WITHDRAWN

MODAL AND SPECTRUM ANALYSIS: DATA DEPENDENT SYSTEMS IN STATE SPACE

MODAL AND SPECTRUM ANALYSIS: DATA DEPENDENT SYSTEMS IN STATE SPACE

SUDHAKAR M. PANDIT
Michigan Technological University

A Wiley-Interscience Publication
JOHN WILEY & SONS, INC.
New York • Chichester • Brisbane • Toronto • Singapore

In recognition of the importance of preserving what has been
written, it is a policy of John Wiley & Sons, Inc., to have books
of enduring value published in the United States printed on
acid-free paper, and we exert our best efforts to that end.

Copyright © 1991 by John Wiley & Sons, Inc.

All rights reserved. Published simultaneously in Canada.

Reproduction or translation of any part of this work
beyond that permitted by Section 107 or 108 of the
1976 United States Copyright Act without the permission
of the copyright owner is unlawful. Requests for
permission or further information should be addressed to
the Permissions Department, John Wiley & Sons, Inc.

Library of Congress Cataloging in Publication Data:
Pandit, Sudhakar M. (Sudhakar Madhavrao)
  Modal and spectrum analysis: data dependent systems in state space/Sudhakar M. Pandit.
      p.   cm.
  Includes bibliographical references.
  ISBN 0-471-63705-X
  1. Modal analysis. 2. Spectrum analysis. 3. State-space methods.
I. Title.
  TA654.15.P36 1991
  620'.004—dc20   90-25164
                                                                               CIP

Printed in the United States of America

10 9 8 7 6 5 4 3 2 1

To my wife

MANEESHA

who, with sons Milind and Devavrat,
had to endure my preoccupation with the manuscript
and yet was the first to suggest that
I write this book!

मायां तु प्रकृतिं विद्यान्मायिनं तु महेश्वरम् ।
तस्यावयवभूतैस्तु व्याप्तं सर्वमिदं जगत् ॥१०॥

"Know Nature to be *Māyā*, and the great God to be the lord of *Māyā*. This whole universe is pervaded by Him through beings which form His parts."

Śvetāśvatara Upaniṣad (IV. 10)

PREFACE

This book is intended for scientists and engineers dealing with real data—experimental or operational. A background typical of a bachelor's degree in science or engineering is assumed. It is customary to include the philosophy, unique features, goals, and contents of a book in its preface; but in this book these will be found in the Introduction, discussed as objectively and impersonally as possible. Here I give a brief account—necessarily subjective, personal, and perhaps naïve—of how I was led to this philosophy, and express my indebtedness to the people who were helpful in this undertaking.

When I first became acquainted with the theory of relativity over 25 years ago, I was not only fascinated and charmed by its grandeur and beauty, but also puzzled and dismayed by its inherent and pervasive determinism. Even without the benefit of quantum mechanics, a rudimentary knowledge of Indian philosophy shows that any description of causation in a physical reality in space-time is necessarily relative *and* uncertain or indeterminate. A modern exposition of this ancient philosophy by Swami Vivekananda in the pre-relativity era, which I happened to read at the same time, made this quite clear: since the ultimate reality is one and undivided by (or beyond) space-time, its divided physical appearance in space-time must be causally related by a mathematical description expressing both relativity and uncertainty. In the description of an ideally isolated system, when boundary conditions result in deterministic effects dominating the uncertainty, e.g., systems in classical celestial mechanics, relativistic mechanics with its consequent determinism alone may suffice as an approximation. Otherwise the uncertainty of quantum mechanics prevails.

The apparently divided physical reality is perceived by means of sensory data, which must reflect all the laws of its space-time relation captured by the sensory mechanism. The classical scientific method conjectures such laws, expresses them in mathematical models, and establishes their validity by verifying the agreement of their consequences with the observed data. Modern physics modifies these laws and models to get better agreement with more and more refined data, thus reducing the uncertainty.

It is also possible to reverse the method, deriving the mathematical models directly from the data and using them in applications such as design, prediction, and control. I have called a system represented by such a model from data alone as "Data Dependent System," and the methodology of deriving and applying such models as "Data Dependent Systems" methodology. The immense practical utility of an approach of this kind is obvious and is illustrated in this book. But it also has theoretical utility in that it provides a rational method of decomposing the possibly complicated models obtained from data into simple parts, which can be utilized to confirm, deny, or modify well-known laws and even to conjecture new laws for the behavior of the underlying system. In this sense the approach is fully complementary to the classical scientific method.

The conceptual framework of the methodology is contrasted with the classical scientific method in Chapter 1 and elaborated with the help of a detailed chapter outline. Its mathematical formalism is evolved in Chapter 2. Its foundation in linear algebra and algebraic geometry is clarified in Chapter 3 and utilized for model decomposition in Chapter 4. Applications to the practice of modal analysis and spectrum analysis are elaborated in Chapters 5 and 6. Comparison and contrast with the prevalent methods of data and signal analysis based on the Fast Fourier Transform (FFT) have been provided wherever necessary. Statistical preliminaries are reviewed in an appendix at the end.

A first draft of this book grew out of lecture notes for a course entitled "Multivariate Data Dependent Systems," which I have been teaching at Michigan Technological University since 1977. Many of the assignments have been incorporated in the body of the text as exercises for the reader—ranging from simple to open-ended problems. In 1982, I challenged the graduate students in the class by showing how the theory in Chapters 2–4 provides a far more satisfactory alternative to the prevalent practice of modal analysis and spectrum analysis. Three students accepted the challenge: Steve Dabrowski developed the univariate and multivariate spectrum examples; whereas, using a robust computer program written by Bill Wittig, Nakul Mehta worked out the univariate modal examples. These examples from the resultant papers and theses growing into Chapters 5–6, and continually revised drafts of the basic Chapters 2–4, were used in subsequent years and amplified by the work of many students, notably: Eric Jacobson, Anthony Okafor, Rock Helsel, Alexandra Ma, Zhiquan Hu, Yingxian Yao, and Chris Weber. The technical contribution of these students can be seen from our joint papers listed in the references. Here, I would like to express my personal thanks for their help through many discussions, disputes, challenges, and counterchallenges of the usual advising process. I have also benefited from the corrections and critical comments of other students, notably: Brian Boudreau and Ghanashyam Joshi.

I would like to express my hearty thanks to a dear friend and professional colleague, Harold A. Evensen, for his constant encouragement and support in this endeavor. A superb experimentalist, Harold was already aware of the limitations of the prevalent FFT based methods; with an equally good grip on the theory, he could immediately see the advantages of the new method. Many parts of this book have greatly benefited from the lively discussions I had with him.

Finally, I appreciate the quiet patience of Margaret Harsh in retyping again and again the innumerable revisions of the class notes from my early disks. Although the computer has simplified the tedious task of correction, now the secretaries have to put up with the idiosyncrasies of both author and word processor!

SUDHAKAR M. PANDIT

Houghton, Michigan
March 1991

The familiar expression of the Indian seers is: "I have *seen*: I have *realized*: and you also can *see*; you also can *realize*." In this approach belief becomes provisional; the great teachers, while imparting their knowledge to their students, ask them to treat this knowledge only as a working hypothesis. The students are asked to validate it by personal experience. This finds clear and powerful expression in Buddha's famous address to the Kālāmas delivered a few months before his passing away. (*The Aṅguttara Nikāya*, Pali Publication Board, 1960 Edition, Nālanda-Devanagari, Pali Series, Vol. I, 3.7.5):

"This I have said to you O Kālāmas, but you may accept it, not because it is a report, not because it is a tradition, not because it is so said in the past, not because it is given from the scripture, not for the sake of discussion, not for the sake of a particular method, not for the sake of careful consideration, not for the sake of forbearing with wrong views, not because it appears to be suitable, not because your preceptor is a recluse, but if you yourselves understand that this is so meritorious and blameless, and, when accepted, is for benefit and happiness, then you may accept it."

SWAMI RANGANATHANANDA

THE MESSAGE OF THE UPANISADS
Bharatiya Vidya Bhavan, Bombay, 1968, pp. 178–179 (2nd Edition, 1980)

CONTENTS

1. **Introduction**   1

    1.1 Classical System Modeling—The Scientific Method, 2
    1.2 The Data Dependent Systems Philosophy, 3
    1.3 The DDS Methodology, 6
    1.4 State Space Formalism for the DDS Approach, 7
    1.5 Modal and Spectrum Analysis, 9
    1.6 Chapter Outline, 11

2. **Data Dependent Systems in State Space**   13

    2.1 First-Order Discrete Scalar Model AR(1), 14

        2.1.1 Deterministic/Homogeneous Difference Equation: Transient Response, 15
        2.1.2 Stochastic/Nonhomogeneous Difference Equation, 16
        2.1.3 Parameter Estimates, 17
        2.1.4 Deterministic as a Limiting Case of Stochastic, 18

    2.2 Solution of the AR(1) Model as a Difference Equation and Its Characteristics, 18

        2.2.1 Solution and Decomposition, 19
        2.2.2 Deterministic System, 20
        2.2.3 Stochastic System, 20
        2.2.4 Stationary and Nonstationary Systems, 21
        2.2.5 Mean and Variance for the AR(1) Model, 21

- 2.2.6 Limiting Case: Stationary AR(1) Model, 22
- 2.2.7 Nonstationary AR(1) Model, 23

2.3 First-Order Differential Equation, 24

- 2.3.1 Homogeneous Solution: Deterministic Part, 24
- 2.3.2 Nonhomogeneous Solution: Deterministic and Stochastic Parts, 25

2.4 Adequate Modeling of DDS in State Space, 26

- 2.4.1 DDS Modeling Strategy in State Space, 26
- 2.4.2 State Models, 27
- 2.4.3 Equivalence of Bivariate ARV(1) with Univariate ARMA(2,1): Need for Moving Average, 31
- 2.4.4 Univariate ARMA$(n, n-1)$, 32
- 2.4.5 Multivariate ARMAV$(n, n-1)$, 33

2.5 Illustration of the DDS Modeling Approach, 35

- 2.5.1 Independent Data: The "AR(0)" Model, 35
- 2.5.2 The AR(1) Model, 37
- 2.5.3 The ARMA(2,1) Model, 37
- 2.5.4 Stopping Criterion—Stochastic Systems: F-Test, 39
- 2.5.5 Stopping Criterion—Deterministic System, 43

2.6 Conventional Modeling in Mechanics: Origin of State Space, 44

- 2.6.1 Definition of State and State Space, 45
- 2.6.2 Lagrange and Hamilton Equations of Motion, 45
- 2.6.3 State Model for Undamped Free One-Degree-of-Freedom System, 46
- 2.6.4 State Model for Damped Forced One-Degree-of-Freedom System, 49
- 2.6.5 State Model for n-Degree-of-Freedom Multivariate System Illustrated by $n = 2$, 50
- 2.6.6 State Model for n-Degree-of-Freedom Univariate System, 51
- 2.6.7 Right-Hand Side Dynamics: Need for Moving Average, 53
- 2.6.8 Right-Hand Side Dynamics as Moving Average: Misnomer of a Misnomer, 53

2.7 Effect of Mean Estimation, 55

- 2.7.1 Constant Deterministic Part (DC offset) and Average $\bar{X}$, 55
- 2.7.2 Exponential Deterministic Part and Average $\bar{X}$, 57

2.8 DDS Modeling as Exponential Expansion, 59

- 2.8.1 Sum of Two Exponentials—Discrete Time: Difference Equation, 60

2.8.2 Sum of Two Exponentials—Continuous Time: Differential Equation, 60
2.8.3 Exponential Expansion: Difference/Differential Equations, 61
2.8.4 Deterministic and Stochastic Components: ARMA Model, 63

2.9 Fourier Series, Fourier Transform and FFT Versus DDS, 64

2.9.1 Fourier Series, 64
2.9.2 Fourier Transform, 65
2.9.3 DDS as Generalized Laplace Transform, 65
2.9.4 Parsimony at the Expense of Orthogonality, 67
2.9.5 Computational Efficiency of Fourier Methods: Fast Fourier Transform, 67
2.9.6 Discrete Fourier Series as the Basis of FFT, 68
2.9.7 DDS Requires Less Computation for Better Final Results, 71
2.9.8 Data with Noise: Random Function, 72
2.9.9 FFT for Qualitative, DDS for Quantitative Results, 75

3. **Fundamentals of Data Dependent System Analysis**   78

3.1 Vector Space, 80

3.1.1 Definition of a Vector Space and Subspace, 80
3.1.2 Basis and Dimension of a Vector Space, 80
3.1.3 Inner Product and Norm, 83
3.1.4 Gram–Schmidt Orthogonalization and Triangular Reduction, 84

3.2 Illustrative Applications to DDS in State Space, 86

3.2.1 Deterministic AR Models, 86
3.2.2 Stochastic AR Models, 92
3.2.3 Extension to ARMA Models, 94

3.3 Applications of Orthogonal Decomposition, 95

3.3.1 Orthogonal Decomposition, 95
3.3.2 Orthogonal Projection, 96
3.3.3 Method of Least Squares: Normal Equations, 97
3.3.4 Forecasting and Control, 98

3.4 Transformations, 102

3.4.1 Change of Basis, 102
3.4.2 Application to AR(n) Parameter Estimation, 103
3.4.3 Similarity Transformation, 104
3.4.4 Unitary (Orthogonal) Transformation, 105
3.4.5 Projection, 106
3.4.6 Reflection: Householder Transformation, 107

xviii   CONTENTS

       3.4.7 Gram–Schmidt, Householder, and Cholesky Methods for Least Squares, 110

  3.5 Matrix Decomposition, 111

       3.5.1 Eigenvalues and Eigenvectors, 111
       3.5.2 Spectral Representation of a Matrix vis-a-vis Conventional and DDS Approaches to Modal and Spectrum Analysis, 116
       3.5.3 Singular Values and Singular Value Decomposition (SVD), 119

  3.6 Spectral/Modal Decomposition of Scalar (Univariate) Systems, 125

       3.6.1 Continuous Time Systems, 125
       3.6.2 Discrete Time Systems, 135

  Appendix A3.1: Proof of $d[\mathscr{S}] + d[\mathscr{M}] = n$ (3.1.2), 141

4. Modal Decomposition of Vector (Multivariate) Systems   142

  4.1 Modal Decomposition of Continuous-Time Systems, 142
  4.2 Illustrative Example: Undamped Two-Degree-of-Freedom System, 146
  4.3 Modal Decomposition of Discrete-Time Systems, 152
  4.4 The ARV(1) Model, 153
  4.5 ARMAV(2,1) and ARMAV(n,m) Models, 161
  4.6 Transient Response—Identifiability, Controllability, and Observability, 165
  4.7 Justification of ARMA($n, n-1$) Strategy for a General State Model, 167

5. Modal Analysis   173

  5.1 Brief Review and Limitations of FFT-Based Approaches, 174

       5.1.1 Damped Single-Degree-of-Freedom (SDOF) System, 176
       5.1.2 Circle-Fit Method and Structural Damping, 182
       5.1.3 Undamped Multiple-Degree-of-Freedom (MDOF) System, 184
       5.1.4 Damped Multiple-Degree-of-Freedom (MDOF) System, 190
       5.1.5 Multiple-Degree-of-Freedom (MDOF) Curve Fitting, 197
       5.1.6 Limitations of the FFT, 204

  5.2 The DDS Modeling for Modal Analysis, 216

       5.2.1 Modeling, 217
       5.2.2 Estimation, 219
       5.2.3 Computation Methods, Their Speeds and Accuracies, 225

5.3 Modal and System Parameter Identification, 230

    5.3.1 Modal Model/Parameters: Frequency, Damping, and Mode Shapes, 231
    5.3.2 Selection of Modes, 234

5.4 Spatial Model: Mass, Stiffness, and Damping Matrices, 240

    5.4.1 The General Approach, 241
    5.4.2 Impulse and Random Excitation, 243
    5.4.3 Sinusoidal Excitation, 245
    5.4.4 Step Excitation, 247

5.5 Illustrative Examples—Scalar Models, 247

    5.5.1 Nonclassically Damped System, 247
    5.5.2 Closely Spaced Modes, 249
    5.5.3 Spatial Model—Mass, Stiffness and Damping Matrices, 255
    5.5.4 Modal Parameters of a Disc–Brake Rotor—Structural Modification, 260

5.6 Illustrative Examples—Vector Models, 266

    5.6.1 Modal Model—Simulated and Real Systems, 267
    5.6.2 Modal and Spatial Model—An Experimental Torsional System, 268
    5.6.3 Modal Analysis of a V-6 Engine, 277

6. Spectrum Analysis   284

6.1 The Univariate Spectrum, 286

    6.1.1 White Noise, 287
    6.1.2 Spectrum Versus Sample Spectrum, 287
    6.1.3 The AR(1) Model, 288
    6.1.4 The ARMA(2,1) Model, 292
    6.1.5 The ARMA(n,m) Model, 294

6.2 Frequency Response Function (FRF) or Transfer Function, 296

    6.2.1 Single-Input–Single-Output (SISO) System, 298
    6.2.2 Multiple-Input–Single-Output (MISO) System, 300
    6.2.3 The Multiple-Input–Multiple-Output (MIMO) System, 301

6.3 The Multivariate Spectrum, 304

    6.3.1 The ARMAV(n,m) Model and Its Modal Decomposition, 306
    6.3.2 Decomposition by White Noise Variance–Covariance, 308
    6.3.3 Partial Correlation/Coherence and Conditioned Spectra, 309

6.4 Two- and Three-Channel SISO and MISO Modeling with Examples, 315

    6.4.1 Extended ARMA Models and Their Transformation to ARMAV Models, 316
    6.4.2 Two-Channel SISO System, 317
    6.4.3 Three-Channel MISO System Illustrating Conditioned Spectra, 322
    6.4.4 Feedback Transfer Function Identification of a Machine Tool, 330

6.5 Multichannel Modeling Example—Forge Hammer Noise, 335

    6.5.1 Modeling Strategy, 335
    6.5.2 Illustration by Forge Hammer Noise Data, 336

Appendix A: Statistical Preliminaries   359

A.1 Mean, Variance, Covariance, and Correlation of Random Variables, and Their Estimators, 359
A.2 Mean and Variance–Covariance of Estimators in the Linear Least Squares Model and Sample Auto Correlations, 364
A.3 Probability Distributions—Normal, Chi-Square, and F, 366

Table A: Percentage Points of the F-Distribution, 372

Appendix B: Data Listing   378

B.1 % Silicon Content of the Blast Furnace Output, 378
B.2 Forge Hammer Accelerations and Sound Pressure Levels, 379

References   403

Index   407

MODAL AND SPECTRUM ANALYSIS: DATA DEPENDENT SYSTEMS IN STATE SPACE

1

INTRODUCTION

We identify and understand the world surrounding us by discerning certain patterns or modes in space and time from sensory perception data, and referring them to the ones in our memory or subconscious, discerned at earlier times and possibly at different places. This process of knowledge or understanding, which may broadly be called "modal analysis," may proceed by two distinct but not necessarily mutually exclusive ways. One is to use our prior knowledge to form certain prejudices and assumptions and then fit the data to their consequences. The other is to objectively analyze and decompose the data into certain basic modes and then select the ones based on some predetermined criteria, including but not limited to similarity with the modes expected from prior knowledge. This book adopts the latter point of view, calling it the data dependent systems (DDS) approach to distinguish it from the former. The approach is best learned in the context of modal analysis of temporal data from mechanical vibrations due to their elegant mathematical formulation and powerful illustrative potential. However, it can be and has been applied to data from such diverse fields as economics, machined surfaces, and image processing.

In practice, even the best data we can get contain not one but many modes mixed together. So we need some mathematical machinery to separate or differentiate them. The DDS methodology uses the machinery of difference/differential equations to accomplish this separation or decomposition. Although single series of data using scalar difference/differential equations can be studied by means of characteristic polynomials, it is simpler, more elegant, and computationally more efficient to use matrix methods; these so-called state-space methods are almost indispensable for high-order scalar and/or vector equations for multiple series of data. It should, however, be emphasized that the state-space formalism adopted in this book follows more closely its origin in classical mechanics than its abstract version common in modern control theory. In particular, inputs are not distinguished from the outputs in the equations; this distinction is used only at the later analysis and interpretation stage.

In reality, of course, the modes or patterns often do not manifest themselves clearly in a deterministic fashion, but are hazy or stochastic in nature. The prevalent methods require that such stochastic data be analyzed to get a transfer function by spectrum analysis first, before extracting the modal information. The DDS approach can accomplish the modal analysis of stochastic data directly and performs modal and spectrum analysis simultaneously.

This introductory chapter erects the conceptual framework and provides the chapterwise outline showing how the book is built around it. Minimum mathematical formalism is used without proof, just to elucidate the logic of these concepts. Their detailed development will follow in the subsequent chapters.

1.1 CLASSICAL SYSTEM MODELING—THE SCIENTIFIC METHOD

An appreciation of the DDS philosophy or point of view will be facilitated by contrasting it with the present methods of system modeling, which are based on the classical scientific method. In his classic account of the decline of mechanism and the rise of quantum mechanics, d'Abro (1939) informs us that the scientific method, devised by Galileo and Newton, may be applied in three successive stages: (1) observational, (2) experimental, and (3) theoretical or mathematical. The second experimental stage, when possible, supplements the first by producing precise quantitative data under tightly controlled conditions. The availability of such precise data in early physics made it possible to apply the mathematical procedure of the third stage with outstanding success. This success of the third stage embodied in precise general laws, and well-developed mathematical models expressing these laws, led to further refined instruments of observation. These instruments became available to physics as well as other physical sciences and engineering, leading to their explosive growth in the subsequent centuries. But the outstanding success of physics in the third stage could not be duplicated to the same degree in other sciences. Why?

Because the third stage requires the investigator to make assumptions of a more or less speculative nature that are crucial for its success. These assumptions must capture the essence of the system, abstract, simplify, and idealize it. They must enable the investigator to conjecture laws of such an idealized system behavior that will lead to tractable mathematical equations, usually differential equations, and also provide manageable solutions. Finally, these solutions must make predictions that match with the observed data. If the assumptions are restrictive, controlled conditions must be created in a laboratory for eliminating all the extraneous influences not permitted by the assumptions to validate the theory.

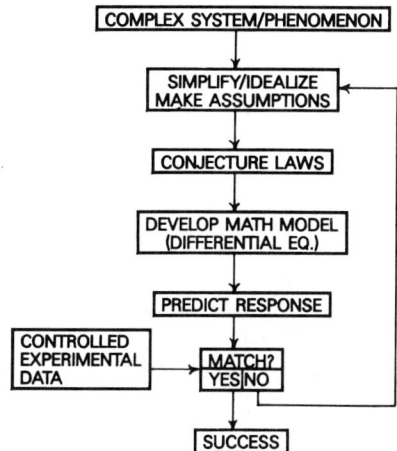

Figure 1.1 Classical system modeling.

No wonder it required the genius of Newton to initiate the third stage at the beginning of the theory of gravitation and of Einstein to culminate it in the theory of relativity! Making the assumptions which will lead to success requires intuition bordering on genius, and for most modern complex systems, system modeling by the classical scientific method could become the endless loop depicted in Figure 1.1 unless one possesses such intuition to get out of it to success. Even for well-abstracted and simplified systems, when the observed data do not match the prediction, a considerable intuitive ingenuity is required to determine how the assumptions should be altered so that the discrepancy is reduced. For complex systems, such looping based on trial and error could become a stupendous task. Moreover, alterations of the assumptions and the resultant modifications in the model to force its predictions to match the observed data could produce artificial and physically meaningless constructions, such as the well-known Ptolemaic epicycle theory of planetary motion.

1.2 THE DATA DEPENDENT SYSTEMS PHILOSOPHY

A thorough understanding of the system is required before its mathematical model can be formulated by the scientific method. However, many of the modern systems are too complex to idealize or to conjecture general laws about their microscopic or macroscopic behavior. Modeling such systems becomes a problem.

Not many of us possess the genius of Newton or Einstein to conjecture such laws. But, thanks to their pioneering work, we do possess three significant advantages over them. The *electronic revolution* ushered in by their work with others has given us instruments capable of providing far more precise and abundant data than ever before; such precise data can be collected even under uncontrolled real-life conditions and no longer necessarily require a controlled laboratory environment. *Mathematics* has developed to an unprecedented degree in our time. And finally, we have *the computer*, perhaps nearest to the replacement of a genius, which can digest and exploit the first two advantages.

Although lack of understanding of the physical mechanism prevents us from formulating a mathematical model of the system, data properly recorded contain this missing knowledge relevant to the measured variables. Therefore, a systematic analysis of the observed data should lead us to this missing knowledge. System equations, which could not be derived from the unknown physical mechanism, should be obtainable from the observed data alone. The system, as represented by the equations derived from and dependent upon the data, may be called a data dependent system (DDS), and the methodology for obtaining such equations or models from the data alone and using them for analysis, prediction, control, and design will be called the data dependent systems (DDS) methodology.

The DDS is invariably stochastic in nature. The stochastic element enters in two different ways. One is through the presence of superimposed random or unpredictable noise such as the error of measuring instruments or experimental error. In such cases, if the effect of noise can be appreciably reduced, say by refined measurement or experimental procedure, then the DDS can be approximated by a deterministic system, corresponding to the macroscopic physics of the terrestrial world tending toward relativistic mechanics. On the other hand, the system may be subject to random influences which affect the temporal or spatial behavior of the system. Here the stochastic element is not merely a superimposed noise, but plays a more fundamental role in the evolution of the system. Then the DDS cannot be approximated by a deterministic system but is *inherently* stochastic, corresponding to the microscopic physics of the subatomic world tending toward quantum mechanics.

As the development of the DDS methodology unfolds in later chapters, the reader will be able to see that the deterministic and random aspects coexist harmoniously in the models, there is no need to assume either aspect a priori. When one of the two aspects dominates, it will become clear at the analysis stage. Then the other can be ignored in further applications. When both aspects are equally dominant, this will also be indicated by the analysis, though the applications now become harder because one of the two aspects must be carefully separated by appropriate computation, depending upon the nature of application.

Interestingly enough, it is the irreducible uncertainty, manifested in the residuals of the adequate model fitted to the data, that allows us to separate the deterministic and random components of the DDS. Boundary values in general, and initial values in particular, will provide the deterministic component and the rest of the residuals that are patternless, colorless, or "white" will provide the source or nucleus of the random component. Although the same modes and hence the same model governs both deterministic and random parts, when one of them dominates, it is possible to select only those modes most influential in that part and simplify the model. Such a simplified model may be related to the one conjectured by the classical scientific method.

This leads to the concept of modal and model decomposition that is central to the DDS philosophy. This concept is not so crucial in the usual system analysis based on the classical scientific method because restrictive assumptions and/or controlled experiments are designed to ensure that only the modes of interest enter the data. The DDS methodology avoids the trial and error inherent in such restrictions and assumptions. Information not contained in the data, even though it may allow us to postulate a simple model, is *not* used in the modeling stage. Only after the model obtained by the DDS methodology using the data alone is decomposed, is this information used to see if the parts of the model are consistent with it and if the other parts are negligible. If so, the simpler model based on such information is accepted. Otherwise, either the information and the assumptions based on it are wrong, or the data has been improperly measured and collected.

Thus, not requiring a priori assumptions before modeling is a strength of the DDS methodology in two ways. First, it relieves the investigator of the tedious task of guessing such assumptions from the necessarily inadequate information available and thus avoids not only trial and error but also prejudicing of interpretations. Second, it utilizes the observed data and other information more effectively by employing them as evidence from independent sources during the analysis stage, rather than mixing them before modeling to make modeling easier. The DDS methodology possesses a simple, straightforward, and rational modeling procedure using data alone, it does not need assumptions to guess and simplify the modeling process. Moreover, the analysis stage can suggest assumptions, which, if confirmed by other evidence, can often improve estimation, computational efficiency, and interpretation, and eventually provide simpler models for applications.

How does the DDS philosophy provide a clear separation and accurate estimation of the modes and arrive at the adequate model based on data alone? By successively driving down the source of uncertainty in the data. The order or complexity of the model is increased until the residuals or the discrepancy between the model and the data contains no more patterns or modes worth modeling. A deterministic system model would be obtained as a limit of the stochastic DDS, when the random irreducible residuals can be driven down to a level negligible compared to the data. The accuracy of modal estimation would of course be limited by the accuracy of the data.

The stochastic nature of DDS is not merely a nuisance to be grudgingly reckoned with because of the possibly random nature of the observed data. It is a fundamental postulate of the DDS philosophy that the world we sense is stochastic; that its randomness is useful and necessary even in arriving at the deterministic as a limiting case. The DDS philosophy is founded on an ancient dictum: Uncertainty, and determinism with its consort relativity, are two sides of the same coin. The world we sense is uncertain because it is relative. Relativity and uncertainty are not only the price but also the tribute we pay for the privilege of knowing the world. Consequently, there is no conflict between the deterministic and the random in the DDS philosophy and methodology; both arise simultaneously and harmoniously in its analysis process.

1.3 THE DDS METHODOLOGY

The initial attempt at the formulation of the DDS methodology in Pandit (1970) was modest in that it was merely proposed to update the parameters of a given model using observed data. The methodology was greatly extended to its present form, with the requisite mathematical foundation and with many illustrative applications in Pandit (1973, 1977a, b). Some of the discussion in the preceding and this section as well as Figure 1.2 have originated from these references.

As depicted diagrammatically in Figure 1.2, the DDS methodology begins with observed data, sampled uniformly from one or many variables, ordered over time or space. The data may be collected under either controlled experimental conditions or uncontrolled operating conditions. Since the modes or patterns manifest themselves as the dependence in the successive observations, the data vector is allowed to express this dependence in a linear vector space of increasing dimension. This is equivalent to fitting difference equation models of increasing order. The increase in order is continued until it fails to significantly reduce the residuals of the model fitted to the data. This provides a statistically adequate model from the data directly, without requiring any other information or assumptions. The least-square or L_2 norm has been used so far and will also be used in this book for estimating model parameters, but recent research

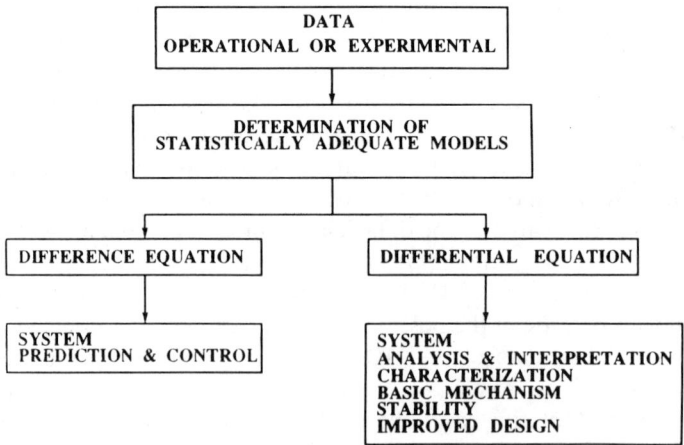

Figure 1.2 The data dependent systems methodology.

indicates that other norms, such as minimum sum of absolute values or L_1, may also be beneficial.

When the observed data "truly" reflects the behavior of the underlying physical system, the DDS provides its "true" representation and can be used for system analysis, prediction, and control in much the same way as conventional methods. In particular, transfer functions of multi-input multi-output systems can be readily obtained from the difference equations and used for optimal control even in the presence of high noise of unknown character in the observed data from real-life or operating systems.

On the other hand, characteristic roots or eigenvalues of the difference equations can be transformed readily into those of the corresponding differential equations. Once the differential equations of the system become available, they can be utilized for the usual applications such as characterization, interpretation and identification, study of the basic mechanism, stability analysis, and design improvements. Information not contained in the data can be brought to bear on the analysis and understanding of the system at this stage. In fact, the third stage of the scientific method discussed earlier can be conducted now much more effectively and rationally to build a mathematical theory. We now have a detailed mathematical description of all the modes and their characteristics obtained rigorously and objectively from the factual data, rather than the vague and often confusing quantitative data from which the classical scientific method must begin.

Not only is the DDS methodology equally applicable to deterministic as well as stochastic systems, but it is also applicable to stable as well as unstable, stationary as well as nonstationary, and single as well as multiple input–output systems, with or without feedback. No prior assumptions regarding these characteristics are necessary. On the contrary, the methodology will itself reveal at the analysis stage whether the system is stable or unstable, with or without feedback, and so on. The paucity of initial assumptions is indeed one of the most desirable features of the method.

1.4 STATE SPACE FORMALISM FOR THE DDS APPROACH

The simplest model relating two successive observations of data, say X_{t-1} and X_t, is

$$X_t = \phi_1 X_{t-1} + a_t$$

where t refers to the sequence index of observations, $t = 0, 1, 2, \ldots$, and a_t expresses the residual or leftover part after expressing the dependence. This part a_t forms the source or nucleus of randomness in the system and is characterized by a mean of zero and a variance of σ_a^2. It arises because the dependence is not perfect or the system is not deterministic. For an ideally deterministic system governed by this model, the dependence would be perfect and σ_a^2, and therefore a_t, would be identically zero.

8   INTRODUCTION

To clarify the deterministic and random parts of the stochastic system we need initial values in particular or boundary values in general. Since we will be primarily concerned in the book with mechanical vibratory systems, albeit for their illustrative potential for studying nonmechanical systems as well, we will restrict ourselves to initial values only and rewrite the model as

$$X_t = \phi_1 X_{t-1} + a_t, \quad X_t = X_0 \quad \text{at} \quad t = 0 \quad (1.1)$$

where X_0 is the initial value. We will also consider t as time, although it could also be a space variable.

By a straightforward recursive substitution, it can be seen that the evolution of the system in time is described by the solution of the difference equation (1.1):

$$X_t = \phi_1^t X_0 + \sum_{j=0}^{t-1} \phi_1^j a_{t-j} \quad (1.2)$$

The first term on the right-hand side is the deterministic part and the second term is the stochastic part. Both parts evolve by a mode or pattern characterized by the single exponential ϕ_1^t. When $|\phi_1| > 1$ the system is unstable and nonstationary. When $|\phi_1| < 1$ the system is stable and, for large enough t, stationary.

If we are primarily interested in modal analysis and hence would like to concentrate on the deterministic part, the differential equation for the deterministic part can be written as

$$\frac{dx}{dt} = \alpha x, \quad \alpha = \frac{1}{\Delta} \ln(\phi_1) \quad (1.3)$$

where Δ is the sampling interval.

If we are primarily interested in the random part, it can be characterized by the autocovariance γ_k, or its Fourier transform, autospectrum $S(\omega)$, provided the stationarity condition $|\phi_1| < 1$ is satisfied. Then, these characteristics are immediately obtained as

$$\gamma_k = \frac{\sigma_a^2}{(1 - \phi_1^2)} \phi_1^k \quad (1.4)$$

and

$$S(\omega) = \frac{\Delta \sigma_a^2}{2\pi} \frac{1}{|e^{i\omega \Delta} - \phi_1|^2} \quad (1.5)$$

How can we decide whether the model (1.1) is adequate, so that a_t's have no more modes or patterns in them and are uncorrelated or white (with flat autospectrum)? We can do so by placing the data vector in a higher dimensional "state" space and then modeling it. For a single series of data, this can be accomplished by forming the state vector $\mathbf{X}_t$ as $[X_t, X_{t-1}]^T$,

$[X_t, X_{t-1}, X_{t-2}]^T, \ldots$, which is equivalent to fitting scalar models of increasingly higher order. For multiple series of the data, this is accomplished by forming the state vector $\mathbf{X}_t$ as $[X_{1t}, X_{2t}, \ldots, X_{pt}, X_{1t-1}, X_{2t-1}, \ldots, X_{pt-1}]^T$, $[X_{1t}, X_{2t}, \ldots, X_{pt}, X_{1t-1}, X_{2t-1}, \ldots, X_{pt-1}, X_{1t-2}, X_{2t-2}, \ldots, X_{pt-2}]^T \ldots$ successively, where $X_{1t}, X_{2t}, \ldots, X_{pt}$ could consist of input force or excitation as well as output response data. Such a model may be written as

$$\mathbf{X}_t = \boldsymbol{\phi}\mathbf{X}_{t-1} + \boldsymbol{\theta}\mathbf{a}_t, \qquad \mathbf{X}_t = \mathbf{X}_0 \quad \text{at} \quad t = 0 \tag{1.6}$$

where $\mathbf{X}_t, \mathbf{a}_t$ are vectors of data and residuals, $\boldsymbol{\phi}$ is the discrete state matrix, and $\boldsymbol{\theta}$ is a matrix parameter. The dimension of the state space, or the order of the model, is increased until the residuals are small enough or until they fail to reduce significantly. The details of such DDS modeling in state space will be worked out in Chapter 2, basics of estimation will be provided in Chapter 3, and estimation algorithms for modal analysis applications will be developed in Chapter 5. Although (1.6) appears to be a linear difference equation, it is capable of providing stochastic linearization of nonlinear systems, as discussed in Pandit (1977b), and further elaborated in Chapter 6.

1.5 MODAL AND SPECTRUM ANALYSIS

It is clear that (1.1) is a scalar or univariate special case of (1.6). The evolution of the system (1.6) in time can thus be described by the multivariate analog of (1.2):

$$\mathbf{X}_t = \boldsymbol{\phi}^t \mathbf{X}_0 + \sum_{j=0}^{t-1} \boldsymbol{\phi}^j \boldsymbol{\theta} \mathbf{a}_{t-j} \tag{1.7}$$

Again the first term on the right-hand side is the deterministic part and the second term is the stochastic part. Both parts evolve as a mixture of modes characterized by linear combinations of exponentials λ_i^t, where λ_i are the possibly complex eigenvalues of the discrete state matrix $\boldsymbol{\phi}$. When $|\lambda_i| > 1$ for any i, the system is unstable and nonstationary. When $|\lambda_i| < 1$ for all i, the system is stable and, for large enough t, stationary.

When we are interested in modal analysis, we concentrate on the deterministic part. It will be shown that the modal analysis can be accomplished far more accurately by the DDS than by other methods based on the Fast Fourier Transform (FFT) because the noise is appropriately modeled and separated naturally by the stochastic part, rather than our guessing the nature of noise and then attempting to filter or average it out. The eigenvectors of $\boldsymbol{\phi}$ provide the modal vectors containing the amplitudes of modes, whereas the complex eigenvalues λ_i provide their natural frequencies and damping ratios, delineating their behavior in time. The basic eigenvalue–eigenvector decomposition, often called the spectral decomposition of a matrix in the mathematical literature, is reviewed and illustrated for the modal decomposition of scalar models in Chapter 3 and extended to vector models in Chapter 4.

Using the information provided by the DDS approach via modal decomposition and other a priori knowledge of the system, we can select appropriate modes of the physical (mechanical) system which provided the data. The eigenvalues λ_i of such selected modes can be transformed to the eigenvalues μ_i of the differential equation by the simple transformation

$$\mu_i = \frac{1}{\Delta}\ln(\lambda_i) \qquad (1.8)$$

Employing the corresponding eigenvectors we can construct the continuous time state matrix **A** governing the differential equation of the system, analogous to (1.3):

$$\dot{\mathbf{x}} = \mathbf{A}\mathbf{x} \qquad (1.9)$$

As will be shown in Chapter 5, such a state matrix obtained from response or output data provides what is commonly known as the modal model in experimental modal testing. It provides scaled mode shapes (relative amplitudes), natural frequencies, and damping ratios, which are enough in many modal analysis applications. Repeating the same analysis for the forced response data, together with the force data, resolves the correct scaling of modal (eigen-) vectors and yields the physical or spatial model of the system consisting of mass, damping, and stiffness matrices **m**, **c**, and **k**:

$$\mathbf{m}\ddot{\mathbf{x}} + \mathbf{c}\dot{\mathbf{x}} + \mathbf{k}\mathbf{x} = \mathbf{f} \qquad (1.10)$$

In particular, when the forcing function is random, the DDS methodology provides a natural method for characterizing and analyzing random vibrations based on response data.

The spectral decomposition of the state matrix, when applied to the stochastic component, provides us with information in much clearer and far more precise quantitative form than is possible by the FFT-based spectrum analysis. As will be shown in Chapter 6, this becomes possible because once an adequate model has been obtained, so that the residuals a_t are white, their spectrum is constant and hence smooth. Moreover, in the DDS approach, the operation of Fourier transformation is carried out on a difference/differential equation (which was the original intent of Fourier transform) and hence yields analytic functions for characteristics such as the frequency response (transfer) function, the auto- and cross-spectrum, and the coherence. Analytic functions are inherently smooth. On the other hand, the FFT of a sample of white noise always remains violently erratic no matter how large a sample we take. Cutting a large record of white noise into pieces and averaging the FFT is equivalent to windowing the original large record and hence introduces bias and other distortions. The FFT of random data, which is nothing but a convolution of such white noise, inherits these problems in spite of skillful "window carpentry."

Second, the deterministic component of the data is appropriately modeled and separated in the DDS approach. Hence it does not affect the spectrum obtained from the stochastic part except perhaps to a small extent caused by the parameter estimation errors. On the other hand, the conventional FFT-based spectrum analysis introduces large errors by including the FFT of the deterministic part as well. Although good practice demands that deterministic parts be removed beforehand, one usually lacks the a priori information needed to filter them out effectively.

It should be emphasized that the common practice of modal analysis, using random or pseudo-random input via auto- and cross-spectra, is not necessary in the DDS approach. The difference equation models obtained by the DDS approach provide all the quantitative modal information via the state matrix, as well as characteristics such as frequency response function if desired for graphical illustration. Only one data record suffices, since no averaging is necessary.

Modal and spectrum analysis is treated in this book as a natural consequence of the spectral decomposition of the state matrix, directly synthesized from the observed data by the DDS approach. Because the deterministic and the stochastic (noise) parts are appropriately modeled and analytically separated, they do not affect each other, unless the data record is so short (compared to even a single record of the FFT-based approach) that the inadequacy of information causes large parameter estimation errors in the model. Even for such inadequate data, the DDS approach does provide signal to noise indicators for the modes, as discussed in Chapter 5.

1.6 CHAPTER OUTLINE

The basics of the DDS approach are developed in Chapters 2 and 3 to show how the state matrix is synthesized from data, Chapter 4 shows how the spectral decomposition of the state matrix yields the modal decomposition of the system, and Chapters 5 and 6 apply this modal decomposition to modal analysis and spectrum analysis, respectively. A more detailed conceptual outline of the chapters is elaborated below.

Chapter 2 formulates the DDS philosophy in the framework of state space and evolves its modeling strategy. Starting from the simple first-order scalar stochastic difference equation and the decomposition of its solution in deterministic and stochastic parts, more elaborate scalar and vector models called autoregressive moving average (ARMA) models are introduced and shown to form a special case of the general state model. Vector differential equations follow by analogy and are used to trace the origins of state space in classical mechanics. Fourier series (transforms), invented originally to solve such equations, are then compared and contrasted with the DDS approach.

Chapter 3 delineates the linear algebra and geometry underlying the state model. The eigenvalue–eigenvector or spectral decomposition of the state matrix is shown to be the crux of modal and spectrum analysis and illustrated by applying it to scalar continuous and discrete-time systems. The Gram–Schmidt, Householder, Cholesky, and singular value decomposition methods of least squares estimation are reviewed for both explaining the rationale of the modeling approach and developing computationally efficient algorithms for parameter estimation. This chapter develops and illustrates all the mathematical tools needed in subsequent chapters.

Chapter 4 extends the modal decomposition to the multivariate case. Both continuous and discrete-time systems are considered and the decomposition is applied to their characteristics, such as impulse and step response from the deterministic part and the Green's function from the stochastic part. The concepts of identifiability, controllability, and observability are briefly discussed and the justification for the modeling strategy from the control theory viewpoint is indicated.

Chapter 5 applies the modal decomposition to the practice of modal analysis. After presenting a brief review of the existing methods of modal analysis (mostly based on the FFT), it illustrates the basic limitations of the FFT in modal analysis. The modeling procedure and estimation algorithms specific to modal analysis applications are outlined. The theory of the DDS approach to modal analysis is developed to show how the state matrix provides the scaled mode shapes, natural frequencies, damping ratios, and finally the mass, damping, and stiffness matrices. Both simulated and experimental examples illustrating the procedure are given.

Chapter 6 briefly reviews the current methods of spectrum analysis, showing how many of the problems associated with this method, such as bias and random error, are avoided in the DDS approach. Analytical expressions are derived for obtaining auto-spectra from scalar models, and auto- and cross-spectra as well as coherence from vector models. These are illustrated using data from real systems. Additional quantitative and graphical information, not available from the existing methods, is extracted and illustrated. The conditioned spectra and partial coherences (useful in source identification and noise reduction) are discussed and illustrated, together with feedforward and feedback frequency response functions.

2

DATA DEPENDENT SYSTEMS IN STATE SPACE

In this chapter we will provide the motivation for the state space approach to data dependent systems (DDS). Although this approach is particularly useful for multivariate models, such as the vector autoregressive moving average (ARMAV) models, it also provides new insight into the higher-order scalar models.

We will introduce a first-order discrete scalar model considering a real set of data in Section 2.1. This model is in the form of a first-order difference equation. If the model is adequate, we can decompose it by a simple method of recursive substitution to understand its implications, as discussed in Section 2.2. This simple recursive method of substitution directly provides us with the "solution" of the difference equation. The solution, by analogy, leads us to the first-order continuous scalar model in the form of a differential equation discussed in Section 2.3. The solution of this first-order differential equation is also discussed. The adequacy of the first-order model in fact implies that the system underlying the data "resides" in a state space of dimension one. If such a model is not adequate, the system requires a state space of dimension greater than one, and hence must be described by vector models with matrix parameters discussed in Section 2.4, which lays the foundation for DDS in state space. The resultant modeling approach is illustrated in Section 2.5, whereas the vector differential equation models are exemplified in Section 2.6 using classical mechanics that originated the concept of state space. Section 2.7 clarifies the model parameter estimation, particularly in relation to the mean. The chapter closes by representing the DDS modeling as a sum of an increasing number of exponentials in Section 2.8 and briefly indicating its similarity and contrast with Fourier series (transform) in Section 2.9, later elaborated in Chapter 5.

14   DATA DEPENDENT SYSTEMS IN STATE SPACE

2.1   FIRST-ORDER DISCRETE SCALAR MODEL AR(1)

Figure 2.1 shows 200 consecutive observations of the silicon percentage in the hot metal of each cast from a blast furnace (listed in Appendix B.1 at the end of the book). The data appears to be random, but on closer inspection one sees some overall patterns. Once it starts going up or down, it continues to do so for two or three consecutive observations. This suggests that the data is statistically dependent, that is, each observation depends on and is influenced by the preceding one. We cannot model the data by conventional techniques such as linear regression because they are based on independent observations. (See Appendix A at the end of the book for a brief review of statistical preliminaries.)

Let us try to express the dependence by a mathematical model since this dependence characterizes the data apart from its mean and variance. An equation adequately describing this dependence of the present observation at time t, say X_t, on the past observations $X_{t-1}, X_{t-2}, \ldots$, should provide a complete model of the system necessary for prediction, control, and analysis. As we will show, a model expressing the dependence in time also characterizes the mean and the variance.

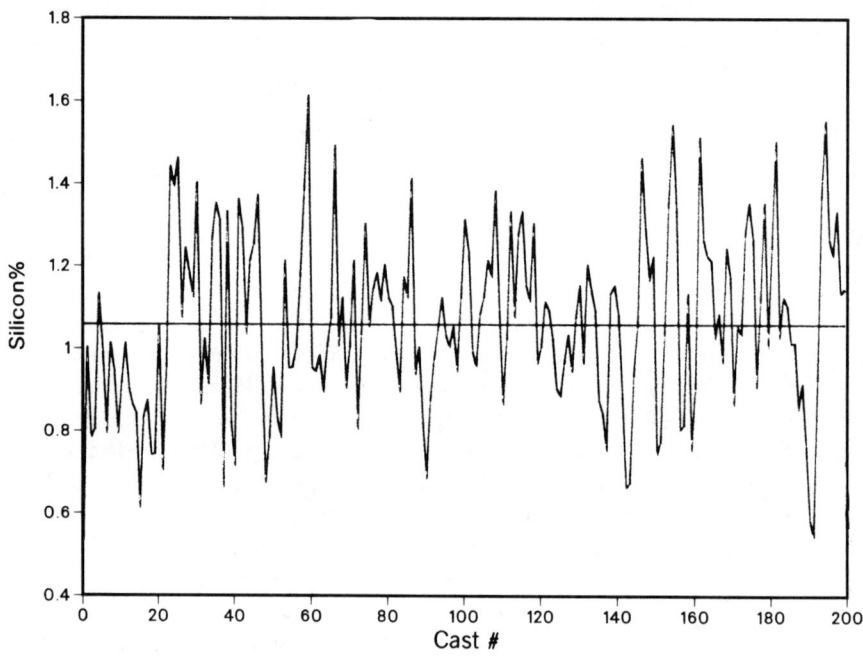

Figure 2.1   Percent silicon content of the blast furnace output.

2.1.1 Deterministic/Homogeneous Difference Equation: Transient Response

The simplest way to get a model for describing the dependence of X_t on X_{t-1} is to plot the observations X_t versus X_{t-1}. This plot for the average subtracted silicon percentage data in Figure 2.2 is seen to be scattered around a straight line. If we ignore the random deviations from the straight line, then, since the line passes through the origin, its equation or model is given by

$$X_t = \phi_1 X_{t-1} \tag{2.1.1}$$

Since we have ignored the random deviations, this model is a *first-order deterministic difference equation*. This is because knowing any one value of X_t, say at $t = 0$, all other values at subsequent times (and also previous times, which we are not presently considering) are completely determined. Thus we have

$$X_1 = \phi_1 X_0$$
$$X_2 = \phi_1 X_1 = \phi_1^2 X_0$$
$$X_3 = \phi_1 X_2 = \phi_1^3 X_0$$

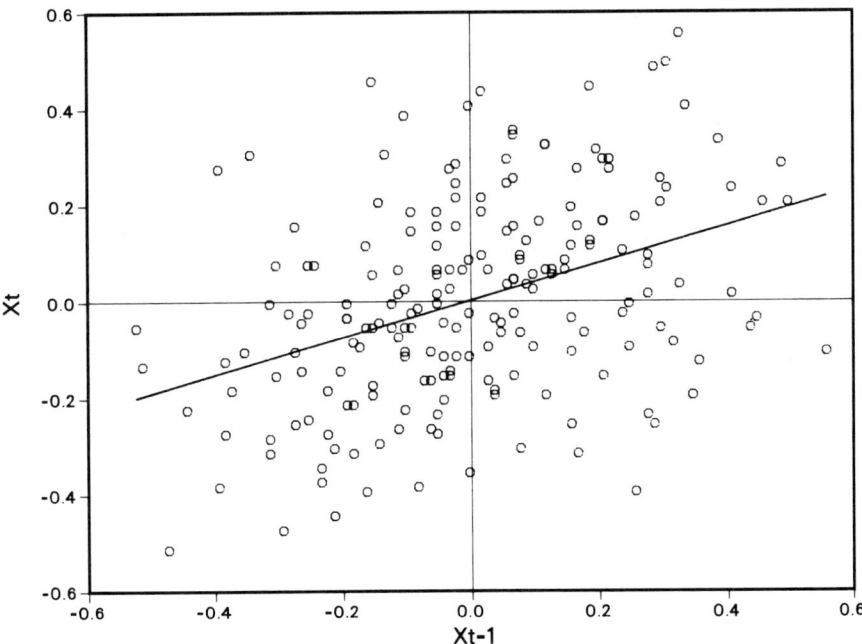

Figure 2.2 Plot of average subtracted blast furnace data X_t versus X_{t-1}.

and in general

$$X_t = \phi_1^t X_0 \tag{2.1.2}$$

which is a deterministic discrete exponential starting at X_0 as expected.

Equation (2.1.2) represents a solution of the homogeneous first-order linear difference equation (2.1.1), because it can be verified that the discrete exponential (2.1.2) satisfies the difference equation (2.1.1). To show this more clearly in analogy with differential equations, note that (2.1.1) can be rewritten in the form of a *homogeneous linear difference equation* with the initial condition

$$X_t - \phi_1 X_{t-1} = 0, \quad \text{given } X_t = X_0 \quad \text{at} \quad t = 0 \tag{2.1.1a}$$

which is satisfied by the solution (2.1.2), also called the transient response. Note that X_0, as an initial condition, is a given parameter and not one of the response data $X_1, X_2, \ldots, X_N$.

2.1.2 Stochastic/Nonhomogeneous Difference Equation

The solution (2.1.2) cannot represent the data (after subtracting the average) in Figure 2.1, because the data is fluctuating randomly around a fixed mean or average, whereas (2.1.2) represents a deterministic discrete exponential. Therefore, the deterministic difference equation (2.1.1) is not an appropriate model for the silicon percentage data. What went wrong? Obviously we ignored the random fluctuations around the straight line in the plot of X_t versus X_{t-1} in Figure 2.2.

Let us denote the random fluctuation at time t by a_t. Then the random variable a_t (or more precisely, a random function of time t) should have zero mean since the random fluctuations a_t's are evenly distributed around the line in Figure 2.2 and we expect their average to be zero. The histogram of all a_t's plotted in Figure 2.3 is seen to be roughly normal. If the a_t's at different times t are dependent (just like X_t's), our modeling effort will be incomplete or inadequate because we need another model for expressing the dependence of a_t's. If, however, the a_t's are independent, there is no more dependence left to model and the complete or adequate model is given by

$$X_t = \phi_1 X_{t-1} + a_t \tag{2.1.3}$$

where the a_t's are normally independently distributed with mean zero and variance, say σ_a^2, which we will briefly denote by

$$a_t \sim \text{NID}(0, \sigma_a^2) \tag{2.1.4}$$

The zero mean assumption can be checked from the computed average of residual a_t's, and the normal distribution can be checked by a histogram of residual a_t's. The crucial assumption, however, is the independence of a_t's. If this

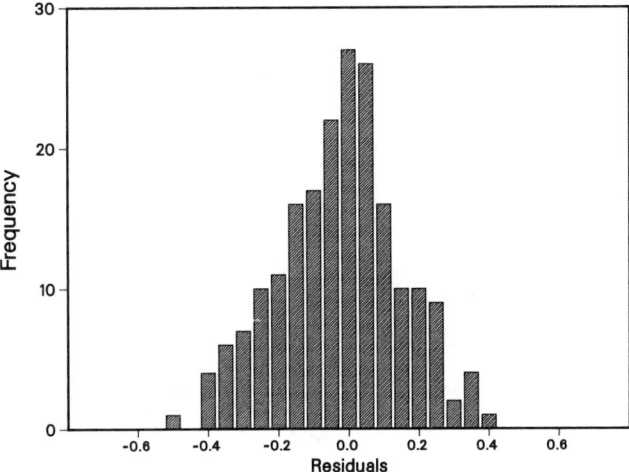

Figure 2.3 Histogram of the residuals—AR(1) model, blast furnace data.

assumption is not satisfied, the stochastic difference equation model (2.1.3) is inadequate or incomplete and we need to find ways of completing it.

2.1.3 Parameter Estimates

Let us suppose that the stochastic difference equation model (2.1.3) is adequate. Then the estimate of the unknown parameters ϕ_1 and σ_a^2 can be obtained by the usual linear least squares methods by minimizing the sum of squares of the residuals a_t's:

$$\text{RSS} = \sum_{t=1}^{N} a_t^2 = \sum_{t=1}^{N} (X_t - \phi_1 X_{t-1})^2 \tag{2.1.5}$$

where N is the number of observations in addition to X_0. To find an estimate of ϕ_1, say $\hat{\phi}_1$, we can differentiate (2.1.5) with respect to ϕ_1 and equate it to zero. This gives the least squares estimate of ϕ_1 that minimizes the residual sum of squares (RSS) in (2.1.5) as

$$\hat{\phi}_1 = \Sigma X_t X_{t-1} / \Sigma X_{t-1}^2 \tag{2.1.6}$$

where the summation is from $t = 1$ to N and the estimate of the residual variance σ_a^2 is given by

$$\hat{\sigma}_a^2 = \frac{\text{RSS}}{N} \tag{2.1.7}$$

(For an unbiased estimate N is replaced by the number of residuals minus the number of estimated parameters.)

18 DATA DEPENDENT SYSTEMS IN STATE SPACE

For the silicon percentage data in Figure 2.1, the parameter estimates are

$$\hat{\phi}_1 = 0.39, \quad \text{RSS} = 7.755$$

and since $N = 199$,

$$\hat{\sigma}_a^2 = 7.755/199 = 0.039$$

where we have taken the first data point in Figure 2.1 after subtracting the average as X_0 for convenience. Since the variance of the original series X_t is 0.0458, we have accounted for about 15% of the variability in X_t by the simple first-order model (2.1.3), expressing the dependence on its previous value only. Since X_t depends or regresses on its own past value at lag 1, this model is called an *autoregressive model of order 1*, abbreviated as AR(1).

2.1.4 Deterministic as a Limiting Case of Stochastic

The AR(1) model (2.1.3) is also a difference equation, similar to the first-order deterministic difference equation model (2.1.1). In fact, if a_t is a deterministic function, the difference equation (2.1.3) reduces to (2.1.1) when a_t is identically zero; in other words, the two models (2.1.1) and (2.1.3) would be representing practically the same data and the underlying system if the a_t's are negligibly small compared to the X_t's. Since a_t is a random function and already has zero mean, its magnitude is measured by its variance. If its variance σ_a^2 is small, a_t's would tend to be small with very large probability. For example, under the normal distribution assumption which we have made, a_t's should be between $-1.96\sigma_a$ and $+1.96\sigma_a$ with 95% probability. Therefore, when a_t is a random function, models (2.1.1) and (2.1.3) represent practically the same data and the same underlying system if the variance of a_t is negligible compared to the sample variance or mean-squared value of X_t (computed assuming a constant mean, since the actual mean may be changing with time). This will be illustrated at the end of Section 2.5.

In contrast, for the silicon data discussed above, the a_t's are not at all negligible compared to X_t's, because the variance of a_t is 0.0388, whereas that of X_t is 0.0457. Therefore, the deterministic difference equation (2.1.1) would indeed be a very poor model to represent the data, as we observed earlier from the solution of (2.1.1) given by the discrete exponential (2.1.2). We will be able to see this even more clearly by obtaining the solution of (2.1.3) by following the same method of substitution used for obtaining the solution (2.1.2) from (2.1.1); this is done in the next section.

2.2 SOLUTION OF THE AR(1) MODEL AS A DIFFERENCE EQUATION AND ITS CHARACTERISTICS

Another way of categorizing models (2.1.1)–(2.1.3) is to realize that the difference equation (2.1.1) is the fundamental homogeneous equation. If we specify a given initial condition X_0, as in (2.1.1a), we get the homogeneous solution or transient response (2.1.2) that is deterministic because X_0 is deterministic. On the other hand, instead of specifying the initial condition X_0 and tying down the response to a specific time origin $t = 0$, if we specify the forcing function a_t for all t, we get the nonhomogeneous, forced model (2.1.3) that becomes stochastic when a_t is a random function specified by (2.1.4).

It is then clear that if we specify *both* the initial condition X_0 and the (random) forcing function a_t, we should get a response that contains both transient and forced, homogeneous and particular, and deterministic and stochastic components. We now proceed to find this response by solving the difference equation (2.1.3) with the specified initial condition X_0. Note that in finding a solution or response we always take the parameters such as ϕ_1, σ_a^2, and X_0 as given, even though they are estimated from the limited data.

2.2.1 Solution and Decomposition

The assumption that a_t is a random function is immaterial as far as finding the solution of the difference equation (2.1.3) is concerned. We can still use the method of substitution as in the case of the deterministic model (2.1.1). Given the sequence a_t and $X_t = X_0$ starting at $t = 0$, we can recursively substitute in (2.1.3) to get, for $a_0 = 0$,

$$X_1 = \phi_1 X_0 + a_1$$
$$X_2 = \phi_1 X_1 + a_2 = \phi_1^2 X_0 + \phi_1 a_1 + a_2$$
$$X_3 = \phi_1 X_2 + a_3 = \phi_1^3 X_0 + \phi_1^2 a_1 + \phi_1 a_2 + a_3$$

and hence in general we can write

$$X_t = \phi_1^t X_0 \;+\; \sum_{j=0}^{t-1} \phi_1^j a_{t-j} \qquad (2.2.1)$$

homogeneous	particular
transient	forced
deterministic	stochastic
signal	noise

which is the complete solution of the nonhomogeneous linear difference equation (2.1.3) given the initial condition X_0 and the discrete function a_t, irrespective of whether a_t is a deterministic or random function.

Again, to make the analogy with differential equation more transparent, note that the AR(1) model (2.1.3) can be written in the form of a nonhomogeneous difference equation as

$$X_t - \phi_1 X_{t-1} = a_t, \qquad \text{given } X_t = X_0 \quad \text{at} \quad t = 0 \qquad (2.2.2)$$

and (2.2.1) is its complete solution, given the initial condition $X_t = X_0$ at $t = 0$

and the input or the forcing function a_t. The first part of (2.2.1) is the homogeneous solution, which is the same as (2.1.2) and the second part may be called a particular solution (or particular sum, instead of particular integral in the case of differential equations). In the present case, since a_t is a random function or sequence, the first part of the solution (2.2.1) is the deterministic component and the second is the stochastic component.

2.2.2 Deterministic System

Moreover, if a_t is identically zero, the AR(1) model (2.2.2) reduces to the deterministic model (2.1.1a), and so does the solution (2.2.1) to its deterministic part (2.1.2). In practice, if the a_t's are negligible compared to X_0, the deterministic part dominates and the system is very closely represented by a deterministic model. When a_t is random, as in our case, their standard deviation σ_a must be negligibly small compared to the initial value X_0 in order for the deterministic model (2.1.1) to be adequate. Otherwise, the stochastic AR(1) model (2.1.3) is required.

Thus, in practice, it is always preferable to fit the AR(1) model to the data. If the system underlying the data is deterministic, this will be indicated by the standard deviation σ_a being very small compared to the data X_0. Otherwise, the system will have both the deterministic as well as stochastic components as shown by (2.2.1).

2.2.3 Stochastic System

When will the system and its response X_t be purely stochastic, with no deterministic component? The obvious answer, from (2.2.1), is when X_0 is zero. In practice, this means that X_0 must be negligibly small compared to σ_a. However, we can see another possibility by considering the silicon data used earlier.

The average of the silicon data in Figure 2.1 is 1.05, which we can take as the mean. Since the first observation is 0.53, the average or mean subtracted observation is $0.53 - 1.05 = -0.52$, which we have taken as X_0. On the other hand, $\sigma_a = 0.197$. Thus X_0 is not negligible compared to σ_a. Then why don't we see the deterministic component in the data in Figure. 2.1, and why does the series appear to be almost purely stochastic?

To answer this question, let us write solution (2.2.1) for the silicon data using the estimated parameters as the "true" parameters. This gives

$$X_t = (0.39)^t(-0.52) + \sum_{j=0}^{t-1} (0.39)^j a_{t-j}$$

The values of the deterministic component at $t = 0, 1, 2, 3, 4, 5$ are $-0.52, -0.20, -0.08, -0.03, -0.01, -0.004$, respectively. It is now easy to see that the deterministic component becomes negligible compared to $\sigma_a = 0.197$ after about 3 lags and certainly after 4 lags. Moreover, the value $X_0 = -0.52$ itself is not very large compared to $\sigma_a = 0.19$ (about 2.5 times, which is within the 99% probability limits on a_t). Hence the deterministic component, even if present, can hardly be distinguished from the stochastic component at the beginning of the series, and is so small after about 3 lags that the series remains almost purely stochastic.

Thus the system and its response X_t can be considered to be almost purely stochastic when X_0 is negligibly small compared to σ_a or when ϕ_1 is so small that $X_0 \phi_1^t$ is negligible in very few lags compared to the number of data points. Although we have illustrated this for a first-order model, we will show later in Chapter 4 that similar results are true for an arbitrary order. The DDS methodology always treats the system as stochastic. After fitting the model, we can decide if either the deterministic or stochastic component can be ignored. Moreover, when the stochastic component is large but is undesirable, like disturbance or noise, the DDS methodology will provide a natural method of filtering it out and recovering the deterministic "signal."

2.2.4 Stationary and Nonstationary Systems

Since X_t consists of both deterministic and stochastic parts, it is a random variable for a given t and a stochastic process or random function of t in general. If a_t is assumed to have normal distribution, since a linear combination of normal variables has a normal distribution, X_t also has a normal distribution for a given t. This distribution is completely defined by its mean and variance. When these two properties do not explicitly depend upon the time origin, that is, when they are independent of t, the stochastic process or the random function is said to be *wide-sense stationary* or simply *stationary*, otherwise the stochastic process is said to be *nonstationary*.

2.2.5 Mean and Variance for the AR(1) Model

Recall that the mean of a random variable is its expected value, which may be thought of as the average over an infinite sample. If we denote the expectation operator by E, then the operator is linear, that is, the expectation of a linear combination of random variables is the same linear combination of the expectation of variables. Thus if a and b are deterministic and X and Y are random variables,

$$\text{Mean}(aX + bY) = E(aX + bY) = aE(X) + bE(Y) = a\,\text{Mean}(X) + b\,\text{Mean}(Y)$$

A deterministic variable can be considered to be a *degenerate* or limiting case of a random variable with the same mean and *zero* variance; for example,

$$E(a) = a, \ \text{Var}(a) = 0; \quad E(10.1) = 10.1, \ \text{Var}(10.1) = 0$$

Since the random variable a_t has zero mean, its expectation is zero. Hence the mean of X_t is obtained from (2.2.1) as

$$E(X_t) = E(\phi_1^t X_0) + E\left(\sum_{j=0}^{t-1} \phi_1^j a_{t-j}\right)$$

$$= \phi_1^t X_0 + \sum_{j=0}^{t-1} \phi_1^j E(a_{t-j})$$

Thus,

$$E(X_t) = \phi_1^t X_0 \qquad (2.2.3)$$

Recall also that the variance of a random variable is the expected value of its squared deviation from the mean,

$$\text{Var } X_t = E[X_t - E(X_t)]^2$$

and therefore, the variance of a zero mean random variable is the expected value of its square. Moreover, the variance of a linear combination of random variables is the same linear combination of the variances with the coefficients squared *if the variables are independent*. In other words,

$$\text{Var}(aX + bY) = a^2 \text{Var}(X) + b^2 \text{Var}(Y)$$

when X and Y are independent. Note that the independence is not required for the mean but is required for the variance.

Thus the variance of X_t is the variance of the stochastic part given by the second term in (2.2.1), and since the a_t's are independent, it is

$$\text{Var } X_t = \sum_{j=0}^{t-1} \phi_1^{2j} \text{Var}(a_{t-j})$$

$$= (1 + \phi_1^2 + \phi_1^4 + \phi_1^6 + \cdots + \phi_1^{2(t-1)}) \sigma_a^2 \qquad (2.2.4a)$$

$$= \frac{(1 - \phi_1^{2t})}{(1 - \phi_1^2)} \sigma_a^2 \qquad (2.2.4)$$

Notice that the variance of X_t given by (2.2.4) is a function of time.

2.2.6 Limiting Case: Stationary AR(1) Model

Considering the mean and variance given by (2.2.3) and (2.2.4), it is seen that the AR(1) model (2.2.2) and its solution (2.2.1) represent a nonstationary stochastic process or random-function X_t since both the mean and the variance are functions of time t and hence depend explicitly on the time origin. However, for

$$|\phi_1| < 1, \qquad \phi_1^t \to 0, \quad \text{and} \quad \phi_1^{2t} \to 0, \quad \text{as} \quad t \to \infty$$

SOLUTION OF THE AR(1) MODEL AS A DIFFERENCE EQUATION

Hence, when ϕ_1 is small and/or when X_0 is small, this dependence on t vanishes for relatively small values of t and from then on the series can be treated as stationary. We can make this statement a little more precise: For $|\phi_1| < 1$, the series becomes stationary beyond t so large that

$$\phi_1^{2t} \ll 1 \quad \text{and} \quad |\phi_1^t X_0| \ll \sigma_a \tag{2.2.5}$$

For such large t we can substitute the limits given above in (2.2.3) and (2.2.4) to get the mean and variance of a stationary AR(1) stochastic process as

$$E(X_t) = 0, \quad \text{Var}(X_t) = \sigma_a^2/(1 - \phi_1^2) \tag{2.2.6}$$

and then, under the stationarity conditions (2.2.5), the general AR(1) model (2.2.2) is the same as the stationary model (2.1.3) since the effect of the initial value X_0 is indistinguishable from a_t's. Note that the stationarity condition $|\phi_1| < 1$ not only ensures that the variance given by (2.2.6) does not involve t, but also ensures that it remains finite.

2.2.7 Nonstationary AR(1) Model

On the other hand, (2.2.3) and (2.2.4a) show that for

$$|\phi_1| = 1, \quad E(X_t) = X_0 \text{ or } (-1)^t X_0 \quad \text{and} \quad \text{Var}(X_t) = t\sigma_a^2 \tag{2.2.7}$$

Thus the mean remains X_0 for $\phi_1 = 1$ (called random walk) or fluctuates between $+X_0$ and $-X_0$ for $\phi_1 = -1$ and thus depends on the time origin for arbitrarily large t, the variance increases and tends to infinity with t as predicted by (2.2.7), and the series is nonstationary.

When $|\phi_1| > 1$, the mean and variance given by the general expressions (2.2.3) and (2.2.4) are functions of t, both increase and tend to infinity with t, and thus the series is nonstationary. The variance grows exponentially for both $\phi_1 < -1$ and $\phi_1 > 1$, whereas the mean grows exponentially for $\phi_1 > 1$ and in a zigzag manner for $\phi_1 < -1$.

It is thus seen that the AR(1) model is capable of representing stationary or nonstationary data depending upon the value of ϕ_1. However, this is not the only parameter in the model that needs to be estimated from the observed data. The parameters required to completely characterize the first-order discrete model, series, or system are ϕ_1, σ_a^2, and X_0. The parameter X_0 is not important when the stationarity condition (2.2.5) is satisfied; the remaining parameters are then related by the variance expression (2.2.6). Otherwise, all three parameters must be specified to characterize the system fully. To emphasize this general nonstationary nature of the AR(1) model, we can write the model in a slightly different form involving all three parameters:

$$X_t = \phi_1 X_{t-1} + \sigma_a Z_t, \quad \text{given } X_t = X_0 \text{ at } t = 0$$
$$\text{and} \quad Z_t \sim \text{NID}(0,1) \tag{2.2.8}$$

24 DATA DEPENDENT SYSTEMS IN STATE SPACE

for which the solution or the decomposition (2.2.1) takes the form

$$X_t = \phi_1^t X_0 + \sum_{j=0}^{t-1} \phi_1^j \sigma_a Z_{t-j} \qquad (2.2.9)$$

where the last term Z_{t-j} is a given forcing function in the sense that its probability distribution is given and the rest of the parameters ϕ_1, σ_a, and X_0 need to be determined from the data X_t.

2.3 FIRST-ORDER DIFFERENTIAL EQUATION

To appreciate the analogy between the discrete and continuous time systems and also to motivate the development of the state space framework, most of which was historically developed for differential equations, consider the *first-order nonhomogeneous differential equation*

$$\frac{dX(t)}{dt} = \alpha X(t) + bu(t), \quad \text{given } X(t) = X(0) \quad \text{at} \quad t = 0 \qquad (2.3.1)$$

Note the formal similarity of this model with the discrete AR(1) model (2.2.8), and that of the solutions of this model with the corresponding ones in Section 2.2. Although not explicitly stated, the input or the forcing function $u(t)$ must be given in order to find the solution of (2.3.1).

2.3.1 Homogeneous Solution: Deterministic Part

As in Section 2.2 we first solve the homogeneous equation:

$$\frac{dX}{dt} = \alpha X, \quad \text{given } X(t) = X(0) \quad \text{at} \quad t = 0 \qquad (2.3.2)$$

by the usual integration of

$$\int_0^t \frac{dX}{X} = \int_0^t \alpha \, dt$$

as

$$\ln\left(\frac{X(t)}{X(0)}\right) = \alpha t$$

which gives the exponential

$$X(t) = X(0) e^{\alpha t} \qquad (2.3.3)$$

which may be compared with the discrete exponential (2.1.2).

2.3.2 Nonhomogeneous Solution: Deterministic and Stochastic Parts

For solving the nonhomogeneous equation (2.3.1), let us use the shorthand notation of

$$\dot{X} \equiv \frac{dX}{dt}$$

to write it as

$$\dot{X} - \alpha X = bu \tag{2.3.1a}$$

To be able to integrate this equation as before, let us multiply both sides by an arbitrary function k (integrating factor),

$$k\dot{X} - \alpha k X = kbu$$

and suppose that the left-hand side is a complete derivative of say kX, that is,

$$\frac{d}{dt}(kX) = k\dot{X} + \dot{k}X$$

then we must have k such that

$$\dot{k} = -\alpha k, \quad \text{or} \quad k = e^{-\alpha t}$$

Now integrating from 0 to t yields

$$\int_0^t d(kX) = \int_0^t kbu\, ds$$

or

$$kX - k(0)X(0) = \int_0^t k(s)bu(s)ds$$

Solving for $X(t)$,

$$X(t) = k^{-1}(t)k(0)X(0) + \int_0^t k^{-1}(t)k(s)bu(s)ds$$

$$= e^{\alpha t} X(0) + \int_0^t e^{\alpha(t-s)} bu(s)ds$$

$$= e^{\alpha t} X(0) + \int_0^t e^{\alpha v} bu(t-v)dv, \quad \text{where} \quad v = t - s \tag{2.3.4}$$

which may be compared with the discrete difference equation solution (2.2.9).

Again, in the differential equation solution (2.3.4), it does not matter whether $u(t)$ is a deterministic or random function. If, however, it is a random function, the first term of (2.3.4) is the deterministic part of $X(t)$ and the second term is the stochastic part. When $u(t)$ is a continuous time analog of a_t, namely the continuous time white noise, then (2.3.1) represents the continuous time AR(1) model. Obviously, now the stationarity condition is $\alpha < 0$, and the discussion of Section 2.2 regarding nonstationarity is applicable to (2.3.1) with this condition.

If the continuous time process (2.3.1) is sampled at uniform interval Δ, the continuous time t really represents time Δt with $t = 0, 1, 2, \ldots$, and we see that since the deterministic parts of (2.3.4) and (2.2.9) must be the same we have

$$\phi_1 = e^{\alpha \Delta} \tag{2.3.5}$$

$$\alpha = \frac{1}{\Delta} \ln(\phi_1) \tag{2.3.6}$$

which shows how the continuous time parameter α can be recovered by fitting the discrete model to the uniformly sampled data.

2.4 ADEQUATE MODELING OF DDS IN STATE SPACE

We will now consider the adequacy of the AR(1) model (2.2.2) in representing a given set of data. This consideration at once leads to DDS modeling in state space, the general discrete time and continuous time state space models, and their special cases of multivariate and univariate higher-order models. These ideas are introduced informally and succinctly at an abstract, conceptual level in this section. They will be subsequently elaborated with specific examples and illustrations.

2.4.1 DDS Modeling Strategy in State Space

The variable X_t denotes the "state" of the system at time t, and the adequacy of the AR(1) model (2.2.2):

$$X_t = \phi_1 X_{t-1} + a_t$$

implies that a single state X_t completely characterizes the behavior of the system by expressing the dependence of the present state X_t on only one past state X_{t-1}. The remainder a_t has no dependence on the past states since it is an independent sequence for different t; hence a_t is also independent of X_{t-1}, $X_{t-2}, \ldots$, which contain only $a_{t-1}, a_{t-2}, \ldots$ [see (2.2.1)]. Thus the model requires the state space of dimension one.

Another way of understanding this is to realize that since the two terms $\phi_1 X_{t-1}$ and a_t are independent, they can be thought of as orthogonal vectors. The AR(1) model then simply says that the orthogonal projection of the vector X_t on the one-dimensional space of the straight line along X_{t-1} is enough to describe the dependence in the data and hence provides an adequate model for the system. Thus the dependence in the data or the behavior of the system is simple enough to be expressible in a one-dimensional state space.

The adequacy of the first-order model thus implies the adequacy of the one-dimensional state space to describe the system. Note that $\phi_1 X_{t-1}$ being an orthogonal projection of X_t on the space of X_{t-1} says that their difference a_t is the smallest vector joining X_t with any point on the line along X_{t-1}. In other words, the length of the vector a_t measured by σ_a cannot be further reduced. For a given N, using (2.1.7):

$$\hat{\sigma}_a^2 = \frac{\text{RSS}}{N}$$

This says that the sum of squares of residual a_t's cannot be further reduced. Note that a_t is a zero-mean random variable, and its magnitude square is represented by $E(a_t^2) = \sigma_a^2$ and hence its length by σ_a.

How can we check that the model is adequate or that the a_t's are independent, and what can we do if the model is not adequate or the a_t's are not independent? A somewhat subtle answer to both these questions is to allow the model more freedom by allowing it to be expressed in a state space of higher dimension. Such a higher dimension will allow the a_t's, if dependent, to be broken down further by expressing their dependence in the added dimension and projecting X_t in a space of dimension two, three, and so on. If such a freedom into higher dimension of say two does not produce smaller a_t's, that is, smaller σ_a^2 or sum of squares of a_t's, then the one-dimensional AR(1) model is adequate, that is, the a_t's of that model must already be independent. Otherwise, not only would we have shown that the a_t's of the AR(1) model are dependent, but we also would have found a better model in the higher-dimensional state space.

How large a dimension do we need? What should be the stopping criterion? The answer depends upon the kind of system we are dealing with and the application we are aiming at in collecting the data. For dominantly stochastic systems, we can continue to increase the dimension of the state space until the reduction in the residual sum of squares is statistically insignificant. For dominantly deterministic systems, we can continue to increase the dimension until the a_t's are so small that the ratio of the sample standard deviation of a_t to that of X_t, $(\hat{\sigma}_a/\hat{\sigma}_x)$, is smaller than the digital accuracy of the data or of the order of the accuracy of the computational process. In between these two extremes, one may use other application-specific criteria and stop at a lower dimension to obtain simpler models with less computation and ease of interpretation at the expense of lower accuracy and resolution. Such criteria may include a known noise floor below which the dependence in a_t's does not matter any more, or stable estimates of desired characteristics of the system; in practice, instrumentation noise may provide the noise floor, or damping estimate may be the primary emphasis, and so on.

2.4.2 State Models

We now introduce models for a state space of arbitrary dimension. An advantage of the approach to DDS in state space outlined in Section 2.4.1 is that the model for an arbitrary dimension, say p, can be written down and solved exactly like the scalar cases discussed in the two previous sections. Thus if bold-faced $\mathbf{X}_t$ and $\mathbf{Z}_t$ represent p-dimensional column vectors and bold-faced $\boldsymbol{\phi}$ represents a $p \times p$ matrix, the discrete state model that is a p-dimensional vector counterpart of the one-dimensional AR(1) model (2.2.8) may be written as

$$\mathbf{X}_t = \boldsymbol{\phi}\mathbf{X}_{t-1} + \mathbf{b}\mathbf{Z}_t, \quad \text{given} \quad \mathbf{X}_t = \mathbf{X}_0 \quad \text{at} \quad t = 0 \qquad (2.4.1\mathrm{a})^*$$

where $\mathbf{b}$ is a $p \times p$ matrix of constants replacing σ_a. Solution of this equation can be found simply by the method of substitution used in obtaining the solution (2.2.1) as

$$\mathbf{X}_t = \boldsymbol{\phi}^t \mathbf{X}_0 + \sum_{j=0}^{t-1} \boldsymbol{\phi}^j \mathbf{b} \mathbf{Z}_{t-j} \qquad (2.4.2\mathrm{a})^*$$

which is the alternate form of solution (2.2.9) using bold-faced letters and replacing σ_a by $\mathbf{b}$.

Similarly, the continuous-time state model follows from the differential equation (2.3.1) as

$$\dot{\mathbf{x}}(t) = \mathbf{A}\mathbf{x}(t) + \mathbf{B}\mathbf{u}(t), \quad \text{given} \quad \mathbf{x}(t) = \mathbf{x}(0) \quad \text{at} \quad t = 0 \qquad (2.4.3)$$

and its solution is found by exactly following the method of Section 2.3 as

$$\mathbf{x}(t) = e^{\mathbf{A}t}\mathbf{x}(0) + \int_0^t e^{\mathbf{A}(t-s)}\mathbf{B}\mathbf{u}(s)\,ds \qquad (2.4.4)$$

which is of course the solution (2.3.4) in bold-faced letters; the matrix exponential follows from its scalar definition

$$e^{\mathbf{A}t} = \mathbf{I} + \mathbf{A}t + \frac{1}{2!}\mathbf{A}^2 t^2 + \frac{1}{3!}\mathbf{A}^3 t^3 + \cdots \qquad (2.4.5)$$

It is easily verified that the homogeneous and nonhomogeneous parts of (2.4.2a) and (2.4.4) satisfy the corresponding equations from (2.4.1a) and (2.4.3). When $\mathbf{Z}_t$ and $\mathbf{u}(t)$ are random (vector) functions, the two parts of (2.4.2a) and (2.4.4) become the deterministic and stochastic components as before in the scalar case.

The generality of these two models and their solutions cannot be overemphasized. Although elegant, their simplicity of appearance is very deceptive. Considerable matrix and vector space theory discussed in Chapter 3 is required to understand and fully appreciate the various implications of these models. In

---

* The standard forms (2.4.1) and (2.4.2) of the models appear after (2.4.10).

this section, however, we will illustrate the special case of (2.4.1a) for $p = 2$ to motivate the introduction of these models; (2.4.3) will be illustrated by models from classical mechanics in Section 2.6.

Bivariate ARV(1)

Suppose we have two series of data X_{1t} and X_{2t}. Then to test the adequacy of the scalar AR(1) model for X_{1t} alone and X_{2t} alone and also to find a better model in case the scalar AR(1) models are inadequate, we can move into state space of dimension two and consider the discrete state model (2.4.1a) with the state and parameters defined as

$$\mathbf{X}_t = \begin{bmatrix} X_{1t} \\ X_{2t} \end{bmatrix} \quad \boldsymbol{\phi} = \begin{bmatrix} \phi_{11} & \phi_{12} \\ \phi_{21} & \phi_{22} \end{bmatrix} \quad \mathbf{b} = \begin{bmatrix} b_{11} & b_{12} \\ b_{21} & b_{22} \end{bmatrix} \quad \mathbf{Z}_t = \begin{bmatrix} Z_{1t} \\ Z_{2t} \end{bmatrix} \quad (2.4.6)$$

Then the scalar form of the vector state model (2.4.1a) with these matrix parameters becomes, using standard row column multiplication of matrices,

$$\begin{aligned} X_{1t} &= \phi_{11} X_{1t-1} + \phi_{12} X_{2t-1} + b_{11} Z_{1t} + b_{12} Z_{2t} \\ X_{2t} &= \phi_{21} X_{1t-1} + \phi_{22} X_{2t-1} + b_{21} Z_{1t} + b_{22} Z_{2t} \end{aligned} \quad (2.4.7a)*$$

Such a model is known as a vector autoregressive model of order 1, denoted as ARV(1); the added V to AR denotes a multivariate model; here we are considering a bivariate form with two series. To solve this scalar form in terms of Z_{1t} and Z_{2t} by recursively substituting starting from X_{10} and X_{20} would be tedious; fortunately, this has been simply and elegantly accomplished by (2.4.2a). However, even though $\boldsymbol{\phi}$ is given by the matrix shown above, what is $\boldsymbol{\phi}^t$? Obviously it must be a 2×2 matrix, but until we can evaluate it element by element, we cannot express X_{1t} and X_{2t} in terms of X_{10}, X_{20}, Z_{1t} and Z_{2t}. In fact we cannot even determine the stationarity and nonstationarity conditions until we can explicitly evaluate $\boldsymbol{\phi}^t$ in Chapter 3.

Univariate ARMA(2,1)

Now consider the single series AR(1) model such as the one used for silicon data fitted in Section 2.1. How can we determine whether it is adequate, and if not, how can we get a better model? The answer again is to consider a state space of dimension two. Since there is no other series involved at this stage, a logical choice is to consider another state, say X_{t-1} for the two-dimensional state space model. Then our $\mathbf{X}_t$ would consist of the column vector $[X_t, X_{t-1}]^T$, where T denotes the transpose. How would we choose $\boldsymbol{\phi}$ and $\mathbf{b}$? A possible choice is

$$\mathbf{X}_t = \begin{bmatrix} X_t \\ X_{t-1} \end{bmatrix} \quad \boldsymbol{\phi} = \begin{bmatrix} \phi_1 & \phi_2 \\ ? & ? \end{bmatrix} \quad \mathbf{b} = \begin{bmatrix} b_1 & b_2 \\ ? & ? \end{bmatrix} \quad \mathbf{Z}_t = \begin{bmatrix} Z_t \\ Z_{t-1} \end{bmatrix}$$

---

*The standard form of this ARV(1) model (2.4.7) appears at the beginning of Section 2.4.5.

This choice implies that the scalar models corresponding to (2.4.1a) are given by using row column multiplication rule for matrix multiplication as

$$X_t = \phi_1 X_{t-1} + \phi_2 X_{t-2} + b_1 Z_t + b_2 Z_{t-1}$$
$$X_{t-1} = ?X_{t-1} + ?X_{t-2} + ?Z_t + ?Z_{t-1}$$

If the two models for X_t and X_{t-1} are to be consistent, an obvious choice is

$$\mathbf{X}_t = \begin{bmatrix} X_t \\ X_{t-1} \end{bmatrix} \quad \boldsymbol{\phi} = \begin{bmatrix} \phi_1 & \phi_2 \\ 1 & 0 \end{bmatrix} \quad \mathbf{b} = \begin{bmatrix} b_1 & b_2 \\ 0 & 0 \end{bmatrix} \quad \mathbf{Z}_t = \begin{bmatrix} Z_t \\ Z_{t-1} \end{bmatrix} \quad (2.4.8)$$

This choice implies that the scalar models corresponding to (2.4.1a) are given by substituting (2.4.8) as

$$X_t = \phi_1 X_{t-1} + \phi_2 X_{t-2} + b_1 Z_t + b_2 Z_{t-1}, \qquad Z_t \sim \text{NID}(0,1) \qquad (2.4.9)$$
$$X_{t-1} = X_{t-1}$$

Now the second row of the model is true but redundant. Note that the state space representation is not unique; nonuniqueness is a common feature of the state space formulation.

The model (2.4.9) has a dependence on its own past values of order two, on X_{t-1} and X_{t-2}; hence it has autoregressive order 2. It also has the dependence on the past values of Z_t, namely Z_{t-1}, of order 1; this dependence is called moving average. Thus the model is called autoregressive moving average model of order 2 and 1, respectively, denoted by ARMA(2,1). However, similar to (2.2.8), this is not the standard form of the ARMA(2,1) model. In the standard form, the random variable consisting of the last two terms of (2.4.9) is replaced by $(a_t - \theta_1 a_{t-1})$, to be written as

$$X_t = \phi_1 X_{t-1} + \phi_2 X_{t-2} + a_t - \theta_1 a_{t-1}, \qquad a_t \sim \text{NID}(0, \sigma_a^2) \qquad (2.4.10)$$

Thus the standard form of the ARMA(2,1) model may be used to write a modified version of (2.4.1a) as the standard discrete state model:

$$\mathbf{X}_t = \boldsymbol{\phi} \mathbf{X}_{t-1} + \boldsymbol{\theta} \mathbf{a}_t, \qquad \text{given } \mathbf{X}_t = \mathbf{X}_0 \quad \text{at} \quad t = 0 \qquad (2.4.1)$$

with its solution given by (2.4.2a) replacing $\mathbf{b}$ by $\boldsymbol{\theta}$ and $\mathbf{Z}$ by $\mathbf{a}$:

$$\mathbf{X}_t = \boldsymbol{\phi}_1^t \mathbf{X}_0 + \sum_{j=0}^{t-1} \boldsymbol{\phi}^j \boldsymbol{\theta} \mathbf{a}_{t-j} \qquad (2.4.2)$$

Note that the expanded version of (2.4.1) for the two-dimensional case clearly shows the standard form of the ARMA(2,1) model in state space:

$$\begin{bmatrix} X_t \\ X_{t-1} \end{bmatrix} = \begin{bmatrix} \phi_1 & \phi_2 \\ 1 & 0 \end{bmatrix} \begin{bmatrix} X_{t-1} \\ X_{t-2} \end{bmatrix} + \begin{bmatrix} 1 & -\theta_1 \\ 0 & 0 \end{bmatrix} \begin{bmatrix} a_t \\ a_{t-1} \end{bmatrix} \quad a_t \sim \text{NID}(0, \sigma_a^2)$$

(2.4.10a)

of which the first row is (2.4.10) and the second row is true but redundant.

2.4.3 Equivalence of Bivariate ARV(1) with Univariate ARMA(2,1): Need for Moving Average

It may appear strange at first sight that when we go from one- to two-dimensional state space we get the ARV(1) model (2.4.7a) when another series is available, whereas we get the ARMA(2,1) model (2.4.9) when the other series is not available. Why should we get two different models depending upon the availability of an additional series? The answer is that these two apparently different models are actually equivalent and their apparent difference is merely a consequence of the nonuniqueness of state space representation mentioned earlier. It is left as an exercise to show, by using the backshift operator (or z transform) and the transfer function algebra to solve for X_{1t} and X_{2t}, that a bivariate ARV(1) model for two series X_{1t} and X_{2t} implies an ARMA(2,1) model for X_{1t} alone and X_{2t} alone, provided all elements of the ϕ matrix are nonzero. In other words, the moving average part becomes essential when we attempt to model a two-dimensional state space by only one series of data and the moving average is no longer required when we use two different series of data simultaneously in modeling. The moving average part is a consequence of replacing the second state needed in a two-dimensional state space model by the preceding state of the same data rather than using another set of data.

The need for moving average in a univariate model persists even when all but one element of the **b** matrix in (2.4.6) are zero. In other words, as long as there is any randomness or noise in any one variable in the ARV(1) model, both variables, modeled individually, require moving average. When $\mathbf{b} = \mathbf{0}$, then of course the system is purely deterministic and the moving average, which relates only the stochastic part, is not required.

Although we have used a bivariate example for simplicity, it is easy to see that the results are true for any p. Thus a p-variate ARV(1) model for p series of data is equivalent to an ARMA($p, p-1$) model for each individual series. Moreover, a $2p$-variate ARV(1) model is equivalent to an ARMAV(2,1) model for any p of the $2p$ variables, and so on. (ARMAV models will be discussed shortly.) Similar results hold for continuous time, for which the moving average side becomes the right-hand side dynamics when the discrete time random a_t is replaced by a continuous time deterministic forcing function (see Sections 2.6.7 and 2.6.8.)

Autoregressive parameters can be estimated by linear least squares with relatively little computation. Moving average parameters require nonlinear least-squares estimation, generally accomplished by iterative linear least squares, with not only much more extensive computation, but also with other difficulties such as convergence problems, local minima, and subjective convergence tolerance choices. The analysis given above shows that we have to pay the price of such computation if we want to model a system containing at least one noise source by means of a limited number of series of data compared to the proper dimension of the state space.

ARMA$(n, n-1)$ and ARMAV$(n, n-1)$ models can be expressed as AR(∞) and ARV(∞) models, respectively. Thus, if we insist on using only autoregressive models even with limited sets of data to save computation, then the dimension of the state space will be unnecessarily large and this will introduce spurious dynamic modes, which are discussed more fully in Chapters 3 and 4.

This analysis also shows that modeling by pure autoregressive models may be justified without appreciable deterioration of accuracy and resolution when the system is nearly deterministic and/or a large number of data sets are used simultaneously in modeling compared to the proper dimension of the state space. This will be illustrated by modal analysis of simple systems with roughly known degrees-of-freedom in Chapter 5.

2.4.4 Univariate ARMA$(n, n-1)$

How can we check the adequacy of the ARMA(2,1) model and, if inadequate, how can we improve it? In other words, how do we know whether a_t's of the ARMA(2,1) model are independent and we need not do any further modeling, or, if a_t's are dependent, how can we complete the modeling process? Obviously, the answer is to go to a state space of dimension three. This leads us to an ARMA(3,2) model exactly as before, which is also given by the discrete state model (2.4.1) with

$$\mathbf{X}_t = \begin{bmatrix} X_t \\ X_{t-1} \\ X_{t-2} \end{bmatrix} \quad \boldsymbol{\phi} = \begin{bmatrix} \phi_1 & \phi_2 & \phi_3 \\ 1 & 0 & 0 \\ 0 & 1 & 0 \end{bmatrix} \quad \boldsymbol{\theta} = \begin{bmatrix} 1 & -\theta_1 & -\theta_2 \\ 0 & 0 & 0 \\ 0 & 0 & 0 \end{bmatrix} \quad \mathbf{a}_t = \begin{bmatrix} a_t \\ a_{t-1} \\ a_{t-2} \end{bmatrix}$$

In this case the second and the third rows are redundant and the first row gives the scalar forms of the ARMA(3,2) model:

$$X_t = \phi_1 X_{t-1} + \phi_2 X_{t-2} + \phi_3 X_{t-3} + a_t - \theta_1 a_{t-1} - \theta_2 a_{t-2}$$

We can continue to move into higher-dimensional state space and thus arrive at the ARMA$(n, n-1)$ model given by the discrete state model (2.4.1) in the n-dimensional state space with $\mathbf{X}_t = [X_t, X_{t-1}, X_{t-2}, \ldots, X_{t-n+1}]^T$,

$\mathbf{a}_t = [a_t, a_{t-1}, a_{t-2}, \ldots, a_{t-n+1}]^T$, and

$$\boldsymbol{\phi} = \begin{bmatrix} \phi_1 & \phi_2 & \cdots & \phi_{n-1} & \phi_n \\ 1 & 0 & \cdots & 0 & 0 \\ 0 & 1 & \cdots & 0 & 0 \\ \vdots & \vdots & \vdots & \vdots & \vdots \\ 0 & 0 & \cdots & 1 & 0 \end{bmatrix} \quad \boldsymbol{\theta} = \begin{bmatrix} 1 & -\theta_1 & -\theta_2 & \cdots & -\theta_{n-1} \\ 0 & 0 & 0 & \cdots & 0 \\ 0 & 0 & 0 & \cdots & 0 \\ \vdots & \vdots & \vdots & \vdots & \vdots \\ 0 & 0 & 0 & \cdots & 0 \end{bmatrix}$$

(2.4.11)*

the scalar form of the model being given by the first row as

$$X_t = \phi_1 X_{t-1} + \phi_2 X_{t-2} + \phi_3 X_{t-3} + \cdots + a_t - \theta_1 a_{t-1} - \theta_2 a_{t-2}$$
$$- \theta_3 a_{t-3} - \cdots - \theta_{n-1} a_{t-n+1} \tag{2.4.12}$$

2.4.5 Multivariate ARMAV$(n, n-1)$

The same approach can be used for the ARV(1) model (2.4.7a), or its standard form

$$\mathbf{X}_t = \boldsymbol{\phi}_1 \mathbf{X}_{t-1} + \mathbf{a}_t \tag{2.4.7}$$

To check whether it is adequate or not, and to find a better model if it is not adequate, we can move into a state space of double the dimension and arrive at the ARMAV(2,1) model

$$\mathbf{X}_t = \boldsymbol{\phi}_1 \mathbf{X}_{t-1} + \boldsymbol{\phi}_2 \mathbf{X}_{t-2} + \mathbf{a}_t - \boldsymbol{\theta}_1 \mathbf{a}_{t-1} \tag{2.4.13}$$

which, for the bivariate case, forms the top two rows of the discrete state model (2.4.1) with

$$\mathbf{X}_t = [X_{1t}, X_{2t}, X_{1t-1}, X_{2t-1}]^T, \quad \mathbf{a}_t = [a_{1t}, a_{2t}, a_{1t-1}, a_{2t-1}]^T$$

and

$$\boldsymbol{\phi} = \begin{bmatrix} \boldsymbol{\phi}_1 & \boldsymbol{\phi}_2 \\ \mathbf{I} & \mathbf{0} \end{bmatrix} \quad \boldsymbol{\theta} = \begin{bmatrix} \mathbf{I} & -\boldsymbol{\theta}_1 \\ \mathbf{0} & \mathbf{0} \end{bmatrix}$$

$$\boldsymbol{\theta}_1 = \begin{bmatrix} \theta_{111} & \theta_{121} \\ \theta_{211} & \theta_{221} \end{bmatrix} \quad \boldsymbol{\phi}_i = \begin{bmatrix} \phi_{11i} & \phi_{12i} \\ \phi_{21i} & \phi_{22i} \end{bmatrix} \quad i = 1, 2 \tag{2.4.14}$$

---

* This choice of θ, although convenient, makes it singular. An alternative choice, more along the lines of control system state space literature, may be found in Pandit and Mehta (1985).

Thus for an arbitrary multivariate series, we can continue to move into higher and higher dimension and arrive at the ARMAV$(n, n-1)$ model

$$\mathbf{X}_t = \boldsymbol{\phi}_1 \mathbf{X}_{t-1} + \boldsymbol{\phi}_2 \mathbf{X}_{t-2} + \cdots + \boldsymbol{\phi}_n \mathbf{X}_{t-n} + \mathbf{a}_t - \boldsymbol{\theta}_1 \mathbf{a}_{t-1} - \boldsymbol{\theta}_2 \mathbf{a}_{t-2}$$
$$- \boldsymbol{\theta}_3 \mathbf{a}_{t-3} - \cdots - \boldsymbol{\theta}_{n-1} \mathbf{a}_{t-n+1} \qquad (2.4.15)$$

This model is given by the first p rows of the discrete state model (2.4.1), if p is the number of series, with $\mathbf{X}_t$ and $\mathbf{a}_t$ consisting of the column vectors

$$\mathbf{X}_t = [X_{1t}, X_{2t}, \ldots, X_{pt}, X_{1t-1}, X_{2t-1}, \ldots, X_{pt-n+1}]^T$$
$$\mathbf{a}_t = [a_{1t}, a_{2t}, \ldots, a_{pt}, a_{1t-1}, a_{2t-1}, \ldots, a_{pt-n+1}]^T$$

and

$$\boldsymbol{\phi} = \begin{bmatrix} \boldsymbol{\phi}_1 & \boldsymbol{\phi}_2 & \boldsymbol{\phi}_3 & \cdots & \boldsymbol{\phi}_n \\ \mathbf{I} & \mathbf{0} & \mathbf{0} & \cdots & \mathbf{0} \\ \mathbf{0} & \mathbf{I} & \mathbf{0} & \cdots & \mathbf{0} \\ \vdots & \vdots & \vdots & \vdots & \vdots \\ \mathbf{0} & \mathbf{0} & \mathbf{0} & \mathbf{I} & \mathbf{0} \end{bmatrix}$$

$$\boldsymbol{\theta} = \begin{bmatrix} \mathbf{I} & -\boldsymbol{\theta}_1 & -\boldsymbol{\theta}_2 & -\boldsymbol{\theta}_3 & \cdots & -\boldsymbol{\theta}_{n-1} \\ \mathbf{0} & \mathbf{0} & \mathbf{0} & \mathbf{0} & \cdots & \mathbf{0} \\ \mathbf{0} & \mathbf{0} & \mathbf{0} & \mathbf{0} & \cdots & \mathbf{0} \\ \vdots & \vdots & \vdots & \vdots & \vdots & \vdots \\ \mathbf{0} & \mathbf{0} & \mathbf{0} & \mathbf{0} & \cdots & \mathbf{0} \end{bmatrix} \qquad (2.4.16)$$

$$\boldsymbol{\phi}_k = \{\phi_{ijk}\}, \quad k = 1, 2, \ldots, n; \quad \boldsymbol{\theta}_k = \{\theta_{ijk}\}, \quad k = 1, 2, \ldots, n-1$$

where $\{\ \}$ denotes a matrix with a typical element inside, in this case a $p \times p$ matrix. $\mathbf{I}$ and $\mathbf{0}$ denote $p \times p$ identity and zero matrices.

The standard form of the discrete state model (2.4.1) with parameters given by (2.4.16) is so far the most general model, with all the previous discrete models as its special cases. For example, the scalar AR(1) is $p = 1$, $n = 1$; scalar ARMA(2,1) is $p = 1$, $n = 2$; bivariate ARV(1) is $p = 2$, $n = 1$; bivariate ARMAV(2,1) is $p = 2$, $n = 2$; and so on. Therefore, once we learn how to evaluate $\boldsymbol{\phi}^t$ in Chapter 3, the solution for all the ARMAV models will be available from (2.4.2). Then we will be able to understand and analyze the system in its natural deterministic and stochastic components. For multivariate series, this will also provide us with the relation between different variables in both its deterministic and stochastic components. Thus the power of this model is immense.

ILLUSTRATION OF THE DDS MODELING APPROACH   35

The power of the model is further magnified by its relation, as a sampled representation, to the continuous-time state model (2.4.3) and its solution (2.4.4) via the simple relations such as (2.3.5). [Equation (2.3.5) is true in the same form with matrix parameters only for the ARV(1) model; its generalization for the discrete state model (2.4.1) is given in Chapter 5.] The continuous-time state model (2.4.3) is itself very powerful and widely used because it naturally arises in modeling physical systems by the conventional methods illustrated later in Section 2.6.

The discrete state model and its solution (2.4.1) and (2.4.2) represent an initial value problem. This initial value formulation is useful in modeling transient response with unwanted noise in which initial values play an important role, as discussed in Chapter 5. It is also useful in modeling forced response with unwanted noise as in Chapters 5–6, and in spectrum analysis of noise discussed in Chapter 6; initial values are not important in these applications. Further generalization of (2.4.1) and (2.4.3) to a boundary value problem is possible and useful in signal analysis, but is not discussed in this book.

2.5 ILLUSTRATION OF THE DDS MODELING APPROACH

The rationale of the DDS modeling strategy given at the beginning of Section 2.4 was somewhat abstract and may seem esoteric because the rationale was also used to introduce the ARMAV models. Now that we have become acquainted with the scalar ARMA and vector ARMAV models, a more concrete illustration of the modeling strategy can be provided, first by using the familiar three-dimensional space to geometrically illustrate the transition from independent data to AR(1) to ARMA(2,1) model, and then by giving some numerical examples of the models for the blast furnace silicon content data introduced in Section 2.1.

2.5.1 Independent Data: The "AR(0)" Model

We begin with the independent data, represented by the "zero" dimensional origin, as shown in Figure 2.4(a). The model for the system can be thought of as the "AR(0)" model

$$X_t = a_t$$

Thus, this trivial model for independent data simply defines the coordinate axes along time instances $t, t - 1, t - 2, \ldots$. It also shows that the projection of the vector X_t is the origin, which requires a zero-dimensional state space to describe it. Moreover, since X_t's are orthogonal, they can be represented along the orthogonal axes a_t. Note that the unit vectors along the orthogonal axes would be Z_t, but since we have to use the standard form of the model in fitting it to the data, as illustrated in Section 2.3, we can also think of the a_t's as the unit vectors for a given set of data, keeping in mind that their lengths are really σ_a. Actually

36   DATA DEPENDENT SYSTEMS IN STATE SPACE

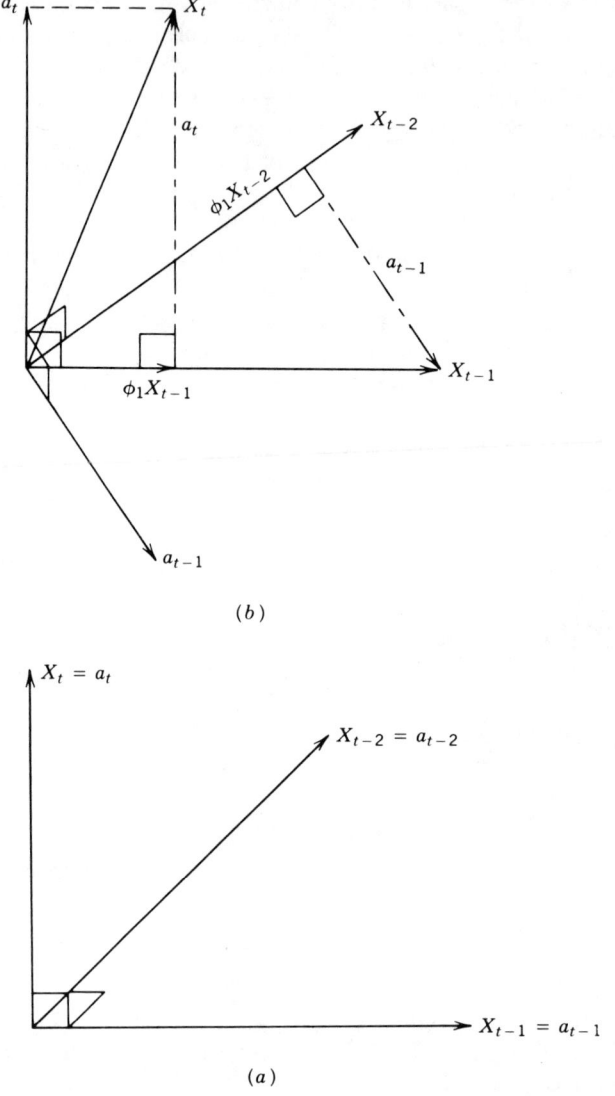

Figure 2.4 State space representation of the AR(1) model. (a) $\phi_1 = 0$, independent data. (b) $\phi_1 > 0$, dependent data.

a_t can take any value as a random variable and σ_a^2 is only their expected squared value, which we use to represent squared length. It should also be noted that although we cannot effectively show them in our familiar three-dimensional view, there are other orthogonal axes $a_{t-3}, a_{t-4}, \ldots$, not shown in the figure; in this sense all the figures used in this section are not strictly correct and have only a pictorial illustrative importance.

2.5.2 The AR(1) Model

If the data is dependent, then $X_t, X_{t-1}, X_{t-2}, \ldots$ are no longer orthogonal and hence cannot be drawn along the orthogonal axes a_t as in Figure 2.4(a). In particular, X_t is no longer along a_t, but makes an angle to it. Therefore, it can be decomposed into two components, one a_t and the other orthogonal to it. How can we get this orthogonal component? It will be the projection of X_t on the "plane" (strictly speaking, hyperplane) spanned by $a_{t-1}, a_{t-2}, \ldots$, or alternatively spanned by $X_{t-1}, X_{t-2}, \ldots$. In fact, this projection $(X_t - a_t)$ is the part of X_t that can be predicted from, and therefore can be expressed in terms of the linear combination of, the past data $X_{t-1}, X_{t-2}, \ldots$.

If X_t can be predicted from X_{t-1} alone and does not require $X_{t-2}, X_{t-3}, \ldots$, then the projection will be along the line X_{t-1}, say $\phi_1 X_{t-1}$, the picture would look like the one shown in Figure 2.4(b) using $\phi_1 > 0$ for convenience, and can be algebraically represented by

$$X_t = \phi_1 X_{t-1} + a_t$$

which is of course the AR(1) model. In other words, the AR(1) model requires that the vectors X_t, X_{t-1}, and the perpendicular a_t from X_t on the space of (not predictable from) $X_{t-1}, X_{t-2}, \ldots$, form a plane or a two-dimensional space. The state space is now a one-dimensional line along X_{t-1}, and the AR(1) model is a restriction that the projection $(X_t - a_t)$ is in this state space. To emphasize that t is only a running index, we have also shown the representation

$$X_{t-1} = \phi_1 X_{t-2} + a_{t-1}$$

in Figure 2.4(b), particularly since it can be done quite easily in this case. Actually it is redundant, and will not be repeated in the subsequent representations, which become slightly more involved. Note that if $\phi_1 = 0$, Figure 2.4(b) reduces to Figure 2.4(a) and the AR(1) model reduces to AR(0) model, as it should.

2.5.3 The ARMA(2,1) Model

If the projection of X_t on the hyperplane spanned by $a_{t-1}, a_{t-2}, \ldots$ or by $X_{t-1}, X_{t-2}, \ldots$ does not fall on the line X_{t-1}, the one-dimensional state space or the AR(1) model would not be adequate. To check this and to find a better model if necessary, we move to the two-dimensional state space shown in Figure 2.5.

Now the triplet X_t, X_{t-1}, and a_t would not form a plane but a three-dimensional space that contains the projection $(X_t - a_t)$. If X_{t-2} and a_{t-1} are also contained in this three-dimensional space, then we can draw a line from the point of projection $(X_t - a_t)$ parallel to a_{t-1}. This line must intersect the plane formed by X_{t-1} and X_{t-2} as shown in Figure 2.5(a) because X_{t-1} and a_{t-1} and X_{t-2} do

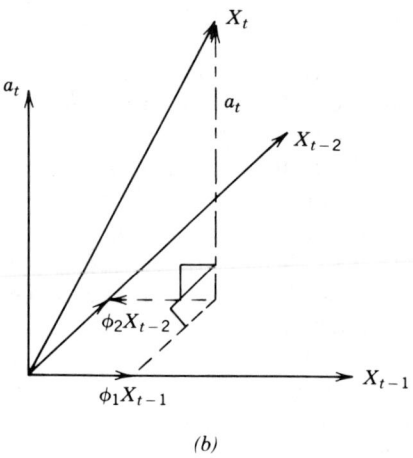

(b)

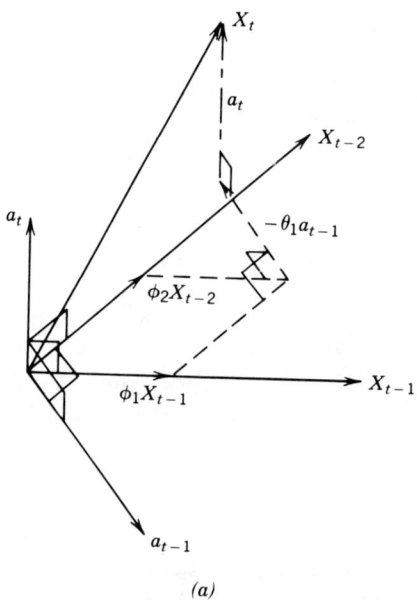

(a)

Figure 2.5 State space representation of the ARMA(2, 1) model. (a) $\theta_1 \neq 0$, (b) $\theta_1 = 0$, AR(2) model.

not form a plane. Denoting this point of intersection $(\phi_1 X_{t-1} + \phi_2 X_{t-2})$ we find that if the projection is in the three-dimensional space spanned by a_t, X_{t-1}, and X_{t-2} which contains a_{t-1}, then the two-dimensional state space is enough and the adequate model is ARMA(2,1):

that is,
$$X_t - a_t = -\theta_1 a_{t-1} + \phi_1 X_{t-1} + \phi_2 X_{t-2}$$

$$X_t = \phi_1 X_{t-1} + \phi_2 X_{t-2} - \theta_1 a_{t-1} + a_t$$

as shown in Figure 2.5(a):

In particular, if the projection is not only in the three-dimensional space spanned by a_t, X_{t-1}, and X_{t-2}, but actually falls on the plane formed by X_{t-1} and X_{t-2}, then we get the AR(2) model as shown in Figure 2.5(b):

$$X_t = \phi_1 X_{t-1} + \phi_2 X_{t-2} + a_t$$

This model is of course a special case of the two-dimensional ARMA(2,1) model with $\theta_1 = 0$.

Note that the dimension of the state space is the same as the AR order in scalar case, and is one less than the dimension needed to show it pictorially. Thus to show the zero-dimensional state space of origin for the AR(0) model, we need a line along a_t in Figure 2.4(a). To show the one-dimensional state space of the line along X_{t-1} for the AR(1) model, we need the plane of a_t and X_{t-1} in Figure 2.4(b). Finally, to show the two-dimensional state space of the plane spanned by X_{t-1} and X_{t-2}, we need the three-dimensional space of a_t, X_{t-1}, and X_{t-2} shown in Figures 2.5(a and b).

For deterministic systems with correct (adequate) model $a_t \equiv 0$ and our definition of the dimension of the state space agrees with the classical one discussed in Section 2.6. Also note that the dimension of the state space is thus defined only based on the data $X_t, X_{t-1}, \ldots$ and not by $a_t, a_{t-1}, \ldots$ manufactured from the data by the fitted model. In the multivariate case for p series, the dimension of the state space would be p times the AR order. For this case as well as the higher-order scalar case, it is difficult to give a pictorial representation of the state space. However, the idea of finding an orthogonal projection from the present state on the past states, with the perpendicular given by a_t when the correct dimension of the state space is reached, will be elaborated in Chapter 3 with the help of the algebraic Gram–Schmidt procedure.

2.5.4 Stopping Criterion—Stochastic Systems: *F*-Test

How long should we continue to increase the dimension of the state space and the AR order of the model? As long as we continue to reduce the length of a_t's, measured by σ_a^2 or residual sum of squares of a_t's, *significantly*. Since the a_t's are random variables, the significance of the reduction in their variance by going from a lower- to a higher-order model with more parameters can be judged by the F-test adapted from linear regression estimation theory in statistics. This is applicable when the sample autocorrelations of the residual a_t's are less than about two times their standard deviations for most lags. (See Chapter 6 or Appendix A at the end of the book for definitions.) The ratio of the sample autocorrelation to its standard deviation is known as unified autocorrelation. When these are less than 2 in absolute value for most lags, the a_t's can be considered to be independent. If we then apply the F-test for the residual sum of squares of the successive models, and the calculated F value is less than the table F value for a specified significance such as 5%, we can stop. The F-test is a test of the hypothesis that some of the parameters in a model are restricted to zero. In the case of linear regression models this is a well-known test for which the criterion may be found in a standard statistics text such as Rao (1965). If the linear regression model has r parameters and we want to test whether s of these are restricted to zero based on N observations, then the criterion is

$$F = \frac{A_1 - A_0}{s} \div \frac{A_0}{N - r} \sim F(s, N - r) \qquad (2.5.1)$$

where A_0 is the (generally smaller) sum of squares of the unrestricted model with r parameters, A_1 is the (generally larger) sum of squares of the restricted model with $(r - s)$ parameters, and $F(s, N - r)$ denotes F distribution with s and $(N - r)$ degrees of freedom.

For the ARMA(n,m) model without mean estimation the above criterion can be used to test the hypothesis that s out of its $(m + n) = r$ parameters are zero. Then, A_0 becomes the residual sum of squares of the ARMA(n,m) model and A_1 that of the model with s parameters dropped out. The justification of the criterion together with its interpretation as a convergence criterion may be found in Pandit (1973).

Stochastic system modeling and the application of the F criterion are illustrated in Table 2.1 for the silicon percentage data. Is the length of the residual a_t's measured by σ_a reduced significantly when we go from a one-dimensional AR(1) model to a two-dimensional ARMA(2,1) model?

Before we answer this question by the F criterion we have to make sure that the absolute values of unified autocorrelations of the a_t's are less than 2. This value of 2 is an approximation of 1.96 and is used because if the sample residual autocorrelations have zero mean and standard deviation σ_a, then there is 95% probability that they will be between $\pm 1.96\sigma_a$ or the ratio of sample autocorrelation with its standard deviation called unified autocorrelation will be between ± 1.96.

Since this limit of 2 for the unified autocorrelation is based on 95% probability, there is a 5% chance that a unified sample autocorrelation is greater than 2 in absolute value even when the theoretical autocorrelation is zero and a_t's are independent. For this reason, we may ignore one unified autocorrelation greater than 2 in absolute value occurring at lag 10 indicated in the fourth column of Table 2.1 and apply the F-test.

Table 2.1 Illustration of Stochastic System Modeling[a]

ARMA Order	Final Residual Sum of Squares	Mean $\hat{\mu}$ With 95% Confidence Interval	Lags of Unified Autocorrelation > 2	Average of Residuals	Residual Variance $\hat{\sigma}_a^2$	Residual Standard Deviation $\hat{\sigma}_a$	F value
(1,0)	7.4791	1.0594 ± 0.043958	10	0.718981×10^{-7}	0.03758	0.19386	
(2,1)	7.3486	1.0583 ± 0.042457	10	0.768415×10^{-4}	0.03693	0.19217	1.740
(3,2)	6.7690	1.0771 ± 0.037887	10	-0.293362×10^{-3}	0.03402	0.18443	8.306
(4,3)	7.0023	1.0575 ± 0.038187	—	0.14149×10^{-2}	0.03519	0.18758	−3.199
(5,4)	6.2382	1.0056 ± 0.047111	—	0.220365×10^{-1}	0.03135	0.17705	11.637
(6,5)	6.1993	1.0532 ± 0.034868	—	0.43932×10^{-2}	0.03115	0.17650	0.589

[a] Silicon % data: $N = 200$ points; average $\bar{x} = 1.055$; variance $\hat{\sigma}_x^2 = 0.04578$.

The ARMA(2,1) model has $n = 2$, $m = 1$ and, including the additional parameter mean, has

$$r = m + n + 1 = 2 + 1 + 1 = 4$$

parameters. The AR(1) model has

$$1 + 0 + 1 = 2$$

parameters. Thus

$$s = 4 - 2 = 2$$

and from Table 2.1

$N = 200$ (number of observations in the series)
$A_1 = 7.4791$ [sum of squares of AR(1)]
$A_0 = 7.3486$ [sum of squares of ARMA(2,1)]

Hence

$$F = \frac{7.4791 - 7.3486}{2} \div \frac{7.3486}{(200 - 4)}$$

$$= 1.740$$

From the F distribution Table A at the end of the book, $F_{0.95}(2,196) = F_{0.95}(2, \infty) = 3.00$. Since calculated $F < F_{0.95}(s, N - r)$ from the table, the reduction in the length of the residuals σ_a is not statistically significant and we may conclude that the AR(1) model is adequate.

If we, however, choose not to ignore the unified autocorrelation greater than 2 in absolute value at lag 10, then, as Table 2.1 shows, the adequate model turns out to be ARMA(5,4). The residual sum of squares of ARMA(5,4) is reduced to 6.2382 compared to 7.4791 of the AR(1); this translates into a reduction of $\hat{\sigma}_a$ from 0.19386 for AR(1) to 0.17705 for the ARMA(5,4).

Since the F-test is essentially a test to check overmodeling, it requires that a_t's of both models be independent or uncorrelated, that is, their unified autocorrelations be less than 2 in absolute value. Therefore, in Table 2.1, the F-test is theoretically valid only between ARMA(4,3) to (5,4) where it is found to be significant, indicating that ARMA(4,3) is not adequate and ARMA(5,4) is a better model, and between ARMA(5,4) to (6,5) where it is found to be insignificant, indicating that ARMA(5,4) is the adequate model and ARMA(6,5) is overmodeling.

The calculated F value is generally large when the state space dimensions are much lower than that of the adequate one, thus correctly indicating that we should increase the dimension even though the test itself is not theoretically valid. This may be verified by calculating the F value for AR(0) to AR(1) as 43.2 using $A_1 = (N - 1) \times$ Variance $= 9.1281$; since $r = 2$, $s = 1$, and $F_{0.999}(1, \infty) = 10.83$, the zero-dimensional AR(0) model is inadequate.

The limitations of the nonlinear least squares minimization routines may, however, give false indications, as seen in Table 2.1 for ARMA(3,2) to (4,3). Thus an insignificant F value should be used to stop increasing the dimension only when the unified autocorrelations are also less than 2 in absolute value. Many more examples illustrating the modeling procedure and the F criterion for single as well as multiple time series may be found in Pandit and Wu (1983).

2.5.5 Stopping Criterion—Deterministic System

As the example given above illustrates, the F criterion is often too stringent, and may require unnecessarily high-order models. In such cases we may be able to stop at a reasonable order model using other criteria. This is particularly true for systems with dominant deterministic components. If the length of the a_t's measured by σ_a is of the order of magnitude considered insignificantly small in the particular problem, we can stop at a lower order model, disregarding the F-test.

To illustrate such a case, consider the data generated by the model

$$X_t = 100(0.9)^t, \quad t = 1, 2, \ldots, 200$$

If the AR(1) model with the mean as an additional parameter is estimated using the computer program from Pandit and Wu (1983), we have the following estimates with their 95% confidence intervals. (Results without mean estimation are given in Section 2.7.)

$$\text{Mean} = 0.13823 \times 10^{-8} \pm 0.17317 \times 10^{-7}$$

$$\phi_1 = 0.90000 \pm 0.13564 \times 10^{-8}$$

Residual sum of squares $= 0.4043 \times 10^{-11}$
Average of the residuals $= -0.107213 \times 10^{-7}$
Absolute value of unified autocorrelations > 2, at lags 1, 12, and 16

This gives us the estimate of the residual standard deviation as

$$\sigma_a = \sqrt{0.4043 \times 10^{-11}/199} \approx 1.4 \times 10^{-7}$$

Now, since the data itself is single precision, it is accurate only up to seventh decimal place because the eighth decimal place may contain roundoff errors. Hence, by moving to a higher dimensional space by fitting the ARMA(2,1) model, we would be simply modeling the dynamics of roundoff and other computer operations with the help of the same computer! Thus we have already reached the "noise floor" for this particular problem, and any further improvement, even if possible, is of no practical importance. Therefore, we can stop. Note that the estimates obtained are also extremely accurate and do not require any refinement. Since the mean is practically zero we see that the model is given by the deterministic difference equation

$$X_t = 0.9 X_{t-1}$$

as expected. In fact, further modeling shows that an ARMA(2,1) modeling gives less accurate estimates and higher residual sum of squares. Although this should not happen in theory, it may occur in practice because the minimization routines used in the computer for minimizing the residual sum of squares of a_t's have their own limitations.

We can formally quantify this argument to develop a general criterion for determining whether the system is deterministic from the results of the fitted model. If A is the digital accuracy or the noise floor of the data (e.g., 10^{-8} for single precision data simulated on the UNIVAC 1100/80 computer used for the examples in this chapter), then A times the sample standard deviation of the data can be taken as the lower limit for the "practical zero" of this data. Therefore, if

$$\sigma_a^2 < A^2 \times \text{Sample Data Variance} \tag{2.5.2a}$$

or

$$\text{Residual Sum of Squares} < NA^2 \times \text{Sample Data Variance} \tag{2.5.2}$$

then the system may be considered to be deterministic. It is left as an exercise to show that for the AR(1) model with $|\phi| < 1$ and large N this criterion reduces to

$$\text{Residual Sum of Squares} < A^2 X_0^2 \phi_1^2 / (1 - \phi_1^2) \tag{2.5.3}$$

and that it is satisfied in the example given above.

2.6 CONVENTIONAL MODELING IN MECHANICS: ORIGIN OF STATE SPACE

State space concepts evolved out of the classical dynamics of particles and rigid bodies. In this section we will briefly review this foundation of state space techniques and then develop the model for an n-degree-of-freedom system in the modern state space framework with a threefold purpose. First, the motivation and rationale for defining the dimension of state space in the previous section will be clarified. Second, the important application of DDS to modal analysis discussed in subsequent chapters is founded on such a model. Third, even for systems from fields such as business and economics, which are not mechanical, such a model helps in their interpretation and becomes an aid in arriving at policy decisions based on the readily available economic data. In view of this specific aim, we will not venture into state space analysis of modern control theory for which many good texts are available, for example, Ogata (1967) and Brogan (1985).

2.6.1 Definition of State and State Space

The position of a moving particle in space may be specified using Cartesian coordinates. Then, for a system of k particles, $3k$ coordinates are required. However, there may be constraints on the system so that not all of these $3k$ coordinates are linearly independent. If there are s constraints on the system, then we can define $(3k - s)$ linearly independent variables and express all the coordinates in terms of these variables. Lagrange used such linearly independent variables, called generalized coordinates q_i, to describe the motion of a conservative system. The number of independent quantities required to specify the position of a system uniquely is called the number of degrees of freedom of that system. Thus the number of generalized coordinates is equal to the number of degrees of freedom.

For example, if a system consisting of a set of n particles (n masses) is constrained to move only in one direction, say horizontal, then $s = 2n$. Thus the position of the system is uniquely specified by $3n - 2n = n$ generalized coordinates $q_1, q_2, \ldots, q_n$, which in this simple case can be chosen to be the horizontal position of each of the masses, say $x_1, x_2, \ldots, x_n$. The system then has n degrees of freedom. Note that any set of quantities by which the configuration of a particle or a system can be uniquely specified forms a set of generalized coordinates.

Similarly, a *state* of a deterministic dynamic system is the smallest set of numbers which must be specified at some initial time in order to be able to predict uniquely the behavior of the system at subsequent times for a given input. Such numbers are called state variables. If at least n *state variables*, say $x_1, x_2, \ldots, x_n$, are required to describe the behavior of a system, then they may be considered as n components of a vector **x** called a *state vector*. A *state space* is then defined as an n-dimensional space with coordinates $x_1, x_2, \ldots, x_n$. State variables need not be measurable or even physically meaningful, although for most applications in this book we take the state vector as a vector of measurements. This somewhat arbitrary choice of state variable is responsible for the nonuniqueness of the state space formulation and also for its mathematical elegance and tractability.

2.6.2 Lagrange and Hamilton Equations of Motion

The mechanical state of a system is completely determined if the positions q_i and velocities $\dot{q}_i$ at some instant are simultaneously specified. Equations of motion then specify these quantities at subsequent times. Lagrange showed that the equations of motion of a free conservative system are given by

$$\frac{d}{dt}\left(\frac{\partial L}{\partial \dot{q}_i}\right) - \frac{\partial L}{\partial q_i} = 0 \qquad i = 1, 2, \ldots, n \qquad (2.6.1)$$

where L is the Lagrangian defined by

$$L = T(\dot{q}, q) - V(q) \qquad (2.6.2)$$

and T denotes the kinetic energy, which is a function of position and velocity only, and V denotes the potential energy, which is a function of position only.

Hamilton showed that by introducing a new function, now called Hamiltonian, that can be simply defined for a conservative system as

$$H = T + V \tag{2.6.3}$$

and defining new coordinates called generalized momentum coordinates:

$$p_i = \frac{\partial T}{\partial \dot{q}_i} \qquad i = 1, 2, \ldots, n$$

the equations of motion can be simplified to what are now known as Hamilton's equations of motion

$$\dot{q}_i = \frac{\partial H}{\partial p_i}$$

$$\dot{p}_i = -\frac{\partial H}{\partial q_i} \qquad i = 1, 2, \ldots, n \tag{2.6.4}$$

These are two sets of *first-order* homogeneous differential equations because the right-hand sides are functions of p and q only and do not involve $\dot{p}$ and $\dot{q}$. Only the kinetic energy T in the Hamiltonian H defined by (2.6.3) is a function of $\dot{q}$ and $\partial T/\partial \dot{q}_i$ is now replaced by p_i. Thus (2.6.4) can be formulated as the general first-order state equation (2.4.3) [with $\mathbf{u}(t) = \mathbf{0}$, since we are dealing with homogeneous equations] and readily solved using the solution (2.4.4) for arbitrary n. On the other hand, the original Lagrange equation (2.6.1) generally leads to second-order equations, as we illustrate below, and is therefore not so easy to solve. This lays the foundation of the state space approach in reducing arbitrary order differential equations to the state model in the form of a vector first-order differential equation, whose solution and the resultant system analysis is succinctly and elegantly accomplished by the matrix theory discussed in Chapter 3.

2.6.3 State Model for Undamped Free One-Degree-of-Freedom System

The simplest illustration of this approach can be given by considering a mass m tied to a spring with spring constant k so that the spring force is kx for displacement x. As shown schematically in Figure 2.6, since the mass is free to move only in the horizontal direction on a frictionless surface, we have a one-degree-of-freedom conservative (no dissipation of energy in friction) free (no external force) system. Hence Lagrange's equation (2.6.1) can be applied. Since there is only one coordinate

$$q = x, \qquad \dot{q} = \dot{x}$$

CONVENTIONAL MODELING IN MECHANICS   47

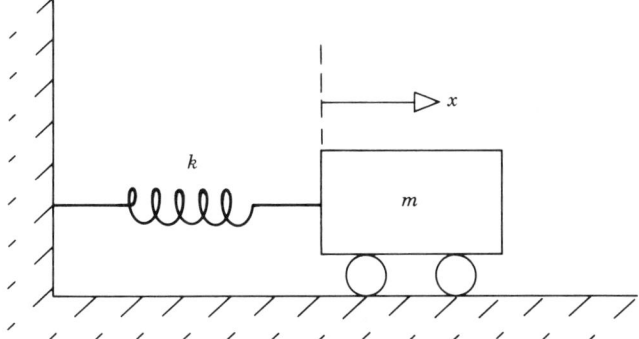

Figure 2.6 One-degree-of-freedom conservative system.

and the kinetic and potential energies are respectively given by

$$T = \tfrac{1}{2}m\dot{x}^2, \qquad V = \tfrac{1}{2}kx^2$$

Substituting in Lagrange's equation (2.6.1) gives

$$\frac{d}{dt}(m\dot{x}) + kx = 0, \quad \text{that is,} \quad m\ddot{x} + kx = 0 \quad \text{or} \quad \ddot{x} + \frac{k}{m}x = 0. \qquad (2.6.5)$$

which is the equation of motion of an undamped one-degree-of-freedom vibratory system with natural frequency $\sqrt{k/m}$. It is a second-order homogeneous differential equation, as suggested above.

On the other hand, Hamilton's approach introduces the additional coordinate

$$p = m\dot{x}$$

the Hamiltonian is

$$H = \frac{1}{2m}p^2 + \frac{1}{2}kq^2$$

and substituting in Hamilton's equations of motion (2.6.4) gives the pair of first-order homogeneous differential equations for the same system as

$$\dot{q} = \frac{1}{m}p$$

$$\dot{p} = -kq$$

(2.6.6a)

Alternatively, the last form of (2.6.5)

$$\ddot{x} + \frac{k}{m} x = 0$$

can be directly reduced to a slightly different form of (2.6.6a) by the choice of the state variables

$$x = q, \quad \dot{x} = p$$

to

$$\dot{q} = p$$

$$\dot{p} = -\frac{k}{m} q \tag{2.6.6}$$

This pair of equations can now be written in the standard form of the continuous-time state equation (2.4.3) as

$$\dot{\mathbf{x}} = \mathbf{A}\mathbf{x}$$

where the state vector $\mathbf{x}$ and the 2×2 matrix $\mathbf{A}$ are given by

$$\mathbf{x} = \begin{bmatrix} q \\ p \end{bmatrix} = \begin{bmatrix} x \\ \dot{x} \end{bmatrix} \quad \mathbf{A} = \begin{bmatrix} 0 & 1 \\ -\frac{k}{m} & 0 \end{bmatrix} \tag{2.6.7}$$

where it is seen by the standard row column multiplication of the matrix $\mathbf{A}$ and the state vector $\mathbf{x}$ that the first row is redundant and the second row is the same as the main equation given by the last form of (2.6.5).

Now given $\mathbf{x}(0) = (x(0), \dot{x}(0))^T$, i.e. the initial position and the initial velocity, the solution of the differential equation is directly obtained from (2.4.4) as

$$\mathbf{x}(t) = e^{\mathbf{A}t} \mathbf{x}(0)$$

and it is seen from the definition of the state vector (2.6.7) that the top element of this vector gives the position as a function of time and the bottom element specifies the velocity as a function of time. Thus, even though the top row of the state equation is redundant, both rows of the solution give physically meaningful results.

It is now clear that one need not explicitly use Hamilton's equations of motion (requiring the additional definition of H) in order to develop the standard form of state equations. Once the equations of motion are known in any form, they can be directly converted to the standard form of vector first-order differential equations by suitably defining the state variables, as illustrated above. This simple and direct route is generally followed in the modern state

2.6.4 State Model for Damped Forced One-Degree-of-Freedom System

Rayleigh extended Lagrange's equations to nonconservative systems by introducing the dissipation energy function D. With the inclusion of dissipation function D and an external force Q_i on the ith particle, Lagrange's equations generalize to

$$\frac{d}{dt}\left(\frac{\partial T}{\partial \dot{q}_i}\right) - \frac{\partial T}{\partial q_i} + \frac{\partial D}{\partial \dot{q}_i} + \frac{\partial V}{\partial q_i} = Q_i \qquad (2.6.8)$$

for a nonconservative forced system.

To illustrate, let us modify the system in Figure 2.6 by adding a damper providing force proportional to velocity, say $c\dot{x}$, as shown in Figure 2.7, and an external force $u(t)$. Since the system is still free to move only in the horizontal direction, it still has one degree of freedom, and hence needs only one coordinate $q = x$. Using

$$T = \tfrac{1}{2}m\dot{x}^2, \qquad V = \tfrac{1}{2}kx^2, \qquad \text{and} \quad D = \tfrac{1}{2}c\dot{x}^2$$

and substituting in (2.6.8) gives

$$m\ddot{x} + c\dot{x} + kx = u$$

or

$$\ddot{x} + \frac{c}{m}\dot{x} + \frac{k}{m}x = \frac{1}{m}u \qquad (2.6.9)$$

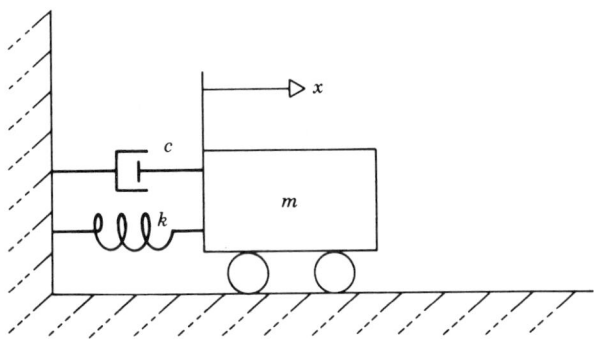

Figure 2.7 One-degree-of-freedom nonconservative system.

which is the equation of motion for a damped forced one-degree-of-freedom vibratory system.

Its state equation in the standard form

$$\dot{\mathbf{x}} = \mathbf{A}\mathbf{x} + \mathbf{B}\mathbf{u}$$

can be found by the obvious choice

$$\mathbf{x} = \begin{bmatrix} x \\ \dot{x} \end{bmatrix}, \quad \mathbf{A} = \begin{bmatrix} 0 & 1 \\ -\dfrac{k}{m} & -\dfrac{c}{m} \end{bmatrix}, \quad \mathbf{B} = \begin{bmatrix} 0 & 0 \\ \dfrac{1}{m} & 0 \end{bmatrix}, \quad \text{and} \quad \mathbf{u} = \begin{bmatrix} u \\ 0 \end{bmatrix} \quad (2.6.10)$$

which reduces to (2.6.7) when c and u are zero, and the system is undamped and free.

2.6.5 State Model for n-Degree-of-Freedom Multivariate System Illustrated by $n = 2$

For a general n-degree-of-freedom system with generalized coordinates given by the column vector

$$\mathbf{q} = (q_1, q_2, q_3, \ldots, q_n)^T,$$

the kinetic, potential, and dissipation energies are given by the quadratic forms

$$
\begin{aligned}
T &= \frac{1}{2} \sum_{i,j=1}^{n} m_{ij} \dot{q}_i \dot{q}_j \\
&= \tfrac{1}{2} \dot{\mathbf{q}}^T \mathbf{m} \dot{\mathbf{q}} \\
V &= \frac{1}{2} \sum_{i,j=1}^{n} k_{ij} q_i q_j \qquad (2.6.11) \\
&= \tfrac{1}{2} \mathbf{q}^T \mathbf{k} \mathbf{q} \\
D &= \frac{1}{2} \sum_{i,j=1}^{n} c_{ij} \dot{q}_i \dot{q}_j \\
&= \tfrac{1}{2} \dot{\mathbf{q}}^T \mathbf{c} \dot{\mathbf{q}}
\end{aligned}
$$

where

$$\mathbf{m} = \{m_{ij}\}, \qquad \mathbf{k} = \{k_{ij}\}, \qquad \text{and} \quad \mathbf{c} = \{c_{ij}\}$$

are $n \times n$ mass, stiffness, and damping matrices; note that when the energies are positive, the corresponding quadratic forms, and hence the matrices, are positive definite. Let the external forces Q_i acting on the masses be given by the elements of the column vector

$$\mathbf{u} = (u_1, u_2, u_3, \ldots, u_n)^T$$

Then a straightforward application of the Lagrange's equation of motion (2.6.8) yields

$$\mathbf{m\ddot{q} + c\dot{q} + kq = u}$$

or

$$\mathbf{\ddot{q} + m^{-1}c\dot{q} + m^{-1}kq = m^{-1}u} \qquad (2.6.12)$$

and its state equation is obtained by the choice of

$$\underset{2n \times 1}{\mathbf{x} = \begin{bmatrix} \mathbf{q} \\ \mathbf{\dot{q}} \end{bmatrix}}, \quad \underset{2n \times 2n}{\mathbf{A} = \begin{bmatrix} \mathbf{0} & \mathbf{I} \\ -\mathbf{m^{-1}k} & -\mathbf{m^{-1}c} \end{bmatrix}}, \quad \underset{2n \times n}{\mathbf{B} = \begin{bmatrix} \mathbf{0} \\ \mathbf{m^{-1}} \end{bmatrix}} \qquad (2.6.13)$$

as

$$\mathbf{\dot{x} = Ax + Bu}$$

To illustrate, it is left as an exercise to show that for the simple two-degree-of freedom system sketched in Figure 2.8

$$\mathbf{x} = \begin{bmatrix} x_1 \\ x_2 \\ \dot{x}_1 \\ \dot{x}_2 \end{bmatrix}, \quad \mathbf{A} = \begin{bmatrix} 0 & 0 & 1 & 0 \\ 0 & 0 & 0 & 1 \\ \dfrac{-(k+k_1)}{m_1} & \dfrac{k}{m_1} & \dfrac{-(c+c_1)}{m_1} & \dfrac{c}{m_1} \\ \dfrac{k}{m_2} & \dfrac{-(k+k_2)}{m_2} & \dfrac{c}{m_2} & \dfrac{-(c+c_2)}{m_2} \end{bmatrix} \qquad (2.6.14)$$

2.6.6 State Model for n-Degree-of-Freedom Univariate System

It is also left as an exercise to show, using the simple D operator for derivative (or by using Laplace transform), that the differential equation for x_1 alone or x_2

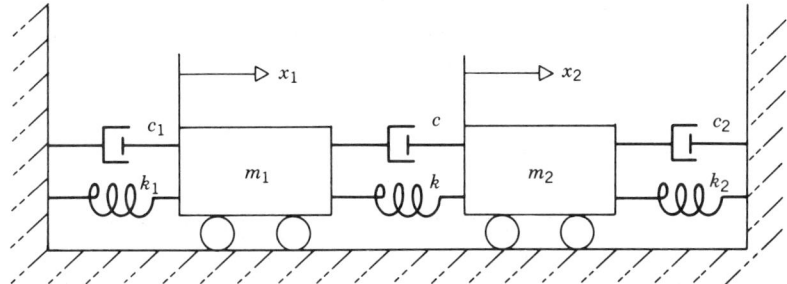

Figure 2.8 Two-degrees-of-freedom nonconservative system.

alone takes the form

$$(D^4 + \alpha_3 D^3 + \alpha_2 D^2 + \alpha_1 D + \alpha_0)x = (b_2 D^2 + b_1 D + b_0)u \quad (2.6.15)$$

where

$$D \equiv \frac{d}{dt}, \quad D^2 \equiv \frac{d^2}{dt^2}, \quad \text{and so on}$$

In general, the motion of a single mass in a system of $n/2$ masses is given by the nth order linear differential equation

$$\begin{aligned}(D^n + \alpha_{n-1}D^{n-1} + \alpha_{n-2}D^{n-2} + \cdots + \alpha_1 D + \alpha_0)x \\ = (b_{n-2}D^{n-2} + b_{n-3}D^{n-3} + \cdots + b_1 D + b_0)u\end{aligned} \quad (2.6.16)$$

which can be reduced to the standard form

$$\dot{\mathbf{x}} = \mathbf{A}\mathbf{x} + \mathbf{B}\mathbf{u}$$

by letting

$$\mathbf{x} = \begin{bmatrix} x \\ Dx \\ D^2 x \\ \vdots \\ D^{n-1} x \end{bmatrix}, \quad \mathbf{A} = \begin{bmatrix} 0 & 1 & 0 & \cdots & 0 \\ 0 & 0 & 1 & \cdots & 0 \\ \vdots & \vdots & \vdots & \vdots & \vdots \\ 0 & 0 & 0 & \cdots & 1 \\ -\alpha_0 & -\alpha_1 & -\alpha_2 & \cdots & -\alpha_{n-1} \end{bmatrix},$$

$$\mathbf{B} = \begin{bmatrix} 0 & 0 & 0 & \cdots & 0 & 0 \\ 0 & 0 & 0 & \cdots & 0 & 0 \\ \vdots & \vdots & \vdots & \vdots & \vdots & \vdots \\ 0 & 0 & 0 & \cdots & 0 & 0 \\ b_0 & b_1 & b_2 & \cdots & b_{n-2} & 0 \end{bmatrix}, \quad \mathbf{u} = \begin{bmatrix} u \\ Du \\ \vdots \\ D^{n-2} u \\ 0 \end{bmatrix} \quad (2.6.17)$$

The matrix $\mathbf{A}$ is often called the companion matrix of the nth order linear differential operator in the differential equation (2.6.16). Note the resemblance of this matrix with the discrete ϕ matrix given by (2.4.11) for the ARMA$(n, n-1)$ model. This resemblance is not accidental; indeed, the eigenvalues of the two matrices are related by

$$\lambda = e^{\mu \Delta}$$

which can be used to recover the differential operator parameters of not only this particular system but any other system governed by the general continuous time state model.

2.6.7 Right-Hand Side Dynamics: Need for Moving Average

The multivariate model (2.6.12) does not have derivatives of the forcing function on the right-hand side; however, the univariate models (2.6.15) and (2.6.16), restricted to only one of the elements of the multivariate vector, do have derivatives of the forcing function on the right-hand side. The derivation of (2.6.15) and (2.6.16) from (2.6.12) explains how and why the so-called right-hand side dynamics involving the derivatives of the forcing function arise. Note that the highest derivative order on the right-hand side of the differential equations (2.6.15) and (2.6.16) is two less than that on the left-hand because the basic model (2.6.12) is second order. If (2.6.12) were first order, the right-hand order in (2.6.16) would be $n - 1$; it will never exceed $n - 1$.

If the forcing functions u in the differential equations (2.6.12, 2.6.15, and 2.6.16) are random, say continuous time white noise, then these models are the continuous-time vector autoregressive model denoted as AV(2) and autoregressive moving average models denoted as AM(4,2) and AM($n,n-2$), respectively. The derivation of (2.6.15) and (2.6.16) from (2.6.12) then mirrors the derivation showing the need for moving average in Section 2.4.3 from discrete time into continuous time. Again, the multivariate model (2.6.12) does not have a moving average, but the univariate models (2.6.15) and (2.6.16) for each of its series alone do. Also, the result is not restricted for multivariate to univariate models discussed here for simplicity; a 2p-variate multivariate system without moving average would require moving average when only p of its variables are modeled by a multivariate model. Thus the moving average or the right-hand side dynamics arises because we are trying to model the system with the number of variables less than those required by the dimension of the state space.

2.6.8 Right-Hand Side Dynamics as Moving Average: Misnomer of a Misnomer

The similarity in the mathematical forms of the right-hand side dynamics and the moving average side of the ARMA model has caused a considerable confusion in the literature. The popularity of ARMA models in the 1970s has lead to the incorrect use of the term moving average for the right-hand side dynamics even with deterministic and known inputs, first in the control system literature and then in the modal analysis literature. Such inappropriate use of the term moving average for the right-hand side dynamics is particularly tempting, because when the ARMA model is formally expressed in the transfer function form, the moving average term appears in the denominator just like the right-hand side dynamics term of a deterministic system.

The correct use of the term moving average should be reserved for the case of random, uncorrelated, and lagged terms such as $\theta_1 a_{t-1}$, $\theta_2 a_{t-2}$, and so on, where a_t's are random residuals and *not* actual inputs at all! As further explained in Section 6.1.3, since the residuals are obtained from the available data X_t using the ARMA model, the transfer function expressing the ratio of the output X_t and the input a_t in the frequency domain is formal and fictitious; moreover X_t and a_t, both being random functions, do not have Fourier, Laplace, or z transforms.

Interestingly enough, the use of the term moving average, for say $(a_t - \theta_1 a_{t-1})$, is itself somewhat of a misnomer. The correct meaning of the term moving average refers to the average of a fixed number of current and past observations as we move in time. No such moving average arises in any MA model! The only coincidental connection between such proper moving average and ARMA is that the forecast from a special ARMA(1,1) model with $\phi_1 = 1$ happens to be an exponentially weighted moving average! This misnomer, common in statistical literature, is now being converted into another misnomer by applying it to the right-hand side dynamics or the numerator of a transfer function in the engineering literature!

One may ask: Is there any harm in such a misnomer of a misnomer? The answer is, unfortunately, yes. In spite of their tempting formal similarity, there is a fundamental difference in the way in which a transfer function between an output and a genuine input is obtained, compared to that between an output and a pseudo-input computed using that very output and the transfer function. When both the output and input are given, the estimation of the transfer function is linear and one-shot. When only output is known and the fictitious input must be computed using the transfer function, its estimation is nonlinear and necessarily recursive. For this reason, we will restrict the use of the term moving average only in dealing with lagged residual terms.

Nevertheless, the similarity of the moving average side with the right-hand side dynamics is important because it explains the way moving average side arises. In the statistical literature the moving average terms are treated just like autoregressive terms and therefore the moving average order can exceed the autoregressive order. Its analogy with the right-hand side dynamics, however, shows that the moving average side arises *from* the autoregressive side under some lack of information. Thus Section 2.4.3 showed that the univariate model for only one element of a vector with multivariate autoregressive model requires moving average. It can be shown that a sum of arbitrary number of variables each with an AR models requires an ARMA model and that a continuous time stochastic process governed by a second- or higher-order autoregressive model, when sampled at uniform intervals, requires an ARMA model, see, for example, Pandit and Wu (1983, pp. 44–45). In such cases the moving average order can never be equal to or exceed the autoregressive order. This supports the rationale for the ARMA($n, n - 1$) modeling strategy developed using the state-space argument in Section 2.4.

It is thus interesting to note that in all these cases both the moving average and the right-hand side dynamics arise from the autoregressive side as a result of

1. Lack of information, such as vector to scalar or continuous to discrete, *and*
2. Presence of a forcing function, that is random for moving average and deterministic for the right-hand side dynamics.

2.7 EFFECT OF MEAN ESTIMATION

Recall from Section 2.1 that the AR(1) model was originally developed after plotting the average subtracted silicon data X_t versus X_{t-1} and seeing that it is a straight-line plot. Since the fitted model eventually indicated that the data was almost purely stochastic and stationary, its mean was constant, and therefore the average $\bar{X}$ was a good estimator of the mean.

The estimation procedure always takes the data minus the mean as X_t in minimizing the sum of squares calculated by using the model. One may, however, choose to take the average as the mean as was done for the silicon data, or estimate the mean as an additional parameter as was done for the exponentially generated deterministic data in Section 2.5.5. In the latter case, the estimated mean turned out to be almost zero, actually 0.13823×10^{-8}, which is of the order of roundoff error, as pointed out in Section 2.5.5.

To clarify the role played by the mean estimation in modeling, let us denote the deterministic and stochastic parts of the data by D_t and S_t, respectively. Then recall from the decomposition (2.2.1) or its generalization (both scalar as well as vector, but we will consider here only the scalar case for ease of illustration) that since the stochastic part S_t is a finite linear combination of zero mean random variables a_t, the mean of S_t is zero. Thus the mean of the data is its deterministic part

$$X_t = D_t + S_t$$
$$\text{Mean} = E(X_t) = D_t$$

a special case of which was illustrated in deriving the mean (2.2.3) of the AR(1) model.

2.7.1 Constant Deterministic Part (DC Offset) and Average $\bar{X}$

When the deterministic part or the mean is only a constant (often called DC offset), say μ, so that it is not dependent upon the time origin, the time series is stationary in the mean. For such a series the average $\bar{X}$ is an unbiased estimator of the mean because

$$E(X_t) = D_t = \mu$$

and
$$E(\bar{X}) = E(1/N)(X_1 + X_2 + \cdots + X_N)$$
$$= (1/N)(\mu + \mu + \cdots + \mu)$$
$$= \mu$$

Moreover, the average subtracted data $(X_t - \bar{X})$ has a zero mean since

$$E(X_t - \bar{X}) = (\mu - \mu) = 0$$

and is therefore purely stochastic and can be used in place of $(X_t - \mu)$ in the model. For such a series, even if the mean is estimated by the estimation program, it will be very close to the data average.

Thus, in fitting the AR(1) model the program always uses the form

$$(X_t - \mu) = \phi_1(X_{t-1} - \mu) + a_t \tag{2.7.1}$$

to calculate the residual sum of squares wherein X_t denotes the actual data. When we do not estimate the mean, it replaces μ by $\bar{X}$. Note that the actual data, therefore, has the model

$$X_t = (1 - \phi_1)\mu + \phi_1 X_{t-1} + a_t, \quad \text{given } X_t = X_0 \quad \text{at } t = 0 \tag{2.7.2}$$

$$= (1 - \phi_1^t)\mu + \phi_1^t X_0 + \sum_{j=0}^{t-1} \phi_1^j a_{t-j} \tag{2.7.2a}$$

which can be proved by recursive substitution as in Section 2.2.1 and in fact reduces to the solution (2.2.1) when the mean μ is zero as assumed there.

Moreover,

$$E(X_t) = D_t = (1 - \phi_1^t)\mu + \phi_1^t X_0 \tag{2.7.3}$$

which, comparing with the mean (2.2.3) derived in Section 2.2.5, shows that a function of time multiplied by the nonzero mean μ has now been added to the deterministic part. Continuing further as in Section 2.2.5, it can be seen that the variance of X_t is still given by (2.2.4):

$$\text{Var } X_t = \frac{(1 - \phi_1^{2t})}{(1 - \phi_1^2)} \sigma_a^2$$

Thus both the mean and the variance are functions of time, explicitly depend upon the time origin, and the time series is in general nonstationary in the mean as well as variance or simply nonstationary.

If, however, the stationarity condition $|\phi_1| < 1$ is satisfied, then for large enough t (more precisely stated by Eq. 2.2.5), the series can be considered stationary (in the mean and variance) and its constant mean and variance are

given by
$$E(X_t) = \mu, \quad \text{Var}(X_t) = \sigma_a^2/(1 - \phi_1^2) \qquad (2.7.4)$$

which is (2.2.6) with zero mean replaced by μ as expected.

2.7.2 Exponential Deterministic Part and Average $\bar{X}$

Now consider the data generated by the deterministic exponential
$$X_t = X_0(\phi)^t \qquad (2.7.5)$$
for which
$$X_t = \phi X_{t-1}, \quad \text{given} \quad X_t = X_0 \quad \text{at} \quad t = 0 \qquad (2.7.5a)$$

What happens if we try to fit an AR(1) model to this data allowing mean estimation? Since the estimation program will use the model (2.7.1), comparing it with (2.7.5a) shows that theoretically, $\phi_1 = \phi$, $\mu = 0$, and $a_t \equiv 0$. Actually, since the roundoff error and other computer noise introduces the stochastic part, we would expect the mean to be of the order of 10^{-8}, which is the precision of the generated single precision data and the magnitude of a_t measured by σ_a about 10^{-7}, as illustrated in Section 2.5.5.

The theoretical relation between the estimated mean and σ_a^2 or the residual sum of squares of a_t's can be easily obtained by noting that the residual sum of squares
$$\text{RSS} = \sum_{t=2}^{N} a_t^2 = \sum_{t=2}^{N} [(X_t - \phi_1 X_{t-1}) - \mu(1 - \phi_1)]^2 \qquad (2.7.6a)$$
$$= (N - 1)\mu^2(1 - \phi)^2 \qquad (2.7.6)$$
when $\phi_1 = \phi$.

If we do not estimate the mean but take the average as the mean and fit an AR(1) model, the same relation should hold with μ replaced by the average $\bar{X}$ given by
$$\bar{X} = (1/N)(X_0\phi + X_0\phi^2 + \cdots + X_0\phi^N)$$
$$= (X_0\phi/N)(1 - \phi^N)/(1 - \phi) \qquad (2.7.7)$$

To illustrate, let us consider the data generated by (2.7.5) with $X_0 = 100$, $\phi = 0.9$, and $N = 150$. Using the computer program with mean estimation we find that the estimated values are:
$$\hat{\phi}_1 = 0.9, \hat{\mu} = -0.43 \times 10^{-5}, \quad \text{and} \quad \text{RSS} = 0.22 \times 10^{-10}$$

whereas (2.7.6) gives
$$\text{RSS} = 149(-0.43)^2(0.1)^2 \times 10^{-10} = 0.28 \times 10^{-10}$$

which agree very closely. Since

$$\sigma_a = \sqrt{RSS/(N-1)} = 3.8 \times 10^{-7}$$

we have reached the noise floor of the data and can stop modeling.

On the other hand, if the mean is not estimated and μ is replaced by the average given by (2.7.7) as

$$\bar{X} = (100 \times 0.9/150)(1 - 0.9^{150})/(1 - 0.9) = 6.0$$

then using (2.7.6)

$$RSS = 149(6.0)^2(1 - 0.9)^2 = 53.64$$

which is what the program gives. This residual sum of squares is obviously too large and therefore the model is not acceptable.

What went wrong? Why is the program unable to reduce the residual sum of squares further? Answers to these questions can be easily obtained by examining the residual sum of squares expression (2.7.6a). In view of the nature of the data given by (2.7.5a), the first term $(X_t - \phi_1 X_{t-1})$ can be made small and actually reduced to zero by choosing $\phi_1 = \phi$ to reduce the RSS to (2.7.6). If the parameter μ is estimated, it can be freely chosen, and therefore a small value of the order of the digital accuracy of the data is chosen as the estimate, resulting in the smallest possible residual sum of squares and smallest possible "perpendicular" a_t measured by the magnitude $\sigma_a = 3.8 \times 10^{-7}$ already near the noise floor of the data in a state space of dimension one.

If the mean is not estimated and μ must be taken as the average $\bar{X}$, the state space of dimension one does not have enough freedom for the data to reduce the sum of squares further from $(N-1)\bar{X}^2(1-\phi)^2$, which is still quite large, 53.64, in the present case. To reduce it further, we have to go to a state space of dimension two, which requires fitting an ARMA(2,1) model. For this model the residual sum of squares is given by

$$RSS = \sum_{t=3}^{N} [(X_t - \phi_1 X_{t-1} - \phi_2 X_{t-2}) - \bar{X}(1 - \phi_1 - \phi_2) + \theta_1 a_{t-1}]^2 \quad (2.7.8)$$

Now, although $\bar{X}$ is fixed, there are two parameters (ϕ_1 and ϕ_2) to choose and they can be chosen to reduce the two terms in parentheses in (2.7.8) to zero, which would then give $\theta_1 \simeq 1$. Thus ϕ_1 and ϕ_2 are determined by the conditions

$$X_t - \phi_1 X_{t-1} - \phi_2 X_{t-2} = 0 \quad (2.7.9)$$

$$1 - \phi_1 - \phi_2 = 0 \quad (2.7.10)$$

The first of these equations, in common with the solution of homogeneous differential equations, can best be solved by operator methods. Using the backshift operator $BX_t = X_{t-1}$ or in general $B^j X_t = X_{t-j}$, (2.7.9) can be alternatively written as

$$(1 - \phi_1 B - \phi_2 B^2)X_t = (1 - \phi_1 B - \phi_2 B^2)X_0(\phi^t) = 0 \quad (2.7.9a)$$

If λ_1 and λ_2 are the characteristic roots of the difference operator with the characteristic polynomial in B:

$$(1 - \phi_1 B - \phi_2 B^2) = (1 - \lambda_1 B)(1 - \lambda_2 B) = [1 - (\lambda_1 + \lambda_2)B + \lambda_1 \lambda_2 B^2] = 0 \quad (2.7.11)$$

then (2.7.9a) will be satisfied if one of these roots is ϕ since

$$(1 - \phi B)X_0(\phi)^t = X_0(\phi)^t - \phi X_0(\phi)^{t-1} = 0$$

and (2.7.10) shows that the second root must be 1 because its left-hand side is the left-hand polynomial in (2.7.11) with $B = 1$. Therefore, (2.7.11) shows that

$$\phi_1 = \lambda_1 + \lambda_2 = 1 + \phi \quad \text{and} \quad \phi_2 = -\lambda_1 \lambda_2 = -\phi$$

which may be verified from the actual program results.

In fact (2.7.9) and (2.7.9a) show that an AR(2) model rather than an ARMA(2,1) would be better for such purely deterministic data. However, the data in practice is rarely so deterministic and therefore ARMA(2,1) is usually better for most data sets, even with slight noise or stochastic part.

2.8 DDS MODELING AS EXPONENTIAL EXPANSION

If we ignore the stochastic part and therefore eliminate the moving average operator from our consideration, the argument in the preceding section can be easily generalized to an n-dimensional state space. When the mean is not estimated, the average subtracted data is given by

$$X_t = X_0(\phi)^t - \bar{X}$$
$$= X_0(\phi)^t - \bar{X}(1)^t \quad (2.8.1)$$

The first discrete exponential is reduced to zero by the difference operator $(1 - \phi B)$ and the second by the operator $(1 - B)$. Note that the operator $(1 - B)$ reduces any constant to zero. Therefore, the model for this data is given by

$$(1 - \phi B)(1 - B)X_t = [1 - (1 + \phi)B + \phi B^2]X_t = 0 \quad (2.8.2)$$

given $X_t = (X_0 - \bar{X})$ at $t = 0$, and $X_t = [(X_0/\phi) - \bar{X}]$ at $t = -1$, which is fitted by an AR(2) model with $\phi_1 = (1 + \phi)$ and $\phi_2 = -\phi$ and will give residual sum of squares to yield σ_a of the order of eighth decimal place accuracy used in generating single precision data. It is now clear that using $\bar{X}$ in place of μ (which should be zero) introduced a step in the deterministic part which was accounted for by the characteristic root 1 in the model. Note that the $\bar{X}$ needed in the data (2.8.1) enters the model (2.8.2) only via initial conditions.

2.8.1 Sum of Two Exponentials—Discrete Time: Difference Equation

Clearly, the second exponential in (2.8.1) need not arise from the root 1 in general. Hence the data generated by the sum of two exponentials

$$X_t = c_1(\lambda_1)^t + c_2(\lambda_2)^t \tag{2.8.3}$$

is modeled with mean estimation by

$$(1 - \lambda_1 B)(1 - \lambda_2 B)X_t = [1 - (\lambda_1 + \lambda_2)B + \lambda_1\lambda_2 B^2]X_t = 0 \tag{2.8.4}$$

given $X_t = (c_1 + c_2)$ at $t = 0$, and $X_t = [(c_1/\lambda_1) + (c_2/\lambda_2)]$ at $t = -1$, which is fitted by an AR(2) model with $\phi_1 = (\lambda_1 + \lambda_2)$, $\phi_2 = -\lambda_1\lambda_2$, and the estimated mean nearly zero. If the mean is not estimated and μ is replaced by $\bar{X}$, the data used for modeling is

$$X_t = c_1(\lambda_1)^t + c_2(\lambda_2)^t - \bar{X} \tag{2.8.5}$$

which is modeled by

$$(1 - \lambda_1 B)(1 - \lambda_2 B)(1 - B)X_t = [1 - (\lambda_1 + \lambda_2 + 1)B \\ + (\lambda_1\lambda_2 + \lambda_1 + \lambda_2)B^2 - \lambda_1\lambda_2 B^3]X_t = 0 \tag{2.8.6}$$

given

$X_t = (c_1 + c_2 - \bar{X})$ at $t = 0$,
$X_t = [(c_1/\lambda_1) + (c_2/\lambda_2) - \bar{X}]$ at $t = -1$, and
$X_t = [(c_1/\lambda_1^2) + (c_2/\lambda_2^2) - \bar{X}]$ at $t = -2$,

fitted by an AR(3) model with $\phi_1 = (\lambda_1 + \lambda_2 + 1)$, $\phi_2 = -(\lambda_1\lambda_2 + \lambda_1 + \lambda_2)$, and $\phi_3 = \lambda_1\lambda_2$. Again the step introduced by $\bar{X}$ is accounted for or reduced to zero by the extra root 1.

2.8.2 Sum of Two Exponentials—Continuous Time: Differential Equation

The analogy of this development in continuous time with differential equations is obvious. Consider the second-order differential equation (2.6.9) for a one-degree-of-freedom system written in the form

$$(D^2 + \alpha_1 D + \alpha_2)x(t) = 0 \quad (2.8.7)$$

If the characteristic roots of the differential operator are defined by a solution of the polynomial in D

$$(D^2 + \alpha_1 D + \alpha_2) = (D - \mu_1)(D - \mu_2) = 0 \quad (2.8.8)$$

then the solution of (2.8.7), for distinct roots $\mu_1 \neq \mu_2$, is given by

$$x(t) = c_1 e^{\mu_1 t} + c_2 e^{\mu_2 t} \quad (2.8.9)$$

and for equal roots $\mu_1 = \mu_2$ by

$$x(t) = (c_1 + c_2 t)e^{\mu_1 t} \quad (2.8.10)$$

where c_1 and c_2 can be determined by the initial conditions $\dot{x}(0)$ and $x(0)$.

If this response is sampled at uniform intervals Δ, then denoting the relation between continuous time t (used in parentheses) and discrete time t (used in subscripts) = 0, 1, 2, ..., by

$$\text{Continuous } t = \Delta t \text{ discrete} \quad (2.8.11)$$

the sampled data is given by

$$X_t = c_1(e^{\mu_1 \Delta})^t + c_2(e^{\mu_2 \Delta})^t$$
$$= c_1(\lambda_1)^t + c_2(\lambda_2)^t$$

which is (2.8.3), and is therefore modeled by (2.8.4) when the mean is estimated and by (2.8.6) when the average is taken as the mean. The relation between the discrete characteristic roots λ_i and continuous characteristic roots μ_i is given by

$$\lambda_i = e^{\mu_i \Delta}, \quad i = 1, 2 \quad (2.8.12)$$

2.8.3 Exponential Expansion: Difference/Differential Equations

Note that although (2.8.3) and (2.8.9) are exponential in form, they can represent a widely different variety of data sets. For real negative μ_i ($\lambda_i < 1$), they represent a combination of decaying exponentials; for zero μ_i ($\lambda_i = 1$), they represent a straight line [using (2.8.10)]; for real positive μ_i ($\lambda_i > 1$), they represent unstable exploding exponentials; for complex (conjugate) μ_i (λ_i), they represent decaying or damped sine waves when the real part of μ_i is negative ($|\lambda_i| < 1$), undamped sine waves when the real part is zero or μ_i are purely imaginary ($|\lambda_i| = 1$), and finally exploding sine waves when the real part is positive ($|\lambda_i| > 1$).

Hence the generalization of (2.8.3) can be used to represent almost any deterministic data of practical interest:

$$X_t = c_1(\lambda_1)^t + c_2(\lambda_2)^t + \cdots + c_n(\lambda_n)^t \tag{2.8.13}$$

which is modeled with mean estimation by

$$(1 - \phi_1 B - \phi_2 B^2 - \cdots - \phi_n B^n)X_t = 0 \tag{2.8.14}$$

given

$$x_{-i} = c_1(\lambda_1)^{-i} + c_2(\lambda_2)^{-i} + \cdots + c_n(\lambda_n)^{-i}, \quad i = 0, 1, \ldots, n-1$$

and can be modeled by an AR(n) model with ϕ_i given by

$$(1 - \phi_1 B - \phi_2 B^2 - \cdots - \phi_n B^n) \equiv (1 - \lambda_1 B)(1 - \lambda_2 B)\cdots(1 - \lambda_n B) \tag{2.8.15}$$

When the mean is not estimated but replaced by $\bar{X}$, we need an AR($n+1$) model with the additional root equal to 1.

The data (2.8.13) could be thought of as being sampled from a continuous time system governed by the nth order differential equation

$$(D^n + \alpha_{n-1}D^{n-1} + \alpha_{n-2}D^{n-2} + \cdots + \alpha_0)x(t) = 0 \tag{2.8.16}$$

so that the data is a uniformly sampled version of its solution

$$x(t) = c_1 e^{\mu_1 t} + c_2 e^{\mu_2 t} + \cdots + c_n e^{\mu_n t} \tag{2.8.17}$$

Hence the AR(n) model (2.8.14) can represent a combination of decaying and/or exploding exponentials and/or sine waves as well as polynomials; whereas the ARMA($n, n-1$) model can represent any such deterministic data superimposed by noise. The DDS approach therefore is capable of modeling stationary or nonstationary data that is stochastic, deterministic, or a mixture of the two.

The DDS modeling approach, in effect, then provides us with the discrete exponential expansion representation (2.8.13) of the data with its interpolation in continuous time given by the exponential expansion (2.8.17), the former being the solution of the difference equation (2.8.14), which can be readily fitted to the data by the least-squares method, and the latter being the solution of the differential equation (2.8.16), which can be readily interpreted by analogy to physical systems such as mechanical vibrations. Thus, difference/differential equation formulation (2.8.14 and 2.8.16) is used to express the data as a linear combination of exponentials (2.8.13 and 2.8.17) with the coefficients c_i determined by the initial conditions.

2.8.4 Deterministic and Stochastic Components: ARMA Model

We have generalized only the deterministic component using $AR(n)$ models for easy understanding. However, it is not difficult to see that in using $ARMA(n,n-1)$ models, the noise present in the data gets expressed as response to white or uncorrelated noise, with the impulse response or Green's function [generalizing, for example, ϕ^j of the $AR(1)$ model in the stochastic part of (2.2.1), elaborated in Section 3.6.2] as a different linear combination of the *same* exponentials, the new coefficients being determined by the moving average parameters θ_i.

For example, if the deterministic component is a sum of m exponentials and the stochastic part (now assumed to contain no deterministic part without loss of generality) has a Green's function that is a sum of $(n-m)$ exponentials generalizing the single exponential $AR(1)$ case, then the theoretical representation of the data is

$$\hat{X}_t = c_1 \lambda_1^t + \cdots + c_m \lambda_m^t + \sum_{j=0}^{t-1} (g_{m+1} \lambda_{m+1}^j + \cdots + g_n \lambda_n^j) a_{t-j} \quad (2.8.18)$$

Now an ARMA model must be used and ignoring other computational noise, $ARMA(n,n-1)$ model would be adequate. This $ARMA(n,n-1)$ model will have the decomposition

$$X_t = c_1 \lambda_1^t + \cdots + c_m \lambda_m^t + c_{m+1} \lambda_{m+1}^t + \cdots + c_n \lambda_n^t$$
$$+ \sum_{j=0}^{t-1} (g_1 \lambda_1^j + \cdots + g_m \lambda_m^j + g_{m+1} \lambda_{m+1}^j + \cdots + g_n \lambda_n^j) a_{t-j} \quad (2.8.19)$$

If the deterministic part is strong enough and/or we take large enough number of observations before it decays to zero, we would expect $c_{m+1}, c_{m+2}, \ldots, c_n$ to be small, instead of their theoretical zero values, and $c_1, c_2, \ldots, c_m$ to be large. Hence the $AR(1)$ argument can be generalized to

$$c_i \gg \sigma_a, \quad i = 1, 2, \ldots, m$$
$$c_i < \sigma_a, \quad i = m+1, \ldots, n \quad (2.8.20)$$

We should also expect to see $g_1, g_2, \ldots, g_m$ small compared to $g_{m+1}, g_{m+2}, \ldots, g_n$; in practice this may not hold because the estimation error from a finite number of data points may cause them to be large. This argument will be generalized to vector models in Chapter 5, and an example illustrating these results will be given in Section 5.5.2 (see Table 5.5).

An exact $ARMA(n,n-1)$ model and initial conditions, which yield the decomposition (2.8.18), are left as an exercise. This exercise shows that for a special case of (2.8.18) when the noise is white, we have an $ARMA(m,m)$ model with equal autoregressive and moving average parameters, that is, $\lambda_i = v_i$, $i = 1, 2, \ldots, m$; the parameters $\phi_i = \theta_i$ can then be readily estimated by total least squares (TLS) method, see Section 3.3.3.

2.9 FOURIER SERIES, FOURIER TRANSFORM AND FFT VERSUS DDS

The exponential expansion representation (2.8.17) of the data reminds us of Fourier series. The analogy is more than casual. In this section we will explore this analogy further to elucidate the similarity and differences between the Fourier series, Fourier transform, their computationally efficient implementation by an algorithm well known as FFT, and DDS.

2.9.1 Fourier Series

Fourier series or Fourier transform are particular cases of the general mathematical technique of eigenfunction expansion. Recall that under certain broad (Dirichlet) conditions, a bounded periodic function of period T [i.e., $f(t+T) = f(t)$] can be expressed as the "complex Fourier series" or "exponential Fourier series":

$$f(t) = \sum_{n=-\infty}^{n=+\infty} c_n e^{i\omega_n t}, \quad \text{with} \quad \omega_n = \frac{2n\pi}{T}, \quad n = 0, \pm 1, \pm 2, \ldots, \quad i = \sqrt{-1} \quad (2.9.1)$$

The eigenfunctions $e^{i\omega_n t}$ used in this expansion are the solutions of the differential equation

$$D^2 x = -\omega^2 x \quad \text{or} \quad (D^2 + \omega^2)x(t) = 0 \quad (2.9.2)$$

with boundary conditions

$$x\left(-\frac{T}{2}\right) = x\left(\frac{T}{2}\right), \quad \dot{x}\left(-\frac{T}{2}\right) = \dot{x}\left(\frac{T}{2}\right) \quad (2.9.2a)$$

Thus $e^{i\omega_n t}$ is an eigenfunction corresponding to the eigenvalue $-\omega_n^2$. Note that the eigenvalues are discrete because of the periodic nature of the eigenfunctions specified by the boundary conditions (2.9.2a) with finite nonzero period T. Hence the expansion is a sum rather than an integral because of the periodic nature of the function $f(t)$.

Once the period T is known, the eigenfunctions are completely determined and hence the coefficients c_n can be calculated. Moreover, the computation of c_n is simplified by the most important property of the eigenfunctions—that they are orthogonal and can be made orthonormal by suitable division (by T):

$$\frac{1}{T} \int_{-T/2}^{T/2} e^{i\omega_m t} e^{-i\omega_n t} \, dt = 0, \quad \text{for} \quad m \neq n$$
$$= 1, \quad \text{for} \quad m = n \quad (2.9.3)$$

which can be readily verified by the simple integration of the exponential function using the periodic boundary conditions (2.9.2a). Therefore, c_n can be computed by

$$c_n = \frac{1}{T} \int_{-T/2}^{T/2} f(t) e^{-i\omega_n t} dt, \qquad n = 0, \pm 1, \pm 2, \ldots \qquad (2.9.4)$$

Thus the orthogonality property (2.9.3) allows each c_n to be calculated separately; in calculating c_n, we do not have to worry about all other c_m, $m \neq n$. Therefore $c_0, c_1, c_2, c_3, \ldots$, can be calculated one by one, which is ideally suited to computers.

2.9.2 Fourier Transform

When the function $f(t)$ is not periodic, its Fourier series does not exist; in other words, it is no longer possible to represent the function as a linear combination of a countably infinite number of complex exponentials (sines and cosines) with discrete multiple frequencies ω_n and coefficients c_n. If we still insist on using the orthogonal eigenfunctions $e^{i\omega t}$, then under suitable conditions such as the absolute value of the function $f(t)$ having a finite integral over the real axis, the expansion (2.9.1) takes the form of integral with continuously varying frequencies ω and the density function $F(\omega)$, called the Fourier transform of $f(t)$, replacing the discrete coefficients c_n. Fourier series relations (2.9.1 and 2.9.4) are now replaced by Fourier transform relations (2.9.5 and 2.9.6), respectively:

$$f(t) = \frac{1}{2\pi} \int_{-\infty}^{\infty} F(\omega) e^{i\omega t} d\omega \qquad (2.9.5)$$

$$F(\omega) = \int_{-\infty}^{\infty} f(t) e^{-i\omega t} dt \qquad (2.9.6)$$

2.9.3 DDS as Generalized Laplace Transform

Fourier methods were originally used for dealing with differential equations, with $f(t)$ representing a known mathematical function. A solution of a differential equation would generally need integration, but the exponential function reduces the operation of differentiation and integration to multiplication and division respectively, thus transforming the problem from calculus to algebra. This advantage of the exponential function is available for arbitrary exponents and hence is shared by both DDS, which uses general complex exponents, and by Fourier methods, which use purely imaginary exponents, as depicted in Figure 2.9.

Why do the Fourier methods restrict the exponent to the imaginary axis in the complex plane, the entire imaginary axis for Fourier transform and discrete points on the axis equally spaced at $2\pi/T$ for Fourier series? They do so to take

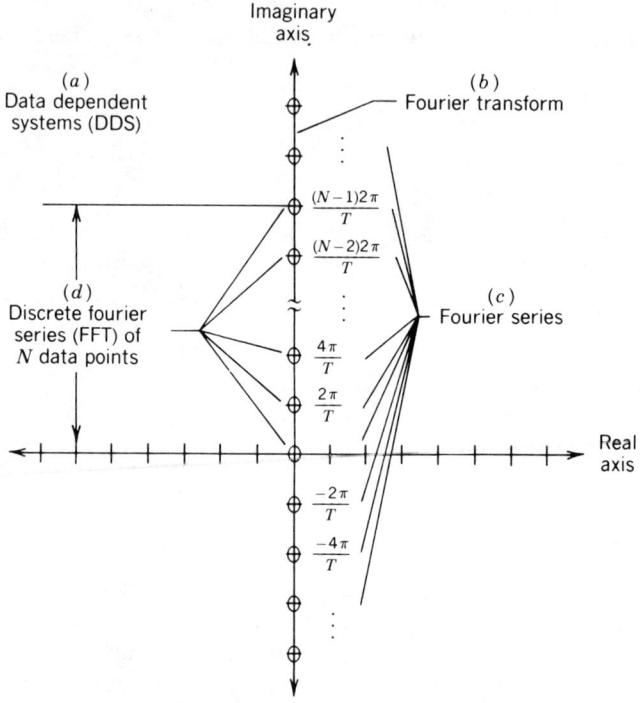

Figure 2.9 DDS versus Fourier methods in the complex plane. Successive narrowing of exponents from: (a) data dependent systems—entire complex plane, to (b) Fourier transform—imaginary axis; (c) Fourier series—discrete points on imaginary axis equally spaced at $2\pi/T$, and (d) discrete Fourier series (FFT)—only N discrete points on imaginary axis equally spaced at $2\pi/T$.

advantage of the orthogonality defined by (2.9.3). This orthogonality provides a mathematical tractability that has made Fourier methods a rich field of research in both applied and pure mathematics.

In contrast, a pair of exponentials with arbitrary exponents are *not* necessarily orthogonal since they may not satisfy (2.9.3). In other words, the finite exponential expansion (2.8.17) provided by DDS is generally not orthogonal, whereas the infinite exponential expansion (2.9.1) of the Fourier series and (2.9.5) of the Fourier integral are orthogonal. As shown in Figure 2.9, broadening the set of exponents from a discrete set $2n\pi i/T$, $n = 0, \pm 1, \pm 2, \ldots$ to the entire imaginary line takes us from Fourier series to Fourier transform without losing orthogonality. The Laplace transform further broadens the set to regions of the complex plane, whereas the DDS method extends it to the entire complex plane (left half for stable systems). Hence the DDS modeling of time series has been dubbed as "generalized Laplace transform" in Pandit (1973), which we follow in the subsequent discussion. Fourier transform as a restriction of the transfer function provided by DDS to the imaginary axis will be further illustrated in Section 5.1.

2.9.4 Parsimony at the Expense of Orthogonality

What have we gained by giving up orthogonality? Parsimony or fewer parameters in a model, that is, fewer exponentials in the expansion. The infinite number of exponentials in the Fourier series or integral is a mathematical idealization that is reduced to an approximation with a finite number of exponentials in practice, as discussed later. Nevertheless, since the choice of exponents is restricted to the imaginary line, a very large number of exponentials may still be required to achieve a requisite degree of approximation. Deliberately widening the choice from a line to a plane (half plane for stable systems) enables DDS usually to achieve the same degree of approximation with only a few exponentials. Because of the abundant choice of exponents from the complex plane, the contribution from all but a few exponentials can be made negligibly small, especially in dealing with a finite amount of data.

For example, the impulse response of a one-degree-of-freedom damped system can be represented by only two exponentials using DDS modeling, whereas Fourier methods require an infinite number in theory and a large number of exponentials with imaginary exponents in practice. An n-degree-of-freedom system can similarly be represented using $2n$ exponentials by DDS and an infinite number of exponentials by Fourier methods.

The only exception is the impulse response from an undamped system. In this case the response is itself a combination of a finite number of sines and cosines and hence can be represented by a sum of a finite number of exponentials using Fourier methods if all the frequencies of the undamped system are harmonic or multiples of the fundamental frequency $2\pi/T$ used in the Fourier series; then the finite Fourier series expansion would be exact and both DDS and Fourier methods would represent the response by the same number of exponentials with imaginary exponents, say $2n\pi i/T$, $n = 0, \pm 1, \pm 2, \ldots, \pm m$. Such a response with a finite Fourier series expansion is known as a trigonometric polynomial:

$$f(t) = \sum_{n=-m}^{m} c_n e^{i\omega_n t}, \qquad \omega_n = \frac{2n\pi}{T} \tag{2.9.7}$$

which is (2.9.1) with $c_n = 0$ for $|n| > m$. Thus, except for trigonometric polynomials, DDS can always represent the response data with a fewer number of exponentials than Fourier methods. Therefore the parsimonious DDS models are usually easier to interpret, understand, and use (especially in analogy with the n-degree-of-freedom system) than a large number of Fourier coefficients c_n.

2.9.5 Computational Efficiency of Fourier Methods: Fast Fourier Transform

What price do we pay for the advantage of parsimonious, easily interpretable, physically meaningful DDS models? Generally, they require more computation than the Fourier methods. Instead of closed-form relations such as (2.9.4) and (2.9.6) in computing the Fourier coefficients, the DDS method requires linear or nonlinear least squares methods to estimate model parameters and root- or eigenvalue-finding routines to recover the exponents and the coefficients of the exponentials. The very broadening of the choice of exponents from the imaginary line to the complex plane that allows DDS to represent the data with fewest exponentials also makes the task of finding these exponents more difficult, mathematically less tractable, and computationally more intensive.

The computational effort is of course minimal for a Fourier series because the exponents are not only restricted to the imaginary axis but are known and countable: $2n\pi/T$, $n = 0, \pm 1, \pm 2, \ldots$. By suitably truncating this countable sequence to a finite arithmetic sequence of multiples of the fundamental frequency $2\pi/T$, the computation can be executed very fast on a computer, yielding the fast Fourier transform (FFT) algorithm that has made Fourier methods more widespread since the 1960s. Its implementation can be further accelerated by parallel processors. Parallel processing is the key to the fast implementation of FFT and is made possible by the orthogonality (2.9.3) that allows c_n given by (2.9.4) to be computed separately and hence in parallel. Moreover, a truncated Fourier series approximation of (2.9.4), similar to (2.9.7), can be further broken down until an individual computation is reduced to two multiplications and an addition. Furthermore, the constants used in the multiplication are sampled values of exponentials with imaginary exponents (sines and cosines) that repeat at periodic intervals, and hence only a limited number over one cycle need to be calculated (which can be done in hardware).

Thus the computational efficiency of the FFT inherently depends upon the assumption that the data is sampled from a truncated Fourier series or a trigonometric polynomial. Since the actual response from which the data has been sampled may not even be periodic, as, for example, a one-degree-of-freedom damped system, the response usually requires a Fourier transform representation, not an infinite Fourier series and certainly not a finite Fourier series or a trigonometric polynomial. Therefore, unless the system is practically undamped and the fundamental frequency used in the FFT is so carefully chosen that all the system frequencies are integral multiples or harmonics of this fundamental frequency, the FFT will introduce not only considerable approximation but also considerable distortion. We will discuss this distortion more fully in Chapter 5 with the necessary illustrations. Here we follow Papoulis (1977) to discuss it briefly and theoretically, for the purpose of comparing the outcome with DDS.

2.9.6 Discrete Fourier Series as the Basis of FFT

How is a nonperiodic response, say $f(t)$, requiring a Fourier transform, made amenable to the FFT, requiring a finite Fourier series? This may be explained in two steps, both of which introduce their own distortions.

FOURIER SERIES, FOURIER TRANSFORM AND FFT VERSUS DDS

In the first step, to reduce a Fourier transform (2.9.5) to a Fourier series (2.9.1), it is assumed that the response function $f(t)$ sampled over a finite period T repeats itself at the same period. The new function, say $\bar{f}(t)$, which is a sum of the original nonperiodic response repeated ad infinitum over time, is now periodic and the Fourier series representation (2.9.1) is justified for this summed response function. This introduces time truncation or windowing distortion called leakage unless the data is sampled from a response with an integral number of cycles over the period T. However, theoretically

$$\bar{f}(t) = \sum_{n=-\infty}^{\infty} f(t + nT) \tag{2.9.8}$$

which will be graphically illustrated in Figure 5.8.

In the second step, it is recognized that although this summed (repeated) response function $\bar{f}(t)$ is now periodic and has a Fourier series representation (2.9.1), its infinite number of coefficients c_n cannot be numerically calculated from a finite number of sampled data points. If N data points are sampled over the interval T so that at sampled points $t = kT/N$, $k = 0, 1, 2, \ldots, N-1$, then

$$\bar{f}\left(\frac{kT}{N}\right) = \sum_{n=-\infty}^{\infty} c_n \exp\left(\frac{i2n\pi}{T} \cdot \frac{kT}{N}\right) \tag{2.9.9}$$

and using

$$n = l + rN, \quad l = 0, 1, \ldots, N-1, \quad r = 0, \pm 1, \pm 2, \ldots \tag{2.9.10}$$

reduces this infinite Fourier series to a finite series called discrete Fourier series (DFS), also often called discrete Fourier transform (DFT):

$$\bar{f}\left(\frac{kT}{N}\right) = \sum_{n=-\infty}^{\infty} c_n \exp\left(\frac{i2\pi nk}{N}\right)$$

$$= \sum_{l=0}^{N-1} \sum_{r=-\infty}^{\infty} c_{(l+rN)} \exp\left[\frac{i2\pi(l+rN)k}{N}\right]$$

$$= \sum_{l=0}^{N-1} \exp\left(\frac{i2\pi kl}{N}\right) \sum_{r=-\infty}^{\infty} c_{(l+rN)}, \quad \text{since } e^{irk2\pi} = 1$$

$$\bar{f}\left(\frac{kT}{N}\right) = \sum_{l=0}^{N-1} \bar{c}_l \exp\left(\frac{i2\pi kl}{N}\right), \quad k = 0, 1, \ldots, N-1 \tag{2.9.11}$$

where

$$\bar{c}_l = c_l + c_{l+N} + c_{l-N} + c_{l+2N} + c_{l-2N} + \cdots \tag{2.9.12}$$

which will be graphically illustrated in Figure 5.10.

Thus, instead of calculating c_n by a numerical approximation of the integral (2.9.4), $\bar{c}_l$ are calculated in an FFT using (2.9.11), which requires solving N linear equations in N unknowns. In general, this would require inversion of an $N \times N$

matrix, but because the elements of this matrix from (2.9.11) are discrete exponentials with imaginary exponents, the solution can be written down explicitly. To show this mainstay of the FFT's computational efficiency, let us denote the sampling interval

$$\Delta = \frac{T}{N} \qquad (2.9.13)$$

so that the discrete sampled values of $\bar{f}$ can be denoted by

$$\bar{f}_k = \bar{f}(k\Delta) = \bar{f}\left(\frac{kT}{N}\right)$$

Let the fundamental exponential of (2.9.11) be

$$w_N = \exp\left(\frac{i2\pi}{N}\right), \qquad w_N^N = 1 \qquad (2.9.14)$$

then the N equations (2.9.11) in $\bar{c}_l$ and their solution for $\bar{c}_l$ take the form

$$\bar{f}_k = \sum_{l=0}^{N-1} \bar{c}_l w_N^{kl} \qquad (2.9.15)$$

and

$$\bar{c}_l = \frac{1}{N} \sum_{k=0}^{N-1} \bar{f}_k w_N^{-kl} \qquad (2.9.16)$$

which can be verified either by direct substitution using (finite) geometric series, or by using matrix methods based on the Vandermonde matrix in Chapter 3. The latter method of verification will show that whereas FFT uses an $N \times N$ circulant matrix with eigenvectors forming a Vandermonde matrix of the *known* exponential $w_N^l, l = 0, 1, \ldots, N - 1$ to be able to explicitly solve for eigenvalues by (2.9.16), DDS uses an $n \times n$ Toeplitz matrix with eigenvectors forming a Vandermonde matrix of unknown exponentials to solve for both n (model order or state space dimension) and eigenvalues via the modeling procedure discussed earlier.

The FFT algorithm further expedites the calculation of $\bar{c}_l$ via (2.9.16) by successively splitting the sum for even and odd k. If $N = 2^s$, the computation can be done in $s = \log_2 N$ stages, each requiring N multiplications; thus the FFT requires only $N \log_2 N$ multiplications compared to N^2 for the general $N \times N$ matrix multiplication in (2.9.16). For a typical $N = 1024$ used in commercial Fourier analyzers, this implies 10,240 multiplications for an FFT compared to 1,048,576 for the general multiplication in (2.9.16)! (Additions, requiring far less time, are not usually counted. A floating point add, a floating point multiply, and the necessary subscripting effort is called a *flop*, and the operation count is given as the number of flops required.)

2.9.7 DDS Requires Less Computation for Better Final Results

The discrete Fourier series basis of the computational efficiency of an FFT clearly shows that the FFT algorithm does not provide us with the Fourier transform, not even the approximating Fourier series coefficients c_n, but their disguised aliases $\bar{c}_l$. This phenomenon of aliasing arises because (2.9.12) shows that $\bar{c}_l$ contains not only c_l corresponding to the frequency $\pm 2\pi l/N$, but also $c_{l \pm rN}$ corresponding to frequencies $2\pi(l \pm rN)/N$ which can no longer be distinguished. Aliasing introduces a truncation and consequent distortion in the frequency domain which is the price of reducing an infinite Fourier series to a (finite) discrete Fourier series, just as windowing or the artificial periodicity of T introduces a truncation and consequent distortion in time domain that is the price of reducing a Fourier integral to a Fourier series. Moreover, unless the data contains integral cycles in the window period T, attempts to reduce the distortion in time domain inevitably increase the distortion in the frequency domain, and vice versa. (A mathematical function cannot be nonzero only over a limited band in both time and frequency domains.)

Therefore, an FFT provides at best a time windowed, frequency-aliased image of the time domain sampled points in the frequency domain. Excepting a few applications such as diagnostic monitoring, where one is merely looking for qualitatively large magnitudes over well-known frequency bands and aliasing has been carefully eliminated, this raw FFT image is not sufficient. To recover the underlying Fourier transform (e.g., frequency response in modal analysis of vibration, or autospectrum in noise spectrum analysis) a considerable post-processing of the raw FFT is needed; now the amount of computation starts to approach that of DDS, but the final results are worse than DDS because they are based on the FFT-distorted image of the original time domain data.

The best post-processing of the FFT results commonly followed in modal analysis of vibrations is "curve fitting" or estimating the parameters such as natural frequency, damping ratio, or mode shapes by fitting the appropriate Fourier transform functions to the raw FFT data. The amount of computation required for such curve fitting is the same as that required for obtaining an equivalent DDS model directly from the time domain data. For such results, the DDS requires less computation because the FFT computation is eliminated, and DDS provides better results (clear and more accurate) because the distortion introduced by FFT is eliminated with it!

Time domain truncation and frequency domain aliasing do affect DDS results, but in much less crucial ways than FFT. For example, the accuracy of the DDS model parameters is inversely proportional to the square root of the number of the data points N and hence improves with record length $N(T = N\Delta)$, since the number of parameters or the dimension of the state space depending on the underlying system does not change with N and hence the extra information from additional data adds to the accuracy of the same parameters as long as N substantially exceeds the model order. This advantage is not available in an FFT since N is also the number of frequencies at which the FFT is computed; hence the extra information from the increased data points gets spread over an equally increased number of frequencies and does not improve the accuracy at any given frequency. The commonly used method of averaging the noisy FFT also does not help because, as discussed later for random signals, this method of averaging is actually equivalent to applying a (Bartlett) window that introduces its own distortion.

As illustrated in Chapters 5 and 6, aliasing affects DDS much less than the FFT. To avoid aliasing in FFT, it is essential that the entire Fourier transform or all the coefficients of Fourier series be negligible at frequencies beyond the (Nyquist) sampling frequency $2\pi N/2T$, whereas in DDS it is only necessary that the peak frequency in the Fourier transform (the frequency determined by the imaginary part of the exponent) be less than $2\pi N/2T$. This is again understandable because DDS only works with a few exponents in the complex plane, whereas FFT works with a fixed number of N discrete points on the imaginary axis.

2.9.8 Data with Noise: Random Function

For noisy data, the corresponding function is random, and neither Fourier series nor Fourier transform exist for a random function because it is neither periodic nor does it decay with time so as to be absolutely integrable over an infinite time interval. However, we can always compute an N point FFT of N sample data points even from such a random function or stochastic system. So what does this mathematical nonexistence of the Fourier series or transform of a random function mean in practice?

Recall that one sample from a random function may be quite different from another; the same is true of the Fourier transform of such a sample or FFT of the sampled data. In other words, the noise in the time domain data points simply transforms into the frequency domain points and one cannot make any valid conclusions from the FFT of sampled data. Thus the FFT of a single noisy record is virtually useless in practice. On the other hand, DDS can use a single record to model and characterize the underlying system. Again this is possible for DDS because with an increasing number of data points, more and more information becomes available on the limited number of exponents, thus increasing the accuracy of their estimates; this is not possible for an FFT because the information from the additional data points gets spread over additional frequencies so that each data point in the time domain and each FFT point in the frequency domain remains as noisy as before.

This situation reminds us of an experimental result that is random because of the random experimental error. We know that in this case we can "average out" the random experimental error if the experiment is carefully repeated. If the careful repetition guarantees that each random outcome X_i has the same mean μ

("true" result), the same variance σ^2 caused by the experimental error, and one repetition does not affect the other so that the random variables X_i are independently distributed with mean μ and variance σ^2, then by the procedure of Section 2.7.1 (or Appendix A.1), it is easy to see that the average

$$\bar{X} = \frac{1}{n}(X_1 + X_2 + \cdots + X_n)$$

has mean μ and variance σ^2/n, that is,

$$E(\bar{X}) = \mu, \qquad \text{Var}(\bar{X}) = \frac{\sigma^2}{n} \qquad (2.9.17)$$

Thus the error can indeed be "averaged out," that is, made arbitrarily small (in variance) by taking large enough n.

Can we similarly average out the noise in FFT? The answer is in general no, not unless restrictions analogous to those of the random variables X_i discussed above are satisfied by the response function; but now, because of the added dependence on time, these restrictions become much more stringent and often unrealistic in practice. These restrictions require that the response be stationary, that is, all the probabilistic properties (strictly stationary) or at least mean, variance, and covariance expressing statistical dependence between two response points separated by a fixed interval (wide sense or simply stationary), be the same for all time. In particular, if the response contains deterministic functions, the mean changes with time and the noise cannot be averaged out. Hence a deterministic component must be removed from the data before using FFT. Since such a component is not known a priori, its removal requires considerable trial and error as well as additional computation.

Even with stationary response data, noise cannot be reduced by simply averaging the FFT because a result similar to (2.9.17) is not available in the frequency domain. Stationarity guarantees that a random function is ergodic in the time domain, that is, time averaging is equivalent to ensemble averaging or averaging over time is the same as averaging over different records; hence the noise variance does reduce by averaging in the time domain, analogous to (2.9.17). But the operation of Fourier transformation does *not* preserve ergodicity and hence the average of the FFT records does not converge to the "true" Fourier transform. Although the mathematical reasoning for this is rather deep, it can be intuitively explained with some oversimplification to convey the basic idea. Stationarity of a random function implies ergodicity, which says that information in N/M samples of M data points each is the same as the information in one record of N points. But this is clearly not true in the frequency domain because, for example, an FFT of N points has information on N frequencies compared to only M frequencies in N/M records of M points each.

The only way to reduce noise in a stationary record is to average it out in the time domain. Since the constant mean can be removed by subtracting the average and the constant variance is merely a scaling factor, the time behavior of a stationary random function is summarized by the autocovariance function which decays to zero with increasing lag time and hence has a well-defined Fourier transform called the autospectrum or power spectrum, so called because it represents the average power or square amplitude per unit time in the frequency domain as the sample size and number of samples tend to infinity. Once a DDS model is fitted to the data, assuming it to be the correct model, the theoretical auto- and cross-covariance and its Fourier transform auto- and cross-spectrum are available as continuous functions of time and frequency, respectively. This avoids the computation of an FFT and its pre- or post-processing, and will be illustrated in Chapter 6.

The current practice of modal analysis and spectrum analysis, however, is based on sample covariance (auto and cross) and its Fourier transform spectrum (auto and cross). The sample spectrum can be computed either by averaging in the time domain first to get the covariance and then taking its FFT, or by first taking the FFT of the record and then averaging at a given frequency the squared amplitudes for the autospectrum and products of amplitudes from two different records for the cross-spectrum. Note that FFT amplitudes themselves should not be averaged for reasons discussed earlier. Both these procedures yield the same result if the weighting functions used in averaging form a Fourier transform pair, but the latter is followed in modern Fourier or spectrum analysis as it is computationally more efficient. Different weighting functions used in weighted averaging are called windows because they effectively block out or modify the undesirable part of the record in time as well as frequency domain in averaging and thereby effectively smooth the spectrum or reduce its variance, thus averaging out the noise. The simple (uniformly weighted) averaging of spectra of N/M records of M points each turns out to be a particular window called a Bartlett window.

Unfortunately, such averaging, smoothing, or windowing introduces bias, that is, the smoothed spectrum systematically differs from the "true" or theoretical spectrum. In fact, for a given N the more the variance reduces, the more the bias increases, and vice versa. If N is the total number of points and M is the time domain width of the window, such as the Bartlett window mentioned earlier, then, for an estimate $\hat{S}(\omega)$ of the "true" spectrum $S(\omega)$, with most commonly used windows, the variance of $\hat{S}(\omega)$ is

$$\text{Var}\,[\hat{S}(\omega)] \propto \frac{M}{N} S^2(\omega) \qquad (2.9.18)$$

and the bias of $\hat{S}(\omega)$ is

$$\text{Bias}\,[\hat{S}(\omega)] \propto -\frac{1}{M} h^{(1)}(\omega) \qquad \text{for Bartlett} \qquad (2.9.19a)$$

$$\propto \frac{1}{M^2} \ddot{S}(\omega) \qquad \text{for all others} \qquad (2.9.19b)$$

[see Priestley, 1981 for $h^{(1)}(\omega)$]. $\ddot{S}(\omega)$ denotes the second derivative of the true spectrum. In particular, note that both the variance and bias are large at a sharp peak in the spectrum which is generally of most interest.

If the time domain window width that determines the number of neighboring covariances used in averaging is M, the frequency domain window width that determines the number of neighboring spectra used in averaging turns out to be proportional to $1/M$. Therefore, for a given N, taking M to be very small (e.g., increasing number of samples N/M in the example given above) to reduce the variance and smooth the spectrum, will increase the frequency domain window width, possibly averaging neighboring distinct peaks. Hence, in practice, the compromise between variance and bias also turns out to be a compromise between smoothness and the ability to resolve closely spaced frequency peaks. Thus, averaging out the noise by applications of smoothing windows requires not only much additional calculation over and above that of a raw FFT, but also considerable trial and error in determining smoothing windows and parameters such as M and N. Moreover, to obtain the best estimates requires correct knowledge of the spectrum we are trying to estimate! To know the variance we need the spectral value and to know the bias we need the second derivative of the spectrum, as seen from (2.9.18) and (2.9.19).

The root cause of all these difficulties, pointed by Kendall (1945) long before the advent of the FFT, is that the sample autocovariance is a very poor estimator of the theoretical autocovariance. It has a large variance and neighboring estimates are highly correlated, thus presenting a distorted version of true autocovariance. However, because of the ergodicity guaranteed by the assumed stationarity, it is at least consistent, that is, it converges to the true autocovariance as $N \to \infty$. Its Fourier transform, the sample spectrum, loses even ergodicity, since the Fourier transformation does not preserve ergodicity. So the sample spectrum is not a consistent estimator and does not converge to the true spectrum; as shown by Grenander (1951), it does not even converge to a limiting random variable as $N \to \infty$. The raw sample spectrum is erratic and fluctuates wildly; no wonder attempts to smooth it may end up combining neighboring peaks and leaving some spurious ones caused merely by large variance intact, particularly when the knowledge of the underlying theoretical spectrum is lacking! Note that these discouraging statistical properties of the raw sample spectrum persist even when it is obtained directly from the FFT of the data rather than from the FFT of the sample covariance, since the two methods of computation are mathematically equivalent.

2.9.9 FFT for Qualitative, DDS for Quantitative Results

There is one advantage of the sample spectrum, besides its computational efficiency via FFT, that has made it a popular data analysis technique in spite of its drawbacks. The advantage is that, irrespective of the (unknown) nature of statistical dependence in the original set of N point data, the N point spectral estimates at the appropriate discrete frequencies (called periodogram in the early statistical literature) are statistically independent. Moreover, FFT of the data is a finite linear transformation, and therefore the spectrum at each discrete frequency has a well-defined distribution (Chi-square if the data is normal). Smoothing is again a linear operation, so the statistical properties of smoothed spectra can be obtained and optimized assuming a known theoretical spectrum in a mathematically tractable way. There exists, therefore, a rich and abundant body of literature on statistical spectral estimation.

A comprehensive account of this literature has been given by Priestley (1981), who points out that in spite of its drawbacks, periodogram (sample spectrum) analysis remains entirely appropriate for the purpose for which it was designed, namely, the analysis of processes *with discrete spectra*. This should be clear even without statistics and is applicable also to the use of FFT on deterministic data, since, as pointed out in Section 2.9.3, the basic discrete (finite) Fourier series assumes the data to be a sample from a trigonometric polynomial, consisting of a sum of exponentials with imaginary exponents containing discrete frequencies that are multiples of a fundamental frequency. These FFT-based techniques, designed for a finite number of discrete frequencies, becomes inappropriate when applied to a system requiring continuous functions of frequency, such as the frequency response or transfer function of even a one-degree-of-freedom damped system, and random functions with continuous spectra. No (necessarily) finite amount of averaging or smoothing will solve the problem, it will merely lead to dichotomies such as a compromise between aliasing and resolution, or between variance and bias, discussed earlier. To put it differently, all these problems arise from a basic difficulty: a Fourier series does not exist for a nonperiodic function and certainly not for a random function!

It is for such systems whose frequency response, transfer functions or spectrum are functions of continuous frequencies that the DDS method becomes appropriate and necessary in spite of its increased computation and the FFT may provide at best, under fortunate circumstances, a qualitative visual indicator. Notwithstanding its computational efficiency, the FFT is really useful only for detecting the presence of certain suspected discrete frequency peaks.

Moreover, for most applications of system analysis in practice, a characteristic such as frequency response, transfer function, or spectrum is only an intermediate step. One likes to use them subsequently for design, prediction, or control. For such quantitative analysis DDS is invaluable. It will be shown in Chapter 6 that for visual inspection, diagnostic monitoring, and other applications, DDS can provide much more quantitative information and insight, as well as plots of these characteristics that are far clearer and easier to interpret than the FFT.

As a final note on the similarity and differences between DDS and FFT, it is interesting to note that for noisy data both DDS and FFT attempt to reduce statistically dependent data to independent data. DDS does so in time domain by reducing it to a_t's by means of a nonlinear least-squares procedure that recovers the parameters and the few necessary exponents that immediately provide a physically meaningful description and hence requires extensive computation. FFT does so by Fourier transforming the autocovariance function at known multiples of frequencies by linear operations that naturally require less computation but also provide less information on the exponents that characterize the system; to recover them would require the same additional computation as DDS. Thus the additional computation of DDS is the price we have to pay for getting a clear, accurate, and physically meaningful system characterization from the response data!

3

FUNDAMENTALS OF DATA DEPENDENT SYSTEM ANALYSIS

In Chapter 2 we formulated the data dependent systems (DDS) philosophy in the framework of state space and evolved the DDS modeling strategy. In its essence, this strategy amounts to giving more and more freedom to the data by letting it express its dependence in a higher and higher dimensional state space until further freedom fails to reduce the "unexplained" residuals significantly or until the unexplained residuals are negligible compared to the scale of numbers in which we are interested. Once the model and its solution are introduced and illustrated in Chapter 2, our next task is to delineate the linear algebra and geometry underlying such a data dependent model and to analyze the implications of its solution.

We can understand the solutions of the *scalar* (univariate) first- and second-order models quite well. The first-order model produces an exponential response, and although we have not yet discussed the response of a second-order system, we can guess from the simplest one-degree-of-freedom system consisting of a spring and a mass studied in Chapter 2 that its response may be oscillatory or sinusoidal. The response of a higher-order scalar or first-order vector system has not yet been discussed. Even though we know its mathematical solution in terms of the matrix exponential $e^{\mathbf{A}t}$ or its discrete version ϕ^t, the nature of this exponential itself is not yet clear. In this chapter we will develop the mathematical tools for tackling this problem. We will show that the response of any linear system can be expressed as a combination of first- and second-order systems.

As explained in Chapter 2, the state of a system is a vector and the space it occupies for different values of t is a vector space. Since its behavior for different t is governed by linear differential or difference equations, the state space is a linear vector space. The algebraic geometry of such a linear vector space will be reviewed in Section 3.1 and applied to the DDS modeling in state space by means of autoregressive models in Section 3.2. The orthogonal decomposition resulting from this algebraic geometry is then illustrated in Section 3.3 with reference to the modeling strategy, estimation by least squares, and forecasting and control. The data vector has N (number of observations) components, and modeling is an attempt to "explain" it with as small residuals as possible using a state space of the smallest possible dimension n.

The central state equation

$$\dot{\mathbf{x}} = \mathbf{A}\mathbf{x} \quad \text{or} \quad \mathbf{X}_t = \phi \mathbf{X}_{t-1}$$

specifies that the derivative of the state vector is a transformation of the state $\mathbf{x}(t)$ by the state matrix $\mathbf{A}$ [or that the state vector at the current time is a transformation of the state vector at previous (discrete) time by the state matrix ϕ] in the linear vector space. Therefore, we will study transformation of vectors represented by matrices in Section 3.4 of this chapter.

If $\mathbf{A}(\phi)$ is a diagonal matrix with say $a_i(\phi_i)$ as its diagonal elements, the state equation would simply reduce to n *separate* scalar equations:

$$\dot{x}_i = a_i x_i \quad \text{or} \quad X_{it} = \phi_i X_{it-1}$$

for which we already know the solution. In general, when a matrix is not diagonal, it can be represented as a product of transformation matrices and a diagonal matrix. For most square matrices of interest, the elements of the diagonal matrix turn out to be the eigenvalues of the matrix and the transformation matrix has columns which are the corresponding eigenvectors; the representation is then called spectral decomposition. Hence this chapter aims at the decomposition of the state space solutions in terms of the first- and second-order solutions via the eigenvalue–eigenvector decomposition of the state matrix $\mathbf{A}$ or ϕ in the linear vector space spanned by $\mathbf{x}$ or $\mathbf{X}_t$, discussed in Section 3.5. Such a decomposition of the vector space of solutions follows from the spectral decomposition or representation of the state matrix, which is shown to be the crux of modal and spectrum analysis in Section 3.5; this underscores the basic difference between the conventional and DDS approaches to modal and spectrum analysis. For nonsquare matrices, the diagonal matrix consists of singular values and the transforming matrices have columns called singular vectors, also discussed in Section 3.5. The resultant singular value decomposition is a useful tool in determining the dimension of the state space.

The spectral decomposition is illustrated by application to univariate continuous and discrete time systems in Section 3.6, confirming the familiar results such as impulse response and step response. These results can also be obtained without matrix methods, quite easily in low-order (or dimension) cases to illustrate how the matrix methods are working, and somewhat tediously in high-order (or dimension) cases to illustrate the elegance and power of matrix methods. These illustrations pave the way in Chapter 4 for their application to multivariate systems, for which these matrix methods are indispensible.

3.1 VECTOR SPACE

Since state space is a vector space, we will review briefly the definition of a vector space from linear algebra with the related concepts of basis, dimension, inner product, and norm. A matrix and its rank evolve naturally from these concepts.

3.1.1 Definition of a Vector Space and Subspace

A linear vector space $\mathscr{X}$ is a set whose elements are called vectors and in which two operations, called addition and scalar multiplication, are defined with the following familiar algebraic properties:

To every pair of vectors **x** and **y** in $\mathscr{X}$ there corresponds a vector $\mathbf{x} + \mathbf{y}$ in $\mathscr{X}$ such that $\mathbf{x} + \mathbf{y} = \mathbf{y} + \mathbf{x}$ and $\mathbf{x} + (\mathbf{y} + \mathbf{z}) = (\mathbf{x} + \mathbf{y}) + \mathbf{z}$; $\mathscr{X}$ contains a unique vector $\mathbf{0}$ (the null vector of origin of $\mathscr{X}$) such that $\mathbf{x} + \mathbf{0} = \mathbf{x}$ for every **x** in $\mathscr{X}$; and to each **x** in $\mathscr{X}$ there corresponds a unique vector $-\mathbf{x}$ such that $\mathbf{x} + (-\mathbf{x}) = \mathbf{0}$.

If c is a scalar (from the field of real or complex numbers, or in general any field in which addition, subtraction, multiplication, and division are suitably defined), there is a vector $c\mathbf{x}$ in $\mathscr{X}$ such that $1\mathbf{x} = \mathbf{x}$, $c_1(c_2\mathbf{x}) = (c_1 c_2)\mathbf{x}$, and such that the vector and scalar distributive laws $c(\mathbf{x} + \mathbf{y}) = c\mathbf{x} + c\mathbf{y}$ and $(c_1 + c_2)\mathbf{x} = c_1\mathbf{x} + c_2\mathbf{x}$ hold.

A subspace or linear manifold in a vector space $\mathscr{X}$ is any subset of vectors $\mathscr{M}$ closed under addition and scalar multiplication, that is, if **x** and **y** are in $\mathscr{M}$, $c_1\mathbf{x} + c_2\mathbf{y}$ is in $\mathscr{M}$ for any scalars c_1 and c_2. Any such subset $\mathscr{M}$ is itself a vector space in the finite dimensional vector spaces we need.

3.1.2 Basis and Dimension of a Vector Space

A set of vectors $\mathbf{x}_1, \mathbf{x}_2, \ldots, \mathbf{x}_n$ is linearly *dependent* if there exist scalars $c_1, c_2, \ldots, c_n$ not all simultaneously zero such that

$$c_1\mathbf{x}_1 + c_2\mathbf{x}_2 + \cdots + c_n\mathbf{x}_n = \mathbf{0}$$

otherwise it is linearly *independent*; we will generally drop the adjective "linearly" unless it is to be emphasized.

The null vector $\mathbf{0}$, or any set of vectors containing a null vector, is a dependent set. A set of non-null vectors is dependent when and only when a member in the set is a linear combination of its predecessors. A linearly independent subset of vectors in a vector space $\mathscr{X}$, generating or spanning $\mathscr{X}$, is called a basis of $\mathscr{X}$. If $\mathbf{a}_1, \mathbf{a}_2, \ldots, \mathbf{a}_m$ and $\mathbf{\beta}_1, \mathbf{\beta}_2, \ldots, \mathbf{\beta}_n$ are two alternate choices for a basis, then $m = n$, which is called the dimension or rank of $\mathscr{X}$. Every vector has a unique representation in terms of a given basis.

Let $\mathbf{a}_1, \mathbf{a}_2, \ldots, \mathbf{a}_n$ be n given vectors in an arbitrary vector space and consider the linear equation in the scalars $x_1, x_2, \ldots, x_n$:

$$x_1\mathbf{a}_1 + x_2\mathbf{a}_2 + \cdots + x_n\mathbf{a}_n = \mathbf{0} \tag{3.1.1}$$

The null vector with all x_i simultaneously zero is a trivial solution of (3.1.1). The necessary and sufficient condition that (3.1.1) has a nontrivial solution is that $a_1, a_2, \ldots, a_n$ be dependent. The solutions $\mathbf{x} = (x_1, x_2, \ldots, x_n)$ constitute a linear vector space.

Let $\mathscr{S}$ be the subspace of solutions and $\mathscr{M}$ be the subspace spanned by vectors $a_1, a_2, \ldots, a_n$. Then, if d denotes the dimension of a vector space, we have the important result (see Appendix A.3.1 for proof):

$$d[\mathscr{S}] + d[\mathscr{M}] = n \qquad (3.1.2)$$

The nonhomogeneous equation corresponding to the homogeneous equation (3.1.1),

$$x_1 a_1 + x_2 a_2 + \cdots + x_n a_n = \mathbf{y} \qquad (3.1.3)$$

has a solution if and only if $\mathbf{y}$ is dependent on $a_1, a_2, \ldots, a_n$, that is, the dimension or rank of a vector space spanned by $a_1, a_2, \ldots, a_n, \mathbf{y}$, is the same as that spanned by $a_1, a_2, \ldots, a_n$; then the equations are said to be consistent and we have either one unique solution or an infinite number of solutions.

A general solution of the nonhomogeneous equation is the sum of a particular solution of the nonhomogeneous equation and a general solution of the homogeneous equation. This follows since the difference of any two nonhomogeneous solutions is a homogeneous solution, as seen by substituting them in (3.1.3) and subtracting the two resultant equations.

Obviously, a nontrivial or nonnull solution of the homogeneous equation is not unique. Hence the solution of a nonhomogeneous equation is unique when the corresponding homogeneous equation has the null vector as the only solution, that is, $d[\mathscr{S}] = n - d[\mathscr{M}] = 0$ or $d[\mathscr{M}] = n$, that is, when $a_1, a_2, \ldots, a_n$ are independent.

To write the nonhomogeneous equations in a more familiar and succinct form, let us introduce matrices. If the column vectors

$$a_i = (a_{1i}, a_{2i}, \ldots, a_{mi})^T, \qquad i = 1, 2, \ldots, n$$

then (3.1.3) is a set of m linear equations in n unknowns:

$$\begin{aligned} a_{11}x_1 + a_{12}x_2 + \cdots + a_{1n}x_n &= y_1 \\ a_{21}x_1 + a_{22}x_2 + \cdots + a_{2n}x_n &= y_2 \\ &\vdots \\ a_{m1}x_1 + a_{m2}x_2 + \cdots + a_{mn}x_n &= y_m \end{aligned} \qquad (3.1.4a)$$

which may be written in a succinct form as

$$\mathbf{Ax} = \mathbf{y} \qquad (3.1.4)$$

where $\mathbf{A}$ is an $m \times n$ matrix with columns $\mathbf{a}_1, \mathbf{a}_2, \ldots, \mathbf{a}_n$, and $\mathbf{x}$ and $\mathbf{y}$ are the column vectors $\mathbf{x} = (x_1, x_2, \ldots, x_n)^T$ and $\mathbf{y} = (y_1, y_2, \ldots, y_m)^T$. The subspace $\mathscr{S}$, which is the set of vectors $\mathbf{x}$ such that $\mathbf{Ax} = \mathbf{0}$, is then called the null space of $\mathbf{A}$, denoted by $\mathbf{N(A)}$. The subspace $\mathscr{M}$ spanned by columns of $\mathbf{A}$ is called the *range* of $\mathbf{A}$, denoted by $\mathbf{R(A)}$. The fundamental relation (3.1.2) now takes the form

$$d[\mathbf{N(A)}] + d[\mathbf{R(A)}] = n \qquad (3.1.2a)$$

Let only r rows of $\mathbf{A}$ be independent and let $k = d[\mathscr{M}(\mathbf{a})] = d[\mathbf{R(A)}]$, be the dimension or rank of the vector space (and of matrix $\mathbf{A}$) spanned by the column vectors. If we retain only the r independent rows, the homogeneous solution space $\mathscr{S}$ will still be the same. Since $d[\mathscr{S}] = d[\mathbf{N(A)}] = n - k$, some k columns in the reduced set of equations are independent. But now the column vector has only r elements, so that it belongs to an r-dimensional space and therefore the number of independent column vectors cannot be greater than r. Therefore $r \geq k$. Reversing the roles of columns and rows in the argument, $k \geq r$. Hence $r = k$. In other words, the rank of a matrix $\mathbf{A}$ is the number of independent rows or columns and remains the same for $\mathbf{A}^T$, $\bar{\mathbf{A}}$, and $\bar{\mathbf{A}}^T$, where the bar denotes complex conjugate. Thus

$$\text{Rank}(\mathbf{A}) = d[\mathscr{M}(\mathbf{a})] = d[\mathbf{R(A)}] = d[\mathbf{R(\bar{A})}] = d[\mathbf{R(A}^T)] \qquad (3.1.5)$$

$$d[\mathscr{S}] = d[\mathbf{N(A)}] = n - d[\mathscr{M}(\mathbf{a})] = n - d[\mathbf{M}(\boldsymbol{\beta})] = n - \text{rank}(\mathbf{A}) \qquad (3.1.2b)$$

Therefore, the solution of (3.1.4) exists, that is, the nonhomogeneous equation is consistent if

$$\text{Rank}(\mathbf{A}) = \text{rank}(\mathbf{A}|\mathbf{y}) \leq n$$

where $(\mathbf{A}|\mathbf{y})$ denotes the matrix $\mathbf{A}$ augmented by the column vector $\mathbf{y}$.

If $\text{rank}(\mathbf{A}) = n$ so that $d[\mathscr{S}] = 0$, the null vector is the only solution of the homogeneous equation, hence the nonhomogeneous equation has a unique solution; in particular, for a square matrix $\mathbf{A}$

$$\mathbf{x} = \mathbf{A}^{-1}\mathbf{y}$$

If $\text{rank}(\mathbf{A}) < n$, there are an infinite number of solution vectors generated by arbitrary linear combinations of $[n - \text{rank}(\mathbf{A})]$ independent solutions. Hence the dimension of the solution space, $[n - \text{rank}(\mathbf{A})]$, or $(n - r_A)$, is often called the degeneracy of $\mathbf{A}$, denoted by

$$q_A = n - r_A \qquad (3.1.2c)$$

3.1.3 Inner Product and Norm

Inner product, written as $\langle \mathbf{x}, \mathbf{y} \rangle$, is any scalar valued function satisfying (bar denotes complex conjugate):

1. $\langle \mathbf{x}, \mathbf{y} \rangle = \overline{\langle \mathbf{y}, \mathbf{x} \rangle}$
2. $\langle \mathbf{x}, \alpha \mathbf{y}_1 + \beta \mathbf{y}_2 \rangle = \alpha \langle \mathbf{x}, \mathbf{y}_1 \rangle + \beta \langle \mathbf{x}, \mathbf{y}_2 \rangle$
3. $\langle \mathbf{x}, \mathbf{x} \rangle \geq 0$ for all $\mathbf{x}$ and $\langle \mathbf{x}, \mathbf{x} \rangle = 0$ if and only if $\mathbf{x} = \mathbf{0}$

Combining 1 and 2 gives

4. $\langle \alpha \mathbf{x}_1 + \beta \mathbf{x}_2, \mathbf{y} \rangle = \bar{\alpha} \langle \mathbf{x}_1, \mathbf{y} \rangle + \bar{\beta} \langle \mathbf{x}_2, \mathbf{y} \rangle$

The most commonly used inner product is $\bar{\mathbf{x}}^T \mathbf{y}$.

The length or norm of a vector is defined by

$$\|\mathbf{x}\| = \sqrt{[\langle \mathbf{x}, \mathbf{x} \rangle]}$$

The following Cauchy–Schwartz inequality connecting the inner product and the norm is quite useful in practice

$$|\langle \mathbf{x}, \mathbf{y} \rangle| \leq \|\mathbf{x}\| \cdot \|\mathbf{y}\|$$

with equality if and only if $\mathbf{x}$ and $\mathbf{y}$ are linearly dependent or colinear.

A unit vector can be obtained from an arbitrary vector by

$$\hat{\mathbf{x}} = \frac{\mathbf{x}}{\|\mathbf{x}\|}$$

The norm of the difference $\|\mathbf{x} - \mathbf{y}\|$ can be used as a metric or distance measure.

The angle θ between the two vectors $\mathbf{x}$ and $\mathbf{y}$ is defined by

$$\langle \mathbf{x}, \mathbf{y} \rangle = \|\mathbf{x}\| \, \|\mathbf{y}\| \cos \theta$$

Hence the two vectors are orthogonal if

$$\langle \mathbf{x}, \mathbf{y} \rangle = 0$$

and orthonormal if in addition

$$\|\mathbf{x}\| = \|\mathbf{y}\| = 1$$

3.1.4 Gram–Schmidt Orthogonalization and Triangular Reduction

Given a set of n linearly independent vectors $y_1, y_2, \ldots, y_n$ in a vector space, an *orthonormal basis* (*o.n.b.*) can be constructed by the Gram–Schmidt orthogonalization process. Let $v_1 = y_1$ and find v_2 by subtracting from y_2 a component in the direction of v_1 so that v_1, v_2 are orthogonal. Let $v_2 = y_2 - av_1$, then to find a so that v_1, v_2 are orthogonal

$$0 = \langle v_1, v_2 \rangle = \langle v_1, y_2 \rangle - a \langle v_1, v_1 \rangle$$

so that

$$v_2 = y_2 - \frac{\langle v_1, y_2 \rangle}{\langle v_1, v_1 \rangle} v_1$$

Similarly

$$v_3 = y_3 - \frac{\langle v_1, y_3 \rangle}{\langle v_1, v_1 \rangle} v_1 - \frac{\langle v_2, y_3 \rangle}{\langle v_2, v_2 \rangle} v_2$$

and so on. Finally normalize to get

$$\hat{v}_i = \frac{v_i}{\|v_i\|}, \qquad i = 1, 2, \ldots, n$$

For the orthonormal basis

$$\langle \hat{v}_i, \hat{v}_j \rangle = \delta_{i-j} = 0 \qquad \text{for } i \neq j$$
$$\qquad\qquad\qquad = 1 \qquad \text{for } i = j$$

The Gram–Schmidt orthogonalization of an arbitrary basis $y_1, y_2, \ldots, y_n$ to an orthogonal basis $[v_1, v_2, \ldots, v_n]$ can be succinctly represented by means of the upper triangular matrix with elements

$$\begin{aligned} t_{ij} &= 0, & i > j \\ &= 1, & i = j \\ &= \frac{\langle v_i, y_j \rangle}{\langle v_i, v_i \rangle}, & i < j \end{aligned} \qquad (3.1.6)$$

as

$$\begin{aligned} v_1 &= y_1 \\ v_2 &= y_2 - t_{12} v_1 \\ v_3 &= y_3 - t_{13} v_1 - t_{23} v_2 \\ &\vdots \\ v_n &= y_n - t_{1n} v_1 - t_{2n} v_2 \cdots - t_{(n-1)n} v_{n-1} \end{aligned} \qquad (3.1.7)$$

that is,

$$[\mathbf{y}_1, \mathbf{y}_2, \ldots, \mathbf{y}_n] = [\mathbf{v}_1, \mathbf{v}_2, \ldots, \mathbf{v}_n] \begin{bmatrix} 1 & t_{12} & t_{13} & \cdots & t_{1n} \\ 0 & 1 & t_{23} & \cdots & t_{2n} \\ \vdots & \vdots & \vdots & \vdots & \vdots \\ 0 & 0 & 0 & \cdots & 1 \end{bmatrix} \quad (3.1.8)$$

and to an o.n.b. $[\hat{\mathbf{v}}_1, \hat{\mathbf{v}}_2, \ldots, \hat{\mathbf{v}}_n]$ as

$$[\mathbf{y}_1, \mathbf{y}_2, \ldots, \mathbf{y}_n]$$
$$= [\hat{\mathbf{v}}_1, \hat{\mathbf{v}}_2, \ldots, \hat{\mathbf{v}}_n] \begin{bmatrix} t_{11}\|\mathbf{v}_1\| & t_{12}\|\mathbf{v}_1\| & t_{13}\|\mathbf{v}_1\| & \cdots & t_{1n}\|\mathbf{v}_1\| \\ 0 & t_{22}\|\mathbf{v}_2\| & t_{23}\|\mathbf{v}_2\| & \cdots & t_{2n}\|\mathbf{v}_2\| \\ \vdots & \vdots & \vdots & \vdots & \vdots \\ 0 & 0 & 0 & \cdots & t_{nn}\|\mathbf{v}_n\| \end{bmatrix} \quad (3.1.9)$$

This expression (3.1.9) is also known as QR factorization that expresses an arbitrary matrix as a product of a matrix **Q** with orthonormal columns and an upper triangular matrix **R**. If $\mathbf{y}_1, \mathbf{y}_2, \ldots, \mathbf{y}_n$ are not linearly independent, then

$$\mathbf{v}_i = \hat{\mathbf{v}}_i = \mathbf{0} \quad \text{for some } i$$

Then supplementing (3.1.6) with

$$t_{ij} = 0, \quad \text{if } \mathbf{v}_i = \mathbf{0} \quad (3.1.6a)$$

extending the nonzero $\hat{\mathbf{v}}_i$'s to a complete set of n orthonormal vectors, and replacing the zero $\hat{\mathbf{v}}_i$ vectors with the additional ones so obtained provides the QR factorization of an arbitrary (not necessarily full rank) matrix.

Although (3.1.8) is, strictly speaking, not a QR decomposition because $\mathbf{v}_i$'s are orthogonal but not orthonormal, one can get the QR decomposition (3.1.9) immediately by dividing $\mathbf{v}_i$ by their lengths and multiplying the ith row of t_{ij}'s by the same length. Hence we will refer to both (3.1.8) and (3.1.9) as QR decomposition.

Every vector **x** in the vector space then has a unique decomposition or expansion in terms of orthogonal or orthonormal basis vectors

$$\mathbf{x} = \sum_{i=1}^{n} a_i \mathbf{v}_i$$

$$= \sum_{i=1}^{n} \frac{\langle \mathbf{v}_i, \mathbf{x} \rangle}{\langle \mathbf{v}_i, \mathbf{v}_i \rangle} \mathbf{v}_i \quad \text{orthogonal decomposition} \quad (3.1.7a)$$

$$= \sum_{i=1}^{n} \langle \hat{\mathbf{v}}_i, \mathbf{x} \rangle \hat{\mathbf{v}}_i \quad \text{orthonormal decomposition} \quad (3.1.7b)$$

FUNDAMENTALS OF DATA DEPENDENT SYSTEM ANALYSIS

If the basis is not orthonormal, then, as an alternative to orthogonalization, we can find the so-called *reciprocal basis* $\{\mathbf{r}_1, \mathbf{r}_2, \ldots, \mathbf{r}_n\}$ to satisfy

$$\langle \mathbf{r}_i, \mathbf{v}_j \rangle = \delta_{ij}, \quad \text{that is,} \quad \mathbf{RB} = \mathbf{I} \quad (3.1.10)$$

where the $n \times n$ matrices

$$\mathbf{B} = \{\mathbf{v}_1, \mathbf{v}_2, \ldots, \mathbf{v}_n\} \text{ and } \mathbf{R} = \{\bar{\mathbf{r}}_1, \bar{\mathbf{r}}_2, \ldots, \bar{\mathbf{r}}_n\}^T. \quad (3.1.11)$$

Then

$$\mathbf{x} = \sum_{i=1}^{n} \langle \mathbf{r}_i, \mathbf{x} \rangle \mathbf{v}_i \quad (3.1.12)$$

3.2 ILLUSTRATIVE APPLICATIONS TO DDS IN STATE SPACE

We will now illustrate the vector space concepts reviewed in Section 3.1 by applying them to AR models. These applications will help clarify the approach to DDS in state space initiated in Chapter 2 and provide a geometrical basis for their subsequent extension to ARMA models. They will illustrate with simple examples how the DDS modeling strategy in state space outlined in Section 2.4.1 is actually implemented in the linear vector space framework starting from the response data.

3.2.1 Deterministic AR Models

Consider four data points: 90, 81, 72.9, and 65.61. What is the dimension of the state space occupied by this data vector? For deterministic data, the answer is the number of linearly independent data vectors that can be constructed from the data taken at lag 1, 2, ..., and so on. Therefore, we consider

$$\begin{bmatrix} x_1 & x_2 \\ x_2 & x_3 \\ x_3 & x_4 \end{bmatrix} = \begin{bmatrix} 90.0 & 81.00 \\ 81.0 & 72.90 \\ 72.9 & 65.61 \end{bmatrix}$$

If these vectors are linearly dependent, then the rank of the matrix and the dimension of the state space is 1 and the model is AR(1). If these vectors are linearly independent, then the matrix has a full rank of 2, the dimension of the state space is at least 2, and the AR(1) model would be inadequate for this data.

We can use the Gram–Schmidt procedure to check the dependence since it essentially removes the linearly dependent part of one vector from another; if the remainder $\mathbf{v}_i$ are zero, the vectors are linearly dependent, otherwise independent. Using (3.1.6)

$$t_{11} = t_{22} = 1, \qquad t_{12} = \frac{x_1 x_2 + x_2 x_3 + x_3 x_4}{x_1^2 + x_2^2 + x_3^2}$$

$$= \frac{90 \times 81 + 81 \times 72.90 + 72.9 \times 65.61}{90^2 + 81^2 + 72.9^2}$$

$$= 0.9$$

Therefore, using (3.1.7) we have

$$\mathbf{v}_1 = \begin{bmatrix} 90.0 \\ 81.0 \\ 72.9 \end{bmatrix}, \quad \mathbf{v}_2 = \begin{bmatrix} 81.00 \\ 72.90 \\ 65.61 \end{bmatrix} - 0.9 \begin{bmatrix} 90.0 \\ 81.0 \\ 72.9 \end{bmatrix} = \begin{bmatrix} 0 \\ 0 \\ 0 \end{bmatrix}$$

showing that the vectors are linearly dependent, the matrix is rank-deficient, and the model is AR(1).

This can be seen even from three data points that give the 2×2 matrix

$$\begin{bmatrix} x_1 & x_2 \\ x_2 & x_3 \end{bmatrix} = \begin{bmatrix} 90.0 & 81.0 \\ 81.0 & 72.9 \end{bmatrix}$$

whose columns are linearly dependent, so the matrix is singular or rank-deficient with rank 1, and a similar Gram–Schmidt calculation again yields $t_{11} = t_{22} = 1$ and $t_{12} = 0.9$. The data is thus modeled by the deterministic AR(1) model:

$$x_t = 0.9 x_{t-1}$$

The formula for calculating $t_{12} = \phi_1$ by the Gram–Schmidt procedure agrees with (2.1.6) obtained for linear regression. Note that $t_{12} = \phi_1$ is true not only for the AR(1) model but also for higher-order models in a slightly different form.

Thus, the deterministic AR(1) modeling of the data is the Gram–Schmidt or QR decomposition without implementing (3.1.6a)

$$\begin{bmatrix} 90.0 & 81.0 \\ 81.0 & 72.9 \end{bmatrix} = \begin{bmatrix} 90.0 & 0 \\ 81.0 & 0 \end{bmatrix} \begin{bmatrix} 1 & 0.9 \\ 0 & 1 \end{bmatrix}$$

FUNDAMENTALS OF DATA DEPENDENT SYSTEM ANALYSIS

or

$$\begin{bmatrix} 90.0 & 81.0 & 72.90 \\ 81.0 & 72.9 & 65.61 \end{bmatrix} = \begin{bmatrix} 90 & 0 & 0 \\ 81 & 0 & 0 \end{bmatrix} \begin{bmatrix} 1 & 0.9 & 0.81 \\ 0 & 1 & 0.9 \\ 0 & 0 & 1 \end{bmatrix}$$

or

$$\begin{bmatrix} 90.00 & 81.00 \\ 81.00 & 72.90 \\ 72.90 & 65.61 \end{bmatrix} = \begin{bmatrix} 90.00 & 0 \\ 81.00 & 0 \\ 72.90 & 0 \end{bmatrix} \begin{bmatrix} 1 & 0.9 \\ 0 & 1 \end{bmatrix}$$

Now consider these data points: 130, 89, 63.7, 47.21, and 35.89. Applying Gram–Schmidt procedure to the 4 × 2 matrix of this data,

$$\begin{bmatrix} 130.00 & 89.00 \\ 89.00 & 63.70 \\ 63.70 & 47.21 \\ 47.21 & 35.89 \end{bmatrix}$$

shows that the two columns are not linearly dependent. Using columns of 3 as before for illustration gives $t_{11} = 1$ and $t_{12} = 0.7011$. Now, by (3.1.7)

$$\mathbf{v}_1 = \begin{bmatrix} 130.0 \\ 89.0 \\ 63.7 \end{bmatrix}, \quad \mathbf{v}_2 = \begin{bmatrix} 89.00 \\ 63.70 \\ 47.21 \end{bmatrix} - 0.7011 \begin{bmatrix} 130.0 \\ 89.0 \\ 63.7 \end{bmatrix} = \begin{bmatrix} -2.14 \\ 1.30 \\ 2.55 \end{bmatrix}$$

Thus column vectors of the matrix

$$\begin{bmatrix} 130.00 & 89.00 \\ 89.00 & 63.70 \\ 63.70 & 47.21 \end{bmatrix}$$

are *not* linearly dependent, or the rank of the matrix is 2. Therefore, state space of dimension 1 and the AR(1) model are inadequate.

In the state space of dimension 2 we consider the matrix

$$\begin{bmatrix} x_1 & x_2 & x_3 \\ x_2 & x_3 & x_4 \\ x_3 & x_4 & x_5 \end{bmatrix} = \begin{bmatrix} 130.00 & 89.00 & 63.70 \\ 89.00 & 63.70 & 47.21 \\ 63.70 & 47.21 & 35.89 \end{bmatrix}$$

We still have $t_{11} = t_{22} = 1$, $t_{12} = 0.7011$, and therefore $\mathbf{v}_2$ as before, but need in addition

$$t_{13} = \frac{130 \times 63.7 + 89 \times 47.21 + 63.7 \times 35.89}{130^2 + 89^2 + 63.7^2} = 0.5114$$

$$t_{23} = \frac{-2.14 \times 63.7 + 1.30 \times 47.21 + 2.55 \times 35.89}{(-2.14)^2 + 1.30^2 + 2.55^2} = 1.2977$$

so that by (3.1.7)

$$\mathbf{v}_3 = \begin{bmatrix} 63.70 \\ 47.21 \\ 35.89 \end{bmatrix} - 0.5114 \begin{bmatrix} 130.00 \\ 89.00 \\ 63.70 \end{bmatrix} - 1.2977 \begin{bmatrix} -2.14 \\ 1.30 \\ 2.55 \end{bmatrix} = \begin{bmatrix} -0.005 \\ 0.008 \\ 0.005 \end{bmatrix} \simeq \mathbf{0}$$

showing that the three column vectors of the matrix are linearly dependent. Therefore, the matrix is rank-deficient, the rank of the matrix and the dimension of the state space is 2, and the AR(2) model is adequate.

Thus the deterministic AR(2) modeling of the data is the Gram–Schmidt or QR decomposition

$$\begin{bmatrix} 130.00 & 89.00 & 63.70 \\ 89.00 & 63.70 & 47.21 \\ 63.70 & 47.21 & 35.89 \end{bmatrix}$$
$$= \begin{bmatrix} 130.00 & -2.14 & -0.005 \\ 89.00 & 1.30 & 0.008 \\ 63.70 & 2.55 & 0.005 \end{bmatrix} \begin{bmatrix} 1 & 0.7011 & 0.5114 \\ 0 & 1 & 1.2977 \\ 0 & 0 & 1 \end{bmatrix}$$

As the dependences are removed by the Gram–Schmidt procedure, the columns become smaller; when the number of columns just exceeds the correct order, the last column jumps to a very small value, which is theoretically supposed to be zero, but in practice would be of the order of round-off errors, as seen in the example given above. Using the next data value of 27.78, it is easy to verify that the application of Gram–Schmidt procedure will make the last two columns zero, or the rank of the 3 × 4 data matrix would still be 2 and the matrix would be rank-deficient.

Once the dimension of the state space or the order of the model is known, the parameters ϕ_1, ϕ_2 can be determined using the equation

$$x_t = \phi_1 x_{t-1} + \phi_2 x_{t-2}$$

with the given data. In fact, only the first four data points are enough to solve for ϕ_1, ϕ_2 by

$$\begin{bmatrix} 130.0 & 89.0 \\ 89.0 & 63.7 \end{bmatrix} \begin{bmatrix} \phi_2 \\ \phi_1 \end{bmatrix} = \begin{bmatrix} 63.70 \\ 47.21 \end{bmatrix}$$

as $\phi_1 = 1.3$ and $\phi_2 = -0.4$.

Is the value $t_{23} = 1.2977$ an approximate version of $\phi_1 = 1.3$, just as $t_{12} = \phi_1$ was for the AR(1) model before? Indeed, after studying transformations in Section 3.4, we will show that ϕ_1, ϕ_2 can be directly obtained from

$$\begin{bmatrix} 1 & t_{12} \\ 0 & 1 \end{bmatrix} \begin{bmatrix} \phi_2 \\ \phi_1 \end{bmatrix} = \begin{bmatrix} t_{13} \\ t_{23} \end{bmatrix}$$

For the present case

$$\begin{bmatrix} 1 & 0.7011 \\ 0 & 1 \end{bmatrix} \begin{bmatrix} \phi_2 \\ \phi_1 \end{bmatrix} = \begin{bmatrix} 0.5114 \\ 1.2977 \end{bmatrix}$$

which gives $\phi_1 = 1.2977$ and

$$\phi_2 + 0.7011\phi_1 = 0.5114, \qquad \text{that is,} \qquad \phi_2 = -0.3984$$

The round-off error illustrated above is characteristic of the poor numerical stability of the original Gram–Schmidt procedure and can be considerably improved by modified Gram–Schmidt and Householder transformations discussed later.

The fifth data point 35.89 was needed to determine the model order by exhibiting the rank deficiency of the 3×3 data matrix. Additional data points will repeat this by returning near zero columns beyond 2 in any Gram–Schmidt orthogonalized matrix of two or more rows and more than two columns.

The generalization to arbitrary order AR models should now be obvious. Given a set of N data values $x_1, x_2, \ldots, x_N$, to determine the dimension of the state space in which the data vector x_t resides, we form matrices of increasing number of columns and decreasing number of rows from the N data values until the Gram–Schmidt procedure starts returning zero columns. This is equivalent to increasing the dimension of the state space until the data vector x_t is a part of the state space and hence is its own projection in the state space and the perpendicular is zero. This perpendicular, in fact, is the last column of the Gram–Schmidt orthogonalized matrix with one more column than the correct dimensions. This procedure is also equivalent to fitting increasing-order AR models until the residuals are reduced to zero.

ILLUSTRATIVE APPLICATIONS TO DDS IN STATE SPACE

If n is the correct order or dimension, then the first time a zero column will appear as

$$\begin{bmatrix} x_1 & x_2 & \cdots & x_{n+1} \\ x_2 & x_3 & \cdots & x_{n+2} \\ \vdots & \vdots & & \vdots \\ x_{N-n} & x_{N-n+1} & \cdots & x_N \end{bmatrix} = \begin{bmatrix} x_1 & v_{12} & \cdots & v_{1n} & 0 \\ x_2 & v_{22} & \cdots & v_{2n} & 0 \\ \vdots & \vdots & & \vdots & \vdots \\ x_{N-n} & v_{(N-n)2} & \cdots & v_{(N-n)n} & 0 \end{bmatrix}$$

$(N-n) \times (n+1)$ $\qquad\qquad$ $(N-n) \times (n+1)$

$$\times \begin{bmatrix} 1 & t_{12} & \cdots & t_{1n} & t_{1(n+1)} \\ 0 & 1 & \cdots & t_{2n} & t_{2(n+1)} \\ \vdots & \vdots & & \vdots & \vdots \\ 0 & 0 & \cdots & 1 & t_{n(n+1)} \\ 0 & 0 & \cdots & 0 & 1 \end{bmatrix} \qquad (3.2.1)$$

$(n+1) \times (n+1)$

and the parameters $\phi_1, \phi_2, \ldots, \phi_n$ of the correct model

$$x_t = \phi_1 x_{t-1} + \phi_2 x_{t-2} + \cdots + \phi_n x_{t-n}$$

can be recovered from the following relation (proved later in Section 3.4.2):

$$\begin{bmatrix} 1 & t_{12} & \cdots & t_{1n} \\ 0 & 1 & \cdots & t_{2n} \\ \vdots & \vdots & \vdots & \vdots \\ 0 & 0 & \cdots & 1 \end{bmatrix} \begin{bmatrix} \phi_n \\ \phi_{n-1} \\ \vdots \\ \phi_1 \end{bmatrix} = \begin{bmatrix} t_{1(n+1)} \\ t_{2(n+1)} \\ \vdots \\ t_{n(n+1)} \end{bmatrix} \qquad (3.2.2)$$

which, by backward substitution, reduces to the recursive computation

$$\phi_{n-i+1} = t_{i(n+1)} - \sum_{j=i+1}^{n} t_{ij} \phi_{n-j+1}, \quad i = n, n-1, \ldots, 1 \qquad (3.2.3)$$

so that in particular for $i = n$ we have (sums from large to small indices are omitted)

$$\phi_1 = t_{n(n+1)} \qquad (3.2.3\text{a})$$

which was verified for AR(1) and AR(2) above. The Gram–Schmidt procedure will continue to return zero columns at the end if the number of columns is increased to $n + 2, n + 3, \ldots$. Note that x_t in the deterministic $AR(n)$ model can represent both the scalar data value at t as well as the data vector $(x_t, x_{t-1}, \ldots, x_{t-N+n+1})^T$; this $(N - n)$ component vector resides in a state space of dimension n.

3.2.2 Stochastic AR Models

In the AR(1) example above, the simplest Gram–Schmidt computation made the second column of the data matrix zero as expected; but even in the next simple AR(2) example, the third column differed from the expected zero value by the roundoff error "residuals" caused by the limited number of digits retained in the data, estimated parameters, and the computational process in general. In view of the discussion in Sections 2.1.4 and 2.2.2 that a deterministic system is a limiting case of a stochastic system when the residual a_t's are negligible compared to the data, it is easy to realize that for a stochastic system with $AR(n)$ model:

$$X_t = \phi_1 X_{t-1} + \phi_2 X_{t-2} + \cdots + \phi_n X_{t-n} + a_t$$

the data matrix would not be rank-deficient even if the correct order is exceeded because the $n + 1, n + 2, \ldots$ columns will contain the *nonzero* residuals when the Gram–Schmidt procedure is applied. However, the residual sum of squares is indeed the squared length of each of these column vectors that would not significantly reduce after the correct order or dimension is exceeded. The statistical significance of the reduction in squared length of this vector can be decided by the F criterion (2.5.1) as discussed in Section 2.5.4. If the squared length of this column length drops below a noise-floor criterion such as (2.5.2), we conclude that the system is deterministic and stop.

Once we decide the stop either by using F-test (2.5.1) and unified correlations for stochastic systems, or (2.5.2) for deterministic systems, equation (3.2.1) is valid if the zero last column is replaced by appropriate residuals. Therefore, the parameters can be computed by (3.2.3). Of course, unlike the deterministic systems discussed earlier, N must be very large compared to n in order to get reasonable results. For example, the standard deviation of $\hat{\phi}_1$ of the AR(1) model is approximately $\sqrt{(1 - \phi_1^2)/N} \approx 0.1$ for small ϕ_1 and $N = 100$; therefore, even with 100 data points, true ϕ_1 lies within $\hat{\phi}_1 \pm 0.2$ with 95% probability and considering that $|\phi_1| < 1$ for a stable system, this accuracy is barely enough. (See Appendix A for the proof of these results.)

Hence it would be inappropriate to illustrate the Gram–Schmidt procedure for stochastic data using only 4 or 5 data points, as was done earlier for deterministic AR models. However, for large N (3.2.1)–(3.2.3) remain valid, with the last column of zeros replaced by residual a_t's. If the average subtracted silicon percentage data discussed in Section 2.1 is used for illustrative purposes, then the AR(1) modeling of this data is equivalent to the Gram–Schmidt or QR decomposition (3.2.1) with $n = 1$ and the last column of zeros on the right-hand side replaced by the residuals:

$$\begin{bmatrix} x_1 & x_2 \\ x_2 & x_3 \\ \vdots & \vdots \\ x_{N-1} & x_N \end{bmatrix} = \begin{bmatrix} -0.525 & -0.055 \\ -0.055 & -0.275 \\ \vdots & \vdots \\ 0.075 & 0.085 \end{bmatrix}$$

$$= \begin{bmatrix} -0.525 & 0.1476 \\ -0.055 & 0.2537 \\ \vdots & \vdots \\ 0.075 & 0.0562 \end{bmatrix} \begin{bmatrix} 1 & 0.386 \\ 0 & 1 \end{bmatrix}$$

where

$$t_{12} = \frac{x_1 x_2 + x_2 x_3 + \cdots + x_{N-1} x_N}{x_1^2 + x_2^2 + \cdots + x_{N-1}^2} = \phi_1 = 0.386$$

in agreement with the least-square formula (2.1.6) and the results of Section 2.1.3. Note that unlike the deterministic AR(1) case, the second column of the matrix on the right-hand side is *not* zero and there is *no rank deficiency*.

This illustrates clearly that for data from an AR(1) model

$$X_t = \phi_1 X_{t-1} + a_t$$

the Gram–Schmidt decomposition yields

$$\begin{bmatrix} X_1 & X_2 \\ X_2 & X_3 \\ \vdots & \vdots \\ X_{N-1} & X_N \end{bmatrix} = \begin{bmatrix} X_1 & a_2 \\ X_2 & a_3 \\ \vdots & \vdots \\ X_{N-1} & a_N \end{bmatrix} \begin{bmatrix} 1 & \phi_1 \\ 0 & 1 \end{bmatrix} \quad (3.2.4)$$

and the F-test applied to the two columns by using $\Sigma X_t^2 = A_1, \Sigma a_t^2 = A_0, r = 1 = s$ in (2.5.1) would indicate statistically significant reduction. This will indicate that the zero-dimensional AR(0) model is inadequate and at least a one-dimensional AR(1) model is required. (See Section 2.5.4 for numerical verification.)

What will happen if the correct model is AR(1) and we attempt the Gram–Schmidt decomposition (3.2.1) with $n = 2$? It is easy to verify that for large N we should get

$$\begin{bmatrix} X_1 & X_2 & X_3 \\ X_2 & X_3 & X_4 \\ \vdots & \vdots & \vdots \\ X_{N-2} & X_{N-1} & X_N \end{bmatrix} = \begin{bmatrix} X_1 & a_2 & a_3 \\ X_2 & a_3 & a_4 \\ \vdots & \vdots & \vdots \\ X_{N-2} & a_{N-1} & a_N \end{bmatrix} \begin{bmatrix} 1 & \phi_1 & \phi_1^2 \\ 0 & 1 & \phi_1 \\ 0 & 0 & 1 \end{bmatrix} \quad (3.2.5)$$

Again, unlike the deterministic AR(1) case, the last two columns of the matrix on the right-hand side are not zero. Hence there is no rank deficiency to indicate that the rank of the data matrix is one and hence the AR(1) model should be adequate. However, the squared lengths of the last two columns given by the residual sums of squares are practically the same, and an F-test based on these sums of squares would indicate statistically insignificant improvement, concluding AR(1) as the adequate model. Note that for this test also we need large N.

There is no need to increase the dimension further for modeling purposes. However, it is interesting to note theoretically that if we started the data matrix in (3.2.5) with X_0 instead of X_1 and increased the number of columns and N simultaneously, the Gram–Schmidt decomposition represents the orthogonal decomposition

$$X_t = \phi_1^t X_0 + \sum_{j=0}^{t-1} \phi_1^j a_{t-j}$$

discussed extensively in Chapter 2. Since

$$\begin{aligned} t_{ij} &= \phi^{j-i} & j \geq i \\ &= 0 & j < i \end{aligned} \quad (3.2.5a)$$

the increase of j beyond 2 is redundant.

The generalization of (3.2.4) for AR(n) models is obvious and was indicated earlier. The generalization of (3.2.5) for the single-series AR(n) models and further to multiple series models is left as an interesting and nontrivial exercise.

3.2.3 Extension to ARMA Models

The Gram–Schmidt procedure is a linear vector space operation and therefore breaks down for the nonlinear ARMA models. An ARMA(n,m) model can be expressed as an AR(∞) model if the characteristic roots v_i of the moving average operator

$$(1 - \theta_1 B - \theta_2 B^2 - \cdots - \theta_m B^m) \equiv (1 - v_1 B)(1 - v_2 B) \cdots (1 - v_m B)$$

are all less than 1 in absolute value. For such an AR(∞) model, its parameters, called inverse function, eventually decay to zero, hence an AR(p) model with large enough p provides a good approximation to AR(∞) model parameters and its residuals. This AR(p) model can be fitted by the method of Section 3.2.2 and often suffices for some applications such as forecasting. If desired, the parameters of the original ARMA(n,m) model can be estimated from this approximate AR(p) model by two methods.

The first method utilizes the parameters of the AR(p) model by treating them as inverse function coefficients to get the initial guess values of the ARMA model parameters. A nonlinear least squares search routine is then used to minimize the residual sum of squares and to estimate the final parameters. The necessary parametric relationships and illustrative examples may be found in Pandit and Wu (1983), and they readily extend to ARMAV models.

The second method utilizes the residuals of the AR(p) model. Treating these residuals as approximate a_t's of the ARMA model, the corresponding Gram–Schmidt procedure is carried out using both x_t and a_t columns. Such iterative Gram–Schmidt procedure readily generalizes to ARMAV models.

3.3 APPLICATIONS OF ORTHOGONAL DECOMPOSITION

The concept of large N used in Section 3.2.2 can be made rigorous by declaring the inner product not as a finite sum of products but as an infinite average or expectation of products. For this expectation to exist and be finite we need the stationarity assumption.

3.3.1 Orthogonal Decomposition

For ARMA models of stationary series, assuming zero mean we can declare the inner product by the expectation

$$\langle X_t, X_{t-j} \rangle = E(X_t X_{t-j})$$

Then a_t's form an orthogonal set since

$$\langle a_i, a_j \rangle = E(a_i a_j) = \delta_{(i-j)} \sigma_a^2$$

and hence the decomposition such as the one given by (2.2.1) for the AR(1) model

$$X_t = a_t + \phi_1 a_{t-1} + \phi_1^2 a_{t-2} + \cdots + \phi_1^{t-1} a_1$$

is an orthogonal decomposition. In general the modeling strategy outlined in Section 2.5 can be viewed as an explicit method of obtaining such an orthogonal set of a_t's from the data. The resultant model then provides coefficients of a_t's in the expansion. The coefficients are ϕ_1^j for the AR(1) model, those for the general ARMA model can be obtained by the subsequent developments in this chapter.

3.3.2 Orthogonal Projection

Let $\mathcal{W}$ be a finite dimensional subspace and a a vector not in $\mathcal{W}$, that is, $a \notin \mathcal{W}$. Then there exist two vectors Γ and β such that

$$a = \beta + \Gamma, \quad \beta \in \mathcal{W}, \quad \Gamma \perp \mathcal{W}, \quad \Gamma \neq 0$$

in other words, a is decomposed into two orthogonal components: β in $\mathcal{W}$ and the nonzero vector Γ orthogonal to $\mathcal{W}$.

To prove this, let $y_1, y_2, \ldots, y_r$ be a basis of $\mathcal{W}$. Then construct $v_1, v_2, \ldots, v_{r+1}$ from $y_1, y_2, \ldots, y_r, a$ by Gram–Schmidt orthogonalization so that

$$\begin{aligned} a &= a_1 v_1 + a_2 v_2 + \cdots + a_r v_r + a_{r+1} v_{r+1} \\ &= \quad\quad\quad\quad \beta \quad\quad\quad + \Gamma \end{aligned}$$

Since $\mathcal{W} = \mathcal{M}(y_1, y_2, \ldots, y_r) = \mathcal{M}(v_1, v_2, \ldots, v_r)$, $\beta \in \mathcal{W}$ and v_{r+1} is by construction orthogonal to each of $v_1, v_2, \ldots, v_r$ and therefore to $\mathcal{W}$. To prove uniqueness, if $a = \beta_1 + \Gamma_1$, then

$$0 = (\beta - \beta_1) + (\Gamma - \Gamma_1)$$

Taking the inner product on both sides we see that

$$\langle (\beta - \beta_1), (\beta - \beta_1) \rangle = 0 = \langle (\Gamma - \Gamma_1), (\Gamma - \Gamma_1) \rangle$$

which shows that $\beta = \beta_1$ and $\Gamma = \Gamma_1$.

Γ is said to be the perpendicular from a to $\mathcal{W}$ and β the (orthogonal) projection of a on $\mathcal{W}$. $\Gamma = 0$ if $a \in \mathcal{W}$. The length of Γ is the shortest distance between a and the space $\mathcal{W}$, that is,

$$\|\Gamma\| = \inf \|a - x\|, \quad x \in \mathcal{W}$$

which can be seen by taking the inner product

$$\begin{aligned} \langle (x - a), (x - a) \rangle &= \langle (x - \beta - \Gamma), (x - \beta - \Gamma) \rangle \\ &= \langle (x - \beta), (x - \beta) \rangle + \langle \Gamma, \Gamma \rangle \\ &\geq \|\Gamma\|^2 \\ &= \|\Gamma\|^2 \quad \text{when } x = \beta \end{aligned}$$

APPLICATIONS OF ORTHOGONAL DECOMPOSITION

The set of all vectors orthogonal to a given set of vectors S in a vector space $\mathscr{X}$ is a subspace of $\mathscr{X}$ denoted by $S^\perp$. Also

$$d[\mathscr{M}(S)] + d[S^\perp] = d[\mathscr{X}]$$

Moreover, every vector $\mathbf{x}$ in $\mathscr{X}$ can be uniquely expressed as

$$\mathbf{x} = \mathbf{u} + \mathbf{v}, \quad \mathbf{u} \in \mathscr{M}(S), \quad \mathbf{v} \in S^\perp, \quad \text{and} \quad \|\mathbf{x}\|^2 = \|\mathbf{u}\|^2 + \|\mathbf{v}\|^2$$

Then $\mathscr{X}$ is said to be the direct sum of $\mathscr{M}(S)$ and $S^\perp$ written as

$$\mathscr{X} = \mathscr{M}(S) \oplus S^\perp$$

3.3.3 Method of Least Squares: Normal Equations

When the number of equations $m > n$, the number of unknowns in the equations $\mathbf{y} = \mathbf{A}\mathbf{x}$, then rank $(\mathbf{A}) \leq n < m$, and rank$(\mathbf{A}|\mathbf{y}) \leq n + 1 \leq m$, and therefore the equations may be inconsistent, that is, there may not be an exact solution. In such cases we look for an approximate solution minimizing the error

$$\mathbf{e} = \mathbf{y} - \mathbf{A}\mathbf{x}$$

The solution minimizing the usual norm of $\mathbf{e}$, assuming real vector space for simplicity,

$$\|\mathbf{e}\|^2 = \mathbf{e}^T\mathbf{e} = (\mathbf{y} - \mathbf{A}\mathbf{x})^T(\mathbf{y} - \mathbf{A}\mathbf{x})$$

is called the least squares estimate of $\mathbf{x}$. The vectors $\mathbf{y}$, $\mathbf{A}\mathbf{x}$, and $\mathbf{e}$ belong to an m-dimensional vector space, say $\mathscr{V}_m$, but $\mathbf{A}\mathbf{x}$ belongs to $\mathbf{M}(a)$, the column space of $\mathbf{A}$. Then

$$\mathscr{V}_m = \mathscr{M}(a) + \mathscr{M}(a)^\perp$$
$$\mathbf{y} - \mathbf{A}\mathbf{x} = \mathbf{e} = \mathbf{e}_1 + \mathbf{e}_2$$
$$\|\mathbf{e}\|^2 = \|\mathbf{e}_1\|^2 + \|\mathbf{e}_2\|^2$$

Since $\mathbf{y}$ is given and $\mathbf{A}\mathbf{x} \in \mathscr{M}(a)$, the choice of $\mathbf{x}$ cannot affect $\mathbf{e}_2$. The least squares solution vector $\hat{\mathbf{x}}$ is therefore the one for which $\|\mathbf{e}_1\|^2 = 0$ or $\mathbf{e}_1 = 0$, so that

$$\mathbf{y} - \mathbf{A}\hat{\mathbf{x}} = \mathbf{e} = \mathbf{e}_2 \in \mathscr{M}(a)^\perp$$

that is,
$$a_i^T \mathbf{e} = 0, \quad i = 1, 2, \ldots, n$$
$$\mathbf{A}^T\mathbf{e} = \mathbf{A}^T(\mathbf{y} - \mathbf{A}\hat{\mathbf{x}}) = 0 = \mathbf{A}^T\mathbf{y} - \mathbf{A}^T\mathbf{A}\hat{\mathbf{x}}$$

or
$$\mathbf{A}^T\mathbf{A}\hat{\mathbf{x}} = \mathbf{A}^T\mathbf{y}$$

which are known as *normal equations* for the least squares problem. If the square matrix $\mathbf{A}^T\mathbf{A}$ is nonsingular, then

$$\hat{\mathbf{x}} = (\mathbf{A}^T\mathbf{A})^{-1}\mathbf{A}^T\mathbf{y}$$

which is the usual least squares formula. For weighted least squares, the norm is taken as $\mathbf{e}^T\mathbf{Q}\mathbf{e}$ and yields

$$\hat{\mathbf{x}} = (\mathbf{A}^T\mathbf{Q}\mathbf{A})^{-1}\mathbf{A}^T\mathbf{Q}\mathbf{y}$$

As an example, for the AR(1) model the $(N-1) \times 1$ matrix $\mathbf{A}$ is the column vector $[X_1, X_2, \ldots, X_{N-1}]^T$, $\mathbf{y} = [X_2, X_3, \ldots, X_N]^T$, and the least squares formula gives the expression (2.1.6) for ϕ_1, alternatively derived by the Gram–Schmidt procedure in Section 3.2.1.

If $\mathbf{A}^T\mathbf{A}$ is singular, its usual inverse does not exist. Then the least squares estimates can be obtained by using pseudo (or generalized) inverse with the help of singular value decomposition (SVD) discussed in Section 3.5.3. The standard least squares solution of $\mathbf{y} = \mathbf{A}\mathbf{x}$ essentially allows errors in the "observation" $\mathbf{y}$ but not in the "design" or "independent variable" matrix $\mathbf{A}$. The SVD can also be used to find the total least squares (TLS) solution of $\mathbf{y} = \mathbf{A}\mathbf{x}$ when errors are present in both $\mathbf{y}$ and $\mathbf{A}$; see for example Golub and Van Loan (1987).

3.3.4 Forecasting and Control

Using the orthogonal decomposition of the ARMA model, described above for the AR(1) special case illustrating the modeling strategy as Gram–Schmidt orthogonalization, we can easily solve the problem of forecasting and control. Again we will illustrate the AR(1) case, but the general case follows along the same lines, once the coefficient function (called Green's function) analogous to ϕ_1^j for the AR(1) case, is evaluated later for the general ARMA models.

Forecasting
The problem of forecasting is that knowing the data $X_1, X_2, \ldots, X_t$ up to and including time t, we would like to forecast the value of $X_{t+\ell}$ ℓ steps ahead. Since the mean is either constant or a known deterministic function of time, it can be forecasted without error, therefore we can subtract it from the data and consider the zero mean series. If $\mathcal{M}_t$ denotes the space spanned by the data $X_1, X_2, \ldots, X_t$, it is also the space spanned by the orthogonal vectors $a_1, a_2, \ldots, a_t$, that is,

$$\mathcal{M}_t = \mathcal{M}(X_1, X_2, \ldots, X_t) = \mathcal{M}(a_1, a_2, \ldots, a_t)$$

Note that subtracting a constant mean has merely the effect of translating the origin to $\mathbf{0}$, from the original nonzero origin. Since $X_{t+\ell} \notin \mathcal{M}_t$, the space of known values, its best forecast is the vector in $\mathcal{M}_t$ closest to it, which is naturally the projection of $X_{t+\ell}$ in $\mathcal{M}_t$. The perpendicular, which is the difference between $X_{t+\ell}$ and its forecast or projection, is then the forecast error. The orthogonal projection guarantees that this error has the minimum length compared to all vectors in $\mathcal{M}_t$. Since the squared norm or the squared length have been chosen in the orthogonal decomposition to be the variance or the mean-squared norm, the forecast so obtained by the orthogonal projection is the minimum mean-squared error or the minimum variance forecast.

To find the forecast and the error, we can express the future observation $X_{t+\ell}$ into two parts, one in $\mathcal{M}_t$ and the other orthogonal to $\mathcal{M}_t$. This can be accomplished exactly like the decomposition of α into β and Γ discussed earlier using the orthogonal decomposition of $X_{t+\ell}$ in terms of the orthogonal vectors $a_1, a_2, \ldots, a_t$:

$$X_{t+\ell} = \phi_1^{t+\ell-1} a_1 + \phi_1^{t+\ell-2} a_2 + \cdots + \phi_1^{\ell} a_t + \phi_1^{\ell-1} a_{t+1} + \cdots + a_{t+\ell}$$
$$= \hat{X}_t(\ell) \qquad\qquad + e_t(\ell)$$

where $\hat{X}_t(\ell)$ denotes the projection in $\mathcal{M}_t$ or forecast at time t of the future observation ℓ steps ahead, and $e_t(\ell)$ is the corresponding forecast error

$$e_t(\ell) = X_{t+\ell} - \hat{X}_t(\ell) = \phi_1^{\ell-1} a_{t+1} + \cdots + a_{t+\ell}$$

The (minimum) squared length or variance of this forecast error is given by its norm found from the definition of the inner product as

$$E[e_t(\ell)]^2 = \langle e_t(\ell), e_t(\ell) \rangle = \sigma_a^2 (\phi_1^{2(\ell-1)} + \cdots + \phi_1^2 + 1)$$
$$= \frac{(1-\phi_1^{2\ell})}{(1-\phi_1^2)} \sigma_a^2$$

This is also loosely called the variance of the forecast, although, strictly speaking, the forecast is not a random variable. A probability limit at the desired probability level, such as say 95%, can then be found by the forecast $\pm 1.96 \sqrt{\text{(variance)}}$. This is the probability limit on the mean subtracted observation. To get the actual forecast limit we need to add the mean at the end.

The forecast error variance given above does not require the actual calculation of a_t's but the forecast seems to require the calculation of $a_1, \ldots, a_t$. Note that the a_t's can be recursively calculated using

$$a_t = e_{t-1}(1) = X_t - \hat{X}_{t-1}(1)$$

However, the calculation of all the a_t's required in the forecast would be somewhat cumbersome. Can we express the forecast directly in terms of the known data $X_1, \ldots, X_t$? This should be possible since the space spanned by a_t's is also spanned by X_t's. Hence the forecast, which is the projection in this space, must be expressible in terms of X_t's as well. Indeed, this can be accomplished by using not the orthogonal decomposition of X_t derived from the fitted model, but the model itself. Because the model form

$$X_{t+\ell} = \phi_1 X_{t+\ell-1} + a_{t+\ell}$$

tells us that the projection of $X_{t+\ell}$ on $\mathcal{M}_t$ is the sum of the projections of $\phi_1 X_{t+\ell-1}$ and $a_{t+\ell}$ on $\mathcal{M}_t$. But since $a_{t+\ell}, a_{t+\ell-1}, \ldots, a_{t+1} \perp \mathcal{M}_t$, their projections in $\mathcal{M}_t$ are zero. On the other hand $a_t, a_{t-1}, \ldots, a_1 \in \mathcal{M}_t$, and so also $X_t, X_{t-1}, \ldots, X_1 \in \mathcal{M}_t$, and hence their projections are they themselves. Thus repeated application of this gives

$$\hat{X}_t(\ell) = \phi_1 \hat{X}_t(\ell - 1)$$
$$= \phi_1^2 \hat{X}_t(\ell - 2)$$
$$\vdots$$
$$= \phi_1^\ell \hat{X}_t(0) = \phi_1^\ell X_t$$

which can be verified to be the same as the one obtained above in terms of a_t's by using the orthogonal decomposition of X_t. The method is readily extended to an arbitrary model by using the proper coefficient function (Green's function) in place of ϕ_1^j for calculating the error variance and using the appropriate model form for obtaining the recursive relations for computing the forecasts as above.

Optimal Control

The problem of optimal control can be solved by a simple extension of the forecasting by orthogonal projection discussed above. If we consider the one-input–one-output case for illustration, we can first forecast the output at the smallest possible lag between the input and the output, say L, by orthogonal projection. We can then adjust the input to set this forecast at the desired, target, or set-point value of the output. To make the formulation easier, we will assume that the output has been measured from this target value, so that the optimal control strategy is given by setting this forecast to zero. The orthogonal projection then guarantees us that the forecast error, which is also the control error or the deviation of the output from the (zero) target, has the smallest length or variance. Thus we get optimal control in the sense of minimum variance or minimum mean-squared error.

If we consider the simplest AR(1) case for one-input X_{1t}, one-output X_{2t}, the model may be written as

$$X_{2t} = \phi_{21L} X_{1t-L} + \phi_{221} X_{2t-1} + a_{2t}$$

Since the lag or dead time between the input and the output is L, the earliest output that can be affected by the present input X_{1t} is X_{2t+L}. Therefore, we can find the projection of this future output in the space of known past and present inputs and outputs, which gives us the forecast. To find the projection, we need the orthogonal decomposition of X_{2t+L} by recursive substitution as in the single series case:

$$\begin{aligned} X_{2t+L} &= \phi_{21L}X_{1t} + \phi_{221}X_{2t+L-1} + a_{2t+L} \\ &= \phi_{21L}X_{1t} + \phi_{221}\phi_{21L}X_{1t-1} + \phi_{221}^2 X_{2t+L-2} + \phi_{221}a_{2t+L-1} + a_{2t+L} \\ &\vdots \\ &= \text{linear combination of } (X_{1t}, X_{1t-1}, \ldots, X_{11}, X_{2t}, X_{2t-1}, \ldots, X_{21}) \\ &\quad + \phi_{221}^{L-1} a_{2t+1} + \phi_{221}^{L-2} a_{2t+2} + \cdots + \phi_{221} a_{2t+L-1} + a_{2t+L} \\ &= \hat{X}_{2t}(L) + e_{2t}(L) \end{aligned}$$

In other words, we have X_{2t+L} expressed as the sum of its forecast or projection on the space $\mathcal{M}_t = \mathcal{M}(X_{1t}, X_{1t-1}, \ldots, X_{11}, X_{2t}, X_{2t-1}, \ldots, X_{21})$ and the forecast (control) error or perpendicular $e_{2t}(L)$. Since the choice of X_{1t} cannot affect $e_{2t}(L)$, the closest we can make the future output X_{2t+L} to its target value zero is by making $\hat{X}_{2t}(L)$ zero. Hence the control strategy is given by writing $\hat{X}_{2t}(L)$ from the model and equating it to zero:

$$\hat{X}_{2t}(L) = 0 = \phi_{21L}X_{1t} + \phi_{221}\hat{X}_{2t}(L-1)$$

In other words we should adjust the input X_{1t} by the rule

$$X_{1t} = \frac{-\phi_{221}}{\phi_{21L}} \hat{X}_{2t}(L-1)$$

If this optimal control strategy is implemented so that the projection $\hat{X}_{2t}(L)$ is zero, then the optimally controlled output is given by only the perpendicular

$$X_{2t+L} = e_{2t}(L) = \phi_{221}^{L-1} a_{2t+1} + \phi_{221}^{L-2} a_{2t+2} + \cdots + \phi_{221} a_{2t+L-1} + a_{2t+L}$$

that is,

$$\begin{aligned} X_{2t} &= e_{2t-L}(L) \\ &= \phi_{221}^{L-1} a_{2t-L+1} + \phi_{221}^{L-2} a_{2t-L+2} + \cdots + \phi_{221} a_{2t-1} + a_{2t} \end{aligned}$$

which is guaranteed to have the minimum squared length or variance given by

$$\begin{aligned} \|X_{2t}\|^2 = \text{Var}(X_{2t}) &= \sigma_{a2}^2 (\phi_{221}^{2(L-1)} + \cdots + \phi_{221}^2 + 1) \\ &= \frac{(1 - \phi_{221}^{2L})}{(1 - \phi_{221}^2)} \sigma_{a2}^2 \end{aligned}$$

This agrees with the corresponding earlier result for a single series.

3.4 TRANSFORMATIONS

A linear transformation has the property that if $\mathbf{x} \to \mathbf{y}$ and $\mathbf{u} \to \mathbf{v}$ then $a\mathbf{x} + b\mathbf{u} \to a\mathbf{y} + b\mathbf{v}$. Consider such a transformation on an n-dimensional Euclidean space E_n and let the columns $\mathbf{e}_i$ of the $n \times n$ identity matrix forming a basis be transformed as

$$\mathbf{e}_1 = (1, 0, 0, \ldots, 0)^T \to (a_{11}, a_{21}, \ldots, a_{n1})^T$$
$$\mathbf{e}_2 = (0, 1, 0, \ldots, 0)^T \to (a_{12}, a_{22}, \ldots, a_{n2})^T$$
$$\vdots$$
$$\mathbf{e}_n = (0, 0, 0, \ldots, 1)^T \to (a_{1n}, a_{2n}, \ldots, a_{nn})^T$$

Multiplying by $x_1, x_2, \ldots, x_n$ and adding we get

$$\mathbf{x} = (x_1, x_2, \ldots, x_n)^T \to (\Sigma a_{1i} x_i, \Sigma a_{2i} x_i, \ldots, \Sigma a_{ni} x_i)^T$$
$$= \mathbf{y} = (y_1, y_2, \ldots, y_n)^T$$

Thus a linear transformation from $\mathbf{x}$ to $\mathbf{y}$ may be written as

$$\mathbf{A}\mathbf{x} = \mathbf{y}$$

and is said to be represented by the $n \times n$ matrix $\mathbf{A}$ where

$$\mathbf{A} = \begin{bmatrix} a_{11} & a_{12} & \cdots & a_{1n} \\ a_{21} & a_{22} & \cdots & a_{2n} \\ \vdots & \vdots & \cdots & \vdots \\ a_{n1} & a_{n2} & \cdots & a_{nn} \end{bmatrix}$$

The transformation is one-to-one only when $\mathbf{A}$ has rank n, that is, it is of full rank or nonsingular so that $\mathbf{A}^{-1}$ exists and the inverse transformation is given by $\mathbf{x} \to \mathbf{A}^{-1}\mathbf{y}$, showing the one-to-one correspondence.

3.4.1 Change of Basis

The matrix $\mathbf{A}$ representing a linear transformation is specific to the basis and will be different for a different basis. The above representation of $\mathbf{A}$ is specific to the basis forming the columns of an identity matrix. From the definition of rank and basis, it is clear that the columns of $\mathbf{A}$ themselves form a basis if and only if $\mathbf{A}$ is of full rank or nonsingular. Then the equation

$$\mathbf{y} = \mathbf{A}\mathbf{x}$$

written in terms of the columns of $\mathbf{A} = \{\mathbf{a}_1, \mathbf{a}_2, \ldots, \mathbf{a}_n\}$

$$\mathbf{y} = x_1 \mathbf{a}_1 + x_2 \mathbf{a}_2 + \cdots + x_n \mathbf{a}_n$$

shows that this vector $\mathbf{y}$ has coordinates $y_1, y_2, \ldots, y_n$ in terms of the "unit" basis consisting of the columns of the identity matrix, whereas it has coordinates $x_1, x_2, \ldots, x_n$ in terms of the basis $\{\mathbf{a}_1, \mathbf{a}_2, \ldots, \mathbf{a}_n\}$; thus a nonsingular matrix $\mathbf{A}$ also represents a change of basis. We can find the coordinates for the new basis by

$$\mathbf{x} = \mathbf{A}^{-1} \mathbf{y}$$

3.4.2 Application to AR(n) Parameter Estimation

As an immediate application of change of basis, consider the Gram–Schmidt decomposition (3.2.1) reproduced below:

$$\begin{bmatrix} x_1 & x_2 & \cdots & x_{n+1} \\ x_2 & x_3 & \cdots & x_{n+2} \\ \vdots & \vdots & & \vdots \\ x_{N-n} & x_{N-n+1} & \cdots & x_N \end{bmatrix} = \begin{bmatrix} x_1 & v_{12} & \cdots & v_{1n} & 0 \\ x_2 & v_{22} & \cdots & v_{2n} & 0 \\ \vdots & \vdots & & \vdots & \vdots \\ x_{N-n} & v_{(N-n)2} & \cdots & v_{(N-n)n} & 0 \end{bmatrix}$$

$(N-n) \times (n+1)$ \qquad\qquad $(N-n) \times (n+1)$

$$\times \begin{bmatrix} 1 & t_{12} & \cdots & t_{1n} & t_{1(n+1)} \\ 0 & 1 & \cdots & t_{2n} & t_{2(n+1)} \\ \vdots & \vdots & \cdots & \vdots & \vdots \\ 0 & 0 & \cdots & 1 & t_{n(n+1)} \\ 0 & 0 & \cdots & 0 & 1 \end{bmatrix} \qquad (3.2.1)$$

$(n+1) \times (n+1)$

To estimate the parameters ϕ_i in the deterministic AR(n) model

$$x_t = \phi_1 x_{t-1} + \phi_2 x_{t-2} + \cdots + \phi_n x_{t-n}$$

we note that this equation expresses the coordinates ϕ_i's of the vector $(x_{n+1}, x_{n+2}, \ldots, x_N)^T$ in terms of the basis consisting of the first n columns of the left-hand side of (3.2.1); this basis is *nonorthogonal*. On the other hand, $t_{i(n+1)}$ express the coordinates of the same vector in terms of the *orthogonal* basis consisting of the first n (nonzero) columns of the first matrix on the right-hand side of (3.2.1).

The change from the orthogonal to the nonorthogonal basis is accomplished by the $n \times n$ triangular matrix of t_{ij}'s. Therefore, the coordinates $(\phi_n, \phi_{n-1}, \ldots, \phi_1)^T$ in terms of the nonorthogonal basis can be obtained from $t_{i(n+1)}$ in terms of the orthogonal basis by

$$\begin{bmatrix} \phi_n \\ \phi_{n-1} \\ \vdots \\ \phi_1 \end{bmatrix} = \begin{bmatrix} 1 & t_{12} & \cdots & t_{1n} \\ 0 & 1 & \cdots & t_{2n} \\ \vdots & \vdots & \cdots & \vdots \\ 0 & 0 & \cdots & 1 \end{bmatrix}^{-1} \begin{bmatrix} t_{1(n+1)} \\ t_{2(n+1)} \\ \vdots \\ t_{n(n+1)} \end{bmatrix}$$

which gives (3.2.2).

3.4.3 Similarity Transformation

If $\{u_1, u_2, \ldots, u_n\}$ and $\{v_1, v_2, \ldots, v_n\}$ are two bases forming the columns of square matrices U and V (representing the change from the unit basis to the respective bases), and if x_u represents the coordinates of a vector in the $\{u_i\}$ basis, then Ux_u represents the coordinates in the unit basis and $V^{-1}Ux_u$ represents the coordinates in $\{v_i\}$ basis, that is,

$$x_v = V^{-1}Ux_u = Bx_u$$

and the nonsingular matrix $B = V^{-1}U$ represents the change of basis from $\{u_i\}$ to $\{v_i\}$.

Now suppose A_u and A_v represent the same linear transformation in basis $\{u_i\}$ and $\{v_i\}$ respectively, that is,

$$A_u x_u = y_u, \qquad A_v x_v = y_v$$

But since $x_v = Bx_u$ and $y_v = By_u$, the second equation gives

$$A_v Bx_u = By_u \qquad \text{or} \qquad B^{-1}A_v Bx_u = y_u$$

and comparing with the first equation,

$$A_u = B^{-1}A_v B$$

or

$$A_v = BA_u B^{-1}$$

This relationship between A_u and A_v is called a *similarity transformation*; any two matrices so related are said to be similar.

3.4.4 Unitary (Orthogonal) Transformation

If both the basis sets $\{\mathbf{u}_i\}$ and $\{\mathbf{v}_i\}$ are, in addition, orthonormal, then by the definition of an orthonormal set

$$\bar{\mathbf{U}}^T \mathbf{U} = \{\bar{\mathbf{u}}_1, \bar{\mathbf{u}}_2, \ldots, \bar{\mathbf{u}}_n\}^T \{\mathbf{u}_1, \mathbf{u}_2, \ldots, \mathbf{u}_n\} = \mathbf{I} = \mathbf{U}\bar{\mathbf{U}}^T$$

$$\bar{\mathbf{V}}^T \mathbf{V} = \mathbf{I} = \mathbf{V}\bar{\mathbf{V}}^T$$

thus

$$\bar{\mathbf{U}}^T = \mathbf{U}^{-1} \quad \text{and} \quad \bar{\mathbf{V}}^T = \mathbf{V}^{-1}$$

and

$$\mathbf{B}\bar{\mathbf{B}}^T = \mathbf{I} \quad \text{or} \quad \bar{\mathbf{B}}^T = \mathbf{B}^{-1}$$

Such matrices are called unitary (orthogonal when real) matrices and the similarity transformation represented by them a unitary (orthogonal) transformation. It is obviously one-to-one and the rows of the matrix also form an o.n.b.; moreover,

$$|\mathbf{B}|^2 = 1 \quad \text{and} \quad |\mathbf{B}| = \pm 1$$

Note that for a unitary similarity transformation, we have the simple relation without requiring an inverse:

$$\mathbf{A}_u = \bar{\mathbf{B}}^T \mathbf{A}_v \mathbf{B} \quad \text{and} \quad \mathbf{A}_v = \mathbf{B}\mathbf{A}_u \bar{\mathbf{B}}^T$$

An example of an orthogonal transformation is pure rotation. In a cartesian coordinate system, a rotation by an angle θ about the x-axis is represented by the matrix

$$\mathbf{A} = \begin{bmatrix} 1 & 0 & 0 \\ 0 & \cos\theta & \sin\theta \\ 0 & -\sin\theta & \cos\theta \end{bmatrix}$$

Note that the transformation is unitary (orthogonal) if and only if the inner product and hence distances and angles are invariant, since

$$\langle \mathbf{x}, \mathbf{y} \rangle = \bar{\mathbf{x}}^T \mathbf{y} = \langle \mathbf{B}\mathbf{x}, \mathbf{B}\mathbf{y} \rangle = \bar{\mathbf{x}}^T \bar{\mathbf{B}}^T \mathbf{B}\mathbf{y}$$

It is often more convenient to talk about real matrices, then we use orthogonal transformation with ordinary transpose, the corresponding results for complex matrices using unitary transformation can be usually obtained by replacing the ordinary transpose by the conjugate transpose; $\mathbf{A}^T$ by $\bar{\mathbf{A}}^T$. On the other hand, general results about unitary transformation can be specialized to orthogonal transformations restricted to the real case by dropping the bar for the complex conjugation.

3.4.5 Projection

All the transformations considered so far represent a change of basis and hence are one-to-one; the matrix representing each is nonsingular and hence invertible. Projection is an example of a linear transformation which is *not* one-to-one, since the projection of a given vector is unique but it may be the projection of many vectors. If $\mathbf{n}$ is a unit vector in a real vector space $\mathscr{X}$, the set of all vectors orthogonal to $\mathbf{n}$ represented by $\mathbf{n}^\perp$ is a subspace of $\mathscr{X}$ (of which $\mathbf{n}$ is a unit normal) and the projection of $\mathbf{x} \in \mathscr{X}$ into this subspace is given by

$$A_p \mathbf{x} = \mathbf{x} - \langle \mathbf{n}, \mathbf{x} \rangle \mathbf{n} = \mathbf{x} - \mathbf{n}^T \mathbf{x} \mathbf{n} = \mathbf{x} - \mathbf{n}\mathbf{n}^T \mathbf{x} = [\mathbf{I} - \mathbf{n}\mathbf{n}^T]\mathbf{x} = [\mathbf{I} - \rangle \mathbf{n}, \mathbf{n}\langle\,]\mathbf{x}$$

So the projection in a subspace with unit normal $\mathbf{n}$ is represented by

$$\mathbf{A}_p = [\mathbf{I} - \mathbf{n}\mathbf{n}^T]$$

Clearly

$$\mathbf{A}_p^2 = [\mathbf{I} - \mathbf{n}\mathbf{n}^T][\mathbf{I} - \mathbf{n}\mathbf{n}^T] = \mathbf{I} - \mathbf{n}\mathbf{n}^T = \mathbf{A}_p$$

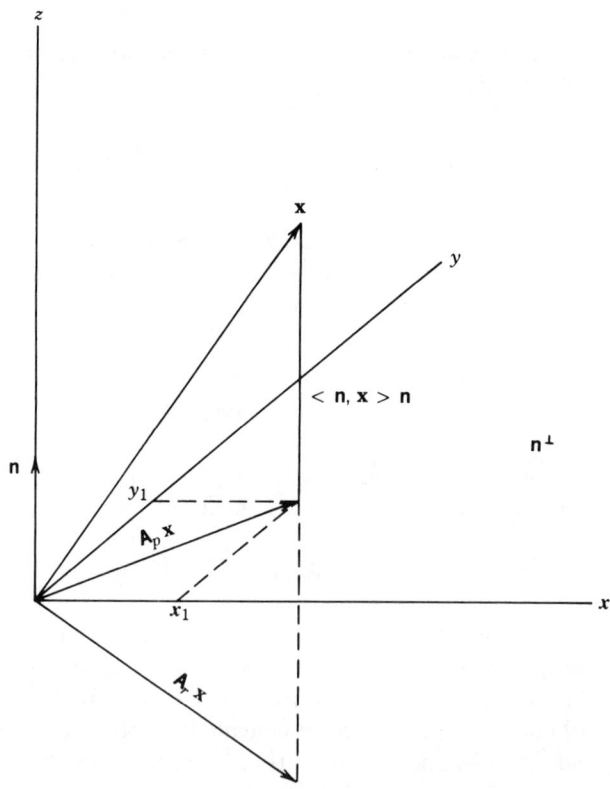

Figure 3.1 Projection and reflection.

Any matrix satisfying $\mathbf{A}^2 = \mathbf{A}$ is called an *idempotent* matrix and it can be shown that its rank is the sum of its diagonal elements. Moreover $\mathbf{A}_p$ is symmetric. A matrix $\mathbf{A}$ represents a projection if and only if it is symmetric and idempotent. Moreover

$$\text{rank}(\mathbf{A}_p) = d[\mathbf{n}^\perp]$$

As an example illustrated in Figure 3.1, consider the usual three-dimensional space with $\mathbf{n} = (0, 0, 1)^T$. Then $\mathbf{n}^\perp$ is the x-y plane and

$$\mathbf{A}_p = \mathbf{I} - \begin{bmatrix} 0 \\ 0 \\ 1 \end{bmatrix}(0, 0, 1) = \begin{bmatrix} 1 & 0 & 0 \\ 0 & 1 & 0 \\ 0 & 0 & 0 \end{bmatrix}$$

If $\mathbf{x} = (x_1, y_1, z_1)^T$ as shown, its projection is given by

$$\mathbf{A}_p \mathbf{x} = \begin{bmatrix} 1 & 0 & 0 \\ 0 & 1 & 0 \\ 0 & 0 & 0 \end{bmatrix}\begin{bmatrix} x_1 \\ y_1 \\ z_1 \end{bmatrix} = \begin{bmatrix} x_1 \\ y_1 \\ 0 \end{bmatrix}$$

Clearly projection, even an orthogonal projection, is a nonorthogonal transformation since it is rank deficient.

3.4.6 Reflection: Householder Transformation

Figure 3.1 suggests that a reflection may be obtained by subtracting from the vector twice its perpendicular as

$$\mathbf{A}_r \mathbf{x} = \mathbf{x} - 2\langle \mathbf{n}, \mathbf{x} \rangle \mathbf{n} = [\mathbf{I} - 2\mathbf{n}\mathbf{n}^T]\mathbf{x}$$

Hence a reflection is represented by the symmetric matrix

$$\mathbf{A}_r = \mathbf{I} - 2\mathbf{n}\mathbf{n}^T$$

which is orthogonal since

$$\mathbf{A}_r^2 = \mathbf{A}_r \mathbf{A}_r^T = \mathbf{I} - 2\mathbf{n}\mathbf{n}^T - 2\mathbf{n}\mathbf{n}^T + 4\mathbf{n}\mathbf{n}^T = \mathbf{I}$$

and is therefore one-to-one, preserving distances, norms, and angles. For the example in the figure

$$\mathbf{A}_r = \begin{bmatrix} 1 & 0 & 0 \\ 0 & 1 & 0 \\ 0 & 0 & -1 \end{bmatrix} \quad \text{and} \quad \mathbf{A}_r \mathbf{x} = \begin{bmatrix} x_1 \\ y_1 \\ -z_1 \end{bmatrix}$$

The unit normal was chosen in the preceeding illustration along the z axis only for ease of illustration. If v is an arbitrary vector, then the unit normal along v is $v/\sqrt{v^T v}$ and the reflection in the hyperplane $v^\perp$ is represented by the matrix

$$H = I - \frac{2v\, v^T}{v^T v}$$

known as Householder matrix (transformation or reflection), which is symmetric and orthogonal with determinant -1. It can be visualized from Figure 3.1 by considering v along the z axis, then the x–y plane represents $v^\perp$. A Householder reflection may be represented as a difference between two projections on the spaces spanned by $v^\perp$ and v

$$H = \left(I - \frac{v\, v^T}{v^T v}\right) - \frac{v\, v^T}{v^T v}$$

$$= P_{v^\perp} - P_{\text{sp}(v)}$$

realized in Figure 3.1 by $A_p x = [I - nn^T]x$ and $\langle n, x\rangle n = n^T x n = nn^T x$, respectively.

Householder transformations provide a computationally efficient and stable alternative to the Gram–Schmidt procedure for obtaining the triangular R matrix of the QR factorization. Note that if only R is needed, then H need not be explicitly computed; Hx can be directly evaluated as

$$Hx = x - \frac{2v^T x}{v^T v} v$$

Given any nonzero x, it is easy to construct v such that Hx is a multiple of e_1, the first column of the identity matrix of the same dimension. Since H preserves norms, $Hx = \pm \|x\| e_1$ and it is easy to verify that $v = x \mp \|x\| e_1$ does the job. If x is already close to a multiple of e_1, the sign is chosen to ensure that the denominator $v^T v$ is not small by selecting

$$v = x + \text{sign}(x_1) \|x\| e_1$$

so that $\|v\| \geq \|x\|$; this is not necessary in our case. (Why?)

To illustrate the application of Householder transformation, consider the $(N-1) \times 2$ data matrix consisting of the column vectors $x_1 = [x_1, x_2, \ldots, x_{N-1}]^T$, $x_2 = [x_2, x_3, \ldots, x_N]^T$, used in applying the Gram–Schmidt procedure to the AR(1) model:

$$X_t = \phi_1 X_{t-1} + a_t$$

The $(N - 1) \times (N - 1)$ Householder matrix $\mathbf{H}_1$ is used to zero all the elements of $\mathbf{x}_1$ except the first one, which is replaced by $\|\mathbf{x}_1\|$. Clearly, for this case $\mathbf{v}_1 = \mathbf{x}_1 - \|\mathbf{x}_1\|\mathbf{e}_1$ and $\mathbf{H}_1 = \mathbf{I} - 2\mathbf{v}_1\mathbf{v}_1^T/\mathbf{v}_1^T\mathbf{v}_1$. It is easy to verify that $\mathbf{H}_1\mathbf{x}_1 = \|\mathbf{x}_1\|\mathbf{e}_1$. Slightly more involved manipulation, using the AR(1) model that implies $\mathbf{x}_2 = \phi_1\mathbf{x}_1 + \mathbf{a}_2$, and since $\mathbf{x}_1$ and $\mathbf{a}_2$ are independent $\mathbf{x}_1^T\mathbf{a}_2 \to 0$ for large N, shows that

$$\mathbf{H}_1\mathbf{x}_2 = \mathbf{a}_2 - a_2\mathbf{e}_1 + \phi_1\|\mathbf{x}_1\|\mathbf{e}_1 + a_2\tilde{\mathbf{x}}_2/(\|\mathbf{x}_1\| - |x_1|)$$

where $\mathbf{a}_2 = [a_2, a_3, \ldots, a_N]^T$ and $\tilde{\mathbf{x}}_2 = (\mathbf{x}_1 - x_1\mathbf{e}_1) = [0, x_2, x_3, \ldots, x_{N-1}]^T$. Letting

$$\hat{\mathbf{a}}_3 = [a_3, a_4, \ldots, a_N]^T + \frac{a_2}{(\|\mathbf{x}_1\| - |x_1|)}[x_2, x_3, \ldots, x_{N-1}]^T$$

$$= [\hat{a}_3, \hat{a}_4, \ldots, \hat{a}_N]^T$$

it is seen that

$$\mathbf{H}_1\begin{bmatrix} x_1 & x_2 \\ x_2 & x_3 \\ \vdots & \vdots \\ x_{N-1} & x_N \end{bmatrix} = \begin{bmatrix} \|\mathbf{x}_1\| & \phi_1\|\mathbf{x}_1\| \\ 0 & \hat{a}_3 \\ \vdots & \vdots \\ 0 & \hat{a}_N \end{bmatrix}$$

Using $\mathbf{x}_1^T\mathbf{a}_2 = 0$, it is left as an exercise to show that $\|\hat{\mathbf{a}}_3\| = \|\mathbf{a}_2\|$.

An $(N - 2) \times (N - 2)$ Householder matrix $\tilde{\mathbf{H}}_2$ can reduce $\hat{\mathbf{a}}_3$ to $\|\hat{\mathbf{a}}_3\|\mathbf{e}_1 = \|\mathbf{a}_2\|\mathbf{e}_1$ by choosing

$$\mathbf{v}_2 = \hat{\mathbf{a}}_3 - \|\hat{\mathbf{a}}_3\|\mathbf{e}_1, \qquad \tilde{\mathbf{H}}_2 = \mathbf{I} - \mathbf{v}_2\mathbf{v}_2^T/\mathbf{v}_2^T\mathbf{v}_2$$

By defining an $(N - 1) \times (N - 1)$ Householder matrix $\mathbf{H}_2$ as

$$\mathbf{H}_2 = \begin{bmatrix} 1 & 0 \\ 0 & \tilde{\mathbf{H}}_2 \end{bmatrix}$$

it is now clear that

$$\mathbf{H}_2\mathbf{H}_1\begin{bmatrix} x_1 & x_2 \\ x_2 & x_3 \\ \vdots & \vdots \\ x_{N-1} & x_N \end{bmatrix} = \begin{bmatrix} \|\mathbf{x}_1\| & \phi_1\|\mathbf{x}_1\| \\ 0 & \|\mathbf{a}_2\| \\ 0 & 0 \\ \vdots & \vdots \\ 0 & 0 \end{bmatrix}$$

which is the normalized version of the QR factorization (3.2.4), similar to (3.1.9) obtained by Gram–Schmidt. Finding a similar normalized version of (3.2.5) is left as an exercise; it confirms that the diagonal elements of **R** continue to be the residual sum of squares after the correct order is exceeded.

If desired, the orthogonal basis **Q** can be recovered from the Householder matrix by

$$\mathbf{Q}^T = \mathbf{H}_2 \mathbf{H}_1$$

and the entire procedure can be generalized to AR(n). However, finding the orthogonal basis by a product of Householder matrices turns out to be twice as inefficient as Gram–Schmidt, although **Q** obtained by Householder is nearer to perfect orthogonality (numerically) than that obtained by Gram–Schmidt. Note that **Q** is a square orthogonal matrix and **R** is a rectangular matrix containing a square upper triangular matrix augmented by zeros.

3.4.7 Gram–Schmidt, Householder, and Cholesky Methods for Least Squares

It is thus seen that the least squares problem $\mathbf{Ax} = \mathbf{b}$ for an $m \times n$ matrix **A** with $m > n$ in general, or ARMA estimation in particular, can be solved with the help of QR factorization by Gram–Schmidt or Householder methods discussed in Sections 3.1.4 and 3.4.6, or by solving normal equations discussed in Section 3.3.3. Normal equations

$$\mathbf{A}^T \mathbf{A} \hat{\mathbf{x}} = \mathbf{A}^T \mathbf{b}$$

can be solved efficiently by implicitly using the QR factorization of **A** as

$$\mathbf{A}^T \mathbf{A} = (\mathbf{QR})^T \mathbf{QR} = \mathbf{R}^T \mathbf{Q}^T \mathbf{QR} = \mathbf{R}^T \mathbf{R} = \mathbf{L} \mathbf{L}^T$$

where **L** is a lower triangular matrix with nonnegative diagonal entries; such a factorization of a nonnegative definite matrix is called Cholesky factorization. The factorization is unique if **A** is nonsingular (why?); then $\mathbf{A}^T \mathbf{A}$ is positive definite and the diagonal entries of **L** are positive. An efficient algorithm is available for finding the **L** matrix. Then, one solves the equations $\mathbf{Ly} = \mathbf{A}^T \mathbf{b}$ by forward substitution and $\mathbf{L}^T \hat{\mathbf{x}} = \mathbf{y}$ by backward substitution, to get the least square estimate $\hat{\mathbf{x}}$ without explicitly inverting the $\mathbf{A}^T \mathbf{A}$ matrix. This requires $(mn^2/2) + (n^3/6)$ operations; roughly $mn^2/2$ for the computation of symmetric $\mathbf{A}^T \mathbf{A}$ and $n^3/6$ for the Cholesky factorization, other terms not involving the product mn are relatively small and hence neglected in the count. The operation counts for (modified) Gram–Schmidt and Householder methods are mn^2 and $(mn^2 - n^3/3)$, respectively. Thus the Cholesky method of solving normal equations is the most computationally efficient procedure for solving the general least squares problem, its operation count reduces to $(mn + n^3/6)$ in the case of AR models when the special (Toeplitz) structure of the $\mathbf{A}^T \mathbf{A}$ matrix is utilized. (Square matrices with the same elements down each diagonal parallel to the main diagonal are called Toeplitz matrices, whereas those with the same elements along each diagonal perpendicular to the main diagonal are called Hankel matrices. Note that all the *square* data matrices of Section 3.2.1 on AR estimation are Hankel, and would be Toeplitz if their column order is reversed. Toeplitz and Hankel matrices can be transformed into each other by left multiplication with "backward identity"—a matrix with 1's on the diagonal perpendicular to the main diagonal and zeros elsewhere.)

The accuracy of the computed normal equations, however, depends on the square of the condition of **A** (since they require the computation of $\mathbf{A}^T\mathbf{A}$), whereas that of the Gram–Schmidt and Householder methods depends on the condition of **A**. Thus, the more ill-conditioned **A** is, the worse is the accuracy of the Cholesky method compared to Householder and Gram–Schmidt methods. For near-singular matrices, such as those from a deterministic AR model, the Cholesky method is more likely to break down before the Householder or Gram–Schmidt methods.

The orthonormal basis **Q** is provided only by Householder and Gram–Schmidt methods. Note that although **Q** provided by the Householder method is $m \times m$, only n columns are needed, whereas **Q** by the Gram–Schmidt method is already $m \times n$, with operation counts of $2(mn^2 - n^3/3)$ and mn^2, respectively. More details of these algorithms may be found in Golub and Van Loan (1987).

3.5 MATRIX DECOMPOSITION

We will now consider the decomposition of a matrix into diagonal form by means of similarity and possibly orthogonal transformation. For a square matrix, when this is possible, the diagonal elements are the eigenvalues and the similarity transform matrix has eigenvectors as its columns. The resultant spectral decomposition, when applied to the state matrix, forms the crux of modal and spectrum analysis.

For a nonsquare matrix, orthogonal similarity transformation to a diagonal matrix is clearly impossible. However, it is possible to diagonalize a nonsquare matrix by means of *different* orthogonal matrices so that the corresponding transformation is no longer similar. Such a decomposition is known as singular value decomposition because the diagonal elements provide a clear measure of the closeness of the original matrix to singularity in the sense that only the number of diagonal elements equal to the rank of the matrix are positive and the rest zero. Recall that the data matrix from deterministic systems has rank n as long as the number of columns is greater than or equal to n. Hence the singular-value decomposition has played a pivotal role in determining the dimension of state space for deterministic systems, although we have chosen to do this more simply by the Gram–Schmidt procedure, as illustrated in Section 3.2.1.

3.5.1 Eigenvalues and Eigenvectors

Let $\mathscr{A}$ be any linear transformation. If, for a nonzero vector $\mathbf{x}_i$ and scalar μ_i

$$\mathscr{A}(\mathbf{x}_i) = \mu_i \mathbf{x}_i$$

then x_i is called an eigenvector and μ_i the corresponding eigenvalue. For finite dimensional vector spaces, $\mathscr{A}$ is represented by an $n \times n$ matrix $\mathbf{A}$ and

or
$$\mathbf{A}x_i = \mu_i x_i$$

$$(\mathbf{A} - \mu_i \mathbf{I})x_i = \mathbf{0}$$

Then the columns of $(\mathbf{A} - \mu_i \mathbf{I})$ are dependent or

$$\text{rank}(\mathbf{A} - \mu_i \mathbf{I}) < n \quad \text{or} \quad |(\mathbf{A} - \mu_i \mathbf{I})| = 0$$

When this determinant is expanded, we get an nth degree polynomial in μ, that is,

$$|\mathbf{A} - \mu \mathbf{I}| = (-1)^n (\mu)^n + c_{n-1} \mu^{n-1} + \cdots + c_1 \mu + c_0 = C(\mu)$$

This is called the characteristic polynomial and its roots are the eigenvalues. If there are $p \leq n$ distinct roots, then

$$C(\mu) = (-1)^n (\mu - \mu_1)^{m_1} (\mu - \mu_2)^{m_2} \cdots (\mu - \mu_p)^{m_p}$$

m_i is called the algebraic multiplicity of μ_i and

$$m_1 + m_2 + \cdots + m_p = n$$

The following results are useful:

1. Since c_i are real, if μ_i is an eigenvalue, so is its complex conjugate $\bar{\mu}_i$.
2. $\text{Trace}(\mathbf{A}) = m_1 \mu_1 + m_2 \mu_2 + \cdots + m_p \mu_p = (-1)^{n+1} c_{n-1}$.
3. $|\mathbf{A}| = \mu_1^{m_1} \mu_2^{m_2} \cdots \mu_p^{m_p} = c_0$.
4. If $\mathbf{A}$ is nonsingular so that it has nonzero eigenvalues μ_i, then $1/\mu_i$ are the eigenvalues of $\mathbf{A}^{-1}$.
5. Similar matrices have the same rank, the same eigenvalues, and hence the same determinant and trace.
   For a Hermitian matrix ($\bar{\mathbf{A}}^T = \mathbf{A}$, that is symmetric when real) it is easy to show that
   i. All the eigenvalues are real and for real $\mathbf{A}$ even eigenvectors can be chosen to be real.
   ii. If $\text{rank}(\mathbf{A}) = r$, then 0 is an eigenvalue of multiplicity $(n - r)$.
   iii. Eigenvectors corresponding to distinct eigenvalues are orthogonal.
   iv. If $\mathbf{A}$ is positive (nonnegative)-definite, that is, $\bar{\mathbf{x}}^T \mathbf{A} \mathbf{x} > 0$ (≥ 0) for arbitrary $\mathbf{x}$, then the eigenvalues are positive (nonnegative), and hence so are the determinant and trace.
   v. For an arbitrary vector $\mathbf{x}$, there exists an eigenvector belonging to $\mathscr{M}(\mathbf{x}, \mathbf{A}\mathbf{x}, \mathbf{A}^2\mathbf{x}, \ldots)$.

The simplest matrix is the identity, next in simplicity is the diagonal matrix. For a diagonal matrix, the diagonal elements are the eigenvalues. Since similar matrices have the same eigenvalues, it would be convenient if we can reduce an arbitrary matrix to a diagonal matrix by similarity transformation, because that diagonal matrix then would contain the eigenvalues as its diagonal elements and therefore the columns of the transforming matrix would be the eigenvectors. When can we reduce a matrix to such diagonal form? The simplest case is when the eigenvalues are distinct.

We first show that there exist at least as many linearly independent eigenvectors as the number of distinct eigenvalues. If there are p distinct eigenvalues and $q < p$ linearly independent eigenvectors corresponding to them, say, $\mathbf{m}_1, \mathbf{m}_2, \ldots, \mathbf{m}_q$, then

$$\mathbf{m}_{q+1} = a_1 \mathbf{m}_1 + a_2 \mathbf{m}_2 + \cdots + a_k \mathbf{m}_k, \qquad k \leq q, \qquad a_i \neq 0,$$

and multiplying by $\mathbf{A}$,

$$\mu_{q+1} \mathbf{m}_{q+1} = a_1 \mu_1 \mathbf{m}_1 + a_2 \mu_2 \mathbf{m}_2 + \cdots + a_k \mu_k \mathbf{m}_k$$

and now multiplying the first equation by μ_{q+1} and subtracting gives

$$\mathbf{0} = a_1(\mu_{q+1} - \mu_1)\mathbf{m}_1 + \cdots + a_k(\mu_{q+1} - \mu_k)\mathbf{m}_k$$

that is,

$$\mu_{q+1} = \mu_1 = \mu_2 = \cdots = \mu_k$$

which is a contradiction, so that at least p eigenvectors must be linearly independent.

The simplest form in which a matrix may be transformed by the similarity transformation may be categorized into three cases:

Case I: n Linearly Independent Eigenvectors—Diagonal Form

One example is a matrix with all eigenvalues distinct, so that $p = n$, and hence $m_i = 1$ for all i. As long as there are n linearly independent eigenvectors

$$\mathbf{A}\mathbf{m}_i = \mu_i \mathbf{m}_i, \qquad i = 1, 2, \ldots, n$$

is

$$\mathbf{AM} = \mathbf{M}\boldsymbol{\mu}$$

where

$$\mathbf{M} = \{\mathbf{m}_1, \mathbf{m}_2, \ldots, \mathbf{m}_n\}$$

is the nonsingular *modal matrix* with linearly independent eigenvectors as its columns and

$$\boldsymbol{\mu} = \mathrm{diag}\{\mu_i\}$$

is a diagonal matrix with eigenvalues as its diagonal elements. Hence, as long as it has n linearly independent eigenvectors (even when this is possible with nondistinct eigenvalues), the matrix $\mathbf{A}$ can be reduced to the diagonal matrix by

$$\mathbf{A} = \mathbf{M}\mu\mathbf{M}^{-1} \quad \text{or} \quad \mu = \mathbf{M}^{-1}\mathbf{A}\mathbf{M}$$

and then the matrix is said to be diagonalizable. Moreover, when all the n eigenvalues are distinct

$$\begin{aligned}\text{rank}(\mathbf{A} - \mathbf{I}\mu_i) &= \text{rank}(\mathbf{M}\mu\mathbf{M}^{-1} - \mathbf{M}\mathbf{I}\mu_i\mathbf{M}^{-1}) \\ &= \text{rank}\,\mathbf{M}(\mu - \mathbf{I}\mu_i)\mathbf{M}^{-1} \\ &= \text{rank}(\mu - \mathbf{I}\mu_i) \\ &= n - 1\end{aligned}$$

and since there can be at most one zero eigenvalue,

$$n - 1 \leq \text{rank}\,\mathbf{A} = \text{rank}\,\mu \leq n$$

Generally, however, if there are n linearly independent eigenvectors but only $p < n$ distinct eigenvalues, the same argument shows that

$$\text{rank}\,(\mathbf{A} - \mathbf{I}\mu_i) = n - m_i$$
$$p - 1 \leq \text{rank}\,\mathbf{A} = \text{rank}\,\mu \leq n$$

When $\mathbf{A}$ is real and symmetric (or complex and Hermitian, $\mathbf{A} = \bar{\mathbf{A}}^T$), then the eigenvalues are real, and if we normalize the eigenvectors to unity then the modal matrix $\mathbf{M}$ is orthogonal for real symmetric $\mathbf{A}$ and unitary ($\mathbf{M}\bar{\mathbf{M}}^T = \bar{\mathbf{M}}^T\mathbf{M} = \mathbf{I}$) for complex Hermitian $\mathbf{A}$. The matrix $\mathbf{A}$ is then said to be unitarily diagonalizable. To prove this for all n distinct eigenvalues

$$\mathbf{A} = \mathbf{M}\mu\mathbf{M}^{-1} = \bar{\mathbf{A}}^T = (\bar{\mathbf{M}}^{-1})^T\bar{\mu}\bar{\mathbf{M}}^T$$

that is,

$$\bar{\mathbf{M}}^T\mathbf{M}\mu(\bar{\mathbf{M}}^T\mathbf{M})^{-1} = \bar{\mu}$$
$$\{\bar{\mathbf{m}}_i^T\mathbf{m}_j\}\,\text{diag}\,\{\mu_i\} = \text{diag}\{\bar{\mu}_j\}\,\{\bar{\mathbf{m}}_i^T\mathbf{m}_j\}$$
$$\{\mu_i\bar{\mathbf{m}}_i^T\mathbf{m}_j\} = \{\bar{\mu}_j\bar{\mathbf{m}}_i^T\mathbf{m}_j\}$$
$$(\mu_i - \bar{\mu}_j)\bar{\mathbf{m}}_i^T\mathbf{m}_j = 0$$

so that for

$i = j$, since $\bar{\mathbf{m}}_i^T\mathbf{m}_i = 1$, $\mu_i = \bar{\mu}_i$, that is, μ_i are real, and for $i \neq j$, $(\mu_i - \bar{\mu}_j)\bar{\mathbf{m}}_i^T\mathbf{m}_j = 0$ implies $\bar{\mathbf{m}}_i^T\mathbf{m}_j = 0$, since $\mu_i \neq \bar{\mu}_j$, showing that **M** is unitary. Therefore $\mathbf{M}^{-1} = \bar{\mathbf{M}}^T$ and the decomposition of the **A** matrix may be alternatively written as

$$\mathbf{A} = \mathbf{M}\,\boldsymbol{\mu}\bar{\mathbf{M}}^T$$

where the bars may be omitted when **A** is real. In other words, for unitary modal matrix **M**, its inverse may be replaced by the conjugate transpose, thus eliminating the computation of inverse.

When one or more eigenvalues is a repeated root of the characteristic equation, a full set of n linearly independent eigenvectors may or may not exist. Since the corresponding eigenvectors are given by

$$(\mathbf{A} - \mu_i \mathbf{I})\mathbf{x} = \mathbf{0}$$

(3.1.2c) shows that the number of linearly independent eigenvectors associated with an eigenvalue μ_i of algebraic multiplicity m_i is equal to the degeneracy q_i of $(\mathbf{A} - \mu_i \mathbf{I})$ given by

$$q_i = n - \text{rank}(\mathbf{A} - \mu_i \mathbf{I}), \qquad 1 \leq q_i \leq m_i$$

which is often referred to as the *geometric* multiplicity of μ_i. Since it is between 1 and m_i, we get three possibilities.

Full Degeneracy, $q_i = m_i$

In this case there exists a full set of m_i linearly independent eigenvectors associated with the eigenvalue μ_i. If this is true for all i, then there exist $m_1 + m_2 + \cdots + m_p = n$ linearly independent eigenvectors for the matrix and hence it can be diagonalized exactly as discussed above. The distinct eigenvalue case is in fact a special case of this with $m_i = 1$.

There exist n not only linearly independent but also orthonormal eigenvectors when the matrix **A** is real symmetric, or complex Hermitian. In general, it can be shown that a matrix **A** has n orthonormal eigenvectors if and only if

$$\mathbf{A}\bar{\mathbf{A}}^T = \bar{\mathbf{A}}^T \mathbf{A}$$

Such a transformation is called *normal* transformation and the matrix representing it is referred to as a *normal matrix*. It can be shown that a matrix being (1) normal, (2) unitarily diagonalizable, and (3) possessing n orthonormal eigenvectors are equivalent conditions. Note that a normal matrix with real eigenvalues is the Hermitian matrix and a Hermitian matrix that is real is the symmetric matrix.

Case II: Less than n Linearly Independent Eigenvectors, Simple Degeneracy, $q_i = 1$, -Jordan Form

In this case there is only one eigenvector associated with each distinct μ_i regardless of its algebraic multiplicity m_i. Then the matrix cannot be diagonalized if $m_i > 1$. The simplest form that $\mathbf{A}$ can be reduced to in this case is called the *Jordan* form. For example, if a 5×5 matrix has root μ_1 repeated three times and μ_2 and μ_3 are the remaining distinct roots then

$$\mathbf{A} = \mathbf{M}\boldsymbol{\mu}\mathbf{M}^{-1}$$

$$\boldsymbol{\mu} = \begin{bmatrix} \mu_1 & 1 & 0 & 0 & 0 \\ 0 & \mu_1 & 1 & 0 & 0 \\ 0 & 0 & \mu_1 & 0 & 0 \\ 0 & 0 & 0 & \mu_2 & 0 \\ 0 & 0 & 0 & 0 & \mu_3 \end{bmatrix}$$

where the top left 3×3 block is known as a Jordan block. Note that in this case $m_1 = 3$, $m_2 = 1$, and $m_3 = 1$, whereas $q_1 = q_2 = q_3 = 1$.

Case III: Less Than n Linearly Independent Eigenvectors, $1 < q_i < m_i$

In this case the reduced matrix may be a mixture of Jordan blocks and diagonal forms even for an individual multiple eigenvalue. Knowledge of m_i and q_i still leaves some ambiguity about the reduced form.

3.5.2 Spectral Representation of a Matrix vis-a-vis Conventional and DDS Approaches to Modal and Spectrum Analysis

The set of all vectors (including the zero vector) satisfying

$$\mathbf{A}\mathbf{x}_i = \mu_i \mathbf{x}_i$$

is called the *eigenspace* of μ_i, which is the same as the null space of $(\mathbf{A} - \mu_i \mathbf{I})$ denoted by $\mathcal{N}_i$. (The null space of a transformation is the set of all vectors which are transformed to zero by the transformation.) $\mathcal{N}_i$ is a q_i dimensional subspace of $\mathcal{X}$ that is invariant under the transformation represented by $\mathbf{A}$, that is, every vector in the subspace is transformed to a vector in the same subspace.

A linear transformation possessing a full set of n linearly independent eigenvectors is said to be *simple*. Note that this is covered by Case I above. In this case the reduced matrix $\boldsymbol{\mu}$ is diagonal irrespective of the algebraic multiplicity m_i of each μ_i. These n linearly independent eigenvectors are the columns of

$$\mathbf{M} = \{\mathbf{m}_1, \mathbf{m}_2, \ldots, \mathbf{m}_n\}$$

that form a basis, whereas conjugate transposes of the rows of $\mathbf{M}^{-1}$ form the reciprocal basis (see (3.1.10–3.1.12))

$$\mathbf{M}^{-1} = \mathbf{R} = \{\bar{\mathbf{r}}_1, \bar{\mathbf{r}}_2, \ldots, \bar{\mathbf{r}}_n\}^T$$

Thus we have a complete and corresponding decomposition of (1) the vector space $\mathscr{X}$, (2) an arbitrary vector $\mathbf{x}$ belonging to it, (3) the transformed vector $\mathbf{Ax}$, and (4) the transformation matrix $\mathbf{A}$:

(1) $\mathscr{X} = \mathscr{N}_1 + \mathscr{N}_2 + \cdots + \mathscr{N}_n$

(2) $\mathbf{x} = \mathbf{M}\mathbf{M}^{-1}\mathbf{x}$
$= \mathbf{m}_1 \bar{\mathbf{r}}_1^T \mathbf{x} + \mathbf{m}_2 \bar{\mathbf{r}}_2^T \mathbf{x} + \cdots + \mathbf{m}_n \bar{\mathbf{r}}_n^T \mathbf{x}$
$= \mathbf{m}_1 \langle \mathbf{r}_1, \mathbf{x} \rangle + \mathbf{m}_2 \langle \mathbf{r}_2, \mathbf{x} \rangle + \cdots + \mathbf{m}_n \langle \mathbf{r}_n, \mathbf{x} \rangle$

(3) $\mathbf{Ax} = \mathbf{A}\mathbf{m}_1 \langle \mathbf{r}_1, \mathbf{x} \rangle + \mathbf{A}\mathbf{m}_2 \langle \mathbf{r}_2, \mathbf{x} \rangle + \cdots + \mathbf{A}\mathbf{m}_n \langle \mathbf{r}_n, \mathbf{x} \rangle$
$= \mu_1 \mathbf{m}_1 \langle \mathbf{r}_1, \mathbf{x} \rangle + \mu_2 \mathbf{m}_2 \langle \mathbf{r}_2, \mathbf{x} \rangle + \cdots + \mu_n \mathbf{m}_n \langle \mathbf{r}_n, \mathbf{x} \rangle$
$= \mu_1 \mathbf{m}_1 \bar{\mathbf{r}}_1^T \mathbf{x} + \mu_2 \mathbf{m}_2 \bar{\mathbf{r}}_2^T \mathbf{x} + \cdots + \mu_n \mathbf{m}_n \bar{\mathbf{r}}_n^T \mathbf{x}$

(4) $\mathbf{A} = \mathbf{M}\mu\mathbf{M}^{-1}$
$= \mu_1 \mathbf{m}_1 \bar{\mathbf{r}}_1^T + \mu_2 \mathbf{m}_2 \bar{\mathbf{r}}_2^T + \cdots + \mu_n \mathbf{m}_n \bar{\mathbf{r}}_n^T$

The last is known as the *spectral representation* of the matrix and of the corresponding linear transformation. When there are p distinct eigenvalues, the n-dimensional state space $\mathscr{X}$ splits into p q_i-dimensional subspaces $\mathscr{N}_i$. Every vector $\mathbf{x}$ can be expressed in terms of the basis consisting of the n independent eigenvectors $\mathbf{m}_i$. For each of these eigenvectors $\mathbf{m}_i$, the operation of the linear transformation or multiplication by the matrix $\mathbf{A}$ is reduced to the multiplication by the scalar eigenvalue μ_i. This is the crux of linear system analysis. A complicated system is reduced to simple subsystems in each of which the system operation can be studied very simply.

If $\mathbf{A}$ is real symmetric or complex Hermitian, the proof given above for the distinct eigenvalue Case I can be extended to show that the n eigenvectors are orthonormal in addition to being linearly independent. Then the modal matrix is unitary (orthogonal when real) so that $\mathbf{R} = \bar{\mathbf{M}}^T$ and the spectral representation takes a particularly simple form:

$$\mathbf{A} = \mathbf{M}\mu\bar{\mathbf{M}}^T$$
$$= \mu_1 \mathbf{m}_1 \bar{\mathbf{m}}_1^T + \mu_2 \mathbf{m}_2 \bar{\mathbf{m}}_2^T + \cdots + \mu_n \mathbf{m}_n \bar{\mathbf{m}}_n^T$$

where the eigenvalues μ_i are real. Again, the bars may be omitted for the real symmetric matrix, since even the eigenvectors can be chosen to be real in that case.

Such spectral representation of a system is at the heart of spectrum analysis and modal analysis, which is discussed in detail in later chapters. But the difference between DDS and the conventional approaches to both spectrum and modal analysis may be pointed out here. The conventional approaches to both attempt to find the eigenvalues from the peaks in the plot of the Fourier transform of the response data or the Fourier transform of its autocovariance function, called the autospectrum or power spectrum; thus both these methods are generally based on frequency domain. Modal analysis in addition tries to find the eigenvectors corresponding to the eigenvalues of interest by curve-fitting in the frequency domain and then attempts to synthesize as much of the matrix **A** as possible. On the other hand, DDS obtains the **A** matrix directly from the data by the least squares method using the modeling strategy briefly discussed in Section 2.4.1; eigenvalues, eigenvectors, and any other desired information then follows naturally and in a straightforward manner by the simple theory of linear operators in a finite dimensional space, as discussed in this chapter.

When a transformation is not simple, it does not have n linearly independent eigenvectors. Then it is not possible to reduce the matrix to diagonal form, it needs a Jordan form or a mixture of Jordan and diagonal form, depending upon whether the degeneracy is full or partial, as seen above. Even with μ taking such a form, the modal matrix **M** needs additional column vectors to make up for the lack of n linearly independent eigenvectors. In other words, we need to complete the basis of n vectors by augmenting the smaller set of linearly independent eigenvectors. Note that otherwise $\mathbf{M}^{-1}$ cannot exist. The additional vectors completing the basis are called *generalized eigenvectors*. Thus we need $(m_i - q_i)$ generalized eigenvectors corresponding to each eigenvalue μ_i to complete the basis.

A generalized eigenvector of rank k is defined as a nonzero vector satisfying

$$(\mathbf{A} - \mu_i \mathbf{I})^k \mathbf{x}_k = \mathbf{0} \quad \text{and} \quad (\mathbf{A} - \mu_i \mathbf{I})^{k-1} \mathbf{x}_k \neq \mathbf{0}$$

If such a vector is found for $k \leq m_i$, then the remaining vectors can be found by the relations

$$\mathbf{x}_{k-1} = (\mathbf{A} - \mu_i \mathbf{I})\mathbf{x}_k, \quad \mathbf{x}_{k-2} = (\mathbf{A} - \mu_i \mathbf{I})\mathbf{x}_{k-1}, \ldots, \mathbf{x}_1 = (\mathbf{A} - \mu_i \mathbf{I})\mathbf{x}_2$$

The last vector $\mathbf{x}_1$ then is a regular eigenvector and hence satisfies

$$(\mathbf{A} - \mu_i \mathbf{I})\mathbf{x}_1 = \mathbf{0}$$

All these vectors belong to the null space of $(\mathbf{A} - \mu_i \mathbf{I})^k$. If this is repeated for each μ_i and the corresponding null space is denoted by $\mathcal{N}_i^{k_i}$ with dimension m_i, then we have the following decomposition of the vector space $\mathcal{X}$:

$$\mathcal{X} = \mathcal{N}_1^{k_1} + \mathcal{N}_2^{k_2} + \cdots + \mathcal{N}_p^{k_p}$$

In each of these we have the usual simple operation for diagonal form

$$\mathbf{A}\mathbf{x}_1 = \mu_1 \mathbf{x}_1$$

and slightly more involved operation

$$\mathbf{A}\mathbf{x}_2 = \mu_1 \mathbf{x}_2 + \mathbf{x}_1$$
$$\mathbf{A}\mathbf{x}_3 = \mu_1 \mathbf{x}_3 + \mathbf{x}_2$$

for a 3×3 Jordan form.

3.5.3 Singular Values and Singular Value Decomposition (SVD)

The dimension of the state space is the same as the dimension of the square state matrix and is equal to the number of its eigenvalues. For deterministic systems, this dimension can be determined from the rank or the number of independent columns of the data matrix, as illustrated in Section 3.2.1. We recommend the QR factorization that can be readily applied with the presence of round-off and computational errors or even for stochastic systems by adapting noise floor or F criterion discussed in Section 3.2.2. Note that for stochastic systems there is no rank deficiency in the data matrix even when the number of columns exceeds the dimension of the state space.

Numerical determination of the rank of an arbitrary (not necessarily square) matrix can be accomplished with some additional computation by an elegant matrix decomposition called singular value decomposition (SVD). Since a large part of literature dealing with deterministic systems uses the SVD to determine the dimension of state space, we will succinctly present the main results of the SVD useful for determination of rank. A valuable aspect of the SVD is that it quantifies the notion of near-rank deficiency.

Recall from Section 3.5.1 that the rank of a nonnegative definite Hermitian matrix is equal to the number of positive eigenvalues. The key to the SVD is the remarkable fact that the rank of an arbitrary (not necessarily square) matrix $\mathbf{A}$ is the same as the rank of the square nonnegative definite Hermitian matrices $\mathbf{\bar{A}}^T\mathbf{A}$ and $\mathbf{A}\mathbf{\bar{A}}^T$.

The central theorem of the SVD [for which the proof and related details may be found in Horn and Johnson (1985)] states that an $m \times n$ matrix $\mathbf{A}$ of rank r can be decomposed as

$$\underset{m \times n}{\mathbf{A}} = \underset{m \times m}{\mathbf{U}} \underset{m \times n}{\text{diag}(\sigma_1, \sigma_2, \ldots, \sigma_p)} \underset{n \times n}{\mathbf{\bar{V}}^T}$$

where $\mathbf{U}$ and $\mathbf{V}$ are unitary (orthogonal when $\mathbf{A}$ is real) and

$$\sigma_1 \geq \sigma_2 \geq \sigma_r > \sigma_{r+1} = \sigma_{r+2} = \cdots = \sigma_p = 0, \quad p = \min(m, n)$$

The real numbers σ_i, called singular values, are the nonnegative square roots of the eigenvalues of $A\bar{A}^T$ and hence are uniquely determined. The columns of U are eigenvectors of $A\bar{A}^T$ and the columns of V are eigenvectors of $\bar{A}^T A$ (arranged in the same order as the corresponding eigenvalues σ_i^2). In other words, the spectral decomposition of $A\bar{A}^T$ and $\bar{A}^T A$ is given by, say, for $m > n$

$$A\bar{A}^T = U \operatorname{diag}(\sigma_1^2, \sigma_2^2, \ldots, \sigma_n^2, 0, \ldots, 0)\, \bar{U}^T, \quad U\bar{U}^T = I$$
$$\underset{m \times m}{} \underset{m \times m}{} \underset{m \times m}{} \underset{m \times m}{}$$

$$\bar{A}^T A = V \operatorname{diag}(\sigma_1^2, \sigma_2^2, \ldots, \sigma_n^2)\, \bar{V}^T, \quad V\bar{V}^T = I$$
$$\underset{n \times n}{} \underset{n \times n}{} \underset{n \times n}{} \underset{n \times n}{}$$

Clearly eigenvalues of A are the same as its singular values only if A is Hermitian and nonnegative definite; if A is positive definite then $U = V$. If $U = (u_1, u_2, \ldots, u_m)$ and $V = (v_1, v_2, \ldots, v_n)$, then u_i and v_i are often referred to as the left and right singular vectors because

$$A v_i = \sigma_i u_i \quad \text{and} \quad \bar{A}^T u_i = \sigma_i v_i, \quad i = 1, 2, \ldots, p$$

The SVD quantifies the rank because, denoting $U_r = (u_1, u_2, \ldots, u_r)$, $V_r = (v_1, v_2, \ldots, v_r)$

$$A = U_r \operatorname{diag}(\sigma_1, \sigma_2, \ldots, \sigma_r)\, \bar{V}_r^T = \sigma_1 u_1 \bar{v}_1^T + \sigma_2 u_2 \bar{v}_2^T + \cdots + \sigma_r u_r \bar{v}_r^T,$$
$$\underset{m \times r}{} \underset{r \times r}{} \underset{r \times n}{} \qquad \bar{U}_r^T U_r = \bar{V}_r^T V_r = I$$

is simply another form of the decomposition given above when the matrix multiplication is carried out, with each matrix $u_i \bar{v}_i^T$ having rank 1. Notice the similarity of this decomposition with the spectral decomposition in Section 3.5.2.

To show rigorously how the SVD quantifies near-rank deficiency or singularity requires definition of matrix norms which we have not introduced since they are not otherwise needed in our work. However, a heuristic argument can be given to see this indirectly. The "most" nonsingular or full-rank matrix is obviously the unitary matrix, since its columns as well as rows are orthonormal and hence linearly independent, and its inverse is the easiest to find since $\bar{A}^T A = I$ implies $A^{-1} = \bar{A}^T$. Then the eigenvalues of $\bar{A}^T A$ and hence the singular values of A are all equal to 1. Therefore the condition number κ of a matrix may be defined as the ratio of its largest and the smallest singular values

$$\kappa(A) = \frac{\sigma_1}{\sigma_p}$$

This condition number is 1 for the "best" unitary matrix for which $\sigma_1 = \sigma_p = 1$ and ∞ for the "worst" singular matrix for which $\sigma_p = 0$ because $r < p$. Thus the larger the condition number, the more ill-conditioned (or computationally difficult to invert) is the matrix. Matrices with small condition numbers are said to be well-conditioned; a unitary (orthogonal when real) matrix with the smallest $\kappa = 1$ is perfectly conditioned. Since the determinant of a singular matrix is zero, it is tempting to use it as a measure of ill-conditioning. The determinant, unfortunately, is a very poor measure of ill-conditioning. A perfectly well-conditioned matrix can have very small determinant: An $n \times n$ matrix diag$(0.1, 0.1, \ldots, 0.1)$ has a determinant of 10^{-n} but $\kappa = 1$ and is of course trivially simple to invert.

The SVD expresses a matrix as a linear combination of rank-one matrices with positive singular values as coefficients that decrease for an ill-conditioned matrix. Therefore, we can find the most well-conditioned n-rank ($n < r$) approximation to an ill-conditioned matrix by retaining the first n terms of this decomposition (with the n largest singular values). The error of approximation by such an n-rank matrix from the SVD is the smallest among all n-rank matrices and is equal to the next singular value σ_{n+1} when the usual squared or 2-norm is used.

SVD and Dimension of State Space

For deterministic systems with guaranteed finite state space of dimension n, Ho and Kalman (1965) proposed an algorithm for determination of a minimal dimensional state matrix (called minimal realization) based on the rank of the pulse response matrix (similar to the data matrix illustrated in Section 3.2.1) that cannot exceed n. Since the numerical determination of rank is difficult in practice even for data from deterministic systems with finite n due to round-off and computational noise, Zeiger and McEwen (1974) proposed the use of SVD in implementing the Ho–Kalman algorithm for treatment of noisy data. This procedure has been adapted by Juang and Pappa (1984) in the eigensystem realization algorithm (ERA) for modal parameter identification of Galileo spacecraft constructed by NASA. It is based on an ad hoc determination of the smallest singular value σ_n below which the singular values will be effectively considered to be zero and omitted in reconstructing the n-rank approximation to the data and state matrix. Their experience indicates that "there can be real difficulties in determining a gap between the computed last nonzero singular value and what should be effectively considered zero when measurement noise is present." Under the ideal conditions of no noise of course σ_n/σ_1 can be taken as the machine precision of the computer.

For stochastic systems, the rank of the data matrix can exceed the dimension of the state space, as illustrated in Section 3.2.2, so the Ho–Kalman algorithm does not apply. There have been several attempts in the time series and control systems literature to circumvent this difficulty by replacing the pulse response output data matrix in the Ho–Kalman algorithm with the covariance matrix. Some of these attempts by Akaike (1976), Pernebo and Silverman (1982), and others have been summarized in Aoki (1987) to propose a scheme of state space modeling of time series based on the SVD. It is indeed true that the *theoretical* covariance matrix of a stationary time series does have a finite rank; in particular, for an ARMA model, the autocovariance at lag greater than the moving average order satisfy the deterministic AR (Yule–Walker) equations. (See Section 6.3.1.) However, this does not mean that the *sample* covariance matrix, that we must necessarily use in practice, will have rank deficiency as the number of rows and columns is increased beyond the dimension of the state space. After all, the sample covariance, that is a random variable as an estimator, may at best satisfy a *stochastic* AR equation and hence the rank of the sample covariance matrix can readily exceed the dimension of the state space as illustrated in Section 3.2.2, so the Ho–Kalman algorithm or its variants do not apply. As rightly pointed out by Kalman, Falb, and Arbib (1969, p. 294), Ho–Kalman algorithm does not merely need but "*hinges*" on the rank deficiency of the pulse response matrix!

Moreover, as pointed out very early in time series literature by Kendall (1945), sample autocovariance is a very poor estimator of the theoretical autocovariance and modeling procedures based on sample autocovariance may have serious pitfalls even for stationary time series. Since the autocovariance does not exist for nonstationary series, they must be transformed to stationary series, which may further confound the autocovariance estimator. For this reason the DDS modeling proposed by Pandit (1973) is *not* based on sample autocovariance and can be applied directly to nonstationary series as well. The DDS modeling procedure only requires the computation of residual sum of squares for successively large n. This may be done by actually fitting the models, as illustrated in Section 2.5.4 (see Pandit and Wu, 1983, for many more examples), or indirectly via the QR factorization, as illustrated simply in Sections 3.2.1 and 3.2.2. The latter method is computationally more efficient, particularly for multivariate data.

Note that the calculation of singular values requires computation far beyond a single QR factorization in the DDS modeling procedure. The most efficient and stable SVD algorithm by Golub and Reinsch (1970) requires an application of QR factorization to obtain a bidiagonal matrix and then several iterations of QR algorithm to obtain the diagonal matrix of singular values. This additional computation may add so much noise to the already noisy covariance matrix that the determination of "small" singular values to be ignored as zero may well nigh be impossible.

To illustrate these difficulties in the use of SVD for determining the dimension of the state space from real data and to emphasize the ad hoc arbitrary decisions it entails, we will give two examples from Aoki (1987, pp. 179–180). For a single series of IBM stock prices, after selecting 5×5 as a large enough dimension for the covariance matrix (replacing the data matrix of Sections 3.2.1 and 3.2.2), the singular values are found to be 11.1, 10.8, 5.2, 3.4, and 2.5. Based on these, $n = 2$ is selected as the dimension of the state space. For two series of Canadian GNP and money data, selecting a 10×10 covariance matrix gives the singular values of $10^{-4} \times$ (4.56, 1.54, 1.43, 1.34, 1.03, 0.72, 0.65, 0.31, 0.16, and 0.01). Again $n = 2$ is selected as the dimension of the state space.

It is thus seen that for real-life data from stochastic systems, including deterministic systems with noise considered in the ERA algorithm for modal analysis discussed above, determination of the dimension of state space by SVD is at best difficult, subjective, ad hoc, and arbitrary. Therefore, the DDS modeling procedure stops at the QR factorization of the data matrix (illustrated in Sections 3.2.1 and 3.2.2) that yields the residual sum of squares on the diagonal of R and then uses the noise floor criteria for deterministic systems and F-test for stochastic systems supplemented by other checks such as residual autocorrelation or stabilization of natural frequency and damping ratio, as desired by the analyst.

SVD and Least Squares

When $m = n$ and the real matrix $\mathbf{A}$ is full rank, that is, $\text{rank}(\mathbf{A}) = n$, then we know that the solution of $\mathbf{Ax} = \mathbf{y}$ is obtained by the usual inverse $\mathbf{A}^{-1}$ as $\mathbf{x} = \mathbf{A}^{-1}\mathbf{y}$. If $m > n$, the problem $\mathbf{Ax} = \mathbf{y}$ is generally overdetermined, without an exact solution; then, as discussed in Section 3.3.3, we can get a least squares solution $\hat{\mathbf{x}} = (\mathbf{A}^T\mathbf{A})^{-1}\mathbf{A}^T\mathbf{y}$ minimizing $\|\mathbf{y} - \mathbf{Ax}\|^2$ if $\text{rank}(\mathbf{A}) = \text{rank}(\mathbf{A}^T\mathbf{A}) = n$ so that the usual inverse of $\mathbf{A}^T\mathbf{A}$ exists. What if the $\text{rank}(\mathbf{A}) = \text{rank}(\mathbf{A}^T\mathbf{A}) < n$? Then the usual inverse of $\mathbf{A}$ or that of $\mathbf{A}^T\mathbf{A}$ does not exist and hence neither the usual linear equation solution, nor the usual linear least squares solution exists. In practice this means that these usual solutions cause numerical difficulties and consequent loss of accuracy when the matrices are nearly rank deficient or ill-conditioned. It is for such (nearly) rank-deficient problems that the SVD is invaluable. (We assume real $\mathbf{A}$ for convenience, results for complex $\mathbf{A}$ are obtained by replacing ordinary transpose by conjugate transpose.)

Using the SVD, we can define the inverse of an arbitrary matrix $\mathbf{A}$ of rank r as

$$\mathbf{A}^\dagger = \mathbf{V}\,\text{diag}(\bar{\sigma}_1^1, \bar{\sigma}_2^1, \ldots, \sigma_r^{-1}, 0, 0, \ldots, 0)\,\mathbf{U}^T = \mathbf{V}_r\,\text{diag}(\bar{\sigma}_1^1, \bar{\sigma}_2^1, \ldots, \bar{\sigma}_r^1)\,\mathbf{U}_r^T$$

$$n \times m \quad n \times n \quad\quad n \times m \quad\quad\quad m \times m \quad n \times r \quad\quad r \times r \quad\quad r \times m$$

$$= (\mathbf{v}_1\mathbf{u}_1^T/\sigma_1) + (\mathbf{v}_2\mathbf{u}_2^T/\sigma_2) + \cdots + (\mathbf{v}_r\mathbf{u}_r^T/\sigma_r)$$

and then the solution of $\mathbf{Ax} = \mathbf{y}$ is

$$\hat{\mathbf{x}} = \mathbf{A}^\dagger \mathbf{y} = \sum_{i=1}^{r} (\mathbf{u}_i^T \mathbf{y}/\sigma_i)\mathbf{v}_i$$

which gives the minimized residual sum of squares (RSS) as

$$\text{RSS} = \sum_{i=r+1}^{m} (\mathbf{u}_i^T \mathbf{y})^2 = \|(\mathbf{I} - \mathbf{A}\mathbf{A}^\dagger)\mathbf{y}\|^2$$

Such $\mathbf{A}^\dagger$ is often called a pseudo-inverse, generalized inverse, or Moore–Penrose inverse. It is a simple but interesting exercise to verify that when $\mathbf{A}$ is square and nonsingular so that $m = n = \text{rank}(\mathbf{A})$, $\mathbf{A}^\dagger = \mathbf{A}^{-1}$ and RSS $= 0$, whereas if $m \geq n$

= rank(A), then $\mathbf{A}^\dagger = (\mathbf{A}^T\mathbf{A})^{-1}\mathbf{A}^T$, in agreement with the preceding solutions; the SVD and the pseudo-inverse generalize the problem to $m \geq n \geq \text{rank}(\mathbf{A})$. Even in the underdetermined case of rank(A) $\leq m < n$, a solution of consistent linear equations [rank(A) = rank(A|y)] $\mathbf{Ax} = \mathbf{y}$ may be obtained as $\mathbf{A}^\dagger \mathbf{y}$, with $\mathbf{A}^\dagger = \mathbf{A}^T(\mathbf{A}\mathbf{A}^T)^{-1}$ for the full rank case of rank(A) = m. Recall from Section 3.1.2 that whenever A is rank-deficient, that is, rank(A) $< p = \min(m, n)$, there are infinite number of ordinary ($m \leq n$) or least squares ($m > n$) solutions as long as the corresponding linear or normal equations are consistent; then $\mathbf{A}^\dagger \mathbf{y}$ provides a solution with minimum norm or the smallest squared length; the reader is referred to Golub and Van Loan (1987) for a more detailed discussion of matrix norms and sensitivity of least squares via the SVD.

The geometry of the SVD in the vector spaces associated with A throws some light on the discussion above. The first $r[= \text{rank}(\mathbf{A})]$ columns of U given by $\mathbf{U}_r = [\mathbf{u}_1, \mathbf{u}_2, \ldots, \mathbf{u}_r]$ form an orthonormal basis (o.n.b.) for the range of A, $R(\mathbf{A})$, or the column space of A; the last $(n - r)$ columns of V given by $[\mathbf{v}_{r+1}, \mathbf{v}_{r+2}, \ldots, \mathbf{v}_n]$ form an o.n.b. for the null space of A, $N(\mathbf{A})$. The remaining columns $[\mathbf{u}_{r+1}, \mathbf{u}_{r+2}, \ldots, \mathbf{u}_m]$ form an o.n.b. for $R(\mathbf{A})^\perp = N(\mathbf{A}^T)$, whereas $\mathbf{V}_r = [\mathbf{v}_1, \mathbf{v}_2, \ldots, \mathbf{v}_r]$ forms an o.n.b. for $N(\mathbf{A})^\perp = R(\mathbf{A}^T)$. Therefore, $\mathbf{A}\mathbf{A}^\dagger = \mathbf{U}_r \mathbf{U}_r^T$ and $\mathbf{A}^\dagger \mathbf{A} = \mathbf{V}_r \mathbf{V}_r^T$ are orthogonal projections onto $R(\mathbf{A})$ and $R(\mathbf{A}^T)$, respectively. The geometry of least squares can then be simply depicted as in Figure 3.2 for a general case. It explains why the least squares estimate $\hat{\mathbf{x}}$ minimizes the residual sum of squares, RSS $= \|(\mathbf{I} - \mathbf{A}\mathbf{A}^\dagger)\mathbf{y}\|^2$, that is the squared length of the residual vector.

The numerical difficulties caused by very small singular values σ_i in inverting a near singular (near rank deficient) matrix are clear from the expression of $\mathbf{A}^\dagger$ that involves a weighted sum of r rank 1 matrices $\mathbf{v}_i \mathbf{u}_i^T$ with weights $1/\sigma_i$. These numerical difficulties and consequent loss of accuracy can be avoided in least squares estimation by proper use of the SVD. For example, r may be chosen such that (σ_r/σ_1) is the smallest ratio above the machine precision; then $\mathbf{A}^\dagger$ will not have the numerical difficulties of the usual least squares method.

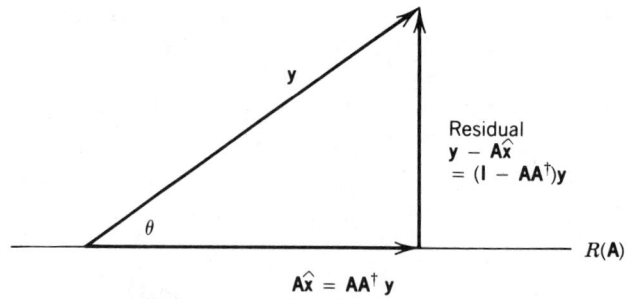

Figure 3.2 Geometry of least squares.

3.6 SPECTRAL/MODAL DECOMPOSITION OF SCALAR (UNIVARIATE) SYSTEMS

The spectral decomposition of the state matrix based on its eigenvalues and eigenvectors will now be illustrated for scalar or univariate systems. It will be seen that in this case the spectral representation yields the same results that are generally obtained by solving operator polynomials or transfer functions; eigenvalues are the roots of the polynomials or poles of the transfer function and the eigenvectors are the residues generally found by partial fractions.

3.6.1 Continuous Time Systems

To illustrate the implications of the spectral representation or decomposition of the linear transformation and its matrix, consider the nth order linear differential equation model:

$$(D^n + \alpha_{n-1} D^{n-1} + \cdots + \alpha_0) x(t) = (b_{n-1} D^{n-1} + \cdots + b_0) u(t)$$

which is a slightly general case of (2.6.16) with the addition of the $b_{n-1} D^{n-1}$ term. It has the state space representation

$$\dot{\mathbf{x}} = \mathbf{A}\mathbf{x} + \mathbf{B}\mathbf{u}$$

This equation involves two linear transformations, one multiplication by a matrix $\mathbf{A}$ that can be analyzed in a finite dimensional vector space, and the other, differentiation, that cannot be analyzed in a finite dimensional vector space. However, the differentiation does not pose any problem for us since we already know its solution:

$$x(t) = e^{\mathbf{A}t} \mathbf{x}(0) + \int_0^t e^{\mathbf{A}(t-s)} \mathbf{B}\mathbf{u}(s)\, ds$$

Therefore, spectral representation of $\mathbf{A}$ should provide us with the complete analysis of the system. This is indeed true as we now show.

The $n \times n$ companion matrix $\mathbf{A}$ is given by

$$\mathbf{A} = \begin{bmatrix} 0 & 1 & 0 & 0 & \cdots & 0 & 0 \\ 0 & 0 & 1 & 0 & \cdots & 0 & 0 \\ \vdots & \vdots & \vdots & \vdots & \vdots & \vdots & \vdots \\ 0 & 0 & 0 & 0 & \cdots & 0 & 1 \\ -\alpha_0 & -\alpha_1 & -\alpha_2 & -\alpha_3 & \cdots & -\alpha_{n-2} & -\alpha_{n-1} \end{bmatrix}$$

Omitting the first column and the last row of the matrix $(\mathbf{A} - \mu_i \mathbf{I})$ leaves an $(n-1) \times (n-1)$ nonsingular lower triangular matrix, showing that rank($\mathbf{A}$

$-\mu_i \mathbf{I}) \geq (n-1)$. On the other hand, its rank is always less than n. Therefore, rank$(\mathbf{A} - \mu_i \mathbf{I}) = (n-1)$ and by (3.1.2c)

$$q_i = n - \text{rank}(\mathbf{A} - \mu_i \mathbf{I}) = 1$$

Thus the degeneracy is simple and the matrix can be reduced to diagonal form for distinct eigenvalues and Jordan form for multiple eigenvalues.

It can be verified that the characteristic polynomial reduces to

$$|\mathbf{A} - \mu \mathbf{I}| = (-1)^n(\mu^n + \alpha_{n-1}\mu^{n-1} + \cdots + \alpha_0)$$
$$= (\mu_1 - \mu)(\mu_2 - \mu) \cdots (\mu_n - \mu)$$

When the characteristic roots or the eigenvalues are all distinct, the eigenvectors forming the columns of the modal matrix $\mathbf{M}$ and its inverse can be explicitly evaluated. Since the eigenvectors are linearly independent, the matrix $\boldsymbol{\mu}$ is diagonal.

$$\mathbf{AM} = \mathbf{M}\boldsymbol{\mu}$$

$$\begin{bmatrix} 0 & 1 & 0 & 0 & \cdots & 0 \\ 0 & 0 & 1 & 0 & \cdots & 0 \\ \vdots & \vdots & \vdots & \vdots & \vdots & \vdots \\ -\alpha_0 & -\alpha_1 & -\alpha_2 & -\alpha_3 & \cdots & -\alpha_{n-1} \end{bmatrix} \begin{bmatrix} m_{11} & m_{12} & \cdots & m_{1n} \\ m_{21} & m_{22} & \cdots & m_{2n} \\ \vdots & \vdots & \vdots & \vdots \\ m_{n1} & m_{n2} & \cdots & m_{nn} \end{bmatrix}$$

$$= \begin{bmatrix} m_{11} & m_{12} & \cdots & m_{1n} \\ m_{21} & m_{22} & \cdots & m_{2n} \\ \vdots & \vdots & \vdots & \vdots \\ m_{n1} & m_{n2} & \cdots & m_{nn} \end{bmatrix} \begin{bmatrix} \mu_1 & & & \\ & \mu_2 & & 0 \\ & 0 & \ddots & \\ & & & \mu_n \end{bmatrix}$$

that is,

$$m_{2i} = \mu_i m_{1i}$$
$$m_{3i} = \mu_i m_{2i}$$
$$\vdots \quad \vdots$$
$$m_{ni} = \mu_i m_{(n-1)i}, \qquad i = 1, 2, \ldots, n$$

Choosing $m_{1i} = 1$, we have

$$\mathbf{M} = \begin{bmatrix} 1 & 1 & \cdots & 1 \\ \mu_1 & \mu_2 & \cdots & \mu_n \\ \mu_1^2 & \mu_2^2 & \cdots & \mu_n^2 \\ \vdots & \vdots & \vdots & \vdots \\ \mu_1^{n-1} & \mu_2^{n-1} & \cdots & \mu_n^{n-1} \end{bmatrix}$$

This matrix is known as a *Vandermonde* matrix. As an example consider the one-degree-of-freedom mass spring dashpot system for which $n = 2$,

$$m\ddot{x} + c\dot{x} + kx = u$$

that is,

$$(D^2 + \alpha_1 D + \alpha_0)x = b_0 u, \quad \alpha_1 = \frac{c}{m}, \quad \alpha_0 = \frac{k}{m}, \quad b_0 = \frac{1}{m}$$

Then the characteristic polynomial is given by

$$\begin{vmatrix} -\mu & 1 \\ -\alpha_0 & -(\alpha_1 + \mu) \end{vmatrix} = \mu^2 + \alpha_1 \mu + \alpha_0 = (\mu - \mu_1)(\mu - \mu_2)$$

So the characteristic roots and the eigenvalues are

$$\mu_{1,2} = \frac{-\alpha_1}{2} \pm \frac{\sqrt{(\alpha_1^2 - 4\alpha_0)}}{2} = \frac{-c}{2m} \pm \left[\left(\frac{c}{2m} \right)^2 - \frac{k}{m} \right]^{1/2}$$

which satisfy

$$\mu_1 + \mu_2 = -\alpha_1 = \text{Trace}(\mathbf{A}) = \text{Trace}(\boldsymbol{\mu})$$

$$\mu_1 \mu_2 = \alpha_0 = |\mathbf{A}| = |\boldsymbol{\mu}|$$

Assuming $\mu_1 \neq \mu_2$, the modal matrix and its inverse are given by

$$\mathbf{M} = \begin{bmatrix} 1 & 1 \\ \mu_1 & \mu_2 \end{bmatrix} \quad \text{and} \quad \mathbf{M}^{-1} = \frac{1}{(\mu_2 - \mu_1)} \begin{bmatrix} \mu_2 & -1 \\ -\mu_1 & 1 \end{bmatrix}$$

An immediate advantage of the spectral representation $\mathbf{A} = \mathbf{M}\boldsymbol{\mu}\mathbf{M}^{-1}$ is in evaluating the matrix exponential $e^{\mathbf{A}t}$. Since it involves the powers of $\mathbf{A}$ in its definition, we first note that

$$\mathbf{A}^2 = \mathbf{M}\boldsymbol{\mu}\mathbf{M}^{-1}\mathbf{M}\boldsymbol{\mu}\mathbf{M}^{-1} = \mathbf{M}\boldsymbol{\mu}^2\mathbf{M}^{-1}$$

Similarly,

$$\mathbf{A}^3 = \mathbf{M}\boldsymbol{\mu}^3\mathbf{M}^{-1}$$

and generally

$$\mathbf{A}^n = \mathbf{M}\boldsymbol{\mu}^n\mathbf{M}^{-1}$$

Therefore,

$$e^{\mathbf{A}t} = \sum_{n=0}^{\infty} \mathbf{A}^n \frac{t^n}{n!} = \mathbf{M} \sum_{n=0}^{\infty} \frac{(\boldsymbol{\mu}t)^n}{n!} \mathbf{M}^{-1} = \mathbf{M}e^{\boldsymbol{\mu}t}\mathbf{M}^{-1}$$

When μ is diagonal, this derivation shows that $e^{\mu t}$ is a diagonal matrix with elements $e^{\mu_i t}$

$$e^{\mu t} = \text{diag}\{e^{\mu_i t}\}$$

Transient Solution or Deterministic Component

As a further example consider the mass-spring-dashpot system without any forcing function, that is, $u = 0$. Now

$$e^{At} = \frac{1}{(\mu_2 - \mu_1)} \begin{bmatrix} 1 & 1 \\ \mu_1 & \mu_2 \end{bmatrix} \begin{bmatrix} e^{\mu_1 t} & 0 \\ 0 & e^{\mu_2 t} \end{bmatrix} \begin{bmatrix} \mu_2 & -1 \\ -\mu_1 & 1 \end{bmatrix}$$

$$= \frac{1}{(\mu_2 - \mu_1)} \begin{bmatrix} \mu_2 e^{\mu_1 t} - \mu_1 e^{\mu_2 t} & e^{\mu_2 t} - e^{\mu_1 t} \\ \mu_1 \mu_2 (e^{\mu_1 t} - e^{\mu_2 t}) & \mu_2 e^{\mu_2 t} - \mu_1 e^{\mu_1 t} \end{bmatrix}$$

Hence the transient solution, or [when $u(t)$ is a random function] the deterministic component, is given by

$$\mathbf{x}(t) = e^{At} \mathbf{x}(0) = e^{At} \begin{bmatrix} x_0 \\ \dot{x}_0 \end{bmatrix}$$

$$= \frac{1}{(\mu_2 - \mu_1)} \begin{bmatrix} (\mu_2 x_0 - \dot{x}_0) e^{\mu_1 t} - (\mu_1 x_0 - \dot{x}_0) e^{\mu_2 t} \\ \mu_1 (\mu_2 x_0 - \dot{x}_0) e^{\mu_1 t} - \mu_2 (\mu_1 x_0 - \dot{x}_0) e^{\mu_2 t} \end{bmatrix}$$

The first row gives the position of the mass at time t and the second row gives the velocity.

Impulse Response

The impulse response of the system is obtained when the initial condition is zero and the input is an impulse function $\delta(t)$. The solution given above yields such an impulse response by setting

$$\mathbf{x}(0) = \mathbf{0}, \quad u(t) = \delta(t)$$

Since the convolution of a function with any derivative of $\delta(s)$ gives the value of the same derivative of the function at $s = 0$, we have, using (2.6.17) and keeping in mind that D operates on elements of e^{At}:

$$\mathbf{x}(t) = e^{At} \mathbf{b}, \quad \mathbf{B} = \mathbf{b} = [0, 0, \ldots, (b_0 + b_1 D + \cdots + b_{n-1} D^{n-1})]^T$$

The impulse response of the simplest first-order scalar system defined by

$$(D + \alpha_0) x = \delta(t)$$

is obtained by setting $A = -\alpha_0$ and $b = 1$ as

$$x(t) = e^{-\alpha_0 t}$$

which, as t increases, approaches zero for $\alpha_0 > 0$ and hence is asymptotically stable, approaches a finite constant 1 for $\alpha_0 = 0$ and hence is stable but not asymptotically stable, and finally, tends to infinity for $\alpha_0 < 0$ and hence is unstable.

The impulse response for the second-order system

$$\ddot{x} + \alpha_1 \dot{x} + \alpha_0 x = \delta(t)$$

is obtained by using $\mathbf{b} = \begin{bmatrix} 0 \\ 1 \end{bmatrix}$ as

$$G(t) = \frac{1}{(\mu_1 - \mu_2)}(e^{\mu_1 t} - e^{\mu_2 t})$$

which is generally called the Green's function of the differential operator.

For the mass-spring-dashpot system under consideration,

$$\ddot{x} + \alpha_1 \dot{x} + \alpha_0 x = \frac{1}{m}\delta(t)$$

the impulse response is obtained by using

$$\mathbf{b} = \begin{bmatrix} 0 \\ 1/m \end{bmatrix}$$

as

$$G(t) = \frac{1}{m(\mu_1 - \mu_2)}(e^{\mu_1 t} - e^{\mu_2 t})$$

It is thus seen that the matrix e^{At} is the crux of the impulse response and can give us the response in any output as a result of an impulse $\delta(t)$ in any input by a proper choice of the matrix $\mathbf{b}$. Thus the matrix exponential e^{At} is the proper matrix generalization of the scalar Green's function. Hence e^{At} is called the (matrix) Green's function or the fundamental matrix of the differential operator, since it is the key to the solution of the differential equation. It provides us with the response of the system, for arbitrary initial conditions and to inputs or forcing functions, by convoluting the given inputs with the Green's function as shown by the general solution.

Step Response

To illustrate, let us find the step response using the (matrix) Green's function. Step response is obtained by setting

$$\mathbf{x}(0) = \mathbf{0} \quad \text{and} \quad \mathbf{u}(t) = u(t) = s(t) = 1, \quad \text{for } t \geq 0$$
$$= 0, \quad \text{for } t < 0$$

and replacing the matrix **B** by the vector **b**. Then the general solution gives the step response

$$\mathbf{x}(t) = \int_0^t e^{\mathbf{A}(t-s)} \mathbf{b}\, ds = -\mathbf{A}^{-1} e^{\mathbf{A}(t-s)} \mathbf{b} \Big|_0^t$$
$$= \mathbf{A}^{-1}(e^{\mathbf{A}t} - \mathbf{I})\mathbf{b} = \mathbf{M}\boldsymbol{\mu}^{-1}(e^{\boldsymbol{\mu}t} - \mathbf{I})\mathbf{M}^{-1}\mathbf{b}$$

For the simplest first-order scalar case

$$(D + \alpha_0)x = u$$

$\mathbf{A} = -\alpha_0$ and $\mathbf{b} = 1$, and therefore we get

$$x(t) = \frac{1}{\alpha_0}(1 - e^{-\alpha_0 t})$$

which asymptotically approaches $1/\alpha_0$ (called the *gain* of the system) for the asymptotically stable case $\alpha_0 > 0$, becomes the ramp $x(t) = t$ for the stable but not asymptotically stable case $\alpha_0 = 0$, and exponentially tends to infinity for the unstable case $\alpha_0 < 0$, as sketched in Figure 3.3a.

For the mass-spring-dashpot system with distinct roots considered above, using $\mathbf{b} = (0, 1/m)^T$ we have

$$\mathbf{x}(t) = \begin{bmatrix} x(t) \\ \dot{x}(t) \end{bmatrix} = \begin{bmatrix} 1 & 1 \\ \mu_1 & \mu_2 \end{bmatrix} \begin{bmatrix} \frac{e^{\mu_1 t} - 1}{\mu_1} & 0 \\ 0 & \frac{e^{\mu_2 t} - 1}{\mu_2} \end{bmatrix} \begin{bmatrix} -1 \\ 1 \end{bmatrix} \frac{1}{m(\mu_2 - \mu_1)}$$

$$= \frac{1}{m(\mu_2 - \mu_1)} \begin{bmatrix} \frac{e^{\mu_2 t} - 1}{\mu_2} - \frac{e^{\mu_1 t} - 1}{\mu_1} \\ e^{\mu_2 t} - e^{\mu_1 t} \end{bmatrix}$$

Thus, in response to a unit step input, the output displacement is given by the first row, whereas the output velocity is given by the second row. The displace-

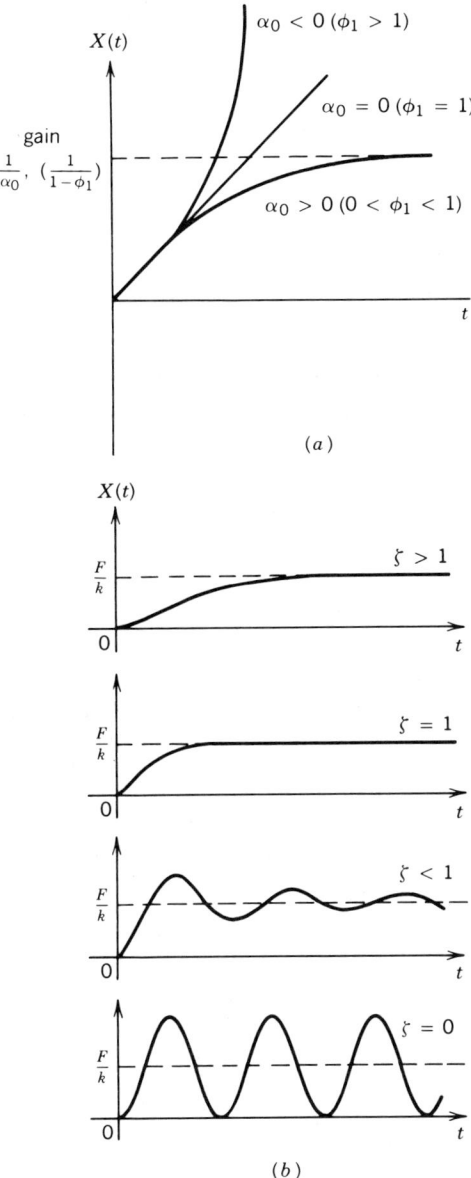

Figure 3.3 Step response of (a) first- and (b) second-order system.

ment response may also be alternatively written as

$$x(t) = \frac{1}{m\mu_1\mu_2}\left[\frac{\mu_1 e^{\mu_2 t} - \mu_2 e^{\mu_1 t}}{\mu_2 - \mu_1} + 1\right] = \frac{1}{k}\left[\frac{\mu_2 e^{\mu_1 t} - \mu_1 e^{\mu_2 t}}{\mu_1 - \mu_2} + 1\right]$$

so that the gain is now $1/k$. When an input force of constant magnitude F is applied, we can get the response by simply replacing $1/m$ by F/m as

$$x(t) = \frac{F}{k}\left[\frac{\mu_2 e^{\mu_1 t} - \mu_1 e^{\mu_2 t}}{\mu_1 - \mu_2} + 1\right]$$

This response for the various values of the damping ratio $\zeta = c/2\sqrt{(km)}$ is sketched in Figure 3.3b.

Determinant and Inverse of the Vandermonde Matrix

To extend these results to an arbitrary nth order scalar differential equation, we need the inverse of the Vandermonde matrix $\mathbf{M}$ given above. The inverse is obtained by dividing the cofactor matrix by the determinant. The determinant of the Vandermonde matrix can be easily verified to be

$$|\mathbf{M}| = \prod_{i>j}^{n}(\mu_i - \mu_j)$$

The cofactor elements of the Vandermonde matrix then can be evaluated as

$$C_{1j} = (-1)^{j+1}\begin{vmatrix} 1 & 1 & \cdots & 1 & 1 & \cdots & 1 \\ \mu_1 & \mu_2 & \cdots & \mu_{j-1} & \mu_{j+1} & \cdots & \mu_n \\ \mu_1^2 & \mu_2^2 & \cdots & \mu_{j-1}^2 & \mu_{j+1}^2 & \cdots & \mu_n^2 \\ \vdots & \vdots & \vdots & \vdots & \vdots & \vdots & \vdots \\ \mu_1^{n-2} & \mu_2^{n-2} & \cdots & \mu_{j-1}^{n-2} & \mu_{j+1}^{n-2} & \cdots & \mu_n^{n-2} \end{vmatrix}$$

$$\cdot \begin{vmatrix} \mu_1 & & & & & & \\ & \mu_2 & & & 0 & & \\ & & \ddots & & & & \\ & & & \mu_{j-1} & & & \\ & 0 & & & \mu_{j+1} & & \\ & & & & & \ddots & \\ & & & & & & \mu_n \end{vmatrix}$$

$$= (-1)^{j+1}\prod_{\substack{i=1 \\ i\neq j}}^{n}\mu_i \prod_{\substack{i>k \\ i,k\neq j}}^{n}(\mu_i - \mu_k)$$

and in general

$$C_{ij} = (-1)^{i+j} \begin{vmatrix} 1 & 1 & \cdots & 1 & 1 & \cdots & 1 \\ \mu_1 & \mu_2 & \cdots & \mu_{j-1} & \mu_{j+1} & \cdots & \mu_n \\ \vdots & \vdots & \cdots & \vdots & \vdots & \cdots & \vdots \\ \mu_1^{i-1} & \mu_2^{i-1} & \cdots & \mu_{j-1}^{i-1} & \mu_{j+1}^{i-1} & \cdots & \mu_n^{i-1} \\ \mu_1^{i+1} & \mu_2^{i+1} & \cdots & \mu_{j-1}^{i+1} & \mu_{j+1}^{i+1} & \cdots & \mu_n^{i+1} \\ \vdots & \vdots & \cdots & \vdots & \vdots & \cdots & \vdots \\ \mu_1^{n-1} & \mu_2^{n-1} & \cdots & \mu_{j-1}^{n-1} & \mu_{j+1}^{n-1} & \cdots & \mu_n^{n-1} \end{vmatrix}$$

$$= (-1)^{i+j} \prod_{\substack{l > k \\ l,k \neq j}}^{n} (\mu_l - \mu_k) \sum_{\substack{i1 < i2 < \cdots < i(n-i) \\ \neq j}}^{n} \mu_{i1} \mu_{i2} \cdots \mu_{i(n-i)}$$

Therefore, denoting $\mathbf{M}^{-1} = \{m^{ij}\}$

$$m^{i1} = (-1)^{i+1} \prod_{\substack{j=1 \\ j \neq i}}^{n} \mu_j \bigg/ \prod_{k=1}^{i-1} (\mu_i - \mu_{i-k}) \prod_{k=1}^{n-i} (\mu_{i+k} - \mu_i)$$

$$m^{in} = (-1)^{i+n} \bigg/ \prod_{k=1}^{i-1} (\mu_i - \mu_{i-k}) \prod_{k=1}^{n-i} (\mu_{i+k} - \mu_i)$$

$$m^{ij} = (-1)^{j+i} \sum_{\substack{i1 < i2 < \cdots < i(n-j) \\ \neq i}}^{n} \mu_{i1} \mu_{i2} \cdots \mu_{i(n-j)} \bigg/ \prod_{k=1}^{i-1} (\mu_i - \mu_{i-k}) \prod_{k=1}^{n-i} (\mu_{i+k} - \mu_i)$$

where a summation or product that runs from high to low indices or contains terms with indices that are not applicable is omitted. Computationally efficient algorithms for computing the term $\mathbf{M}^{-1}\mathbf{b}$ in n^2 flops with high precision are available; see for example Golub and Van Loan (1987).

Solution of the nth-Order Scalar System

We can now write down explicitly the solution of the nth-order scalar system given at the beginning of this section. Denoting the right-hand side of this equation by $bu(t)$ for brevity, that is,

$$bu(t) = (b_{n-1}D^{n-1} + \cdots + b_0)u(t)$$

we have

$$\mathbf{Bu} = (0, 0, \ldots, bu)^T$$

and therefore the solution of the state equation becomes

$$x(t) = Me^{\mu t}M^{-1}x(0) + \int_0^t Me^{\mu(t-s)}M^{-1}Bu(s)\,ds$$

$$= \begin{bmatrix} e^{\mu_1 t} & e^{\mu_2 t} & \cdots & e^{\mu_n t} \\ \mu_1 e^{\mu_1 t} & \mu_2 e^{\mu_2 t} & \cdots & \mu_n e^{\mu_n t} \\ \vdots & \vdots & \vdots & \vdots \\ \mu_1^{n-1} e^{\mu_1 t} & \mu_2^{n-1} e^{\mu_2 t} & \cdots & \mu_n^{n-1} e^{\mu_n t} \end{bmatrix} \begin{bmatrix} \sum_{j=1}^n m^{1j} D^{(j-1)} x(0) \\ \sum_{j=1}^n m^{2j} D^{(j-1)} x(0) \\ \vdots \\ \sum_{j=1}^n m^{nj} D^{(j-1)} x(0) \end{bmatrix}$$

$$+ \int_0^t \begin{bmatrix} e^{\mu_1(t-s)} & e^{\mu_2(t-s)} & \cdots & e^{\mu_n(t-s)} \\ \mu_1 e^{\mu_1(t-s)} & \mu_2 e^{\mu_2(t-s)} & \cdots & \mu_n e^{\mu_n(t-s)} \\ \vdots & \vdots & \vdots & \vdots \\ \mu_1^{n-1} e^{\mu_1(t-s)} & \mu_2^{n-1} e^{\mu_2(t-s)} & \cdots & \mu_n^{n-1} e^{\mu_n(t-s)} \end{bmatrix} \begin{bmatrix} m^{1n} bu(s) \\ m^{2n} bu(s) \\ \vdots \\ m^{nn} bu(s) \end{bmatrix} ds$$

It is now clear that the displacement is given by the first row of this solution and can be succinctly written as

$$x(t) = \sum_{i=1}^n e^{\mu_i t} \left[\sum_{j=1}^n m^{ij} D^{(j-1)} x(0) + \int_0^t e^{-\mu_i s} m^{in} bu(s)\,ds \right]$$

Note that this provides us not only with the solution, but also with the complete analysis or spectral decomposition of the output $x(t)$. The term in the bracket provides us with the contribution due to the eigenvalue μ_i, the first sum resulting from the initial conditions and the second term arising from the accumulated or integrated result of the input or the forcing function applied over time. The jth element m^{ij} of the corresponding column vector $\mathbf{m}^i$ of $\mathbf{M}^{-1}$ determines the weight given to the $(j-1)$st derivative of x evaluated at the initial time 0 for $j = 1, 2, \ldots, n$, whereas m^{in}, the last element of the column vector $\mathbf{m}^i$, controls the weight given in the output to the convolution of the "eigenfunction" with the input.

If μ_i is complex then its complex conjugate must be one of the eigenvalues since α_i's are real. If we let μ_i and μ_{i+1} be a complex conjugate pair of eigenvalues, their contributions can be combined and the sum expressed as a damped sine with amplitude and phase. Then each pair of eigenvalues yields a mode of correspondingly damped vibration and the amplitude of the mode provides an indication of the importance of the particular frequency amongst all other frequencies. Such modal analysis is useful not only in mechanical systems in which the vibration has a clear physical connotation, but also in other systems which are not mechanical but in which the frequencies or periodicities can be associated with tangible phenomena, such as seasonalities in business and economic systems. The analysis can be readily extended to discrete systems modeled by the DDS approach outlined in Chapter 2 and used with similar interpretation as discussed in the next section.

Each component in the spectral decomposition is a solution of a first-order differential equation. In fact, if we denote each component by $x_i(t)$, then it is the solution of the nonhomogeneous differential equation

$$\dot{x}_i(t) = \mu_i x_i(t) + m^{in} bu(t), \quad \text{given} \quad x_i(0) = \sum_{j=1}^{n} m^{ij} D^{(j-1)} x(0)$$

$$i = 1, 2, \ldots, n.$$

Thus the spectral decomposition "uncouples" the system by reducing an nth-order differential equation to n simple first-order equations.

3.6.2 Discrete Time Systems

We can find the spectral decomposition of the corresponding discrete time system governed by the ARMA($n, n-1$) model

$$(1 - \phi_1 B - \phi_2 B^2 - \cdots - \phi_n B^n) X_t = (1 - \theta_1 B - \theta_2 B^2 - \cdots - \theta_{n-1} B^{n-1}) a_t$$

where B represents the backshift operator, $BX_t = X_{t-1}$, $B^n X_t = X_{t-n}$, in a manner completely analogous to the development for the continuous time systems given in Section 3.6.1. Therefore, we only give the key results.

The state equation is

$$\mathbf{X}_t = \boldsymbol{\phi} \mathbf{X}_{t-1} + \boldsymbol{\theta} \mathbf{a}_t, \quad \mathbf{a}_t = 0 \quad \text{for} \quad t < 0, \quad \text{and} \quad \mathbf{X}_t = \mathbf{X}_0 \quad \text{at} \quad t = 0$$

with solution

$$\mathbf{X}_t = \boldsymbol{\phi}^t \mathbf{X}_0 + \sum_{j=0}^{t} \boldsymbol{\phi}^j \boldsymbol{\theta} \mathbf{a}_{t-j}$$

where $\mathbf{X}_t = [X_t, X_{t-1}, \ldots, X_{t-n+1}]^T$, $\mathbf{a}_t = [a_t, a_{t-1}, \ldots, a_{t-n+1}]^T$, and

$$\boldsymbol{\phi} = \begin{bmatrix} \phi_1 & \phi_2 & \phi_3 & \cdots & \phi_{n-1} & \phi_n \\ 1 & 0 & 0 & \cdots & 0 & 0 \\ 0 & 1 & 0 & \cdots & 0 & 0 \\ \vdots & \vdots & \vdots & \vdots & \vdots & \vdots \\ 0 & 0 & 0 & \cdots & 1 & 0 \end{bmatrix}$$

FUNDAMENTALS OF DATA DEPENDENT SYSTEM ANALYSIS

$$\theta = \begin{bmatrix} 1 & -\theta_1 & -\theta_2 & -\theta_3 & \cdots & -\theta_{n-1} \\ 0 & 0 & 0 & 0 & \cdots & 0 \\ 0 & 0 & 0 & 0 & \cdots & 0 \\ \vdots & \vdots & \vdots & \vdots & \vdots & \vdots \\ 0 & 0 & 0 & 0 & \cdots & 0 \end{bmatrix}$$

Note that we have allowed $\mathbf{a}_0$ to be nonzero to permit step and impulse inputs starting at zero. Hence the upper limit in the sum is t rather than the $t-1$ used in Chapter 2.

The characteristic polynomial now takes the form

$$|\phi - \lambda \mathbf{I}| = (-1)^n (\lambda^n - \phi_1 \lambda^{n-1} - \cdots - \phi_n)$$
$$= (\lambda_1 - \lambda)(\lambda_2 - \lambda) \cdots (\lambda_n - \lambda)$$

and the spectral representation of the matrix

$$\phi = \mathbf{L} \lambda \mathbf{L}^{-1}$$

where λ is the diagonal matrix of distinct eigenvalues $\lambda = \{\lambda_i\}$ and

$$\mathbf{L} = \begin{bmatrix} \lambda_1^{n-1} & \lambda_2^{n-1} & \cdots & \lambda_n^{n-1} \\ \lambda_1^{n-2} & \lambda_2^{n-2} & \cdots & \lambda_n^{n-2} \\ \lambda_1^{n-3} & \lambda_2^{n-3} & \cdots & \lambda_n^{n-3} \\ \vdots & \vdots & \vdots & \vdots \\ 1 & 1 & \cdots & 1 \end{bmatrix}$$

which is also a Vandermonde matrix with the determinant given by

$$|\mathbf{L}| = \prod_{i<j}^{n} (\lambda_i - \lambda_j)$$

Now the cofactors can be found as before and the inverse denoted by

$$\mathbf{L}^{-1} = \{\ell^{ij}\}$$

is given by

$$\ell^{in} = (-1)^{i+n} \prod_{\substack{j=1 \\ j \neq i}}^{n} \lambda_j \bigg/ \prod_{k=1}^{n-i} (\lambda_i - \lambda_{i+k}) \prod_{k=1}^{i-1} (\lambda_{i-k} - \lambda_i)$$

$$\ell^{il} = (-1)^{i+1} \bigg/ \prod_{k=1}^{n-i} (\lambda_i - \lambda_{i+k}) \prod_{k=1}^{i-1} (\lambda_{i-k} - \lambda_i)$$

$$\ell^{ij} = (-1)^{j+i} \sum_{\substack{i1 < i2 < \ldots < i(j-1) \\ \neq i}} \lambda_{i1} \lambda_{i2} \ldots \lambda_{i(j-1)} \bigg/ \prod_{k=1}^{n-i} (\lambda_i - \lambda_{i+k})$$

$$\times \prod_{k=1}^{i-1} (\lambda_{i-k} - \lambda_i)$$

Hence the solution of the ARMA$(n, n-1)$ difference equation with the given initial values $X_0, X_{-1}, \ldots, X_{-n+1}$, and the forcing function a_t is given by

$$X_t = \sum_{i=1}^{n} \left[\lambda_i^{n+t-1} \sum_{k=1}^{n} \ell^{ik} X_{1-k} \right] + \left[\sum_{j=0}^{t} \lambda_i^{n+j-1} \ell^{i1} \left(a_{t-j} - \sum_{k=1}^{n-1} \theta_k a_{t-j-k} \right) \right]$$

Note the formal similarity of this spectral decomposition of X_t to that of $x(t)$ given at the end of the preceding Section 3.6.1. In fact the last term in the parentheses may be denoted by θa_t in analogy with the term $bu(t)$, both being the right-hand side of the corresponding difference and differential equations. Hence all the remarks about the decomposition made at the end of Section 3.6.1 apply to this discrete decomposition as well. The only exception is that when a_t is stochastic, expressing the spectral decomposition of X_t as a sum of n random variables X_{it}, each with an AR(1) model with $\phi_1 = \lambda_i$, is conceptually simple, but computationally tedious in establishing the parametric relations. (Why?)

To illustrate, let us consider the case of the ARMA$(n, n-1)$ model obtained by setting $n = 2$. Then the characteristic polynomial is

$$\left\| \begin{bmatrix} \phi_1 & \phi_2 \\ 1 & 0 \end{bmatrix} - \begin{bmatrix} \lambda & 0 \\ 0 & \lambda \end{bmatrix} \right\| = \lambda^2 - \phi_1 \lambda - \phi_2 = (\lambda - \lambda_1)(\lambda - \lambda_2)$$

so that the eigenvalues are the characteristic roots of the autoregressive polynomial

$$\lambda^2 - \phi_1 \lambda - \phi_2 = 0$$

and satisfy

$$\phi_1 = \lambda_1 + \lambda_2$$
$$\phi_2 = -\lambda_1 \lambda_2$$

The elements of $\mathbf{L}^{-1}$ are obtained as

$$\ell^{11} = 1/(\lambda_1 - \lambda_2) \qquad \ell^{12} = -\lambda_2/(\lambda_1 - \lambda_2)$$
$$\ell^{21} = -1/(\lambda_1 - \lambda_2) \qquad \ell^{22} = \lambda_1/(\lambda_1 - \lambda_2)$$

which can easily be verified to be the inverse of the 2 × 2 **L** matrix. The matrix Green's function of the difference (autoregressive) operator is found to be

$$\phi^j = \mathbf{L}\lambda^j\mathbf{L}^{-1}$$

$$= \frac{1}{(\lambda_1 - \lambda_2)}\begin{bmatrix} \lambda_1 & \lambda_2 \\ 1 & 1 \end{bmatrix}\begin{bmatrix} \lambda_1^j & 0 \\ 0 & \lambda_2^j \end{bmatrix}\begin{bmatrix} 1 & -\lambda_2 \\ -1 & \lambda_1 \end{bmatrix}$$

$$= \frac{1}{(\lambda_1 - \lambda_2)}\begin{bmatrix} \lambda_1^{j+1} - \lambda_2^{j+1} & -\lambda_2\lambda_1^{j+1} + \lambda_1\lambda_2^{j+1} \\ \lambda_1^j - \lambda_2^j & -\lambda_2\lambda_1^j + \lambda_1\lambda_2^j \end{bmatrix}$$

$$= \begin{bmatrix} G'_j & \phi_2 G'_{j-1} \\ G'_{j-1} & \phi_2 G'_{j-2} \end{bmatrix}$$

where G'_j denotes the scalar Green's function (given by the top left element) of the AR(2) model for which $\theta_1 = 0$, that is,

$$\theta = \begin{bmatrix} 1 & 0 \\ 0 & 0 \end{bmatrix}$$

Now the transient response or the deterministic component [of both AR(2) and ARMA(2,1) models since it depends only on the left-hand side autoregressive difference operator] can be written as

$$\mathbf{X}_t = \begin{bmatrix} X_t \\ X_{t-1} \end{bmatrix} = \phi^t\mathbf{X}_0 = \phi^t\begin{bmatrix} X_0 \\ X_{-1} \end{bmatrix} = \begin{bmatrix} G'_t X_0 + \phi_2 G'_{t-1} X_{-1} \\ G'_{t-1} X_0 + \phi_2 G'_{t-2} X_{-1} \end{bmatrix}$$

The forced response or the nonhomogeneous stochastic component for the ARMA(2,1) model becomes

$$\begin{bmatrix} X_t \\ X_{t-1} \end{bmatrix} = \sum_{j=0}^{t} \phi^j \theta \mathbf{a}_{t-j} = \sum_{j=0}^{t} \phi^j \begin{bmatrix} a_{t-j} - \theta_1 a_{t-j-1} \\ 0 \end{bmatrix}$$

Hence

$$X_t = \sum_{j=0}^{t} G'_j(a_{t-j} - \theta_1 a_{t-j-1})$$

$$= G'_0 a_t + \sum_{j=1}^{t} (G'_j - \theta_1 G'_{j-1})a_{t-j} - \theta_1 G'_{t-1} a_{-1}$$

$$= \sum_{j=0}^{t} G_j a_{t-j}$$

where we have used $a_{-1} = 0$ and defined

$$G_j = G'_j - \theta_1 G'_{j-1} = [(\lambda_1 - \theta_1)\lambda_1^j - (\lambda_2 - \theta_1)\lambda_2^j]/(\lambda_1 - \lambda_2)$$

as the scalar Green's function of the ARMA(2,1) model. Note that $G'_0 = G_0 = 1$. It is left as an exercise to show that both these scalar Green's functions, G'_j of the AR(2) and G_j of the ARMA(2,1) models, can be directly obtained from the general solution by using the discrete impulse input, that is, $\mathbf{X}_0 = \mathbf{0}$ and $\mathbf{a}_t = (\delta_t, \delta_{t-1})^T$, similar to the continuous time case. Similarly, the impulse response for the general ARMA(n, $n-1$) case can be obtained from

$$G_t = G'_t - \theta_1 G'_{t-1} - \theta_2 G'_{t-2} - \cdots - \theta_{n-1} G'_{t-n+1}$$

where G'_t is the Green's function of the AR(n) model given by the top left element of the $n \times n$ matrix ϕ^t. The explicit closed-form expression for an ARMA(n,m) model with $m < n$ can be readily obtained from this and will be found in Section 6.1.5.

Moreover, since the transient response is of the form $\phi^t \mathbf{X}_0$, it is clear that the impulse response can also be obtained from the transient response by a suitable choice of $\mathbf{X}_0$. This is equivalent to replacing a nonhomogeneous differential equation by a homogeneous equation with suitable initial conditions. Thus the AR(2) and ARMA(2,1) impulse responses can be found by using the transient response with initial conditions $X_0 = 1$, $X_{-1} = 0$, and $X_0 = 1$, $X_{-1} = -\theta_1/\phi_2$, respectively. Finding such initial conditions for the general ARMA($n,n-1$) case is left as a nontrivial exercise.

The step response can be obtained by using $a_t \equiv 1$, $t \geq 0$, and may be written in explicit form for the distinct eigenvalue case as

$$\mathbf{X}_t = \sum_{j=0}^{t-(n-1)} \phi^j \theta(1, 1, \ldots, 1)^T + \sum_{j=t-(n-2)}^{t} \phi^j \theta[\overbrace{1, 1, \ldots, 1}^{t-j+1}, \overbrace{0, \ldots, 0}^{n-t+j-1}]^T$$

$$= \sum_{j=0}^{t-n+1} \phi^j [(1 - \theta_1 - \theta_2 - \cdots - \theta_{n-1}), 0, 0, \ldots, 0]^T$$

$$+ \sum_{i=0}^{n-2} \phi^{t-i} \theta[\overbrace{1, 1, \ldots, 1}^{i+1}, \overbrace{0, \ldots, 0}^{n-i-1}]^T$$

$$= \mathbf{L}\left\{\sum_{j=0}^{t-n+1} \lambda_i^j\right\} \mathbf{L}^{-1}[(1 - \theta_1 - \theta_2 - \cdots - \theta_{n-1}), 0, 0, \ldots, 0]^T$$

$$+ \sum_{i=0}^{n-2} \phi^{t-i} [(1 - \theta_1 - \theta_2 - \cdots - \theta_i), 0, 0, \ldots, 0]^T$$

$$= \mathbf{L}\left\{\frac{1 - \lambda_i^{t-n+2}}{1 - \lambda_i}\right\} \mathbf{L}^{-1}[(1 - \theta_1 - \theta_2 - \cdots - \theta_{n-1}), 0, 0, \ldots, 0]^T$$

$$+ \sum_{i=0}^{n-2} \phi^{t-i} [(1 - \theta_1 - \theta_2 - \cdots - \theta_i), 0, 0, \ldots, 0]^T$$

where the terms in the curly brackets with a single index denote a diagonal matrix as before. Now the desired step response can be taken from the top element of this column vector. It is left as an exercise to derive this using the explicit form of G_j, since the step response is simply $(G_0 + G_1 + \cdots + G_t)$.

To illustrate, consider the simplest first-order case for which $n = L = 1$, and $\lambda_1 = \phi_1$, hence the step response is simply given by

$$X_t = \left(\frac{1}{1 - \phi_1}\right)(1 - \phi_1^{t+1})$$

which is of the same form as $(1 - e^{-\alpha_0 t})/\alpha_0$ obtained for the corresponding continuous time first-order case in Section 3.6.1. As t increases, the step response asymptotically approaches the gain $1/(1 - \phi_1)$ for the asymptotically stable case $|\phi_1| < 1$, becomes a ramp $X_t = (1 + t)$ for the stable but not asymptotically stable case $\phi_1 = 1$, and exponentially tends to infinity for the unstable case $|\phi_1| > 1$, similar to the behavior sketched for the continuous time case in Figure 3.3.

Interestingly, this step response of a first-order system written in a slightly different form

$$X_t = \frac{1}{(1 - \phi_1)}(1)^t - \frac{\phi_1}{(1 - \phi_1)}(\phi_1)^t$$

shows that it is also the transient or homogeneous response of a *second*-order system with roots $\lambda_1 = 1$ and $\lambda_2 = \phi_1$. In fact, using the results for the transient response of a second-order system given above we see that it is the response of

$$X_t - (1 + \phi_1)X_{t-1} + \phi_1 X_{t-2} = 0, \quad \text{given } X_0 = 1 \text{ and } X_{-1} = 0$$

This has practical implications in DDS modeling. We have seen in Section 2.5.5 that when we estimate the mean, the AR(1) model

$$X_t - 0.9X_{t-1} = a_t$$

with zero mean is adequate for the data obtained from

$$X_t = 100(0.9)^t$$

Since the data is an exact deterministic function, the only stochastic part is that due to the computer operation and the computational process itself. The mean therefore refers to the mean of this computer noise and its estimate with 95% confidence interval of $0.13823 \times 10^{-8} \pm 0.17317 \times 10^{-7}$ suggests that it is practically zero. The estimate of σ_a given by 1.4×10^{-7} also confirms that the noise is of the order of computational accuracy.

What will happen if we require the model to take the average as the mean, as is usually done in modeling stationary data? This is essentially forcing the model with a step function since the actual data used for modeling will be

$$100(0.9)^t - \bar{X}, \quad \text{that is,} \quad 100(0.9)^t - \bar{X}(1)^t$$

Therefore we would need at least a second-order model with roots 1 and 0.9 to model this data. Actually the computer noise dynamics requires some more roots and we need ARMA(3,2) or higher-order models to get σ_a of the same order as above. However, all these models invariably contain the roots 1 and 0.9. Moreover, an AR(2) model fitted to the data gives $\phi_1 = 1.9$, $\phi_2 = -0.9$, and σ_a of the order of computer noise as expected. This agrees with the analysis in Section 2.8.

APPENDIX A3.1: Proof of $d[\mathscr{S}] + d[\mathscr{M}] = n$ (3.1.2)

To prove this, let $a_1, a_2, \ldots, a_k$ be independent, that is, $d[\mathscr{M}] = k$, then

$$a_j = a_{j1}a_1 + \cdots + a_{jk}a_k, \quad j = k+1, \ldots, n$$

Hence the vectors

$$\beta_1 = (a_{k+1,1}, \ldots, a_{k+1,k}, -1, 0, \ldots, 0)$$
$$\vdots$$
$$\beta_{n-k} = (a_{n,1}, \ldots, a_{n,k}, 0, 0, \ldots, -1)$$

are in the solution space because they satisfy the homogeneous equation (3.1.1) and they are independent because of the last $(n - k)$ elements. Do they form a basis of the solution space? To answer this, let $\mathbf{y} = (y_1, y_2, \ldots, y_n)$ be a solution and consider the augmented solution

$$\mathbf{y} + y_{k+1}\beta_1 + y_{k+2}\beta_2 + \cdots + y_n\beta_{n-k}$$

which is also a solution and is of the form $(u_1, u_2, \ldots, u_k, 0, 0, \ldots, 0)$ so that

$$u_1 a_1 + u_2 a_2 + \cdots + u_k a_k = \mathbf{0}$$

But this is possible only if $u_1 = u_2 = \cdots = u_k = 0$, so that the augmented solution is a null vector. Thus every solution $\mathbf{y}$ can be expressed in terms of β_i and, therefore, $\beta_1, \beta_2, \ldots, \beta_{n-k}$ is a basis of $\mathscr{S}$ so that

$$d[\mathscr{S}] = n - k$$

which is (3.1.2). Although we generally take a, β, $\mathbf{x}$ as column vectors, we have deliberately taken them here as row vectors following Rao (1965) to emphasize the general nature of vector space concepts.

4

MODAL DECOMPOSITION OF VECTOR (MULTIVARIATE) SYSTEMS

The modal decomposition illustrated in Chapter 3 for the scalar or univariate cases formally extends to the general multivariate case in a straightforward manner, as we will show in this chapter. However, the explicit evaluation of the modal matrix with the eigenvectors as its columns, and especially the evaluation of its inverse, is quite tedious except in a few special cases. Therefore, after developing the multivariate generalization of the spectral decomposition for both continuous- and discrete-time general state space models developed in Chapter 2, we will discuss a few of these simple special cases to provide insight into the results. Based on these results, a DDS approach to what is now commonly known as (experimental) modal analysis and spectrum analysis is developed and illustrated in subsequent chapters.

4.1 MODAL DECOMPOSITION OF CONTINUOUS-TIME SYSTEMS

Consider the continuous-time state space model

$$\dot{\mathbf{x}}(t) = \mathbf{A}\mathbf{x}(t) + \mathbf{B}(t)\mathbf{u}(t)$$

Assuming that $\mathbf{A}$ is diagonalizable, that is, it has a diagonal spectral matrix μ for convenience, and denoting the kth element of the column vector $\mathbf{B}(s)\mathbf{u}(s)$ by $b_k(s)$ temporarily, the solution can be written as [see (2.4.4)]

$$\mathbf{x}(t) = e^{\mathbf{A}t}\mathbf{x}(0) + \int_0^t e^{\mathbf{A}(t-s)}\mathbf{B}(s)\mathbf{u}(s)\,\mathrm{d}s$$

$$= \mathbf{M}e^{\mu t}\mathbf{M}^{-1}\mathbf{x}(0) + \int_0^t \mathbf{M}e^{\mu(t-s)}\mathbf{M}^{-1}\mathbf{B}(s)\mathbf{u}(s)\,\mathrm{d}s$$

$$= \{m_{ij}e^{\mu_j t}\}\left\{\sum_{k=1}^n m^{ik}x_k(0) + \int_0^t e^{-\mu_i s}m^{ik}b_k(s)\,\mathrm{d}s\right\}$$

$$= \sum_{j=1}^n e^{\mu_j t}\left\{\sum_{k=1}^n m_{ij}m^{jk}\left[x_k(0) + \int_0^t e^{-\mu_j s}b_k(s)\,\mathrm{d}s\right]\right\} \quad (4.1.1\mathrm{a})$$

$$= \sum_{j=1}^n \left\{\sum_{k=1}^n m_{ij}m^{jk}\left[x_k(0)e^{\mu_j t} + \int_0^t e^{\mu_j(t-s)}b_k(s)\,\mathrm{d}s\right]\right\} \quad (4.1.1\mathrm{b})$$

Also

$$\mathbf{x}(t) = \{\mathbf{m}_j\}\left\{\underbrace{e^{\mu_i t}\underbrace{\sum_{k=1}^n m^{ik}x_k(0)}_{q_i(0)} + \underbrace{\int_0^t e^{\mu_i(t-s)}\sum_{k=1}^n m^{ik}b_k(s)\,\mathrm{d}s}_{\beta_i(s)}}_{q_i(t)}\right\}$$

$$= \mathbf{M}\mathbf{q}(t)$$

$$= \mathbf{m}_1 q_1(t) + \mathbf{m}_2 q_2(t) + \cdots + \mathbf{m}_n q_n(t) \quad (4.1.1\mathrm{c})$$

where $q_i(t)$ is the solution of

$$\dot{q}_i = \mu_i q_i + \beta_i, \quad \text{given } q_i(t)|_{t=0} = q_i(0), \quad i = 1, 2, \ldots, n$$

and

$$\mathbf{q} = e^{\mu t}\mathbf{M}^{-1}\mathbf{x}(0) + \int_0^t e^{\mu(t-s)}\mathbf{M}^{-1}\mathbf{B}(s)\mathbf{u}(s)\,\mathrm{d}s = \mathbf{M}^{-1}\mathbf{x} \quad (4.1.1\mathrm{d})$$

When the eigenvalues are not distinct, μ is replaced by $\mathbf{J}$, which has diagonal blocks for distinct roots μ_i and Jordan blocks for repeated μ_i. The modal matrix $\mathbf{M}$ then consists of eigenvectors and generalized eigenvectors. We will not discuss such a multiple eigenvalue case, the interested reader is referred to Section 4.4 for the discrete case, and Mehta and Pandit (1989) if necessary.

Equations (4.1.1a and b) provide the most complete decomposition of the system, (4.1.1c and d) merely provide a different interpretation of the same results. Note that the brackets in (4.1.1a and b) represent $n \times 1$ column vectors, whereas in the preceding equation the first curly bracket represents an $n \times n$ matrix and the second bracket an $n \times 1$ column vector. The ith element of the column vector in (4.1.1a and b) gives the decomposition of the ith state variable $x_i(t)$, which, in the special case of scalar equations discussed in Section 3.6.1, happened to be the $(i-1)$st derivative of the single variable $x(t)$, but in general may be other variables and their derivatives, particularly in the multi-input-output case.

The first layer of the decomposition (4.1.1a and b) gives the contribution by the eigenvalue μ_j. For a given i, this contribution is represented by the jth term in the curly brackets. This ijth contribution further breaks down into two parts, one due to the initial conditions or the homogeneous response and the other due to the input forcing function or the nonhomogeneous response; when the forcing function is entirely stochastic, these components are the deterministic and stochastic components, respectively. These two parts are further broken down into components resulting from the initial condition in the kth state variable $x_k(0)$ and that due to the effective input forcing function in the kth state variable represented by $b_k(s)$. The term $b_k(s)$, however, represents the contribution from all the inputs $u_i(t)$ since it is the kth element of the column vector $\mathbf{B}(s)\mathbf{u}(s)$. [Although we use the constant $\mathbf{B}$ case throughout the book, $\mathbf{B}(t)$ is used as a function of t here only to emphasize the generality of results.]

Equation (4.1.1c) gives a slightly different interpretation to the same decomposition. The state vector $\mathbf{x}(t)$ described by a vector differential equation is now expressed in terms of uncoupled "generalized" coordinates $q_i(t)$, each of which is described by a simple scalar first-order differential equation with appropriate initial conditions following (4.1.1c). In fact, (4.1.1c) represents a change of basis from that of the columns of an identity matrix to q_i, and this is accomplished by the modal matrix $\mathbf{M}$ via the similarity transformation of $\mathbf{A} = \mathbf{M}\boldsymbol{\mu}\mathbf{M}^{-1}$, as discussed in Section 3.4.3. This new basis is obtained from the old basis by multiplication of $\mathbf{M}^{-1}$, as the relation $\mathbf{q} = \mathbf{M}^{-1}\mathbf{x}$ shows. The matrix $\mathbf{M}^{-1}$ can be thought of as the decoupling matrix. Thus the entire decomposition (4.1.1a) can be thought of as first decoupling the initial conditions and the inputs by $\mathbf{M}^{-1}$ referring them to the q_i basis, solving the simple first-order nonhomogeneous equation by means of $e^{\mu_i t}$, and then going back to the original basis by multiplying by $\mathbf{M}$.

Since $\mathbf{A}$ is real, the complex eigenvalues μ_i occur in conjugate pairs and so do the eigenvectors $\mathbf{m}_i$. Then the modal or spectral decomposition is more conveniently expressed in terms of sine–cosine functions rather than exponentials, as illustrated in the next section.

Impulse Response

Let the columns of $\mathbf{B}(t)$ be denoted by $\mathbf{b}_j(t)$, that is,

$$\mathbf{B} = \{\mathbf{b}_1, \mathbf{b}_2, \ldots, \mathbf{b}_n\}$$

If there is an impulse input *only* in the kth input, that is,

$$\mathbf{x}(0) = \mathbf{0}, \quad u_k(t) = \delta(t), \quad u_i(t) = 0 \quad \text{for} \quad i \neq k$$

then clearly, since the convolution with a delta function yields the value of a function at the zero argument of the delta function, the impulse response is given by

$$\mathbf{x}(t) = e^{\mathbf{A}t}\mathbf{b}_k(0) \tag{4.1.2}$$

Thus the matrix exponential $e^{\mathbf{A}t}$ is the crux of the impulse response. A constant $\mathbf{B}$ is the most important case, and agrees with the results obtained in Section 3.6.1.

Moreover, the matrix exponential itself provides a characteristic decomposition that is the key to the modal decomposition (4.1.1):

$$e^{\mathbf{A}t} = \mathbf{M}e^{\mu t}\mathbf{M}^{-1} = \{m_{ij}e^{\mu_j t}\}\{m^{ij}\} = \left\{\sum_{j=1}^{n} e^{\mu_j t}m_{ij}m^{jk}\right\} \tag{4.1.3}$$

A typical ikth term of this $n \times n$ matrix is the response of the ith state variable to a single impulse input in the kth variable. This response is then further decomposed into the contribution due to each eigenvalue μ_j: $e^{\mu_j t}$ giving the "shape" of the response and $m_{ij}m^{jk}$ its magnitude. For real μ_j the shape term $e^{\mu_j t}$ represents an exponential and the $m_{ij}m^{jk}$ term the response at $t = 0$. For a pair of complex conjugate eigenvalues μ_j, μ_{j+1}, the two shape terms $e^{\mu_j t}$ and $e^{\mu_{(j+1)} t}$ combine to give a damped sine wave, with its amplitude and phase determined from the corresponding $m_{ij}m^{jk}$ and $m_{i(j+1)}m^{(j+1)k}$ terms by elementary trigonometric calculations. The modal decomposition (4.1.1) then merely shows how these are added over the initial conditions and added and convoluted over the forcing functions from all the state variables.

Step Response
The step response is of interest mainly for the constant $\mathbf{B}$ case. If there is a step input *only* in the kth input, that is,

$$\mathbf{x}(0) = \mathbf{0}, \quad u_k(t) = s(t), \quad u_i(t) = 0 \quad \text{for} \quad i \neq k$$

then similar to Section 3.6.1, the step response is given by

$$\mathbf{x}(t) = \mathbf{M}\mu^{-1}(e^{\mu t} - \mathbf{I})\mathbf{M}^{-1}\mathbf{b}_k \tag{4.1.4}$$

Stability and Stationarity
The decompositions (4.1.1) as well as (4.1.3) show that when all the eigenvalues satisfy

$$\text{Re}(\mu_j) < 0, \quad j = 1, 2, \ldots, n \tag{4.1.5}$$

where Re denotes the real part, the system is asymptotically stable and for random $u(t)$ the stochastic process is stationary (see Sections 2.2.3). If $\text{Re}(\mu_j) > 0$ for at least one j, then the system is unstable and the stochastic process nonstationary. If $\text{Re}(\mu_j) = 0$ and μ_j are distinct, then the system is stable but not asymptotically, since the impulse response tends to a nonzero constant or is an

146   MODAL DECOMPOSITION OF VECTOR SYSTEMS

undamped sine wave, and the stochastic process nonstationary. If $\mu_j = 0$ for more than one j, then the system may be unstable and the stochastic process nonstationary.

4.2 ILLUSTRATIVE EXAMPLE: UNDAMPED TWO-DEGREE-OF-FREEDOM SYSTEM

Consider the two-mass system discussed in Section 2.6.5 without any damping so that the characteristics simplify to

$$\mathbf{x} = \begin{bmatrix} x_1 \\ x_2 \\ \dot{x}_1 \\ \dot{x}_2 \end{bmatrix}, \quad \mathbf{A} = \begin{bmatrix} 0 & 0 & 1 & 0 \\ 0 & 0 & 0 & 1 \\ -\dfrac{(k+k_1)}{m_1} & \dfrac{k}{m_1} & 0 & 0 \\ \dfrac{k}{m_2} & -\dfrac{(k+k_2)}{m_2} & 0 & 0 \end{bmatrix} \quad (4.2.1)$$

Then the characteristic equation is

$$|\mathbf{A} - \mu\mathbf{I}| = \begin{vmatrix} -\mu & 0 & 1 & 0 \\ 0 & -\mu & 0 & 1 \\ -\dfrac{(k+k_1)}{m_1} & \dfrac{k}{m_1} & -\mu & 0 \\ \dfrac{k}{m_2} & -\dfrac{(k+k_2)}{m_2} & 0 & -\mu \end{vmatrix}$$

$$= \begin{vmatrix} \mu^2 + \dfrac{(k+k_1)}{m_1} & -\dfrac{k}{m_1} \\ -\dfrac{k}{m_2} & \mu^2 + \dfrac{(k+k_2)}{m_2} \end{vmatrix} \quad \text{by partitioning of determinants*}$$

$$= \mu^4 + \mu^2 \left(\dfrac{k+k_1}{m_1} + \dfrac{k+k_2}{m_2} \right) + \dfrac{(k+k_1)(k+k_2) - k^2}{m_1 m_2}$$

$$= 0$$

*$\begin{vmatrix} \mathbf{B} & \mathbf{C} \\ \mathbf{D} & \mathbf{E} \end{vmatrix} = |\mathbf{E}||(\mathbf{B} - \mathbf{CE}^{-1}\mathbf{D})| = |\mathbf{B}||(\mathbf{E} - \mathbf{DB}^{-1}\mathbf{C})|$ provided the necessary inverses exist.

which gives the roots

$$\mu^2 = \frac{1}{2}\left\{-\left(\frac{k+k_1}{m_1}+\frac{k+k_2}{m_2}\right) \pm \left[\left(\frac{k+k_1}{m_1}-\frac{k+k_2}{m_2}\right)^2 + \frac{4k^2}{m_1 m_2}\right]^{1/2}\right\} \quad (4.2.2)$$

Since both values of μ^2 are negative, the eigenvalues are purely imaginary. This is expected since there is no damping and therefore the motion will be purely sinusoidal without damping. Hence we may abbreviate the four eigenvalues as

$$\mu = \pm j\omega_1, \pm j\omega_2, \quad j = \sqrt{-1} \quad (4.2.2a)$$

Although in general the frequencies ω_1 and ω_2 are the square roots of the negative of (4.2.2), in the special case of equal masses and equal springs they take a particularly simple form. For $k = k_1 = k_2$ and $m = m_1 = m_2$

$$\mu^2 = \frac{1}{2}\left[-\frac{4k}{m} \pm \frac{2k}{m}\right] = -\frac{k}{m}, -\frac{3k}{m}$$

and therefore the frequencies are simply

$$\omega_1 = \sqrt{k/m} \quad \text{and} \quad \omega_2 = \sqrt{3k/m} \quad (4.2.3)$$

Straightforward algebra using the definition of eigenvectors yields the modal matrix

$$\mathbf{M} = \begin{bmatrix} 1 & 1 & 1 & 1 \\ 1 & 1 & -1 & -1 \\ j\omega_1 & -j\omega_1 & j\omega_2 & -j\omega_2 \\ j\omega_1 & -j\omega_1 & -j\omega_2 & j\omega_2 \end{bmatrix} \quad (4.2.4)$$

for the equal mass and spring case; we will continue to use this simple special case for illustrative purposes in the remainder of this section. Somewhat tedious algebra shows that $|\mathbf{M}| = 16\omega_1\omega_2$; the inverse of the modal matrix is found to be

$$\mathbf{M}^{-1} = \frac{1}{4}\begin{bmatrix} 1 & 1 & -j/\omega_1 & -j/\omega_1 \\ 1 & 1 & j/\omega_1 & j/\omega_1 \\ 1 & -1 & -j/\omega_2 & j/\omega_2 \\ 1 & -1 & j/\omega_2 & -j/\omega_2 \end{bmatrix} \quad (4.2.5)$$

148 MODAL DECOMPOSITION OF VECTOR SYSTEMS

and the matrix exponential is given by

$$e^{At} = \begin{bmatrix} 1 & 1 & 1 & 1 \\ 1 & 1 & -1 & -1 \\ j\omega_1 & -j\omega_1 & j\omega_2 & -j\omega_2 \\ j\omega_1 & -j\omega_1 & -j\omega_2 & j\omega_2 \end{bmatrix} \begin{bmatrix} e^{j\omega_1 t} & & & \\ & e^{-j\omega_1 t} & & 0 \\ & & e^{j\omega_2 t} & \\ & 0 & & e^{-j\omega_2 t} \end{bmatrix}$$

$$\frac{1}{4} \begin{bmatrix} 1 & 1 & -j/\omega_1 & -j/\omega_1 \\ 1 & 1 & j/\omega_1 & j/\omega_1 \\ 1 & -1 & -j/\omega_2 & j/\omega_2 \\ 1 & -1 & j/\omega_2 & -j/\omega_2 \end{bmatrix}$$

$$= \frac{1}{2} \begin{bmatrix} 1 & 1 & 0 & 0 \\ 1 & 1 & 0 & 0 \\ 0 & 0 & 1 & 1 \\ 0 & 0 & 1 & 1 \end{bmatrix} \cos\omega_1 t + \frac{1}{2} \begin{bmatrix} 1 & -1 & 0 & 0 \\ -1 & 1 & 0 & 0 \\ 0 & 0 & 1 & -1 \\ 0 & 0 & -1 & 1 \end{bmatrix} \cos\omega_2 t$$

$$+ \frac{1}{2} \begin{bmatrix} 0 & 0 & 1/\omega_1 & 1/\omega_1 \\ 0 & 0 & 1/\omega_1 & 1/\omega_1 \\ -\omega_1 & -\omega_1 & 0 & 0 \\ -\omega_1 & -\omega_1 & 0 & 0 \end{bmatrix} \sin\omega_1 t$$

$$+ \frac{1}{2} \begin{bmatrix} 0 & 0 & 1/\omega_2 & -1/\omega_2 \\ 0 & 0 & -1/\omega_2 & 1/\omega_2 \\ -\omega_2 & \omega_2 & 0 & 0 \\ \omega_2 & -\omega_2 & 0 & 0 \end{bmatrix} \sin\omega_2 t \qquad (4.2.6)$$

The impulse response can be readily obtained using $e^{At}\mathbf{b}_3$. For an impulse in mass 1, $\mathbf{b}_3 = [0, 0, 1/m, 0]^T$, and therefore the response is found from the third column of the matrices in the matrix Green's function (4.2.6) as

$$x_1(t) = \frac{1}{2m}\left(\frac{1}{\omega_1}\sin\omega_1 t + \frac{1}{\omega_2}\sin\omega_2 t\right)$$

$$x_2(t) = \frac{1}{2m}\left(\frac{1}{\omega_1}\sin\omega_1 t - \frac{1}{\omega_2}\sin\omega_2 t\right)$$

UNDAMPED TWO-DEGREE-OF FREEDOM SYSTEM 149

$$\dot{x}_1(t) = \frac{1}{2m}\left(\cos\omega_1 t + \cos\omega_2 t\right)$$

$$\dot{x}_2(t) = \frac{1}{2m}\left(\cos\omega_1 t - \cos\omega_2 t\right) \qquad (4.2.7a)$$

Of course the velocities in the last two equations also follow directly by differentiating the displacements in the first two equations, rather than using (4.2.6) for that purpose. For an impulse in the second mass, we use $\mathbf{b}_4 = [0, 0, 0, 1/m]^T$ to get the impulse response from the fourth column of the matrices in (4.2.6) as

$$x_1(t) = \frac{1}{2m}\left(\frac{1}{\omega_1}\sin\omega_1 t - \frac{1}{\omega_2}\sin\omega_2 t\right)$$

$$x_2(t) = \frac{1}{2m}\left(\frac{1}{\omega_1}\sin\omega_1 t + \frac{1}{\omega_2}\sin\omega_2 t\right)$$

$$\dot{x}_1(t) = \frac{1}{2m}\left(\cos\omega_1 t - \cos\omega_2 t\right) \qquad (4.2.7b)$$

$$\dot{x}_2(t) = \frac{1}{2m}\left(\cos\omega_1 t + \cos\omega_2 t\right)$$

Similarly, the step response can be obtained using

$$\mathbf{x}(t) = \mathbf{M}\boldsymbol{\mu}^{-1}(e^{\mu t} - \mathbf{I})\mathbf{M}^{-1}\mathbf{b}_3$$

To illustrate, we will evaluate it for the step input in mass 1. Then $\mathbf{b}_3 = [0, 0, 1/m, 0]^T$, $\mathbf{M}^{-1}\mathbf{b}_3$ is obtained from the third column of $\mathbf{M}^{-1}$ in (4.2.5) as

$$\mathbf{M}^{-1}\mathbf{b}_3 = \frac{1}{4m}\begin{bmatrix} -j/\omega_1 \\ j/\omega_1 \\ -j/\omega_2 \\ j/\omega_2 \end{bmatrix}$$

$$\mathbf{M}\boldsymbol{\mu}^{-1}(e^{\mu t} - \mathbf{I}) =$$

$$\begin{bmatrix} \frac{1}{\mu_1}(e^{\mu_1 t} - 1) & \frac{1}{\mu_2}(e^{\mu_2 t} - 1) & \frac{1}{\mu_3}(e^{\mu_3 t} - 1) & \frac{1}{\mu_4}(e^{\mu_4 t} - 1) \\ \frac{1}{\mu_1}(e^{\mu_1 t} - 1) & \frac{1}{\mu_2}(e^{\mu_2 t} - 1) & -\frac{1}{\mu_3}(e^{\mu_3 t} - 1) & -\frac{1}{\mu_4}(e^{\mu_4 t} - 1) \\ (e^{\mu_1 t} - 1) & (e^{\mu_2 t} - 1) & (e^{\mu_3 t} - 1) & (e^{\mu_4 t} - 1) \\ (e^{\mu_1 t} - 1) & (e^{\mu_2 t} - 1) & -(e^{\mu_3 t} - 1) & -(e^{\mu_4 t} - 1) \end{bmatrix}$$

and keeping in mind that the four eigenvalues μ_i are given by (4.2.2a), the step response can be evaluated as

$$x_1(t) = \frac{1}{4m}\left[\frac{1-e^{j\omega_1 t}}{\omega_1^2} + \frac{1-e^{-j\omega_1 t}}{\omega_1^2} + \frac{1-e^{j\omega_2 t}}{\omega_2^2} + \frac{1-e^{-j\omega_2 t}}{\omega_2^2}\right]$$

$$= \frac{1}{2m\omega_1^2\omega_2^2}[\omega_1^2 + \omega_2^2 - \omega_2^2\cos\omega_1 t - \omega_1^2\cos\omega_2 t]$$

$$x_2(t) = \frac{1}{2m\omega_1^2\omega_2^2}[\omega_2^2 - \omega_1^2 - \omega_2^2\cos\omega_1 t + \omega_1^2\cos\omega_2 t] \quad (4.2.8)$$

$$\dot{x}_1(t) = \frac{1}{2m\omega_1\omega_2}[\omega_2\sin\omega_1 t + \omega_1\sin\omega_2 t]$$

$$\dot{x}_2(t) = \frac{1}{2m\omega_1\omega_2}[\omega_2\sin\omega_1 t - \omega_1\sin\omega_2 t]$$

Since the transient response is given by $e^{\mathbf{A}t}\mathbf{x}_0$, it is clear that the impulse response can be obtained as transient response to the initial condition $\mathbf{x}_0 = \mathbf{b}_3$. It is left as an interesting and nontrivial exercise to find the system (i.e., $\mathbf{A}$ matrix) and the initial conditions that will yield the step response given above as the transient response. These are matrix generalizations of the common practice of solving scalar nonhomogeneous differential equations by replacing them with homogeneous equations with suitably chosen initial conditions.

Taking a clue from the discussion at the end of Section 3.6.2, it is not hard to see that if the step response is expressed as the transient response, we will get an additional zero eigenvalue in the system. Similarly, forced response to a sinusoidal input can be expressed as the transient response of a system with a pair of complex conjugate eigenvalues added to the original set of eigenvalues. Thus, unless we know the input, or at least the eigenvalues from the input, it is impossible to determine whether the eigenvalues present in the model come from the system or from its input.

An important application of the impulse response is in ascertaining the relative importance of the various modes indexed by their frequencies present in the system response. Since the response to many other inputs can be readily synthesized from the impulse response, as illustrated above, the relative amplitudes of the various modes in the impulse response may be taken as typical or characteristic of the system in general. For example, in the current undamped system with two equal masses and two equal springs under consideration, it is seen that for mass 1 the amplitudes in the impulse response of two modes with frequencies $\omega_1 = \sqrt{k/m}$ and $\omega_2 = \sqrt{3k/m}$ are $(1/2m\omega_1)$ and $(1/2m\omega_2)$ respectively. Thus the lower frequency is dominant since its amplitude is $\sqrt{3}$ times that of the higher frequency. In the step response these amplitudes are $(1/2m\omega_1^2)$ and $(1/2m\omega_2^2)$, respectively, besides the step increase of magnitude $(1/2m\omega_1^2$

UNDAMPED TWO-DEGREE-OF FREEDOM SYSTEM   151

$+ 1/2m\omega_2^2$). Thus the low-frequency amplitude is now three times that of the high frequency.

Although we have called the matrix **M** as the modal matrix following the common terminology in the state space literature, in the vibration literature the term usually refers to the coefficient matrix of sine, with phase, if necessary, suitably normalized. Thus if we consider the impulse response or transient response to the initial conditions

$$x_1(0) = 1/m, \quad x_2(0) = 0, \quad \dot{x}_1(0) = \dot{x}_2(0) = 0$$

given by the first two equations of (4.2.7a):

$$\begin{bmatrix} x_1 \\ x_2 \end{bmatrix} = \frac{1}{2m\omega_1} \begin{bmatrix} 1 \\ 1 \end{bmatrix} \sin \omega_1 t + \frac{1}{2m\omega_2} \begin{bmatrix} 1 \\ -1 \end{bmatrix} \sin \omega_2 t$$

then the column vector $(1, 1)^T$ is called a *modal vector*. It represents the *relative amplitude* or the *mode shape* of the motions $x_1(t)$ and $x_2(t)$ at the frequency ω_1 with reference to $x_1(t)$. Similarly, the column vector $(1, -1)^T$ is the modal vector representing the relative amplitude or mode shape at frequency ω_2. The matrix

$$\begin{bmatrix} 1 & 1 \\ 1 & -1 \end{bmatrix}$$

is then called the modal matrix of the system. The modal matrix shows that at the lower frequency $\omega_1 = \sqrt{k/m}$ the two masses move together, whereas at the higher frequency $\omega_2 = \sqrt{3k/m}$ their motions oppose each other. This is also true of the impulse response to the impulse in the second mass or the transient response to the initial conditions

$$x_1(0) = 0, \quad x_2(0) = 1/m, \quad \dot{x}_1(0) = 0, \quad \dot{x}_2(0) = 0$$

for which the modal matrix is obtained from (4.2.7b) as

$$\begin{bmatrix} 1 & -1 \\ 1 & 1 \end{bmatrix}$$

The fact that the magnitudes of their relative amplitudes is unity, showing that at each frequency the amplitudes of $x_1(t)$ and $x_2(t)$ are equal, is of course a consequence of the masses and springs being equal. It is not hard to verify that if the initial conditions are changed, the amplitudes of the motions at the two frequencies would change, but not the relative amplitudes, and hence the modal matrix is unchanged. Thus the modal matrix is an inherent property of the system irrespective of the individual initial conditions, hence its importance in characterizing the system.

4.3 MODAL DECOMPOSITION OF DISCRETE-TIME SYSTEMS

Now consider the discrete-time state space model

$$\mathbf{X}_t = \boldsymbol{\phi}\mathbf{X}_{t-1} + \boldsymbol{\theta}\mathbf{a}_t, \qquad \mathbf{X}_t = \mathbf{X}_0 \quad \text{at} \quad t = 0 \qquad (4.3.1)$$

Using the notation of Section 3.6.2, its solution can be decomposed along the same lines as in Section 4.1 for continuous systems. Denoting the kth element of the column vector $\boldsymbol{\theta}\mathbf{a}_{t-j}$ by $\theta_{k(t-j)}$ for convenience,

$$\begin{aligned}
\mathbf{X}_t &= \boldsymbol{\phi}^t \mathbf{X}_0 + \sum_{j=0}^{t} \boldsymbol{\phi}^j \boldsymbol{\theta}\mathbf{a}_{t-j} \\
&= \mathbf{L}\boldsymbol{\lambda}^t\mathbf{L}^{-1}\mathbf{X}_0 + \sum_{j=0}^{t} \mathbf{L}\boldsymbol{\lambda}^j\mathbf{L}^{-1}\boldsymbol{\theta}\mathbf{a}_{t-j} \\
&= \sum_{j=1}^{n} \sum_{k=1}^{n} \left\{ \ell_{ij}\ell^{jk}\left[\lambda_j^t X_{k0} + \sum_{r=0}^{t} \lambda_j^r \theta_{k(t-r)}\right]\right\}
\end{aligned} \qquad (4.3.1\text{a})$$

where again we have allowed $\mathbf{a}_0$ to be nonzero to accommodate impulse and step inputs starting at zero; hence the upper limit in the sum is t rather than $t - 1$.

This form corresponds to (4.1.1a and b) of the continuous case and all the remarks made regarding them in Section 4.1 apply to (4.3.1a). We can also express the state vector in the alternate decoupled coordinates $\mathbf{q}_t = \{q_{it}\}$ similar to (4.1.1c) as

$$\mathbf{X}_t = \mathbf{L}\mathbf{q}_t \qquad (4.3.1\text{b})$$

where

$$\mathbf{q}_t = \boldsymbol{\lambda}^t\mathbf{L}^{-1}\mathbf{X}_0 + \sum_{j=0}^{t} \boldsymbol{\lambda}^j\mathbf{L}^{-1}\boldsymbol{\theta}\mathbf{a}_{t-j} \qquad (4.3.1\text{c})$$

and each q_{it} is the solution of a first-order difference equation or AR(1) model with the autoregressive parameter λ_i with a suitable initial condition and forcing function that can be easily obtained from (4.3.1a).

Denoting the columns of $\boldsymbol{\theta}$ by $\mathbf{b}_j$, that is,

$$\boldsymbol{\theta} = \{\mathbf{b}_1, \mathbf{b}_2, \ldots, \mathbf{b}_n\}$$

the impulse response to an impulse *only* in the kth input is given by

$$\mathbf{X}_t = \boldsymbol{\phi}^t \mathbf{b}_k \qquad (4.3.2)$$

Again the crux of the impulse response or transient response is the matrix exponential $\boldsymbol{\phi}^t$, which itself provides a characteristic decomposition that is the key to the modal decomposition (4.3.1):

$$\boldsymbol{\phi}^t = \mathbf{L}\boldsymbol{\lambda}^t\mathbf{L}^{-1} = \left\{\sum_{j=1}^{n} \lambda_j^t \ell_{ij} \ell^{jk}\right\} \qquad (4.3.3)$$

similar to (4.1.2) for the continuous case. Similarly, following the development in Section 3.6.2, the step response is given by

$$\mathbf{X}_t = \mathbf{L}\left\{\frac{1 - \lambda_i^{t-n+2}}{1 - \lambda_i}\right\}\mathbf{L}^{-1}\sum_{k=1}^{n} \mathbf{b}_k + \sum_{i=0}^{n-2} \boldsymbol{\phi}^{t-i}\sum_{k=1}^{n} s_{ik}\mathbf{b}_k \qquad (4.3.4)$$

where $s_{ik} = 0$ or 1 are the elements of the initial column vectors $\mathbf{s}_i = \mathbf{a}_i$.

Finally, the decomposition (4.3.1) as well as (4.3.3) show that when all the eigenvalues satisfy

$$|\lambda_j| < 1, \qquad j = 1, 2, \ldots, n \qquad (4.3.5)$$

the system is asymptotically stable and for random a_t the stochastic process is stationary. If $|\lambda_j| > 1$ for at least one j, then the system is unstable and the stochastic process nonstationary. If $|\lambda_j| = 1$ and λ_j are distinct, then the system is stable but not asymptotically, since the impulse response tends to a nonzero constant or is an undamped sine wave, and the stochastic process is nonstationary. If $|\lambda_j| = 1$ for more than one j, then the system may be unstable and the stochastic process nonstationary.

When the eigenvalues λ_i are complex they occur in conjugate pairs and so do the eigenvectors. Then the modal or spectral decomposition as well as impulse and step responses are more conveniently expressed in terms of sine–cosine functions rather than exponentials, as illustrated for the continuous-time case in the last section, and the discrete-time case in the next section.

4.4 The ARV(1) MODEL

The simplest illustration of the modal decomposition derived in the preceding section for general discrete systems can be given by considering the bivariate ARV(1) model

$$X_{1t} = \phi_{111}X_{1t-1} + \phi_{121}X_{2t-1} + a_{1t}$$
$$X_{2t} = \phi_{211}X_{1t-1} + \phi_{221}X_{2t-1} + a_{2t} \qquad (4.4.1)$$

so that

$$\mathbf{X}_t = \begin{bmatrix} X_{1t} \\ X_{2t} \end{bmatrix} \qquad \boldsymbol{\phi} = \begin{bmatrix} \phi_{111} & \phi_{121} \\ \phi_{211} & \phi_{221} \end{bmatrix} \qquad \mathbf{a}_t = \begin{bmatrix} a_{1t} \\ a_{2t} \end{bmatrix} \qquad (4.4.1a)$$

If we consider X_{1t} to be the input and X_{2t} to be the output for convenience, then feedback is present when $\phi_{121} \neq 0$ and absent when $\phi_{121} = 0$. Let us consider the simpler no-feedback case first. Then the characteristic equation is

$$|\phi - \lambda \mathbf{I}| = \begin{vmatrix} \phi_{111} - \lambda & 0 \\ \phi_{211} & \phi_{221} - \lambda \end{vmatrix} = (\phi_{111} - \lambda)(\phi_{221} - \lambda) = 0$$

so that the eigenvalues (assumed to be distinct for convenience) are simply

$$\lambda_1 = \phi_{111}, \qquad \lambda_2 = \phi_{221}$$

and the eigenvalue $\lambda_1 = \phi_{111}$ can be clearly associated with the input X_{1t} only and the second eigenvalue $\lambda_2 = \phi_{221}$ associated with the output X_{2t} only.

To find the modal matrix $\mathbf{L} = \{\ell_1, \ell_2\}$ we use the definition of eigenvector

that is,

$$\begin{bmatrix} \phi_{111} & 0 \\ \phi_{211} & \phi_{221} \end{bmatrix} \begin{bmatrix} \ell_{11} \\ \ell_{21} \end{bmatrix} = \phi_{111} \begin{bmatrix} \ell_{11} \\ \ell_{21} \end{bmatrix}$$

$$\phi_{111} \ell_{11} = \phi_{111} \ell_{11}$$
$$\phi_{211} \ell_{11} + \phi_{221} \ell_{21} = \phi_{111} \ell_{21}$$

Since the first equation is an identity, we can choose an arbitrary value for ℓ_{11}. However, if we choose 0, then ℓ_{21} must also be zero to satisfy the second equation and we get the trivial eigenvector $\mathbf{0}$. To avoid this we choose

$$\ell_{11} = 1$$

and then

$$\ell_{21} = \frac{\phi_{211}}{(\phi_{111} - \phi_{221})}$$

Similarly

$$\begin{bmatrix} \phi_{111} & 0 \\ \phi_{211} & \phi_{221} \end{bmatrix} \begin{bmatrix} \ell_{12} \\ \ell_{22} \end{bmatrix} = \phi_{221} \begin{bmatrix} \ell_{12} \\ \ell_{22} \end{bmatrix}$$

that is,

$$\phi_{111} \ell_{12} = \phi_{221} \ell_{12}$$
$$\phi_{211} \ell_{12} + \phi_{221} \ell_{22} = \phi_{221} \ell_{22}$$

which gives

$$\ell_{12} = 0$$

and ℓ_{22} can be arbitrarily chosen as

$$\ell_{22} = 1$$

Therefore,

$$\mathbf{L} = \begin{bmatrix} 1 & 0 \\ \dfrac{\phi_{211}}{(\phi_{111} - \phi_{221})} & 1 \end{bmatrix} \quad \mathbf{L}^{-1} = \begin{bmatrix} 1 & 0 \\ \dfrac{-\phi_{211}}{(\phi_{111} - \phi_{221})} & 1 \end{bmatrix} \quad (4.4.2)$$

The matrix exponential $\boldsymbol{\phi}^j$, which is also the matrix Green's function in this case because there is no moving average parameter, takes the form

$$\mathbf{G}_j = \boldsymbol{\phi}^j = \mathbf{L} \begin{bmatrix} \phi_{111}^j & 0 \\ 0 & \phi_{221}^j \end{bmatrix} \mathbf{L}^{-1}$$

$$= \begin{bmatrix} \phi_{111}^j & 0 \\ \dfrac{\phi_{211}(\phi_{111}^j - \phi_{221}^j)}{(\phi_{111} - \phi_{221})} & \phi_{221}^j \end{bmatrix} \quad (4.4.3)$$

Of course this result can be obtained more directly by the transfer function approach. Using the B operator ($BX_t = X_{t-1}$, $B^2 X_t = X_{t-2}$, and $B^j X_t = X_{t-j}$), the input model for the no-feedback case with $\phi_{121} = 0$ can be written in the form

$$X_{1t} = \phi_{111} X_{1t-1} + a_{1t}$$
$$(1 - \phi_{111} B) X_{1t} = a_{1t}$$

which is just the scalar AR(1) model, so that, assuming zero initial conditions for convenience (including $a_{1t} = 0$ for $t < 0$)

$$X_{1t} = \sum_{j=0}^{t} \phi_{111}^j a_{1t-j}$$
$$= \sum_{j=0}^{t} G_{11j} a_{1t-j} + \sum_{j=0}^{t} G_{12j} a_{2t-j}$$

and therefore

$$G_{11j} = \phi_{111}^j, \quad G_{12j} \equiv 0$$

as given by top row of the $\mathbf{G}_j$ matrix in (4.4.3). Similarly, the output model

$$X_{2t} = \phi_{211} X_{1t-1} + \phi_{221} X_{2t-1} + a_{2t}$$
$$(1 - \phi_{221} B) X_{2t} = \phi_{211} X_{1t-1} + a_{2t}$$

156 MODAL DECOMPOSITION OF VECTOR SYSTEMS

$$X_{2t} = \frac{\phi_{211}}{(1-\phi_{221}B)(1-\phi_{111}B)} a_{1t-1} + \frac{1}{(1-\phi_{221}B)} a_{2t}$$

$$= \frac{\phi_{211}}{(\phi_{111}-\phi_{221})} \sum_{j=0}^{t-1} (\phi_{111}^{j+1} - \phi_{221}^{j+1}) a_{1t-1-j} + \sum_{j=0}^{t} \phi_{221}^{j} a_{2t-j}$$

$$= \frac{\phi_{211}}{(\phi_{111}-\phi_{221})} \sum_{j=1}^{t} (\phi_{111}^{j} - \phi_{221}^{j}) a_{1t-j} + \sum_{j=0}^{t} \phi_{221}^{j} a_{2t-j}$$

$$= \frac{\phi_{211}}{(\phi_{111}-\phi_{221})} \sum_{j=0}^{t} (\phi_{111}^{j} - \phi_{221}^{j}) a_{1t-j} + \sum_{j=0}^{t} \phi_{221}^{j} a_{2t-j}$$

$$= \sum_{j=0}^{t} G_{21j} a_{1t-j} + \sum_{j=0}^{t} G_{22j} a_{2t-j}$$

where we have used the Green's function G'_j of the AR(2) model in the second step, as derived in Section 3.6.2. This gives

$$G_{21j} = \frac{\phi_{211}(\phi_{111}^{j} - \phi_{221}^{j})}{(\phi_{111} - \phi_{221})}$$

$$G_{22j} = \phi_{221}^{j}$$

agreeing with (4.4.3). Note that all these expressions for $j = 0$ yield $\mathbf{G}_0 = \mathbf{I}$ as required by the definition.

As a further special case, when the input X_{1t} is white noise, $\phi_{111} = \phi_{121} = 0$, and we get the Green's function from (4.4.3) by setting $\phi_{111} = 0$ as

$$\mathbf{G}_j = \begin{bmatrix} 0 & 0 \\ \phi_{211}(\phi_{221})^{j-1} & \phi_{221}^{j} \end{bmatrix} \quad j \geq 1, \quad \mathbf{G}_0 = \mathbf{I} \quad (4.4.3a)$$

Now, since $\phi_{111} = \phi_{121} = 0$, the matrix $\boldsymbol{\phi} = \boldsymbol{\phi}_1$ is singular and therefore although $\mathbf{G}_j = \boldsymbol{\phi}^j$, $\mathbf{G}_0$ cannot be obtained from this but must be set separately by definition as $\mathbf{G}_0 = \mathbf{I}$.

The expression (4.4.3) for $\mathbf{G}_j$ is valid only when the eigenvalues are distinct, that is, $\phi_{111} \neq \phi_{221}$. In the case of equal eigenvalues, either by taking the limit in (4.4.3) as $\phi_{221} \to \phi_{111}$ or by assuming $\phi_{111} = \phi_{221}$, which yields the λ matrix in the Jordan form

$$\lambda = \begin{bmatrix} \phi_{111} & 1 \\ 0 & \phi_{111} \end{bmatrix}$$

and then directly using

$$\boldsymbol{\phi} \mathbf{L} = \mathbf{L} \lambda$$

to find the elements of $\mathbf{L}$, it can be shown that for equal eigenvalues the Green's function is given by

$$\mathbf{G}_j = \begin{bmatrix} \phi_{111}^j & 0 \\ j\phi_{211}\phi_{111}^{j-1} & \phi_{111}^j \end{bmatrix} \qquad (4.4.4)$$

When feedback is present and/or when we have more than two series, the method described above, based on explicit evaluation of the modal matrix $\mathbf{L}$ and its inverse $\mathbf{L}^{-1}$, is somewhat tedious. It was given primarily to illustrate the role of eigenvalues and eigenvectors and also to emphasize that the modal or spectral decomposition is essentially carrying out in a systematic manner the substitution procedure illustrated above. A more general method can be devised based on (4.3.3) so that it avoids the explicit evaluation of $\mathbf{L}$ and $\mathbf{L}^{-1}$.

An Alternative Method
We know from (4.3.3) that a general expression for the Green's function is given by

$$\mathbf{G}_j = \mathbf{g}_1 \lambda_1^j + \mathbf{g}_2 \lambda_2^j \qquad (4.4.5)$$

where the two matrices $\mathbf{g}_1$ and $\mathbf{g}_2$ are

$$\mathbf{g}_i = \{\ell_{ki}\ell^{im}\}, \qquad i = 1, 2 \qquad (4.4.6)$$

However, rather than evaluating these matrices from $\mathbf{L}$ and $\mathbf{L}^{-1}$, we can treat them as unknowns and evaluate them from the initial values of $\mathbf{G}_j$ known to be

$$\mathbf{G}_0 = \mathbf{I} = \mathbf{g}_1 + \mathbf{g}_2$$
$$\mathbf{G}_1 = \boldsymbol{\phi}_1 = \lambda_1 \mathbf{g}_1 + \lambda_2 \mathbf{g}_2 \qquad (4.4.7)$$

Note that when $\boldsymbol{\phi}_1$ is singular, the first of these conditions is not true, nor is it needed, since there is only one nonzero eigenvalue.

Equations (4.4.7) can be readily solved as linear simultaneous equations by treating the matrices as scalars to get *formally*

$$\begin{bmatrix} \mathbf{g}_1 \\ \mathbf{g}_2 \end{bmatrix} = \begin{bmatrix} 1 & 1 \\ \lambda_1 & \lambda_2 \end{bmatrix}^{-1} \begin{bmatrix} \mathbf{I} \\ \boldsymbol{\phi}_1 \end{bmatrix}$$

$$= \begin{bmatrix} \dfrac{\lambda_2}{(\lambda_2 - \lambda_1)} & \dfrac{-1}{(\lambda_2 - \lambda_1)} \\ \dfrac{-\lambda_1}{(\lambda_2 - \lambda_1)} & \dfrac{1}{(\lambda_2 - \lambda_1)} \end{bmatrix} \begin{bmatrix} \mathbf{I} \\ \boldsymbol{\phi}_1 \end{bmatrix}$$

It should be emphasized that the multiplication used above is *not* the usual

158 MODAL DECOMPOSITION OF VECTOR SYSTEMS

matrix multiplication, but obtained by treating matrices as scalars. The expressions for $\mathbf{g}_1$ and $\mathbf{g}_2$ are therefore

$$\mathbf{g}_1 = \frac{1}{(\lambda_2 - \lambda_1)}[\lambda_2 \mathbf{I} - \boldsymbol{\phi}_1] \tag{4.4.8a}$$

$$\mathbf{g}_2 = \frac{-1}{(\lambda_2 - \lambda_1)}[\lambda_1 \mathbf{I} - \boldsymbol{\phi}_1] \tag{4.4.8b}$$

It can easily be verified that the results for the no-feedback case discussed above can be obtained by substituting these in (4.4.5) with $\phi_{121} = 0$, $\lambda_1 = \phi_{111}$, and $\lambda_2 = \phi_{221}$.

The method can be easily extended to the case of multiple eigenvalues. For the bivariate case the solution (4.4.5) takes the form

$$\mathbf{G}_j = \mathbf{g}_1 \lambda_1^j + \mathbf{g}_2 j \lambda_1^{j-1} \tag{4.4.9}$$

with initial conditions

$$\mathbf{G}_0 = \mathbf{I} = \mathbf{g}_1$$

$$\mathbf{G}_1 = \boldsymbol{\phi}_1 = \mathbf{g}_1 \lambda_1 + \mathbf{g}_2$$

so that

$$\mathbf{g}_2 = \boldsymbol{\phi}_1 - \lambda_1 \mathbf{I}$$

Using $\lambda_1 = \phi_{111}$ and substituting for $\mathbf{g}_1$ and $\mathbf{g}_2$ in (4.4.9) it can easily be verified that we get (4.4.4).

Although we have used the bivariate case for illustrative purposes, it is clear that the method extends readily to an arbitrary p-variate system. The Green's function then becomes

$$\mathbf{G}_j = \mathbf{g}_1 \lambda_1^j + \mathbf{g}_2 \lambda_2^j + \cdots + \mathbf{g}_p \lambda_p^j \tag{4.4.10}$$

where the $\mathbf{g}_i$'s are obtained by utilizing the p initial values of $\mathbf{G}_j \equiv \boldsymbol{\phi}_1^j$ similar to (4.4.7) for $p = 2$ as

$$\begin{bmatrix} \mathbf{g}_1 \\ \mathbf{g}_2 \\ \vdots \\ \mathbf{g}_p \end{bmatrix} = \begin{bmatrix} 1 & 1 & \cdots & 1 \\ \lambda_1 & \lambda_2 & \cdots & \lambda_p \\ \vdots & \vdots & & \vdots \\ \lambda_1^{p-1} & \lambda_2^{p-1} & \cdots & \lambda_p^{p-1} \end{bmatrix}^{-1} \begin{bmatrix} \mathbf{I} \\ \boldsymbol{\phi}_1 \\ \vdots \\ \boldsymbol{\phi}_1^{p-1} \end{bmatrix} \tag{4.4.11}$$

Here $\mathbf{g}_i$'s and $\boldsymbol{\phi}_i$'s are treated as scalars until we obtain a linear relation in them such as (4.4.8a and b). Thus (4.4.11) is to be formally used by inverting the Vandermonde matrix only to find out what linear combination of $\mathbf{I}$, $\boldsymbol{\phi}_1$,

$\phi_1^2, \ldots, \phi_1^{p-1}$ yields $\mathbf{g}_i$. Recall that all the elements of this Vandermonde matrix inverse have been explicitly evaluated in Section 3.6.2.

When only one of the eigenvalues is zero and therefore the matrix ϕ_1 is singular, we can set $\lambda_1 = 0$. The remaining $(p - 1)$ $\mathbf{g}_i$'s can be obtained from (4.4.10) by considering $j = 1, 2, \ldots, (p - 1)$, that is, by dropping the first row and column in (4.4.11).

Finally, if we have multiple eigenvalues, say the ith eigenvalue λ_i has multiplicity m_i and there are k distinct eigenvalues, then the solution (4.4.10) takes the form

$$\mathbf{G}_j = \sum_{i=1}^{k} \sum_{r=1}^{m_i} \mathbf{g}_{ir} \frac{j!}{(j-r+1)!(r-1)!} \lambda_i^{j-r+1} \quad (4.4.12)$$

and the $\mathbf{g}_{ir}$'s can be evaluated from the initial conditions

$$\mathbf{G}_j = \phi_1^j, \quad j = 0, 1, 2, \ldots, (p-1)$$

when the matrix ϕ_1 is nonsingular, and $j = 1, 2, \ldots, (p - s)$ when there are s zero eigenvalues. Conditions under which (4.4.12) holds, its derivation and that of its continuous time analog using the Jordan form, and the resultant generalization of the discrete and continuous time modal decompositions (4.1.1a) and (4.3.1a) are left as exercises.

Complex Eigenvalues

In the case of complex eigenvalues, directly following the methods described above involves inversion of complex matrices in both cases, since the matrix $\mathbf{L}$ as well as the matrix composed of λ_i's in (4.4.11) are complex. This inversion of complex matrices can be avoided because finally the Green's function is expressible in real terms in the form of sine and cosine. To illustrate, let us consider the bivariate case with $p = 2$. Then there are two complex eigenvalues and they must form a conjugate pair since the matrix ϕ_1 is real. Therefore, the corresponding eigenvectors forming the columns of $\mathbf{L}$ must also be complex conjugate. We can, however, circumvent the explicit evaluation of the eigenvectors by first showing that the Green's function can be expressed in terms of sine and cosine equivalent to (4.4.10) or (4.4.12) and then using the initial conditions to directly evaluate the coefficient matrices $\mathbf{g}_i$ as illustrated in the alternative method above.

Let the eigenvalues be denoted by $(i = \sqrt{-1})$

$$\lambda_1, \lambda_2 = de^{\pm i\beta} = d(\cos \beta \pm i \sin \beta) \quad (4.4.13)$$

where

$$d = |\lambda_1| = |\lambda_2| \quad \text{and} \quad \beta = \tan^{-1}[\text{Im}(\lambda_1)/\text{Re}(\lambda_1)]$$

Hence for the case of $p = 2$ the Green's function of the ARV(1) model with a complex conjugate pair of eigenvalues may be written as

$$G_j = L\lambda^j L^{-1}$$

$$= Ld^j \begin{bmatrix} e^{i\beta} & 0 \\ 0 & e^{-i\beta} \end{bmatrix}^j L^{-1}$$

$$= Ld^j \begin{bmatrix} e^{ij\beta} & 0 \\ 0 & e^{-ij\beta} \end{bmatrix} L^{-1}$$

$$= Ld^j \begin{bmatrix} \cos j\beta + i\sin j\beta & 0 \\ 0 & \cos j\beta - i\sin j\beta \end{bmatrix} L^{-1}$$

$$= g_1 d^j \cos j\beta + g_2 d^j \sin j\beta \tag{4.4.14}$$

Although g_1 and g_2 can be expressed in terms of the elements of matrix L, it is far easier to obtain them directly using the initial conditions at $j = 0, 1$, as

$$G_0 = I = g_1$$

and

$$G_1 = \phi_1 = g_1 d \cos \beta + g_2 d \sin \beta$$

so that

$$g_2 = (1/d \sin \beta)(\phi_1 - Id \cos \beta)$$

It is now clear that the method readily generalizes for arbitrary p. Thus if there are q pairs of complex conjugate eigenvalues and the remaining $(p - 2q)$ eigenvalues are real, then the Green's function is given by

$$G_j = g_1 d_1^j \cos j\beta_1 + g_2 d_1^j \sin j\beta_1 + \cdots + g_{2q-1} d_q^j \cos j\beta_q$$
$$+ g_{2q} d_q^j \sin j\beta_q + g_{2q+1} \lambda_{2q+1}^j + \cdots + g_p \lambda_p^j \tag{4.4.15}$$

where the g_i's are found from the initial conditions

$$G_j = \phi_1^j, \quad j = 0, 1, 2, \ldots, p - 1$$

It should be emphasized that the g_i's of (4.4.14 and 4.4.15) are *not* the same as the g_i's of (4.4.10) although the latter expression is also valid for complex eigenvalues. This is merely to avoid extra notation; we will generally use (4.4.10) for real eigenvalues and (4.4.15) for the complex case to avoid confusion. As an illustrative example, for $p = 3$, there can be at most one pair of complex conjugate eigenvalues and the remaining third eigenvalue must be real. Therefore the coefficient matrices g_1, g_2, and g_3 can be evaluated from

$$\begin{bmatrix} g_1 \\ g_2 \\ g_3 \end{bmatrix} = \begin{bmatrix} 1 & 0 & 1 \\ d_1 \cos \beta_1 & d_1 \sin \beta_1 & \lambda_3 \\ d_1^2 \cos 2\beta_1 & d_1^2 \sin 2\beta_1 & \lambda_3^2 \end{bmatrix}^{-1} \begin{bmatrix} I \\ \phi_1 \\ \phi_1^2 \end{bmatrix}$$

where the matrices $\mathbf{g}_i$ and $\mathbf{I}, \boldsymbol{\phi}_1, \boldsymbol{\phi}_1^2$ are treated as scalars for matrix multiplication purposes similar to the derivation of (4.4.8a and b).

Physical Interpretation
As pointed out earlier, the elements of the $\mathbf{G}_j$ matrix decompose the series X_{1t} and X_{2t} into the contributions from a_{1t} and a_{2t}: G_{11j} and G_{12j} provide the contribution from a_{1t} and a_{2t} to X_{1t} and G_{21j} and G_{22j} provide the contribution from a_{1t} and a_{2t} to X_{2t}, respectively. The $\mathbf{g}_i$'s further break down each of these contributions by the dynamic modes λ_i for real roots and sines and cosines of the frequencies of the corresponding pairs of complex conjugate roots. We will illustrate this breakdown for the no-feedback case in which such a breakdown can also be intuitively explained.

For the no-feedback case the eigenvalues $\lambda_1 = \phi_{111}$ and $\lambda_2 = \phi_{221}$. Therefore, either by substituting these with $\boldsymbol{\phi}_1$ from (4.4.1a) in (4.4.8a and b) or by the direct comparison of (4.4.5) with (4.4.3), it is easy to see that the coefficient matrices $\mathbf{g}_i$'s in this case are given by

$$\mathbf{g}_1 = \begin{bmatrix} 1 & 0 \\ \dfrac{\phi_{211}}{(\phi_{111} - \phi_{221})} & 0 \end{bmatrix} \quad \mathbf{g}_2 = \begin{bmatrix} 0 & 0 \\ \dfrac{-\phi_{211}}{(\phi_{111} - \phi_{221})} & 1 \end{bmatrix}$$

These coefficient matrices show that the only contribution to X_{1t} is from a_{1t} by the first mode ϕ_{111}^j; X_{1t} has no contribution from a_{2t} by the first mode and also no contribution from a_{1t} as well as a_{2t} by the second mode ϕ_{221}^j. This is intuitively clear since it was shown earlier that for the no-feedback case X_{1t} has just an AR(1) model in a_{1t} with parameter ϕ_{111}. Since there is no feedback, neither the white noise a_{2t} generating the output noise nor the dynamic mode of the output noise ϕ_{221}^j enters into X_{1t}.

On the other hand, X_{2t} has equal and opposite contributions by the two modes from a_{1t}. This is explained by the transfer function from the input to the output governed by the dynamic mode ϕ_{221}^j and the dynamic mode ϕ_{111}^j of the input itself. However, a_{2t} contributes to X_{2t} only by the mode ϕ_{221}^j via the output noise; it has no contribution by the first mode ϕ_{111}^j belonging to the input again because there is no feedback. It is left as an interesting exercise for the reader to slowly alter the feedback parameter ϕ_{121} from zero and observe the changes in the eigenvalues and their contributions given by the $\mathbf{g}_1$ and $\mathbf{g}_2$ matrices. In fact, when the parameter ϕ_{121} is large enough, the eigenvalues become complex and illustrate the complex eigenvalue case discussed above.

4.5 ARMAV(2,1) AND ARMAV(n,m) MODELS

Now consider the ARMAV(2,1) model

$$\mathbf{X}_t = \boldsymbol{\phi}_1 \mathbf{X}_{t-1} + \boldsymbol{\phi}_2 \mathbf{X}_{t-2} + \mathbf{a}_t - \boldsymbol{\theta}_1 \mathbf{a}_{t-1} \qquad (4.5.1)$$

where, for the bivariate case $p = 2$, $\mathbf{X}_t = [X_{1t}, X_{2t}]^T$ and $\mathbf{a}_t = [a_{1t}, a_{2t}]^T$. Recall from Section 2.4 that this model, considered in the standard state space form (2.4.1) or (4.3.1), consists of the top two rows with $\mathbf{X}_t = [X_{1t}, X_{2t}, X_{1t-1}, X_{2t-1}]^T$, $\mathbf{a}_t = [a_{1t}, a_{2t}, a_{1t-1}, a_{2t-1}]^T$, and

$$\boldsymbol{\phi} = \begin{bmatrix} \boldsymbol{\phi}_1 & \boldsymbol{\phi}_2 \\ \mathbf{I} & 0 \end{bmatrix} \quad \boldsymbol{\theta} = \begin{bmatrix} \mathbf{I} & -\boldsymbol{\theta}_1 \\ 0 & 0 \end{bmatrix}$$

$$\boldsymbol{\theta}_1 = \begin{bmatrix} \theta_{111} & \theta_{121} \\ \theta_{211} & \theta_{221} \end{bmatrix} \quad \boldsymbol{\phi}_i = \begin{bmatrix} \phi_{11i} & \phi_{12i} \\ \phi_{21i} & \phi_{22i} \end{bmatrix} \quad i = 1, 2 \quad (4.5.2)$$

Hence its formal solution is given by (4.3.1a) and, recalling from the corresponding scalar case in Section 3.6.2, the Green's function of the ARV(2) model will be the top left $p \times p$ matrix element of $\boldsymbol{\phi}^j$. Then the Green's function of the ARMAV(2,1) can be obtained from the term $\boldsymbol{\phi}^j \boldsymbol{\theta} \mathbf{a}_{t-j}$ in (4.3.1a). To evaluate $\boldsymbol{\phi}^j$ and the Green's function, we need the $2p$ eigenvalues given by the determinant of the partitioned matrix (see Section 4.2)

$$|\boldsymbol{\phi} - \lambda \mathbf{I}| = \begin{vmatrix} \boldsymbol{\phi}_1 - \lambda \mathbf{I} & \boldsymbol{\phi}_2 \\ \mathbf{I} & -\lambda \mathbf{I} \end{vmatrix} = |-\lambda \mathbf{I}||\boldsymbol{\phi}_1 - \lambda \mathbf{I} + (1/\lambda)\boldsymbol{\phi}_2|$$

$$= |\lambda^2 \mathbf{I} - \lambda \boldsymbol{\phi}_1 - \boldsymbol{\phi}_2| = 0 \quad (4.5.3)$$

which is the determinant of a second order autoregressive matrix polynomial.

We can evaluate $\boldsymbol{\phi}^j$ and the Green's function by finding the eigenvectors and then using the modal matrix $\mathbf{L}$. However, it is clear from the development in the last section (as well as that of the scalar case in Section 3.6.2) that for distinct eigenvalues the Green's function will have the explicit form

$$\mathbf{G}_j = \mathbf{g}_1 \lambda_1^j + \mathbf{g}_2 \lambda_2^j + \ldots + \mathbf{g}_{2p} \lambda_{2p}^j \quad (4.5.4)$$

where the $p \times p$ matrices $\mathbf{g}_i$'s are obtained by utilizing the $2p$ initial values of $\mathbf{G}_j$. Similarly, $\boldsymbol{\phi}^j$ (required in evaluating the transient response) is of the same form and the corresponding $2p \times 2p$ matrices $\mathbf{g}_i$ can be obtained by utilizing the $2p$ initial values $\mathbf{I}, \boldsymbol{\phi}, \boldsymbol{\phi}^2, \ldots, \boldsymbol{\phi}^{2p-1}$.

The initial values of $\mathbf{G}_j$ can be obtained by using the backshift operator $B^j \mathbf{X}_t = \mathbf{X}_{t-j}$ and the definition of the Green's function

$$\mathbf{X}_t = (\mathbf{G}_0 + \mathbf{G}_1 B + \mathbf{G}_2 B^2 + \cdots)\mathbf{a}_t$$

substituted in the operator form of the ARMA(2,1) model (4.5.1):

$$(\mathbf{I} - \boldsymbol{\phi}_1 B - \boldsymbol{\phi}_2 B^2)\mathbf{X}_t = (\mathbf{I} - \boldsymbol{\theta}_1 B)\mathbf{a}_t$$

to yield the operator identity

$$(\mathbf{I} - \boldsymbol{\phi}_1 B - \boldsymbol{\phi}_2 B^2)(\mathbf{G}_0 + \mathbf{G}_1 B + \mathbf{G}_2 B^2 + \cdots) \equiv (\mathbf{I} - \boldsymbol{\theta}_1 B)$$

Equating the coefficients of equal powers of B on both sides gives the implicit method of evaluating $\mathbf{G}_j$:

0: $\mathbf{G}_0 = \mathbf{I}$
1: $\mathbf{G}_1 - \boldsymbol{\phi}_1 = -\boldsymbol{\theta}_1$ or $\mathbf{G}_1 = \boldsymbol{\phi}_1 - \boldsymbol{\theta}_1$
2: $\mathbf{G}_2 - \boldsymbol{\phi}_1 \mathbf{G}_1 - \boldsymbol{\phi}_2 = 0$ or $\mathbf{G}_2 = \boldsymbol{\phi}_1^2 - \boldsymbol{\phi}_1 \boldsymbol{\theta}_1 + \boldsymbol{\phi}_2$ (4.5.5)
3: $\mathbf{G}_3 - \boldsymbol{\phi}_1 \mathbf{G}_2 - \boldsymbol{\phi}_2 \mathbf{G}_1 = 0$ or $\mathbf{G}_3 = \boldsymbol{\phi}_1^3 - \boldsymbol{\phi}_1^2 \boldsymbol{\theta}_1 + \boldsymbol{\phi}_1 \boldsymbol{\phi}_2$
$\qquad\qquad\qquad\qquad\qquad\qquad\qquad + \boldsymbol{\phi}_2 \boldsymbol{\phi}_1 - \boldsymbol{\phi}_2 \boldsymbol{\theta}_1$

and in general

$$j: \mathbf{G}_j = \boldsymbol{\phi}_1 \mathbf{G}_{j-1} + \boldsymbol{\phi}_2 \mathbf{G}_{j-2}, \quad j \geq 2 \qquad (4.5.6)$$

Thus, for distinct eigenvalues, we can again use the elements of the Vandermonde matrix inverse in Section 3.6.2 to find $\mathbf{g}_i$'s similar to (4.4.11) as

$$\begin{bmatrix} \mathbf{g}_1 \\ \mathbf{g}_2 \\ \vdots \\ \mathbf{g}_{2p} \end{bmatrix} = \begin{bmatrix} 1 & 1 & \cdots & 1 \\ \lambda_1 & \lambda_2 & \cdots & \lambda_{2p} \\ \vdots & \vdots & & \vdots \\ \lambda_1^{2p-1} & \lambda_2^{2p-1} & \cdots & \lambda_{2p}^{2p-1} \end{bmatrix}^{-1} \begin{bmatrix} \mathbf{I} \\ \mathbf{G}_1 \\ \vdots \\ \mathbf{G}_{2p-1} \end{bmatrix} \qquad (4.5.7)$$

For multiple eigenvalues we can use (4.4.12) and evaluate the $\mathbf{g}_{ir}$'s from the necessary number of initial conditions taken from (4.5.5) and (4.5.6).

To illustrate this case consider

$$\boldsymbol{\phi}_1 = \begin{bmatrix} 0 & 0 \\ 0 & 1.5 \end{bmatrix} \quad \boldsymbol{\phi}_2 = \begin{bmatrix} 0 & 0 \\ -0.3 & -0.5 \end{bmatrix} \quad \boldsymbol{\theta}_1 = \begin{bmatrix} 0 & 0 \\ 0 & 0.9 \end{bmatrix}$$

Then the eigenvalues are obtained from

$$|\lambda^2 \mathbf{I} - \lambda \boldsymbol{\phi}_1 - \boldsymbol{\phi}_2| = \begin{vmatrix} \lambda^2 & 0 \\ 0.3 & \lambda^2 - 1.5\lambda + 0.5 \end{vmatrix}$$
$$= \lambda^2(\lambda^2 - 1.5\lambda + 0.5) = 0$$

as $\lambda_1 = 1$, $\lambda_2 = 0.5$, $\lambda_3 = 0$, and $\lambda_4 = 0$.

Although in theory we should use (4.4.12) because of the multiple eigenvalues, since they are zero (4.5.4) can be used for the distinct nonzero eigenvalues to get

$$\mathbf{G}_j = \mathbf{g}_1(1)^j + \mathbf{g}_2(0.5)^j$$

So we need only two initial values of $\mathbf{G}_j$. However, note that because of the zero eigenvalues, the ϕ matrix is singular and the expression for $\mathbf{G}_j$ will not yield $\mathbf{G}_0 = \mathbf{I}$; in other words, the expression given above for $\mathbf{G}_j$ is valid only for $j \geq 1$. Hence we use the initial conditions

$$\mathbf{G}_1 = \boldsymbol{\phi}_1 - \boldsymbol{\theta}_1 = \begin{bmatrix} 0 & 0 \\ 0 & 0.6 \end{bmatrix}$$

$$\mathbf{G}_2 = \boldsymbol{\phi}_1 \mathbf{G}_1 + \boldsymbol{\phi}_2 = \begin{bmatrix} 0 & 0 \\ 0 & 0.9 \end{bmatrix} + \begin{bmatrix} 0 & 0 \\ -0.3 & -0.5 \end{bmatrix} = \begin{bmatrix} 0 & 0 \\ -0.3 & 0.4 \end{bmatrix}$$

$$\mathbf{G}_1 = \mathbf{g}_1 + 0.5\mathbf{g}_2$$
$$\mathbf{G}_2 = \mathbf{g}_1 + 0.25\mathbf{g}_2$$

Subtracting

$$\mathbf{g}_2 = \frac{1}{0.25}(\mathbf{G}_1 - \mathbf{G}_2) = \begin{bmatrix} 0 & 0 \\ 1.2 & 0.8 \end{bmatrix}$$

$$\mathbf{g}_1 = \mathbf{G}_1 - 0.5\mathbf{g}_2 = \begin{bmatrix} 0 & 0 \\ -0.6 & 0.2 \end{bmatrix}$$

It is not hard to see that this $\mathbf{G}_j$ can be obtained directly from the respective models by the transfer function approach *without* using the matrix method.

These results readily generalize to ARMAV(n,m) models. It may be verified that eigenvalues are obtained by generalizing (4.5.3) to the determinant of an nth order autoregressive matrix polynomial

$$|\boldsymbol{\phi} - \lambda \mathbf{I}| = (-1)^{np}|\lambda^n \mathbf{I} - \lambda^{n-1}\boldsymbol{\phi}_1 - \lambda^{n-2}\boldsymbol{\phi}_2 - \cdots - \boldsymbol{\phi}_n| = 0 \quad (4.5.8)$$

Incidentally, this result shows that $|\boldsymbol{\phi}| = |\boldsymbol{\phi}_n|$, so that singularity or nonsingularity of $\boldsymbol{\phi}$ is equivalent to that of $\boldsymbol{\phi}_n$. Equation (4.5.8) is a polynomial of degree np in λ; hence it has np roots or eigenvalues, some of which would be zero if $\boldsymbol{\phi}_n$ is singular.

For distinct eigenvalues the Green's function is given by (4.5.4) and (4.5.7) with $2p$ replaced by np. The initial values needed in (4.5.7) are obtained by comparing the coefficients just like (4.5.5) and (4.5.6) using nth order autoregressive and mth order moving average operators. For multiple eigenvalues we can use (4.4.12). Both these expressions for $\mathbf{G}_j$ are valid for $j \geq 0$ only when all the eigenvalues are nonzero (i.e., $\boldsymbol{\phi}_n$ is nonsingular) and $n > m$. For singular $\boldsymbol{\phi}_n$ and $n > m$ these expressions are valid for $j \geq 1$. For $m \geq n$ they are valid for

$j \geq m - n + 1$. Since $\mathbf{G}_j$ represents the dynamics of the system which we expect generally to be smooth in some sense, we want it to be valid from $j \geq 0$, which provides an added justification for the ARMAV($n, n - 1$) modeling strategy evolved using the geometry of state space in Chapter 2. For complex eigenvalues, they occur in conjugate pairs, and we can use (4.4.15) with p replaced by np and $\mathbf{g}_j$'s evaluated from the initial values in the same way as shown above.

4.6 TRANSIENT RESPONSE—IDENTIFIABILITY, CONTROLLABILITY, AND OBSERVABILITY

In the preceding sections we have concentrated on the function $\mathbf{G}_j$ that includes both the autoregressive and the moving average parameters and is important in the response to forced input, particularly to the random input $\mathbf{a}_t$ giving the stochastic part of the response. In particular, when the elements of $\mathbf{a}_t$ are white noise having flat spectrum over the frequency range limited by the sampling interval (with unlimited range for the continuous time case), the input excites all the modes of the system and therefore all the modes are "identifiable." Therefore, when the stochastic part is dominant, the DDS methodology is capable of identifying the entire dynamics of the system which is completely characterized by $\mathbf{G}_j$.

Can we make the same assertion about the transient part? In other words, if the stochastic part of the response is negligible, can we always identify the entire dynamics of the system from its transient part that is the response to initial conditions? To state it more directly, does every set of initial conditions excite *all* the modes of the system? The answer to this question is obviously no. In fact, a little reflection on the transient solution would show that by a suitable choice of initial conditions we can suppress any mode(s) of the system. The transient part is governed by $e^{\mathbf{A}t}$ and $\boldsymbol{\phi}^t$, which are a part of the Green's function arising from the autoregressive part of the model. Therefore, the system identifiability hinges on the question of whether we can recover $\mathbf{A}$ or $\boldsymbol{\phi}$ from its transient response. This, in turn, is possible if all the eigenvalues of $\mathbf{A}$ or $\boldsymbol{\phi}$ appear in the transient response. It is easy to find initial conditions such that arbitrary eigenvalues we choose vanish from the transient response. This is most easily seen from the modal decomposition of the transient part in (4.1.1d) in the continuous case and (4.3.1c) in the discrete case:

$$\mathbf{q}(t) = e^{\mu t} \mathbf{M}^{-1} \mathbf{x}(0)$$

$$\mathbf{q}_t = \lambda^t \mathbf{L}^{-1} \mathbf{X}_0$$

It is clear from these forms (particularly for the distinct eigenvalues case) that the ith eigenvalue will not appear in the transient response if the ith element of $\mathbf{M}^{-1}\mathbf{x}(0)$ or $\mathbf{L}^{-1}\mathbf{X}_0$ is zero. Recall from Section 3.4.1 that premultiplication by an inverse of a matrix expresses a vector in terms of the basis forming the columns of that matrix. In the present case the columns happen to be the eigenvectors of the system. The ith element of $\mathbf{M}^{-1}\mathbf{x}(0)$ or $\mathbf{L}^{-1}\mathbf{X}_0$ is thus the ith coordinate of the initial value vector along the ith eigenvector, or the projection of the initial value vector along the ith eigenvector. This projection must be nonzero for the ith mode to appear in the transient response. In other words, the initial value vector must be a nontrivial linear combination of the system eigenvectors, each eigenvector appearing with a nonzero coefficient. Thus the initial conditions must excite all the modes of the system so that they can be identified from the transient response.

The condition that each element of the column $\mathbf{M}^{-1}\mathbf{x}(0)$ or $\mathbf{L}^{-1}\mathbf{X}_0$ be nonzero requires the calculation of the eigenvectors and the inverse of the matrix $\mathbf{L}$. An equivalent form of the condition that is easier to evaluate can be expressed directly in terms of ϕ. We consider the discrete case for illustration. Considering only the transient part,

$$\mathbf{X}_1 = \phi \mathbf{X}_0$$
$$\mathbf{X}_2 = \phi \mathbf{X}_1 = \phi^2 \mathbf{X}_0$$
$$\vdots$$
$$\mathbf{X}_n = \phi \mathbf{X}_{n-1} = \phi^n \mathbf{X}_0$$

Therefore the matrix

$$[\mathbf{X}_1, \mathbf{X}_2, \ldots, \mathbf{X}_n] = [\phi \mathbf{X}_0, \phi \mathbf{X}_1, \ldots, \phi \mathbf{X}_{n-1}] = \phi [\mathbf{X}_0, \mathbf{X}_1, \ldots, \mathbf{X}_{n-1}]$$

If ϕ is to be uniquely determined from n measurements of the transient response, then the last matrix multiplying ϕ must be nonsingular. This matrix is obviously the same as

$$\mathbf{B} = [\mathbf{X}_0, \phi \mathbf{X}_0, \ldots, \phi^{n-1} \mathbf{X}_0] \qquad (4.6.1)$$

and the identifiability condition is that this matrix be nonsingular. This matrix is often referred to as the identifiability matrix and is expressed only in terms of ϕ and the initial value vector $\mathbf{X}_0$.

Note that the identifiability condition can be verified only when the system is known or postulated. Hence it is of limited usefulness in the DDS approach when we are trying to model the system from the observed data alone. It, however, does bring out the fact that when the stochastic part is negligible, the observed data and hence the DDS model obtained from it may miss some of the dynamic modes not excited by the initial conditions used in generating the data. For the sake of completeness, we now give two more conditions applicable to systems with known dynamics and known input or control functions. These conditions, called controllability and observability, are usually discussed in modern control theory texts and hence we follow their common notation for matrices so that these conditions can be stated for both discrete and continuous systems at the same time. In this notation the state and observation equations are written as

$$\dot{x}(t) = Ax(t) + Bu(t) \quad (4.6.2)$$

$$y(t) = Cx(t) + Du(t) \quad (4.6.3)$$

$$x_t = Ax_{t-1} + Bu_{t-1} \quad (4.6.4)$$

$$y_t = Cx_t + Du_t \quad (4.6.5)$$

The idea here is that the state x ($n \times 1$ column vector) develops as a response to control input u by the first dynamic equation and is then transformed into the observation y ($m \times 1$ column vector). If there are r inputs, so that u is an $r \times 1$ column vector, then matrix A is $n \times n$, B is $n \times r$, C is $m \times n$, and D is $m \times r$. Using arguments similar to those used for identifiability above, it can be shown that the system is controllable if and only if the $n \times rn$ controllability matrix

$$P = (B, AB, A^2B, A^3B, \ldots, A^{n-1}B) \quad (4.6.6)$$

has rank n. The proof in the discrete case is almost identical to the one given for identifiability when we consider the second term on the right-hand side of (4.6.4) which naturally determines controllability. Similarly, it can be shown that the system is observable, that is, all the modes of the system are present in the observation vector y, if and only if the $mn \times n$ observability matrix

$$Q = \begin{bmatrix} C \\ C\phi \\ C\phi^2 \\ \vdots \\ C\phi^{n-1} \end{bmatrix} \quad (4.6.7)$$

has rank n. Again the proof in the discrete case is obtained as above by considering the first term on the right-hand side of (4.6.5), which determines the observability.

4.7 JUSTIFICATION OF ARMA($n, n-1$) STRATEGY FOR A GENERAL STATE MODEL

The justification of the ARMA($n, n-1$) strategy from the DDS point of view, given only the response data, was outlined in Section 2.4.1 for the scalar case and then extended to the vector case. This model was transformed into the standard state space form by a special choice of ϕ and θ matrices. Of these the choice of the θ matrix was somewhat arbitrary and unimportant since it governs only the initial values; however, the choice of the ϕ matrix was crucial since it governs the dynamics of the system, as shown by the Green's function development discussed in this chapter. In fact, the ARMA$(n,n-1)$ strategy essentially amounts to modeling with this special form of the ϕ matrix in the state space of increasing dimension. Increasing the dimension of the state space means that more and more states are required to describe the dynamics of the system completely. In the DDS approach, X_t was taken as the data and states were simply $X_t, X_{t-1}, \ldots,$ and X_{t-n} with the choice of the ϕ matrix as

$$\phi = \begin{bmatrix} \phi_1 & \phi_2 & \phi_3 & \cdots & \phi_n \\ & & & & 0 \\ & & & & 0 \\ & \mathbf{I} & & & \vdots \\ & & & & 0 \end{bmatrix} \qquad (4.7.1)$$

However, for a system with given inputs, is the ARMA$(n,n-1)$ modeling strategy, that is, the particular choice of the ϕ matrix with increasing n, still justified? In fact, the choice of $X_t, X_{t-1}, \ldots,$ and X_{t-n} as states is itself arbitrary; the system dynamics may be described by many other sets of state variables which may have or require quite different forms of the ϕ matrices even if the system is linear. Therefore, a particular choice of the ϕ matrix implying the ARMA$(n,n-1)$ model for the observed response data on the measured variable needs some justification. Fortunately, modern control theory provides such a justification, which reinforces the ARMA$(n,n-1)$ strategy of the DDS methodology. We will consider a simple version of this argument in this section, more rigorous proofs may be found in advanced control theory texts.

For this purpose, we will use slightly different notation to conform to the usual control theory format of the proof. Let the state vector $\mathbf{x}_t = [x_{t+n-1}, x_{t+n-2}, \ldots, x_t]^T$, so that our standard state model for the ARMA$(n,n-1)$ model can be written in the usual control theory format introduced in the last section as (using $\mathbf{u}_t$ and $\mathbf{B}$ to replace $\mathbf{a}_t$ and θ, respectively)

State: $\qquad \mathbf{x}_t = \phi \mathbf{x}_{t-1} + \mathbf{B}\mathbf{u}_{t-1}$

Measurement: $\quad y_t = \mathbf{C}\mathbf{x}_t, \qquad \mathbf{C} = [0, 0, 0, \ldots, 1] \qquad (4.7.2)$

Here the state has been defined (somewhat arbitrarily) based on the measurement x_t; in fact, in the control theory literature it is defined using both x_t and a_t to clarify the structure of $\mathbf{B}$, which we will not discuss here. However, the actual state vector can be totally different from the measurement and its dynamics may be described by a general ϕ^* matrix quite different from the standard ϕ matrix of the form (4.7.1). Can we still use the ARMA$(n,n-1)$ strategy for modeling the system from the given measurement, which is equivalent to using the model shown above with increasing n, implying a particular form of ϕ and $\mathbf{C}$ given by (4.7.1) and (4.7.2)?

It is indeed fortunate that for a linear system the answer to this question is yes. Following the modern control theory [see Lee (1964)], we will now show that if the system dynamics is described in any state vector z_t by any general ϕ^* matrix, and the state is related to the measurement by any row vector C^*, then (4.7.2) is still the correct model for the system. In other words, we will show that the state model

$$\text{State:} \quad z_t = \phi^* z_{t-1} + B^* u_{t-1}$$

$$\text{Measurement:} \quad y_t = C^* z_t \tag{4.7.3}$$

with arbitrary ϕ^* and C^* is equivalent to (4.7.2). In fact, the linear transformation from the general state z_t of (4.7.3) to the particular measurement-based state x_t of (4.7.2) turns out to be

$$x_t = R z_t \tag{4.7.4}$$

where R is the reverse of the observability matrix Q of (4.6.7):

$$R = \begin{bmatrix} C^* \phi^{*n-1} \\ C^* \phi^{*n-2} \\ \vdots \\ C^* \end{bmatrix} \tag{4.7.5}$$

Note that the reverse matrix R needs to be used rather than the observability matrix Q simply because we have originally defined the state in a "backward" way as $x_t = [x_{t+n-1}, x_{t+n-2}, \ldots, x_t]^T$, leading to the ϕ matrix of the familiar form (4.7.2). In modern control theory literature the state is defined in a forward way as $x_t = [x_t, x_{t+1}, \ldots, x_{t+n-1}]^T$, leading to a ϕ matrix with $(\phi_1, \phi_2, \ldots, \phi_n)$ in the *bottom* row and the identity and zero vectors at the top; then the transformation shown above is accomplished by the observability matrix Q itself. For an observable system Q is nonsingular as is R, so that

$$z_t = R^{-1} x_t \tag{4.7.6}$$

We will now show that with this transformation, ϕ^* and C^* transform into ϕ and C of (4.7.2).

Substituting (4.7.6) in (4.7.3) gives

$$x_t = R \phi^* R^{-1} x_{t-1} + R B^* u_{t-1}$$
$$= \phi x_{t-1} + B u_{t-1}$$
$$y_t = C^* R^{-1} x_t$$
$$= C x_t$$

MODAL DECOMPOSITION OF VECTOR SYSTEMS

Now

$$\phi = \mathbf{R}\phi^*\mathbf{R}^{-1}$$

$$= \begin{bmatrix} \mathbf{C}^*\phi^{*n-1} \\ \mathbf{C}^*\phi^{*n-2} \\ \vdots \\ \mathbf{C}^* \end{bmatrix} \phi^* \begin{bmatrix} \mathbf{C}^*\phi^{*n-1} \\ \mathbf{C}^*\phi^{*n-2} \\ \vdots \\ \mathbf{C}^* \end{bmatrix}^{-1}$$

$$= \begin{bmatrix} \mathbf{C}^*\phi^{*n-1} \\ \hline \mathbf{C}^*\phi^{*n-2} \\ \vdots \\ \mathbf{C}^* \end{bmatrix} \mathbf{I} \begin{bmatrix} \mathbf{C}^*\phi^{*n-2} \\ \mathbf{C}^*\phi^{*n-3} \\ \vdots \\ \hline \mathbf{C}^*\phi^{*-1} \end{bmatrix}^{-1}$$

$$= \begin{bmatrix} \mathbf{d}^T \\ \hline \mathbf{D} \end{bmatrix} \begin{bmatrix} \mathbf{D} \\ \hline \mathbf{e}^T \end{bmatrix}^{-1}$$

$$= \begin{bmatrix} \mathbf{d}^T \\ \hline \mathbf{D} \end{bmatrix} [\mathbf{D}_{-1} | \mathbf{e}_{-1}]$$

$$= \begin{bmatrix} \mathbf{d}^T\mathbf{D}_{-1} & \mathbf{d}^T\mathbf{e}_{-1} \\ & 0 \\ & 0 \\ \mathbf{I} & \vdots \\ & 0 \end{bmatrix}$$

$$= \begin{bmatrix} \phi_1 & \phi_2 & \cdots & \phi_n \\ & & & 0 \\ & & & 0 \\ & \mathbf{I} & & \vdots \\ & & & 0 \end{bmatrix}$$

where, since we have set

$$\begin{bmatrix} \mathbf{D} \\ \hline \mathbf{e}^T \end{bmatrix}^{-1} = [\mathbf{D}_{-1} | \mathbf{e}_{-1}]$$

therefore

$$DD_{-1} = I, \quad De_{-1} = \begin{bmatrix} 0 \\ 0 \\ \vdots \\ 0 \end{bmatrix}$$

$$e^T D_{-1} = [0, 0, \ldots, 0], \quad e^T e_{-1} = 1$$

and

$$C^*R^{-1} = C^* \begin{bmatrix} C^*\phi^{*n-1} \\ C^*\phi^{*n-2} \\ \vdots \\ C^* \end{bmatrix}^{-1}$$

$$= [C^*\phi^{*-1}] \begin{bmatrix} D \\ \hline e^T \end{bmatrix}^{-1}$$

$$= [e^T][D_{-1} | e_{-1}]$$

$$= [0, 0, 0, \ldots, 1]$$

$$= C$$

which proves that (4.7.2) and (4.7.3) are equivalent. Thus, even for systems with deterministic known inputs such as a step, ramp, or sinusoidal, it is enough to concentrate on ARMA$(n, n-1)$ models.

Actually, even when deterministic response is modeled by the computer using the ARMA$(n, n-1)$ strategy, the computer processing noise, including the round-off in the digits truncated in the input data, provides the stochastic part and adds its own modes in addition to the system modes from the main deterministic component. Hence the justification of the ARMA$(n, n-1)$ strategy given in Section 2.4.1 still applies. The modeling procedure simply considers this computer noise as the stochastic part and therefore, after the adequate model has been reached, the a_t's of the model will be of magnitude comparable to this noise, for example 10^{-7} for single-precision data. The proof given above merely shows that even if such a stochastic part is absent in theory, the ARMA$(n, n-1)$ modeling strategy of DDS does not miss anything as long as the system is observable. If the system is not observable, then the measurement data does not contain certain modes of the system, and these will be omitted from the DDS model of the system. The DDS model is as good as the data it is based on!

Note that the observability is necessarily a concept based on a *known* equation of state and *given* measurement. Therefore, the only way to ascertain the observability of a DDS based on some measurement data is to exhibit that a state model such as (4.7.3), obtained a priori without making use of the actual measurement data, is observable with reference to the same measurement data. The results given above then show that the ϕ of the DDS model is related to the ϕ^* of the a priori model by $\phi = \mathbf{R}\phi^*\mathbf{R}^{-1}$, where $\mathbf{R}$ is the matrix used in proving the observability. The ϕ matrix constructed from the DDS model may require selection of modes (eigenvalues) and modal (eigen) vectors as discussed in Chapter 5.

5

MODAL ANALYSIS

The concept of modal decomposition, introduced in Chapter 3 and generalized in Chapter 4, provides a natural foundation to experimental modal analysis. When the data comes from the measured response of a vibrating structure, *relevant* eigenvalues and eigenvectors of the state matrix may be interpreted as those of the underlying vibration system. If the nature of excitation is known but it is not measured, a mathematical description of the system in terms of natural frequency, damping ratio, and scaled mode shapes can be obtained. If, further, the excitation is known or measured, the scaling of mode shapes can be recovered to obtain a full mass-spring-dashpot model of the system from multiple response data.

In contrast to the existing methods of experimental modal analysis, such an approach can be used on "operational" noisy input–output data recorded during the operation of the system under actual working conditions; the data need not be collected under tightly controlled laboratory conditions. If the data is collected under tightly controlled conditions, as required by the current practice of "modal testing," then of course the approach works as well as the existing methods, and provides more accurate results for reasons conceptually discussed in Section 2.9.7 and illustrated in this chapter.

It should be emphasized that although we will use a mechanical or structural lumped parameter system as a concrete model, the method is not restricted to the conventional modal testing problem. Since the data can be operational as well as experimental, the method is applicable to the analysis of any system in which the measured or observed data consists of oscillatory components. For systems other than mechanical, such quantities as natural frequency, damping ratio, scaled mode shapes, and mass, damping, and stiffness matrices need to be suitably interpreted.

174   MODAL ANALYSIS

As most of the popular methods of modal testing and analysis are primarily based on the fast Fourier transform (FFT), we begin with a brief review of these techniques in Section 5.1, elucidating their limitations; more details of the methods may be found in the documentation of commercial Fourier analyzers implementing these techniques. These limitations refer to the FFT without curve fitting, because as discussed in Section 2.9, good curve fitting of real data requires more computation than DDS and yet provides less accurate results. Section 5.2 outlines the DDS modeling procedure and estimation algorithms particularly suitable for modal analysis. Sections 5.3 and 5.4 present the theory for obtaining the modal and spatial models by the DDS approach. Sections 5.5 and 5.6 illustrate modal parameter identification from univariate and multivariate data, respectively; both simulated and experimental examples illustrating the methods are given.

5.1   BRIEF REVIEW AND LIMITATIONS OF FFT-BASED APPROACHES

The frequency domain approach to experimental modal analysis is motivated by the solution of the single-degree-of-freedom (SDOF) undamped system with equation of motion

$$m\ddot{x} + kx = f \tag{5.1.1}$$

Rather than solving it by the "time-domain" method of integration (reviewed in Section 2.3), if we assume a sinusoidal forcing function

$$f(t) = f e^{i\omega t}$$

where $i = \sqrt{-1}$, the response or solution will be of the form

$$x(t) = x e^{i\omega t}$$

Hence, the equation of motion (5.1.1) transforms into

$$(-\omega^2 m + k) x e^{i\omega t} = f e^{i\omega t} \tag{5.1.2a}$$

or

$$(-\omega^2 + m^{-1} k) x e^{i\omega t} = m^{-1} f e^{i\omega t} \tag{5.1.2}$$

which yields the response model or the frequency response function (FRF):

$$\text{Receptance} = \frac{\text{Displacement}}{\text{Force}} = \frac{x}{f} = r(\omega) = \frac{1}{(k - \omega^2 m)} \tag{5.1.3a}$$

$$= \frac{m^{-1}}{(m^{-1} k - \omega^2)} \tag{5.1.3}$$

Receptance is also known as admittance, compliance, or dynamic flexibility. Plotted as a function of the frequency ω (particularly on a log–log scale), it is clear from (5.1.3) that the plot of receptance has a peak reaching infinity at the natural frequency $\omega_n = \sqrt{m^{-1}k}$, is constant at low frequency near $\omega = 0$, providing a rough estimate of the stiffness k, and has negative slope at high frequencies, providing a rough estimate of the mass m. Precise recovery of the natural frequency as the frequency at which peak occurs in the FRF is the key to the frequency domain approach; it is most useful for this purpose when the system is nearly undamped.

The FRF may also be alternatively represented by

$$\text{Mobility} = \frac{\text{Velocity}}{\text{Force}} = \frac{\dot{x}}{f} = i\omega r(\omega) = v(\omega) \tag{5.1.4}$$

$$\text{Accelerance} = \frac{\text{Acceleration}}{\text{Force}} = \frac{\ddot{x}}{f} = -\omega^2 r(\omega) = a(\omega) \tag{5.1.5}$$

Accelerance is also known as inertance, and the reverses of the ratios in (5.1.3–5.1.5), f/x, $f/\dot{x}$, and $f/\ddot{x}$, are sometimes called dynamic stiffness, mechanical impedance, and apparent mass, respectively, for obvious reasons. Mobility is the FRF most commonly plotted in the frequency domain approach, although accelerance is the most readily obtained raw data. The log–log plot of the mobility amplitude has a positive slope determined by the stiffness at low frequencies and a negative slope determined by the mass at high frequencies; for the accelerance amplitude the corresponding slopes are positive and constant, respectively. Although we discuss these plots for their intuitive appeal, such plots and inferences from them are no longer commonly used.

In theory, a plot of an FRF may be obtained by sweeping the system with sinusoidal input excitation force over the range of frequencies over which the computed FRF is nonzero. With the advent of computers and FFT, this may now be accomplished much more effectively by using any input excitation force, as long as its Fourier analysis shows that it is nonzero and covers the frequency range over which the FRF extends.

To see why the frequency domain approach based on FFT is most appropriate for the undamped SDOF system (and less so for all others, as further discussed later), consider the excitation

$$f(t) = fe^{st}$$

where $s = \sigma + i\omega$ is a more general complex variable. Repetition of the preceding steps now yields the more general response model or (Laplace) transfer function replacing (5.1.3):

$$r(s) = \frac{m^{-1}}{(m^{-1}k + s^2)} \tag{5.1.6}$$

176   MODAL ANALYSIS

A comparison with (5.1.3) shows that an FRF is a restriction of the more general transfer function evaluated along the imaginary axis $s = i\omega$. Nevertheless, $r(s)$ as a function of two variables σ and ω in $s = \sigma + i\omega$ has a peak reaching infinity at

$$s^2 = -m^{-1}k$$
$$s = \pm i\sqrt{m^{-1}k}$$

which is on the imaginary axis. Thus Fourier analysis and consequently FFT-based frequency domain methods restricting s to $i\omega$ impose no restrictions on the analysis and hence are quite appropriate for undamped SDOF systems. Recalling the discussion of Section 2.9.3, the reader can surmise that this is no longer the case for damped systems, as we now show.

5.1.1 Damped Single-Degree-of-Freedom (SDOF) System

The equation of motion for a damped SDOF system is

$$m\ddot{x} + c\dot{x} + kx = f \tag{5.1.7}$$

which is often called a spatial model as it describes the physical characteristics of the system such as mass, damper, and spring. Sinusoidal forcing function and solution transform it into

$$(-\omega^2 m + i\omega c + k)xe^{i\omega t} = fe^{i\omega t} \tag{5.1.8a}$$

or

$$(-\omega^2 + i\omega m^{-1}c + m^{-1}k)xe^{i\omega t} = m^{-1}fe^{i\omega t} \tag{5.1.8}$$

and give the receptance

$$\frac{x}{f} = r(\omega) = \frac{1}{[(k - \omega^2 m) + i\omega c]} \tag{5.1.9a}$$

$$= \frac{m^{-1}}{[(\omega_n^2 - \omega^2) + i2\zeta\omega_n\omega]} \tag{5.1.9}$$

with mobility and acceleration defined as $i\omega r(\omega)$ and $-\omega^2 r(\omega)$, respectively, as before.

Unlike the undamped case, all FRF's are now complex and must be plotted either as pairs of functions: (1) real and imaginary parts versus frequency, (2) magnitude and phase versus frequency (Bode plot), or alternatively, (3) imaginary part versus real part plotted for various frequencies (Nyquist plot). Using the natural frequency ω_n and damping factor or ratio ζ of (5.1.9) defined by

$$\omega_n^2 = m^{-1}k, \qquad 2\zeta\omega_n = m^{-1}c \tag{5.1.10}$$

these three functions for receptance can be written as

$$\text{Re}[r(\omega)] = \frac{m^{-1}(\omega_n^2 - \omega^2)}{[(\omega_n^2 - \omega^2)^2 + 4\zeta^2\omega_n^2\omega^2]} \quad (5.1.11a)$$

$$\text{Im}[r(\omega)] = \frac{-m^{-1}2\zeta\omega_n\omega}{[(\omega_n^2 - \omega^2)^2 + 4\zeta^2\omega_n^2\omega^2]} \quad (5.1.11b)$$

$$\text{Magnitude}: |r(\omega)| = \frac{m^{-1}}{[(\omega_n^2 - \omega^2)^2 + 4\zeta^2\omega_n^2\omega^2]^{1/2}} \quad (5.1.12a)$$

$$\text{Phase Angle}: \angle r(\omega) = \tan^{-1}\frac{-2\zeta\omega_n\omega}{(\omega_n^2 - \omega^2)} \quad (5.1.12b)$$

those for accelerance $a(\omega)$ are obtained by multiplying the $r(\omega)$ given above by $-\omega^2$, and for mobility $v(\omega)$ they take the form

$$\text{Re}[v(\omega)] = \frac{m^{-1}2\zeta\omega_n\omega^2}{[(\omega_n^2 - \omega^2)^2 + 4\zeta^2\omega_n^2\omega^2]} \quad (5.1.13a)$$

$$\text{Im}[v(\omega)] = \frac{m^{-1}\omega(\omega_n^2 - \omega^2)}{[(\omega_n^2 - \omega^2)^2 + 4\zeta^2\omega_n^2\omega^2]} \quad (5.1.13b)$$

$$\text{Magnitude}: |v(\omega)| = \frac{m^{-1}\omega}{[(\omega_n^2 - \omega^2)^2 + 4\zeta^2\omega_n^2\omega^2]^{1/2}} \quad (5.1.14a)$$

$$\text{Phase Angle}: \angle v(\omega) = \tan^{-1}\frac{(\omega_n^2 - \omega^2)}{2\zeta\omega_n\omega} = \angle r(\omega) + \frac{\pi}{2} \quad (5.1.14b)$$

The ratio of dynamic to static displacement amplitudes, $k|r(\omega)|$, is called the *magnification factor*.

It is clear from (5.1.11a) that the real part of receptance $r(\omega)$ has a peak and sign reversal from positive to negative at the natural frequency ω_n that causes a 180° phase shift in (5.1.12b), whereas for small damping factor $\zeta \ll 1$, the imaginary part of $r(\omega)$ and its magnitude $|r(\omega)|$ given by (5.1.11b and 5.1.12a) have a peak near ω_n; similar results hold for $v(\omega)$ when the roles of real and imaginary parts are interchanged. However, the higher the damping ratio, the shallower the peak, and the more difficult it becomes to pin down the frequency ω_n.

Does the magnitude $|r(\omega)|$ at least *have* a peak, no matter how shallow, for every vibratory system? The answer is, unfortunately, no. Considering the denominator of $|r(\omega)|$ in (5.1.12a), it is easy to show that its minimum exists only if

$$\zeta < \frac{1}{\sqrt{2}} \quad (5.1.15)$$

178 MODAL ANALYSIS

Thus the peak disappears when the damping ratio is in the range $1/\sqrt{2} < \zeta < 1$ even if the basic system does vibrate, and hence cannot be seen even if the exact Fourier transform $r(\omega)$ is plotted.

In contrast, the DDS method attempts to synthesize the transfer function, obtained by choosing the excitation $f(t) = fe^{st}$, $s = \sigma + i\omega$, as before:

$$(s^2 + m^{-1}cs + m^{-1}k)xe^{st} = m^{-1}fe^{st} \qquad (5.1.16)$$

$$r(s) = \frac{m^{-1}}{(s^2 + m^{-1}cs + m^{-1}k)} = \frac{m^{-1}}{(s^2 + 2\zeta\omega_n s + \omega_n^2)} \qquad (5.1.17)$$

which has a peak, reaching infinity, at

$$s = -\zeta\omega_n \pm \omega_n\sqrt{\zeta^2 - 1}$$
$$= -\zeta\omega_n \pm i\omega_n\sqrt{1 - \zeta^2}, \qquad \zeta < 1$$

that is *not* on the imaginary axis unless $\zeta = 0$ and the system is undamped. Thus $r(\omega)$ obtained by Fourier transform is indeed a restriction of $r(s)$:

$$r(\omega) = r(s)|_{s=i\omega} \qquad (5.1.18)$$

and would provide a distorted and often misleading picture of $r(s)$ when the damping is significantly large. Note that this is the case even if the exact $r(\omega)$ could be obtained by Fourier transform; in practice one has to use FFT, which introduces its own distortions, as discussed in Section 2.9.

This fundamental limitation of the Fourier transform in modal analysis, discussed somewhat abstractly in Section 2.9.3 (see, particularly, Fig. 2.9), is graphically depicted in Figure 5.1. The figure shows $r(s)$ as a function of σ and ω. The relation $r(\omega) = r(s)|_{s=i\omega}$ is depicted as a cut of the surface $r(s)$ by the plane perpendicular to the complex plane along the frequency (ω) or imaginary axis. Although it is drawn for a two-degree-of-freedom system (to be discussed shortly), it can be used to understand the SDOF relations of this section.

The Fourier transform $r(\omega)$ drawn by the thick line is only a slice along the frequency or imaginary axis of the surface $r(\omega)$ shown by light lines. Note that the low-frequency peak with lower damping has survived in the cut, but the high-frequency peak with higher damping has disappeared. If the crest of the mountain is along the frequency or imaginary axis, as in the case of an undamped system, the cut will be fully representative. Otherwise, Fourier transform will be an attempt to scale a mountain from the foothills without climbing it! Its practical implications will be illustrated at the end of Section 5.5.4.

The real and imaginary parts of the surface $r(s)$ and its section $r(\omega)$ by Fourier transform are depicted in Figures 5.2 and 5.3, respectively. The same limitations discussed above are apparent. The cuts are even less representative than in the magnitude plot.

BRIEF REVIEW AND LIMITATIONS OF FFT-BASED APPROACHES

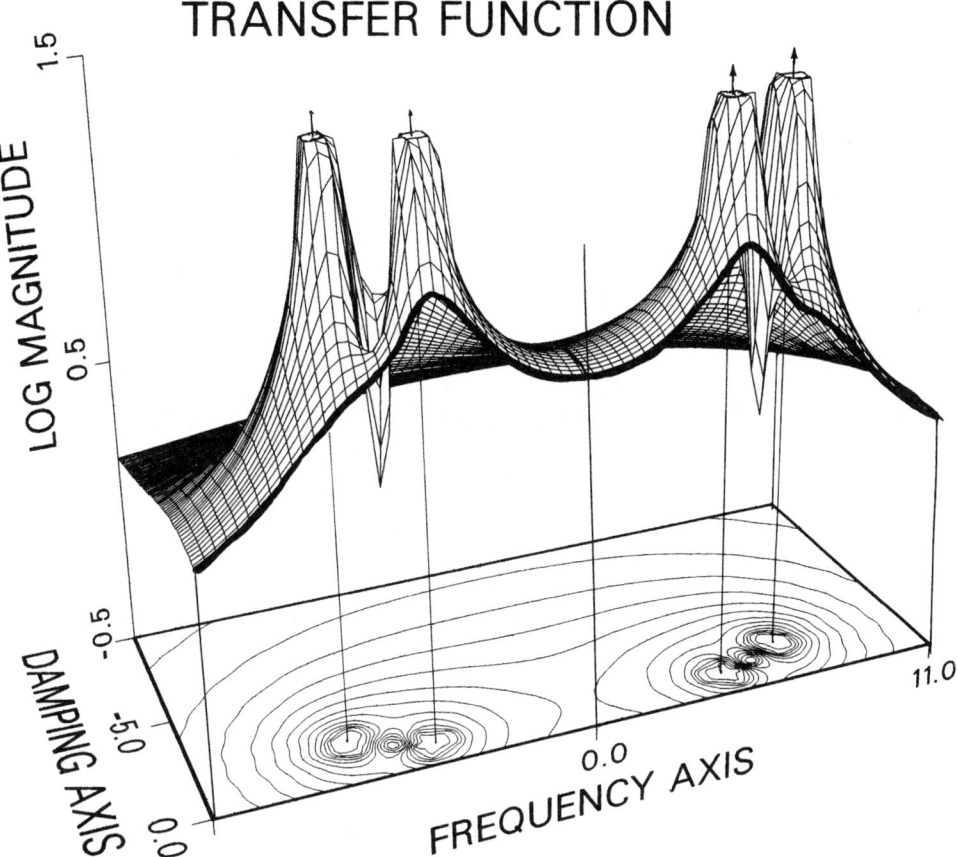

Figure 5.1 Fourier transform as a cut of Laplace transform—magnitude.

Recall that these continuous Fourier transform cuts will be discretized by Fourier series when finite record length is used and aliased by the sampling process when FFT is used in practical implementation. These further distortions due to leakage and aliasing are illustrated in Section 5.1.6.

One may now ask: How is it that the frequency domain techniques based on FFT are so popular in spite of these limitations? Is it because most systems in practice are very nearly undamped? The answer to this, again, is no. In fact, although Fourier transform is ideal for very nearly undamped systems, FFT is not. The limited frequency resolution, imposed by the discretization of Fourier series due to the finite record length, causes estimation difficulties with the rapidly changing slopes of sharp peaks.

There are indeed many mechanical systems in practice with damping not so low as to cause problems owing to frequency resolution of Fourier series, and yet not so high as to cause problems owing to the limitations of Fourier transform. However, it is to the ingenuity of a generation of experimentalists

180 MODAL ANALYSIS

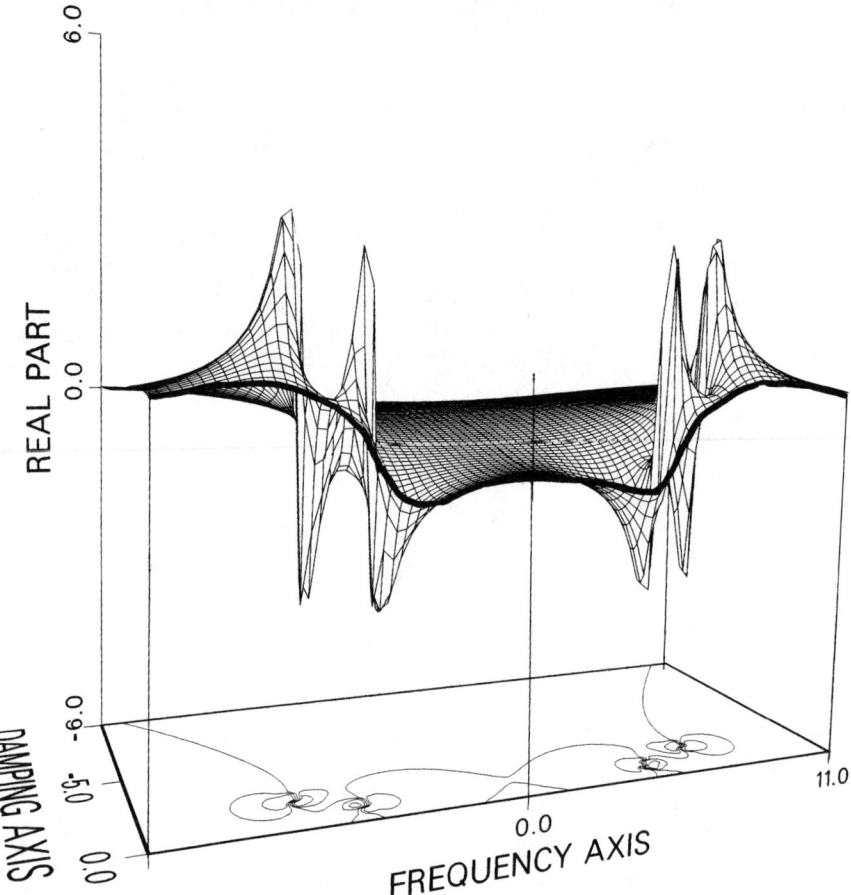

Figure 5.2 Fourier transform as a cut of Laplace transform—real part.

intent upon exploiting the advantage of computers and FFT that we owe the widespread use of FFT in modal analysis. They have provided us with approximations good enough in practice to quickly obtain rough estimates of the modal parameters. Although we will not need them for the more precise time domain DDS method used later, we will review some of these methods for the valuable insight they provide.

The simplest and classical among these is the so called peak-amplitude or peak-picking method that uses the symmetric nature of $|r(\omega)|$ near a resonance peak given by (5.1.12a). Ignoring the small damping, the frequency at which the peak occurs is taken as the natural frequency ω_n. If the "half power" frequencies ω_1 and ω_2 are chosen to the left and right of ω_n such that

$$|r(\omega_1)| = |r(\omega_2)| = \frac{|r(\omega_n)|}{\sqrt{2}} = 0.707|r(\omega_n)| \qquad (5.1.19a)$$

BRIEF REVIEW AND LIMITATIONS OF FFT-BASED APPROACHES **181**

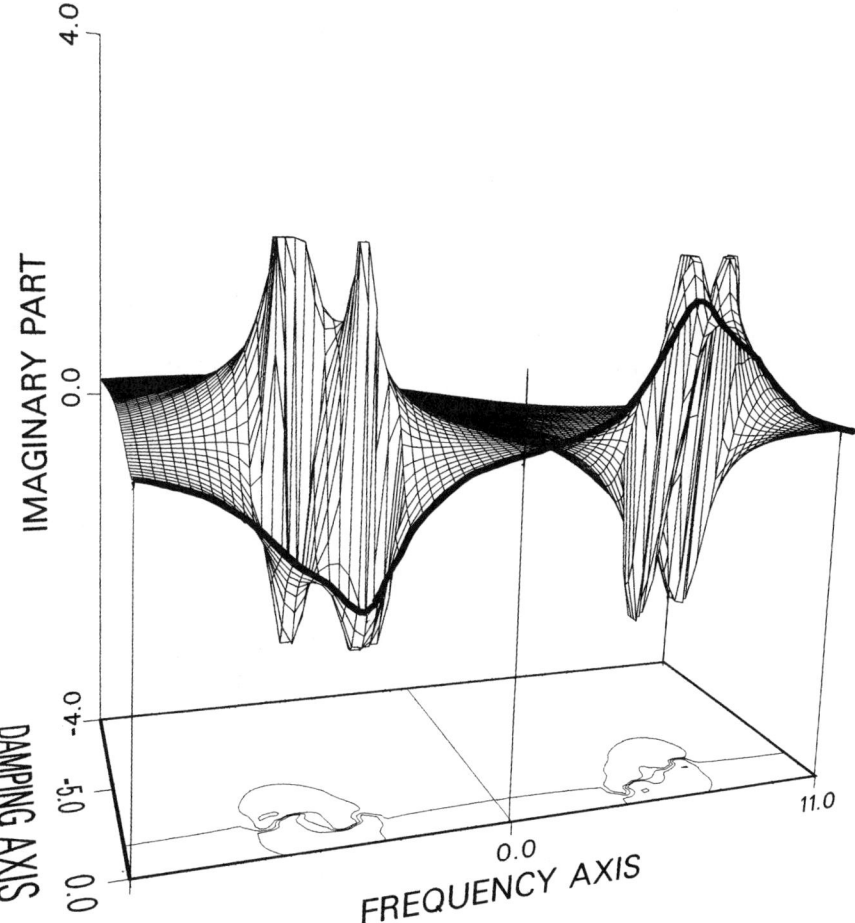

Figure 5.3 Fourier transform as a cut of Laplace transform—imaginary part.

then the damping ratio ζ can be obtained from the "bandwidth" [e.g., see Tse, Morse, and Hinkle (1978)]

$$\frac{\omega_2 - \omega_1}{\omega_n} \simeq 2\zeta \qquad (5.1.19b)$$

Knowing ω_n and ζ, one can use (5.1.12a) to solve for m^{-1} by

$$m^{-1} \simeq 2\zeta\omega_n^2 |r(\omega_n)| \qquad (5.1.20)$$

and the relations $\omega_n = \sqrt{m^{-1}k}$, $2\zeta\omega_n = m^{-1}c$ now yield c and k.

Although simple and intuitively appealing, serious limitations of this method arise from its total reliance on only three points of the FRF in the resonance region where it is most error prone. Apart from the concentration of measurement errors in this region, if damping is low and the resonant peak is sharp, ω_n is accurately determined, but the estimates of the peak height $|r(\omega_n)|$ and ω_1, ω_2 are poor because of the sharp slopes; on the other hand, if damping is high and the peak is shallow, the peak height $|r(\omega_n)|$ is accurately determined but ω_n and hence ω_1 and ω_2 are hard to pin down. Moreover, the method does not extend to a multiple-degree-of-freedom system since the contribution from other modes generally leads to overestimation of damping. A frequency domain method which addresses these limitations by taking into account more points around resonance, and hence naturally involves much more computation, is the circle-fit method.

5.1.2 Circle-Fit Method and Structural Damping

The third, Nyquist plot of FRF of a properly chosen parameter can be shown to be an exact circle; this is true not only of the viscous damping we have considered so far, but also of the so-called structural or hysteretic damping. Recall that the viscous or linear damping force $c\dot{x}$ assumes that the damping coefficient or rate c does not depend upon frequency. A popular alternative for real structures has been a damping rate that varies inversely with frequency, called hysteretic or structural damping:

$$c = \frac{h}{\omega} \qquad (5.1.21)$$

Although the differential equation does not make sense in such a case, the FRF of receptance can be written in contrast to the viscous as

$$\text{Structural:} \quad r(\omega) = \frac{1}{(k - \omega^2 m) + ih} \qquad (5.1.22)$$

$$\text{Viscous:} \quad r(\omega) = \frac{1}{(k - \omega^2 m) + i\omega c} \qquad (5.1.9a)$$

with mobility $v(\omega) = i\omega r(\omega)$ and accelerance or inertance $a(\omega) = -\omega^2 r(\omega)$ defined as before.

With some algebraic manipulation, it can be shown [e.g., see Ewins (1985)] that in the case of structural damping a plot of imaginary versus real parts of receptance for different frequencies is a circle of radius $1/2h$ with center $(0, -1/2h)$, that is,

$$[\text{Re}(r)]^2 + \left[\text{Im}(r) + \frac{1}{2h}\right]^2 = \left(\frac{1}{2h}\right)^2 \qquad (5.1.23)$$

whereas, in the case of viscous damping, a similar relation is true for the mobility plot:

$$\left[\text{Re}(v) - \frac{1}{2c}\right]^2 + [\text{Im}(v)]^2 = \left(\frac{1}{2c}\right)^2 \tag{5.1.24}$$

that is, the plot of imaginary versus real part of mobility for different frequencies is a circle of radius $1/2c$ with center $(1/2c, 0)$.

By selecting a few points on both sides of the resonance peak, a circle which minimizes the least squares error of these points on a Nyquist plot can be fitted. If these points are selected at equally spaced frequencies (as is automatic in FFT), then the center point of the largest arc on the circle between two consecutive frequency points gives the natural frequency; other approximations include choosing frequency of the point on the circle with (1) maximum magnitude (distance from the origin), (2) maximum imaginary part, or (3) zero real part.

Once the natural frequency is determined, damping estimates can be obtained from any pair of points on the circle, with frequencies ω_a above and ω_b below the natural frequency ω_n and subtending angles θ_a and θ_b at the center measured from the radius joining ω_n, by the expressions (see Fig. 5.4)

$$\text{Structural:} \quad \eta = \frac{h}{k} = \frac{(\omega_a^2 - \omega_b^2)}{\omega_n^2[\tan(\theta_a/2) + \tan(\theta_b/2)]} \tag{5.1.25}$$

$$\text{Viscous:} \quad \zeta = \frac{(\omega_a^2 - \omega_b^2)}{2\omega_n[\omega_a\tan(\theta_a/2) + \omega_b\tan(\theta_b/2)]} \tag{5.1.26}$$

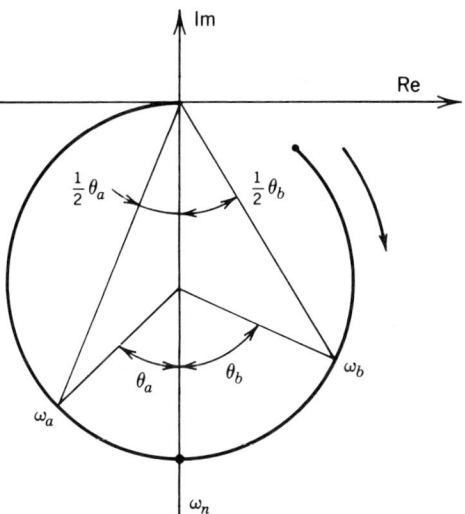

Figure 5.4 Modal circle.

where η is called the structural damping loss factor. These estimates should be the same in theory for any pair, but they are generally different in practice and their average is used as the correct value; the scatter from this average is often used as an indication of the quality of data and analysis. A systematic trend in the scatter may be caused by factors such as experimental error, nonlinearity, and inappropriate assumption.

After determining the natural frequency and damping, the remaining parameters can be determined from the diameter of the circle to yield m, c, and k in the SDOF (and ω_n, ζ, and the modal constant in the multiple-degree-of-freedom) system. Usually the errors in ω_n are small, but even small errors in ω_n can lead to large errors in the other parameters.

An alternative to the circle-fit method called the "inverse method," which is not as widely used, employs the inverse

$$r^{-1}(\omega) = (k - \omega^2 m) + ih \qquad (5.1.27)$$

Now the (Nyquist) plot of imaginary versus real part for different frequencies is a horizontal straight line parallel to the x axis at a distance $h = k\eta$ from the origin. A least squares fitting of the imaginary versus real gives the damping; another least squares fitting of the real parts of the measured FRF from this fitted line then yields m and k. Again the mobility $v(\omega)$ is used for viscous damping.

5.1.3 Undamped Multiple-Degree-of-Freedom (MDOF) System

We will begin with an undamped n-degree-of-freedom system with equations of motion in matrix form

$$\mathbf{m\ddot{x}} + \mathbf{kx} = \mathbf{f} \qquad (5.1.28)$$

where $\mathbf{m}$ and $\mathbf{k}$ and $n \times n$ mass and stiffness (constant) matrices and $\mathbf{x}$ and $\mathbf{f}$ are $n \times 1$ vectors of displacements and forces that are functions of time. Conceptually, the transfer function and FRF can be derived in a way analogous to the SDOF system, but now these will be matrix functions, each element of the matrix relating an element of the response vector with an element of the force vector. Therefore, we will emphasize the matrix theory aspects rather than the physical systems which may be found in standard vibration texts; moreover, the matrix theory aspects are useful in the DDS approach to modal analysis discussed later.

We may try to repeat the development of equations (5.1.1–5.1.3) by assuming a vector of sinusoidal forcing functions, all of the same frequency but possibly different amplitudes and phases:

$$\mathbf{f}(t) = \mathbf{f}e^{i\omega t}$$

This assumption is not restrictive since we can choose any forcing function.

However, the next step, that the response is to be of the same form

$$\mathbf{x}(t) = \mathbf{x}e^{i\omega t}$$

is restrictive, because it assumes that the whole system is capable of vibrating at a single frequency ω at all the points of measurements. If this restrictive assumption is true, then, as before, the equation of motion (5.1.28) transforms into

$$(-\omega^2 \mathbf{m} + \mathbf{k})\mathbf{x}e^{i\omega t} = \mathbf{f}e^{i\omega t} \qquad (5.1.29a)$$

and if $\mathbf{m}$ is nonsingular

$$(-\omega^2 \mathbf{I} + \mathbf{m}^{-1}\mathbf{k})\mathbf{x}e^{i\omega t} = \mathbf{m}^{-1}\mathbf{f}e^{i\omega t} \qquad (5.1.29)$$

so that the matrix FRF can be defined analogous to scalar receptance FRF (5.1.3a) as

$$\mathbf{H}(\omega) = (\mathbf{k} - \omega^2 \mathbf{m})^{-1} \qquad (5.1.30)$$

with the corresponding (Laplace) transfer function

$$\mathbf{H}(s) = (\mathbf{k} + s^2 \mathbf{m})^{-1} \qquad (5.1.31)$$

$$\mathbf{H}(\omega) = \mathbf{H}(s)|_{s=i\omega} \qquad (5.1.32)$$

generalizing (5.1.6) and (5.1.18) to matrix forms.

How can we solve (5.1.29a)? Recall from Section 4.1 that a vector differential equation can be transformed into scalar uncoupled equations by diagonalizing the state matrix. To solve (5.1.29a) in this way requires not only that $\mathbf{m}$ and $\mathbf{k}$ each be diagonalizable (see Section 3.5.1 for definition), but that they be simultaneously diagonalizable by the same matrix, which is possible if and only if $\mathbf{m}$ and $\mathbf{k}$ commute, that is, $\mathbf{mk} = \mathbf{km}$. Is this enough? The answer is no. To see this, let $\mathbf{M}$ diagonalize both $\mathbf{m}$ and $\mathbf{k}$ that commute; then $\mathbf{M}$ also diagonalizes $\mathbf{m}^{-1}\mathbf{k}$:

$$\mathbf{MmM}^{-1} = \text{diag}\{\mu_{mi}\} \qquad (5.1.33a)$$

$$\mathbf{MkM}^{-1} = \text{diag}\{\mu_{ki}\} \qquad (5.1.33b)$$

$$\mathbf{Mm}^{-1}\mathbf{kM}^{-1} = \text{diag}\{\mu_{ki}/\mu_{mi}\} \qquad (5.1.33c)$$

Then, for $\mathbf{f} = \mathbf{0}$, the homogeneous equation from (5.1.28) decouples into n scalar equations in q_i, the elements of $\mathbf{q} = \mathbf{M}^{-1}\mathbf{x}$:

$$\mu_{mi}\ddot{q}_i + \mu_{ki}q_i = 0, \quad i = 1, 2, \ldots, n \qquad (5.1.34)$$

which give the eigenvalue equations corresponding to (5.1.29a and 5.1.29) as

$$-\mu_{mi}\omega^2 + \mu_{ki} = 0, \quad i = 1, 2, \ldots, n \quad (5.1.35a)$$

$$-\omega^2 + \frac{\mu_{ki}}{\mu_{mi}} = 0, \quad i = 1, 2, \ldots, n \quad (5.1.35b)$$

This shows that the squared natural frequencies of the system $\mu_{mi}^{-1}\mu_{ki}$, which are the eigenvalues of $\mathbf{m}^{-1}\mathbf{k}$, must be real and nonnegative, in addition to a diagonalizable nonsingular $\mathbf{m}$ commuting with a diagonalizable $\mathbf{k}$, for the solution to be physically meaningful. Recall that an $n \times n$ matrix is diagonalizable if and only if it has n linearly independent eigenvectors that form the columns of the modal matrix $\mathbf{M}$ and are often called "modal vectors." Being an eigenvector, a modal vector is not unique: a modal vector multiplied by a constant is also a modal vector. Therefore $\mathbf{M}$ is not unique, but μ_{mi} and μ_{ki} are; it is easy to see from (5.1.33c) that the squared frequencies $\omega_i^2 (= \mu_{ki}/\mu_{mi})$ are unique and can be more readily found by the roots of the characteristic equation of (5.1.29):

$$|-\omega^2 \mathbf{I} + \mathbf{m}^{-1}\mathbf{k}| = (\omega_1^2 - \omega^2)(\omega_2^2 - \omega^2) \cdots (\omega_n^2 - \omega^2) = 0 \quad (5.1.35)$$

which is clearly the matrix analog of the scalar equation $\omega_n^2 = m^{-1}k$ for the SDOF system. Moreover, in the present undamped case, the modal vectors as (nonunique) eigenvectors can be chosen to be real because the eigenvalues (squared frequencies) are real.

If the modal/eigenvectors are denoted by $\mathbf{m}_i$ (do not confuse with the unsubscripted mass *matrix* $\mathbf{m}$) so that

$$\mathbf{M} = (\mathbf{m}_1, \mathbf{m}_2, \ldots, \mathbf{m}_n)$$

and

$$\mathbf{M}^{-1} = (\bar{\mathbf{r}}_1, \bar{\mathbf{r}}_2, \ldots, \bar{\mathbf{r}}_n)^T$$

then using (5.1.33a–c) gives the spectral representation (see Section 3.5.2 for review) of the FRF matrix (5.1.30):

$$\mathbf{H}(\omega) = (\mathbf{k} - \omega^2 \mathbf{m})^{-1}$$

$$= \mathbf{M} \operatorname{diag}\{\mu_{ki} - \omega^2 \mu_{mi}\}^{-1} \mathbf{M}^{-1}$$

$$= (\mathbf{m}_1, \mathbf{m}_2, \ldots, \mathbf{m}_n) \operatorname{diag}\left\{\frac{\mu_{mi}^{-1}}{(\omega_i^2 - \omega^2)}\right\} \begin{bmatrix} \bar{\mathbf{r}}_1^T \\ \bar{\mathbf{r}}_2^T \\ \vdots \\ \bar{\mathbf{r}}_n^T \end{bmatrix}$$

$$= \sum_{i=1}^{n} \frac{\mu_{mi}^{-1}}{(\omega_i^2 - \omega^2)} \mathbf{m}_i \bar{\mathbf{r}}_i^T \quad (5.1.36)$$

showing that a typical element $H_{kl}(\omega)$ relating the response at the kth point with the excitation force at the lth point is given by

$$H_{kl}(\omega) = \sum_{i=1}^{n} \frac{\mu_{mi}^{-1} m_{ki} \bar{r}_{li}}{(\omega_i^2 - \omega^2)} \quad (5.1.36a)$$

where the ith *row* of $\mathbf{M}^{-1}$ is given by $\bar{\mathbf{r}}_i^T = (\bar{r}_{1i}, \bar{r}_{2i}, \ldots, \bar{r}_{ni})$. Remembering that μ_{mi}^{-1} is the ith element of diagonalized $\mathbf{m}^{-1}$, this equation compares well with (5.1.3) for the SDOF system. It shows that all the elements of the matrix FRF of an MDOF system are the weighted sums of the corresponding scalar FRF's of SDOF systems with masses and stiffnesses given by the diagonalized $\mathbf{m}$ and $\mathbf{k}$, and the weights controlled by the modal vectors.

Some simplifications are possible in classical mechanics when the equations of motion are obtained by physical considerations, particularly using energy methods generally followed in standard vibration texts. Since the kinetic and potential energies are given by

$$T = \tfrac{1}{2} \dot{\mathbf{x}}^T \mathbf{m} \dot{\mathbf{x}}, \qquad U = \tfrac{1}{2} \mathbf{x}^T \mathbf{k} \mathbf{x}$$

respectively, $\mathbf{m}$ and $\mathbf{k}$ are not only real, but also can be assumed to be symmetric without loss of generality, because if they are *not* symmetric we can replace them by the symmetric matrices $\tfrac{1}{2}(\mathbf{m} + \mathbf{m}^T)$ and $\tfrac{1}{2}(\mathbf{k} + \mathbf{k}^T)$ without altering the energies. (Maxwell's reciprocity theorem may be invoked when the deflections are measured from a fixed position in space.) Moreover, since the kinetic energy is always positive when there is motion, $\mathbf{m}$ is not only real and symmetric, but usually positive definite. For real symmetric $\mathbf{m}$ and $\mathbf{k}$ and positive definite $\mathbf{m}$, they can be simultaneously diagonalized by a simpler transformation using transpose instead of inverse, illustrating the simpler spectral representation of Section 3.5.2.

To see this, recall from Section 3.5.1 that a real symmetric matrix has real eigenvalues and we can choose a real orthogonal eigenvector matrix so that $\mathbf{m}$ has a spectral representation, say

$$\mathbf{m} = \mathbf{R} \mathbf{D}_m \mathbf{R}^T, \qquad \mathbf{R} \mathbf{R}^T = \mathbf{I}$$

Moreover, since $\mathbf{m}$ is positive definite, its eigenvalues forming the elements of the diagonal matrix $\mathbf{D}_m$ are positive, hence

$$\mathbf{m} = \mathbf{R} \sqrt{\mathbf{D}_m} \sqrt{\mathbf{D}_m} \mathbf{R}^T = \mathbf{T} \mathbf{T}^T, \qquad \mathbf{T} = \mathbf{R} \sqrt{\mathbf{D}_m} \quad (5.1.37)$$

In fact, we know from Section 3.4.7 that $\mathbf{T}$ can even be chosen to be triangular.

Now, even if $\mathbf{m}^{-1}$ and $\mathbf{k}$ are symmetric $\mathbf{m}^{-1}\mathbf{k}$ is not, unless $\mathbf{m}$ and $\mathbf{k}$ commute. In spite of this, a real nonsymmetric $\mathbf{m}^{-1}\mathbf{k}$ will have real eigenvalues (squared natural frequencies) if we can find a real symmetric matrix that has the same eigenvalues as $\mathbf{m}^{-1}\mathbf{k}$. A simple calculation,

$$|-\omega^2\mathbf{m} + \mathbf{k}| = |-\omega^2\mathbf{T}\mathbf{T}^T + \mathbf{k}| = 0 = |-\omega^2\mathbf{I} + \mathbf{T}^{-1}\mathbf{k}\mathbf{T}^{-T}|$$
$$= |-\omega^2\mathbf{I} + \mathbf{m}^{-1}\mathbf{k}|$$

shows that $\mathbf{T}^{-1}\mathbf{k}\mathbf{T}^{-T}$ is such a matrix, which therefore has a spectral representation by a real orthogonal matrix

$$\mathbf{T}^{-1}\mathbf{k}\mathbf{T}^{-T} = \mathbf{P}\mathbf{D}_{k/m}\mathbf{P}^T \qquad \mathbf{P}\mathbf{P}^T = \mathbf{I} = \mathbf{P}^T\mathbf{P} \qquad (5.1.38)$$

that is,

$$\mathbf{P}^T\mathbf{T}^{-1}\mathbf{k}\mathbf{T}^{-T}\mathbf{P} = \mathbf{D}_{k/m} \qquad (5.1.38a)$$

Then the modal matrix $\mathbf{S} = \mathbf{T}^{-T}\mathbf{P}$ does the job, as it is easy to verify that

$$\mathbf{S}^T\mathbf{m}\mathbf{S} = \mathbf{I}, \qquad \text{that is,} \qquad \mathbf{m}^{-1} = \mathbf{S}\mathbf{S}^T \qquad (5.1.39a)$$
$$\mathbf{S}^T\mathbf{k}\mathbf{S} = \mathbf{D}_{k/m} = \mathbf{S}^{-1}\mathbf{m}^{-1}\mathbf{k}\mathbf{S} \qquad (5.1.39b)$$

which replace (5.1.33a–c) for the real symmetric $\mathbf{m}$ and $\mathbf{k}$ and positive definite $\mathbf{m}$ case. Note that the second equation of (5.1.39b) is a spectral representation of $\mathbf{m}^{-1}\mathbf{k}$ in that $\mathbf{D}_{k/m}$ is a diagonal matrix of eigenvalues of $\mathbf{m}^{-1}\mathbf{k}$, whereas the first equation of (5.1.39b) and (5.1.39a) are not, because $\mathbf{S}\mathbf{S}^T \neq \mathbf{I}$. In contrast, all three equations (5.1.33a–c) are spectral representations.

If $\mathbf{k}$ is also positive definite, then so is $\mathbf{m}^{-1}\mathbf{k}$, its eigenvalues are then positive and can be denoted by the squared natural frequencies ω_i^2 to rewrite (5.1.39b) as

$$\mathbf{S}^T\mathbf{k}\mathbf{S} = \text{diag}\{\omega_i^2\} = \mathbf{S}^{-1}\mathbf{m}^{-1}\mathbf{k}\mathbf{S} \qquad (5.1.39c)$$

The unique ω_i^2 can be found from the characteristic equation

$$|-\omega^2\mathbf{I} + \mathbf{m}^{-1}\mathbf{k}| = (\omega_1^2 - \omega^2)(\omega_2^2 - \omega^2)\cdots(\omega_n^2 - \omega^2) = 0$$

as before, and the unique eigenvectors $\mathbf{s}_i$, the columns of $\mathbf{S}$ corresponding to eigenvalue ω_i^2, by

$$(-\omega_i^2\mathbf{m} + \mathbf{k})\mathbf{s}_i = \mathbf{0}, \qquad i = 1, 2, \ldots, n$$

normalized by $\mathbf{S}^T\mathbf{m}\mathbf{S} = \mathbf{I}$, which is called mass normalization for obvious reasons. The modal/eigenvector $\mathbf{s}_i$ is called mass-normalized mode shape corresponding to the ith mode of frequency ω_i. Thus $\mathbf{S}$ is the mass-normalized modal matrix. $\mathbf{s}_i$ are also referred to as normal modes and the new coordinates $\mathbf{q}$ defined by $\mathbf{x} = \mathbf{S}\mathbf{q}$ are called normal coordinates.

Other normalizations are possible. Since $\mathbf{m}$ and $\mathbf{k}$ can be simultaneously diagonalized as shown above, we may choose eigenvectors $\mathbf{u}_i$

$$(-\omega_i^2\mathbf{m} + \mathbf{k})\mathbf{u}_i = \mathbf{0}, \qquad i = 1, 2, \ldots, n$$

BRIEF REVIEW AND LIMITATIONS OF FFT-BASED APPROACHES 189

so that $\mathbf{U} = (\mathbf{u}_1, \mathbf{u}_2, \ldots, \mathbf{u}_n)$ diagonalizes $\mathbf{m}$ and $\mathbf{k}$:

$$\mathbf{U}^T \mathbf{m} \mathbf{U} = \text{diag}\{m_i\} \tag{5.1.40a}$$

$$\mathbf{U}^T \mathbf{k} \mathbf{U} = \text{diag}\{k_i\} \tag{5.1.40b}$$

where m_i and k_i are often referred to as modal or generalized mass and stiffness corresponding to mode i. Being a matrix of eigenvectors the modal matrix $\mathbf{U}$ is not unique, neither are m_i and k_i, but their ratio $k_i/m_i = \omega_i^2$ is unique because

$$\mathbf{U}^{-1} \mathbf{m}^{-1} \mathbf{k} \mathbf{U} = \text{diag}\{k_i/m_i\} = \text{diag}\{\omega_i^2\} \tag{5.1.40c}$$

Note the similarity and contrast of (5.1.40a–c) with (5.1.33a–c). The eigenvectors $\mathbf{u}_i$ now may be arbitrarily normalized by choosing the first element unity, or length to be unity, and so on. The corresponding modes or coordinates are referred to as principal modes or coordinates in classical vibration theory. They are related to the normal modes by the simple relation

$$\mathbf{U} = \text{diag}\{\sqrt{m_i}\}\mathbf{S}, \quad \text{that is,} \quad \mathbf{u}_i = \sqrt{m_i}\,\mathbf{s}_i, \quad i = 1, 2, \ldots, n \tag{5.1.41}$$

Finally, just as the spectral representation of $\mathbf{m}$ and $\mathbf{k}$ requiring the inverse of the modal matrix $\mathbf{M}$ in (5.1.33a–c) was used to obtain the spectral representation of the FRF matrix in (5.1.36), the simpler representations (5.1.39a–c) or (5.1.40a–c) requiring only the transposes of the modal matrices $\mathbf{S}$ and $\mathbf{U}$ provide the representation of the FRF matrix (5.1.30) as

$$\mathbf{H}(\omega) = \mathbf{S}\,\text{diag}\{(\omega_i^2 - \omega^2)\}^{-1}\mathbf{S}^T = \sum_{i=1}^{n} \frac{\mathbf{s}_i \mathbf{s}_i^T}{(\omega_i^2 - \omega^2)} \tag{5.1.41a}$$

$$= \mathbf{U}\,\text{diag}\{m_i(\omega_i^2 - \omega^2)\}^{-1}\mathbf{U}^T = \sum_{i=1}^{n} \frac{\mathbf{u}_i \mathbf{u}_i^T}{m_i(\omega_i^2 - \omega^2)} \tag{5.1.41b}$$

with a typical element given by

$$H_{kl}(\omega) = \sum_{i=1}^{n} \frac{s_{ki} s_{li}}{(\omega_i^2 - \omega^2)} \tag{5.1.41c}$$

$$= \sum_{i=1}^{n} \frac{u_{ki} u_{li}}{m_i(\omega_i^2 - \omega^2)} \tag{5.1.41d}$$

Again, comparing with (5.1.3) shows that each FRF of the MDOF system is a weighted sum of SDOF FRF's with the modal mass m_i and weighting factor given by the product of modal vector elements. Note that neither m_i nor u_{ki}, u_{li} are unique, but the ratio $(u_{ki} u_{li}/m_i)$ is unique and equal to $(s_{ki} s_{li})$, variously called the modal constant, modal participation factor, or residue; the last term is used because the relation (5.1.41c–d) can also be derived by applying partial fraction expansion to each element $H_{kl}(\omega)$ that is a ratio of two polynomials in ω^2, the roots of the denominator or "poles" being ω_i^2.

5.1.4 Damped Multiple-Degree-of-Freedom (MDOF) System

A damped n-degree-of-freedom system has the matrix equation of motion

$$\mathbf{m\ddot{x} + c\dot{x} + kx = f} \tag{5.1.42}$$

where, in addition to the mass and stiffness matrices $\mathbf{m}$ and $\mathbf{k}$ as before, $\mathbf{c}$ is the $n \times n$ damping matrix. If the system is capable of vibrating at a single frequency at all points of measurements, then

$$(-\omega^2\mathbf{m} + i\omega\mathbf{c} + \mathbf{k})\mathbf{x}e^{i\omega t} = \mathbf{f}e^{i\omega t} \tag{5.1.43a}$$

Thus, for this system to have n linearly independnent modal/eigenvectors requires that all $\mathbf{m}$, $\mathbf{c}$, and $\mathbf{k}$ be simultaneously diagonalizable, which, in turn, requires that they commute. However, for a nonsingular $\mathbf{m}$, this equation reduces to

$$(-\omega^2\mathbf{I} + i\omega\mathbf{m}^{-1}\mathbf{c} + \mathbf{m}^{-1}\mathbf{k})\mathbf{x}e^{i\omega t} = \mathbf{m}^{-1}\mathbf{f}e^{i\omega t} \tag{5.1.43}$$

and since the identity matrix commutes with all matrices, it suffices that $\mathbf{m}^{-1}\mathbf{c}$ commutes with $\mathbf{m}^{-1}\mathbf{k}$.

The general case can be treated by the spectral representation of the state matrix that will be discussed in Section 5.3.1. The simple approach adopted for the undamped case at the beginning of the last section does not work, because a glance at (5.1.43) shows that when $\mathbf{c} = \mathbf{0}$, we get an eigenvalue problem with eigenvalues $\mu = \omega^2$, whereas when $\mathbf{c} \neq \mathbf{0}$ we do not. Even if we could solve for ω by equating the determinant of the matrix in parenthesis of (5.1.43) to zero, we would have to solve the second-order complex matrix polynomial

$$(-\omega^2\mathbf{I} + i\omega\mathbf{m}^{-1}\mathbf{c} + \mathbf{m}^{-1}\mathbf{k})\mathbf{x} = \mathbf{0}$$

to get the eigenvectors $\mathbf{x}$. In fact, recall from Chapter 2 that the state space approach converts a second-order differential equation into a first-order equation that yields the standard eigenvalue problem involving a linear equation or a first-order polynomial.

When $\mathbf{m}$ and $\mathbf{k}$ are real symmetric and $\mathbf{m}$ positive definite, the classical mechanics approach for the undamped case, using the transpose instead of the inverse of the modal matrix, can be readily extended to the damped case if suitable assumptions are made on $\mathbf{c}$. Two such commonly made assumptions, justifiable on physical grounds, are proportional damping and structural or hysteretic damping. We will review these before giving the standard vibration theory version that still uses the transpose instead of the inverse of the modal matrix.

Proportional Damping
The analysis of the undamped case in the last section, based on real symmetric and positive definite $\mathbf{m}$, $\mathbf{k}$ will simply extend to the damped case if the (real) modal matrix that diagonalizes $\mathbf{m}$ and $\mathbf{k}$ also happens to diagonalize $\mathbf{c}$. Then we have a repetition of (5.1.39a–c) and (5.1.40a–c) with the addition of $\mathbf{c}$:

$$\mathbf{S}^T\mathbf{m}\mathbf{S} = \mathbf{I}, \quad \text{that is,} \quad \mathbf{m}^{-1} = \mathbf{S}\mathbf{S}^T; \quad \mathbf{S}^T\mathbf{k}\mathbf{S} = \mathbf{S}^{-1}\mathbf{m}^{-1}\mathbf{k}\mathbf{S} = \text{diag}\{\omega_i^2\}$$
$$\mathbf{S}^T\mathbf{c}\mathbf{S} = \mathbf{S}^{-1}\mathbf{m}^{-1}\mathbf{c}\mathbf{S} = \text{diag}\{2\zeta_i\omega_i\}$$
(5.1.44)

for mass normalization (normal modes), which, substituting $\mathbf{x} = \mathbf{S}\mathbf{q}$, decouples (5.1.42) into

$$\ddot{q}_i + 2\zeta_i\omega_i\dot{q}_i + \omega_i^2 q_i = (\mathbf{S}^T\mathbf{f})_i, \quad i = 1, 2, \ldots, n \quad (5.1.45)$$

where $(\mathbf{S}^T\mathbf{f})_i$ denotes the ith element of the column $\mathbf{S}^T\mathbf{f}$, and

$$\mathbf{U}^T\mathbf{m}\mathbf{U} = \text{diag}\{m_i\}, \quad \mathbf{U}^T\mathbf{k}\mathbf{U} = \text{diag}\{k_i\}$$
$$\mathbf{U}^T\mathbf{c}\mathbf{U} = \text{diag}\{c_i\}$$
(5.1.46)

for principal modes that decouples (5.1.42) with $\mathbf{x} = \mathbf{U}\mathbf{q}$ into

$$m_i\ddot{q}_i + c_i\dot{q}_i + k_i q_i = (\mathbf{U}^T\mathbf{f})_i, \quad i = 1, 2, \ldots, n \quad (5.1.47)$$

where $(\mathbf{U}^T\mathbf{f})_i$ denotes the ith element of the column $\mathbf{U}^T\mathbf{f}$. Hence the FRF matrix has now the corresponding representations:

$$\mathbf{H}(\omega) = \mathbf{S}\,\text{diag}\{(\omega_l^2 - \omega^2) + i2\zeta_l\omega_l\omega\}^{-1}\mathbf{S}^T$$

$$= \sum_{l=1}^{n} \frac{\mathbf{s}_l\mathbf{s}_l^T}{[(\omega_l^2 - \omega^2) + i2\zeta_l\omega_l\omega]} \quad (5.1.48a)$$

$$= \mathbf{U}\,\text{diag}\{m_l(\omega_l^2 - \omega^2) + ic_l\omega\}^{-1}\mathbf{U}^T$$

$$= \sum_{l=1}^{n} \frac{\mathbf{u}_l\mathbf{u}_l^T}{[(k_l - \omega^2 m_l) + ic_l\omega]} \quad (5.1.48b)$$

$$H_{rt}(\omega) = \sum_{l=1}^{n} \frac{s_{rl}s_{tl}}{[(\omega_l^2 - \omega^2) + i2\zeta_l\omega_l\omega]} \quad (5.1.48c)$$

$$= \sum_{l=1}^{n} \frac{u_{rl}u_{tl}}{[(k_l - \omega^2 m_l) + ic_l\omega]} \quad (5.1.48d)$$

A particular case in which this is possible is when **c** is proportional to **m** and/or **k**:

$$\mathbf{c} = \alpha\mathbf{m} + \beta\mathbf{k} \tag{5.1.49}$$

called proportional damping. Obviously, in this case,

$$2\zeta_i\omega_i = \alpha + \beta\omega_i^2, \quad i = 1, 2, \ldots, n \tag{5.1.50}$$

and

$$c_i = \alpha m_i + \beta k_i, \quad i = 1, 2, \ldots, n \tag{5.1.51}$$

for normal and principal modes, respectively. As before, modal mass and damping rates m_i and c_i are nonunique since they depend upon the normalization, but $m_i^{-1}c_i = 2\zeta_i\omega_i$, being mass-normalized, are unique.

The assumption of proportional damping defined by $\mathbf{c} = \alpha\mathbf{m} + \beta\mathbf{k}$ is motivated primarily by mathematical considerations; it essentially eliminates **c** and leaves only **m**, **k**, thus allowing a simple extension of the undamped system method. In practice, it turns out to be a fairly good approximation when damping is small, so that the error caused by the assumption is small. Also, in many mechanical systems, the damping mechanism is associated with the mass (inertia) elements and stiffness elements, thus providing a physical basis for the assumption.

A slight generalization of proportional damping requires that the matrix that diagonalizes **m** and **k** also diagonalizes **c**, which is possible if and only if **m**, **c**, and **k** commute pairwise. This is difficult to interpret physically, but is useful since it yields real mode shapes $\mathbf{s}_i$ and $\mathbf{u}_i$ for real symmetric **m**, **c**, and **k**. It is conceptually attractive and easy to understand because it allows each element of the MDOF system FRF matrix given by (5.1.48c and d) to be expressed as a weighted sum of SDOF FRF's (5.1.9), the weights being determined by the elements of the modal vectors.

Structural (Hysteretic) Damping

An alternative damping assumption that reduces the diagonalization of three matrices to two is that of structural or hysteretic damping. Recall from the SDOF analysis that in this case the damping is inversely proportional to the frequency, which can be generalized to the matrix form

$$\mathbf{c} = \frac{1}{\omega}\mathbf{h} \tag{5.1.52}$$

Now the differential equation (5.1.42) is not valid, but the forced response model (5.1.43a) takes the form

$$(-\omega^2\mathbf{m} + i\mathbf{h} + \mathbf{k})\mathbf{x}e^{i\omega t} = \mathbf{f}e^{i\omega t} \tag{5.1.53}$$

which is the same as the undamped case with the real matrix **k** replaced by the complex matrix $(i\mathbf{h} + \mathbf{k})$. If, in addition to the real positive definite **m** and **k**, the real damping matrix **h** is also symmetric, then the problem reduces to the simultaneous diagonalization of positive definite **m** and complex symmetric (*not* Hermitian) $(i\mathbf{h} + \mathbf{k})$, which is possible if and only if $\mathbf{m}^{-1}(i\mathbf{h} + \mathbf{k})$ is diagonalizable. Then, by a method similar to the one used in deriving (5.1.39a–c) but more involved, particularly to account for possibly equal eigenvalues, it can be shown [see Horn and Johnson (1985, Theorem 4.5.15)] that using the complex modal matrices **S**, **U**, complex squared frequencies, and complex modal masses and stiffnesses m_l and k_l,

$$\mathbf{S}^T \mathbf{m} \mathbf{S} = \mathbf{I}, \quad \text{that is,} \quad \mathbf{m}^{-1} = \mathbf{S}\mathbf{S}^T; \quad \mathbf{S}^T(i\mathbf{h} + \mathbf{k})\mathbf{S} = \text{diag}\{\omega_l^2(1 + i\eta_l)\} \quad (5.1.54)$$

which mass normalizes and decouples (5.1.53) after substituting $\mathbf{x} = \mathbf{S}\mathbf{q}$ into

$$[-\omega^2 + \omega_l^2(1 + i\eta_l)]q_l e^{i\omega t} = (\mathbf{S}^T \mathbf{f})_l e^{i\omega t}, \quad l = 1, 2, \ldots, n \quad (5.1.55)$$

whereas

$$\mathbf{U}^T \mathbf{m} \mathbf{U} = \text{diag}\{m_l\}, \quad \mathbf{U}^T(i\mathbf{h} + \mathbf{k})\mathbf{U} = \text{diag}\{k_l\} \quad (5.1.56)$$

for arbitrary normalization that decouples (5.1.53) with $\mathbf{x} = \mathbf{U}\mathbf{q}$ into

$$[-\omega^2 m_l + k_l]q_l e^{i\omega t} = (\mathbf{U}^T \mathbf{f})_l e^{i\omega t}, \quad l = 1, 2, \ldots, n \quad (5.1.57)$$

where $(\mathbf{S}^T \mathbf{f})_l$ and $(\mathbf{U}^T \mathbf{f})_l$ denote the *l*th element of columns $\mathbf{S}^T \mathbf{f}$ and $\mathbf{U}^T \mathbf{f}$ as before. The FRF matrix can now be decomposed as

$$\mathbf{H}(\omega) = \mathbf{S}\,\text{diag}\{\omega_l^2(1 + i\eta_l) - \omega^2\}^{-1}\mathbf{S}^T = \sum_{l=1}^{n} \frac{\mathbf{s}_l \mathbf{s}_l^T}{[\omega_l^2(1 + i\eta_l) - \omega^2]} \quad (5.1.58a)$$

$$= \mathbf{U}\,\text{diag}\{k_l - \omega^2 m_l\}^{-1}\mathbf{U}^T = \sum_{l=1}^{n} \frac{\mathbf{u}_l \mathbf{u}_l^T}{(k_l - \omega^2 m_l)} \quad (5.1.58b)$$

$$H_{rt}(\omega) = \sum_{l=1}^{n} \frac{s_{rl} s_{tl}}{[\omega_l^2(1 + i\eta_l) - \omega^2]} \quad (5.1.58c)$$

$$= \sum_{l=1}^{n} \frac{u_{rl} u_{tl}}{(k_l - \omega^2 m_l)} \quad (5.1.58d)$$

Although these results appear to be deceptively similar to the preceding ones, not only complex mode shapes $\mathbf{s}_i$ and $\mathbf{u}_i$, but also complex squared frequencies, modal masses and stiffnesses m_l and k_l, make them physically meaningless. As before, the mode shapes $\mathbf{s}_i$ and $\mathbf{u}_i$ are real if it is assumed that the real modal matrix that diagonalizes real symmetric $\mathbf{m}$ and $\mathbf{k}$ also happens to diagonalize real symmetric $\mathbf{h}$. This is possible, in particular for proportional damping:

$$\mathbf{h} = \alpha \mathbf{m} + \beta \mathbf{k} \tag{5.1.59}$$

so that all the results (5.1.54–5.1.58) are true with real $\mathbf{S}$, $\mathbf{U}$, $\mathbf{s}_i$, $\mathbf{u}_i$, and

$$\omega_l^2 \eta_l = \alpha + \beta \omega_l^2 \tag{5.1.60}$$

Note that the squared frequencies and modal masses and stiffnesses are still complex; it is this physically meaningless phenomenon and the absence of a differential equation description that makes structural damping a questionable assumption. The only saving grace in practice is when the loss factors η_l are so small that they can be ignored and then ω_l^2 treated practically as the undamped natural frequency of the lth mode.

Viscous Damping

As explained at the beginning of this section, for a general viscous damping, it is no longer possible to solve the problem with real eigenvalues $\mu = \omega^2$, the squared frequencies, but it is necessary to consider a problem with complex eigenvalues. In other words, the $2n$th order characteristic polynomial can no longer be expressed as an nth order polynomial in the (undamped) squared frequencies ω^2, but must be solved as a $2n$th order polynomial in complex eigenvalues containing (damped) frequencies. This requires the construction of a $2n \times 2n$ matrix which has such a $2n$th order characteristic polynomial, and correspondingly a $2n$th order response or coordinate vector on which the matrix operates. This $2n \times 1$ vector is the same as the state vector consisting of the displacements $\mathbf{x}$ and velocities $\dot{\mathbf{x}}$:

$$\mathbf{y} = \begin{bmatrix} \mathbf{x} \\ \dot{\mathbf{x}} \end{bmatrix} \tag{5.1.61}$$

which reduces the second-order equation of motion (5.1.42) to the first-order differential equation

$$\mathbf{B}\dot{\mathbf{y}} + \mathbf{K}\mathbf{y} = \mathbf{F} \tag{5.1.62}$$

$$\mathbf{B} = \begin{bmatrix} \mathbf{c} & \mathbf{m} \\ \mathbf{m} & 0 \end{bmatrix}, \quad \mathbf{K} = \begin{bmatrix} \mathbf{k} & 0 \\ 0 & -\mathbf{m} \end{bmatrix}, \quad \mathbf{F} = \begin{bmatrix} \mathbf{f} \\ 0 \end{bmatrix} \tag{5.1.62a}$$

It is easy to see that the first n rows of (5.1.62) are the same as the original equation of motion (5.1.42), whereas the remaining rows are the identity $\mathbf{m}\dot{\mathbf{x}} - \mathbf{m}\dot{\mathbf{x}} = 0$, reminiscent of the state space representation of Section 2.6.5. In fact, it is left as an exercise to show that for a nonsingular **m**, $-\mathbf{B}^{-1}\mathbf{K}$ is the state matrix **A**:

$$-\mathbf{B}^{-1}\mathbf{K} = \begin{bmatrix} 0 & \mathbf{I} \\ -\mathbf{m}^{-1}\mathbf{k} & -\mathbf{m}^{-1}\mathbf{c} \end{bmatrix} = \mathbf{A} \tag{5.1.63}$$

The minor difference between the state-space treatment discussed in Section 5.3.1 and the standard vibration theory treatment reviewed here stems from the assumption of symmetric **m**, **c**, and **k** which allows the use of the transpose of the modal matrix instead of the inverse. If **m**, **c**, and **k** are symmetric, then so are **B** and **K**. Symmetric **B** and **K** can be simultaneously reduced to diagonal forms by the complex modal matrix and its transpose if and only if $\mathbf{B}^{-1}\mathbf{K}$ is diagonalizable and hence we can write

$$\mathbf{S}^T\mathbf{BS} = \mathbf{I}, \quad \text{that is,} \quad \mathbf{B}^{-1} = \mathbf{SS}^T, \quad \mathbf{S}^T\mathbf{KS} = \mathbf{S}^{-1}\mathbf{B}^{-1}\mathbf{KS} = -\mathbf{S}^{-1}\mathbf{AS}$$
$$= \text{diag}\{-\mu_l\} \tag{5.1.64}$$

$$\mu_l, \mu_{l+1} = \omega_l(-\zeta_l \pm \sqrt{\zeta_l^2 - 1}), \tag{5.1.64a}$$

$$= \omega_l(-\zeta_l \pm i\sqrt{1 - \zeta_l^2}), \quad \zeta_l < 1, \quad l = 1, 3, \ldots, (2n-1) \tag{5.1.64b}$$

which is **B**-normalization, similar to mass normalization, and

$$\mathbf{U}^T\mathbf{BU} = \text{diag}\{b_l\}, \quad \mathbf{U}^T\mathbf{KU} = \text{diag}\{k_l\} \tag{5.1.65}$$

$$\mu_l = -\frac{k_l}{b_l}, \quad l = 1, 2, \ldots, 2n \tag{5.1.66}$$

the eigenvalues μ_l being the roots of the characteristics polynomial

$$|i\omega\mathbf{B} + \mathbf{K}| = 0 = |i\omega\mathbf{I} + \mathbf{S}^T\mathbf{KS}| = |i\omega\mathbf{I} - \mathbf{A}| \tag{5.1.67}$$

for arbitrary normalization, so that k_l and b_l are not unique but (k_l/b_l) are.

In view of the structure of (5.1.61) it is clear that the columns of the $2n \times 2n$ matrix **S** and **U** are of the form

$$\mathbf{S} = \begin{bmatrix} \mathbf{s}_1 & \mathbf{s}_2 & \cdots & \mathbf{s}_{2n} \\ \mu_1\mathbf{s}_1 & \mu_2\mathbf{s}_2 & \cdots & \mu_{2n}\mathbf{s}_{2n} \end{bmatrix}, \quad \mathbf{U} = \begin{bmatrix} \mathbf{u}_1 & \mathbf{u}_2 & \cdots & \mathbf{u}_{2n} \\ \mu_1\mathbf{u}_1 & \mu_2\mathbf{u}_2 & \cdots & \mu_{2n}\mathbf{u}_{2n} \end{bmatrix} \tag{5.1.68}$$

where $\mathbf{s}_i$ and $\mathbf{u}_i$ are complex $n \times 1$ modal/eigenvectors as before, and considering that $\mathbf{F} = [\mathbf{f}^T, \mathbf{0}^T]^T$ in (5.1.62a) it is easy to see that the FRF matrix relating the elements of response vector **x** with the elements of the forcing vector **f** is given by

$$\mathbf{H}(\omega) = [\mathbf{s}_1, \mathbf{s}_2, \ldots, \mathbf{s}_{2n}] \operatorname{diag}\{i\omega - \mu_l\}^{-1} \begin{bmatrix} \mathbf{s}_1^T \\ \mathbf{s}_2^T \\ \vdots \\ \mathbf{s}_{2n}^T \end{bmatrix} = \sum_{l=1}^{2n} \frac{\mathbf{s}_l \mathbf{s}_l^T}{(i\omega - \mu_l)} \quad (5.1.69\text{a})$$

$$= [\mathbf{u}_1, \mathbf{u}_2, \ldots, \mathbf{u}_{2n}] \operatorname{diag}\{i\omega b_l + k_l\}^{-1} \begin{bmatrix} \mathbf{u}_1^T \\ \mathbf{u}_2^T \\ \vdots \\ \mathbf{u}_{2n}^T \end{bmatrix} = \sum_{l=1}^{2n} \frac{\mathbf{u}_l \mathbf{u}_l^T}{b_l(i\omega - \mu_l)} \quad (5.1.69\text{b})$$

$$H_{jk} = \sum_{l=1}^{2n} \frac{s_{jl} s_{kl}}{(i\omega - \mu_l)} \quad (5.1.69\text{c})$$

$$= \sum_{l=1}^{2n} \frac{u_{jl} u_{kl}}{b_l(i\omega - \mu_l)} \quad (5.1.69\text{d})$$

Usually mechanical systems are underdamped so that all the μ_l are complex, and then, because the matrices **m**, **c**, **k**, and hence **B**, **K** are real, modal/eigenvectors $\mathbf{s}_i$ and $\mathbf{u}_i$ and eigenvalues μ_i occur in conjugate pairs. Then the above expressions can be expressed in the familiar form

$$H_{jk}(\omega) = \sum_{l=1}^{n} \frac{s_{jl} s_{kl}}{[i\omega - \omega_l(-\zeta_l + i\sqrt{1-\zeta_l^2})]} + \frac{\bar{s}_{jl} \bar{s}_{kl}}{[i\omega - \omega_l(-\zeta_l - i\sqrt{1-\zeta_l^2})]}$$

(5.1.69e)

$$= \sum_{l=1}^{n} \frac{u_{jl} u_{kl}}{b_l[i\omega - \omega_l(-\zeta_l + i\sqrt{1-\zeta_l^2})]} + \frac{\bar{u}_{jl} \bar{u}_{kl}}{b_l[i\omega - \omega_l(-\zeta_l - i\sqrt{1-\zeta_l^2})]}$$

(5.1.69f)

Elementary complex arithmetic can be used to combine the two fractions in (5.1.69e and f) and show that $H_{jk}(\omega)$ takes the form somewhat similar to the proportionally damped case, with the major exception that the numerators are now complex and functions of frequencies.

Although ω_l and $\omega_l\sqrt{1-\zeta_l^2}$ can be considered as the natural frequency and the damped frequency of each mode, they are not simply related to $\mathbf{m}^{-1}\mathbf{k}$ and $\mathbf{m}^{-1}\mathbf{c}$ as in earlier cases. Moreover, even if we could define modal mass, stiffness, and damping parameters using each complex conjugate pair of eigenvalues and modal/eigenvectors, the original vector differential equation (5.1.42) does not

BRIEF REVIEW AND LIMITATIONS OF FFT-BASED APPROACHES 197

decouple into n scalar equations in these parameters; that is why after completing the addition the numerators in (5.1.69e and f) remain not only complex but also become functions of frequency ω.

Nevertheless, the complex mode shapes $\mathbf{s}_i$, $\mathbf{u}_i$ and the residues or modal participation factors $s_{jl}s_{kl}$ and $u_{jl}u_{kl}$ in (5.1.69e and f) can be meaningfully interpreted. They suggest that the amplitudes of two different modes need not be in the same or opposite direction as in the real mode shape case, but may lag or lead by a phase angle different from 0 to 180°. Hence the motion with complex mode shapes may be considered as a traveling wave as opposed to the standing waves in the real modes case. The modes are no longer "normal" or "orthogonal" as they are for the undamped or proportionally damped cases; thus the modal decoupling is more mathematical than physical. It is for this reason that we will follow the more general, yet mathematically simpler state-space format in the DDS approach, even though it loses some of the physical appeal of the normal mode approach reviewed so far.

5.1.5 Multiple-Degree-of-Freedom (MDOF) Curve Fitting

The FFT provides the estimates of FRF at multiples of the fundamental frequency (1/record length) by taking the ratio of the FFT's of the response and the excitation or their spectra. Because of the finite record length, these FRF estimates are necessarily discrete at uniformly spaced frequency intervals (= 1/record length). Although rough estimates of parameters such as natural frequency, damping ratio, and residues or modal constants (mass in the SDOF case) can be obtained for each mode with the help of a few points around resonance, as discussed for the SDOF case earlier, a more systematic determination of these parameters is possible by curve fitting the (continuous) theoretical FRF to these discrete estimates. Recall that these estimates suffer from aliasing, leakage, bias, and variance errors; the curve fitting, in addition, is affected by the limited frequency resolution imposed by the finite record length. Moreover, it is clear from the forms of FRF derived in the last section that such curve fitting amounts to finding the eigenvalues/poles and modal/eigenvectors in the frequency domain. This, in turn, is equivalent to locating the peaks of the mountain $H(s)$ from a cut at the foothills $H(\omega)$, as depicted in Figures 5.1–5.3.

Nevertheless, some of the simpler curve-fitting methods readily available on commercial Fourier analyzers can provide a quick insight into the system and hence we will review them in this section. More complicated nonlinear methods, although more precise, require far more computation than the DDS method for the same accuracy, hence they will be omitted.

SDOF Method

All the MDOF FRF's in the preceding sections were expressed as sums of SDOF FRF's. An SDOF FRF varies rapidly with frequency only around the resonant frequency, and varies so slowly at other frequencies that it may be considered to be constant over any narrow frequency band far away from

198   MODAL ANALYSIS

resonance. Therefore, if the modes are well separated, so that the peaks are far away from each other, and the damping is small enough not only to help separate the modes but also to limit the resonant peak to a narrow frequency band, then the contribution of all other modes can be considered to be a constant around the resonance peak of a given mode. Thus, an MDOF FRF of a system with modes well separated in frequencies and with small enough damping can be written around a given resonant frequency ω_r as

$$\text{MDOF FRF}(\omega)|_{\omega \simeq \omega_r} \simeq \text{SDOF FRF}(\omega)|_{\omega \simeq \omega_r} + \text{constant} \qquad (5.1.70)$$

Therefore, we can apply the circle-fit method of Section 5.1.2 to each peak separately, which yields ω_r, ζ_r, and the residue $s_{rj}s_{rk}$ for the rth mode of FRF relating response point j to excitation point k; all that the added constant contributed by other modes does is to shift the center of the circle by the contribution from the other modes. If this SDOF method is applied to all the peaks identified in the discrete FRF and the sum of the resultant continuous FRF's fits well through these discrete points, then the task of curve fitting is accomplished.

Residual Modes
If the continuous FRF generated using the estimated parameters does not fit well with the discrete version, one reason could be the neglected influence of the modes outside the range used for fitting. These neglected modes are either low frequency—below the fitted range—or high frequency—above the fitted range. Recall from the beginning of Section 5.1 that the FRF is roughly constant at low frequencies near zero and is dominated by stiffness, whereas it varies inversely with the square of frequency at high frequencies and is dominated by mass. Thus the fitted FRF using SDOF method is altered by the residual mass and stiffness specifically dependent on the fitted FRF:

$$\text{FRF}(\omega) \simeq \frac{1}{k_r} + \text{fitted FRF}(\omega) - \frac{1}{\omega^2 m_r} \qquad (5.1.71)$$

Then the residual mass m_r and the residual stiffness k_r are adjusted by trial and error until a good fit is obtained.

Iterative SDOF Method
A comparison of (5.1.71) with (5.1.70) shows that (5.1.71) relaxes the assumption of (5.1.70) that the contribution of all the modes except the ones considered one by one in SDOF fitting is constant; it allows high–frequency variation caused by the residual mass term. If this still does not produce a satisfactory fit, then it implies that the contribution of other modes at the resonance of a given mode is not constant. A simplistic remedy is to repeat the fitting by taking another pass, except that in this pass the contribution of all other modes is obtained by using the estimates in the previous pass. Thus, in SDOF fitting of an ith mode we use,

not the original FRF estimates, but the following refined ones:

$$\text{SDOFFRF}(\omega)|_{\omega \simeq \omega_i} = \text{FRF Data} - \sum_{\omega_r \neq \omega_i} \text{SDOFFRF}(\omega)|_{\omega \simeq \omega_r}$$

$$-\frac{1}{k_r} + \frac{1}{\omega^2 m_r} \qquad (5.1.72)$$

where the parameters needed in subtracting the terms on the right-hand side are obtained from the previous pass. This can be repeated for each mode and then for another pass until a satisfactory fit is obtained.

Lightly Damped Structures

For lightly damped structures the resonant frequencies can be determined at a glance quite accurately from the sharp peaks. If only the modal constants or residues are desired, one can select a few reliable FRF estimates away from resonance and straightaway use the undamped MDOF FRF expression such as (5.1.41c) to solve for the numerator terms since the denominators are known. One can also include the low-frequency residual stiffness and high-frequency residual mass terms. If the same number of frequency points as the number of parameters are chosen, one can get the parameters by solving linear equations, as illustrated by Ewins and Gleeson (1982). Alternatively, one can use more number of frequency points and solve by the linear least-squares method.

Complex Exponential and Polyreference Methods

In contrast to the preceding curve-fitting methods, these are essentially "time domain" methods because they use the impulse response function obtained by taking the inverse Fourier transform of (5.1.69c):

$$h_{jk}(t) = \sum_{l=1}^{2n} s_{jl} s_{kl} e^{\mu_l t} \qquad (5.1.73)$$

If it is assumed that the inverse FFT of an estimated FRF at discrete frequencies is an exact sampled version of (5.1.73), it follows that such a discrete version at sampling interval Δ has the form

$$h_t = \sum_{l=1}^{2n} A_l \lambda_l^t; \qquad A_l = s_{jl} s_{kl}, \qquad \lambda_l = e^{\mu_l \Delta}, \qquad t = 0, 1, 2, \ldots \qquad (5.1.74)$$

Therefore, h_t satisfies the deterministic AR(2n) model (see Chapter 2):

$$(1 - \phi_1 B - \phi_2 B^2 - \cdots - \phi_{2n} B^{2n}) h_t = 0, \qquad B^j h_t = h_{t-j} \qquad (5.1.75)$$

$$(1 - \phi_1 B - \phi_2 B^2 - \cdots - \phi_{2n} B^{2n}) \equiv (1 - \lambda_1 B)(1 - \lambda_2 B) \cdots (1 - \lambda_{2n} B) \qquad (5.1.75a)$$

The complex exponential method of Brown et al. (1979) uses the sampled impulse response h_t to solve for ϕ_l's and then λ_l's by (5.1.75 and 5.1.75a) and is a simplified AR version of DDS. Equations (5.1.74) are then used (Prony method) to solve for the poles or characteristic roots μ_l yielding natural frequencies and damping ratios, and, knowing λ_l, modal constants or residues A_l can be obtained from h_t. Alternatively, once the poles μ_l are found using h_t, residues A_l can also be obtained from the estimated FRF values using (5.1.69c). If $2n$ values of h_t are used, both ϕ_l and A_l can be obtained by solving simultaneous linear equations; if more than $2n$ values of h_t are used, the linear least-squares method is needed.

After estimating the poles and residues using an initial guess of n, the continuous FRF is generated and compared against the estimated FRF to compute the error between the two. The number of DOF n is then increased until this error reaches a minimum. Similar to DDS, the method usually identifies additional "computational" or "noise" modes, which can generally be distinguished and separated by their high damping ratios and/or small residues.

The polyreference method reported by Vold et al. (1982) is a generalization of the complex exponential method. If the vector $h_{jk}(t)$ is considered as a vector for multiple excitation points k, equations (5.1.73) and (5.1.74) can be written as matrix equations. Equation (5.1.75) now becomes a matrix equation that can be solved for matrix ϕ_l's. Solving the matrix polynomial corresponding to (5.1.75a) or forming a state matrix as shown for ARV models in Chapter 2 and finding its characteristic roots, one can solve for the poles. Then the matrix equations (5.1.74) or their frequency domain FRF version can be solved for the residues. Thus the polyreference method is a simplified ARV version of DDS. Note that such a vector of responses must be used if the system has repeated roots/poles, while the state matrix is diagonalizable, that is, it possesses n linearly independent modal/eigenvectors.

Rational Fraction Polynomial Curve Fitter

Jones and Kobayashi (1986) outlined a method of global parameter estimation in the frequency domain based on the rational fraction polynomial representation of the measured frequency response proposed in Adock and Potter (1985). Selecting an arbitrary order for the numerator and denominator, a rational polynomial is fitted by linear least squares to the measured FRF to estimate the coefficients of the polynomials in the numerator and the denominator. This can be done for each set of measurement separately, or all sets of measurements simultaneously. Using the first method gives a measure of scatter around the selected global estimates indicative of problems such as leakage and distortion, but the computational poles different in separate measurements are lost. These computational poles, which compensate for the out-of-band modes, are retained in the second method, but there is no indication of measurement problems.

Ibrahim Time Domain (ITD) Method

This method, initially outlined in Ibrahim and Mikulcik (1977) for a set of free-decay vibration measurements, can also be used for impulse response obtained by inverse FFT from FRF measurements; it then becomes essentially the same as the polyreference method in that the state matrix $\boldsymbol{\Phi}$ is first synthesized from the free-decay or impulse response measurements and then modal/eigenvectors are obtained. Using the notation of Chapter 2 and denoting the state vector of say p responses as $\mathbf{x}_t = (x_{1t}, x_{2t}, \ldots, x_{pt}, x_{1t-1}, x_{2t-1}, \ldots, x_{pt-1})^T$, the ITD method solves for the discrete state matrix $\boldsymbol{\Phi}$ defined by the deterministic version of the AR or discrete time state model

$$\mathbf{x}_t = \boldsymbol{\Phi}\mathbf{x}_{t-1} \tag{5.1.76}$$

by the least-squares formula, using the data matrix say $\mathbf{X}_t = [\mathbf{x}_{t1}, \mathbf{x}_{t2}, \ldots, \mathbf{x}_{tq}]$ as

$$\boldsymbol{\Phi} = \mathbf{X}_t \mathbf{X}_{t-1}^T (\mathbf{X}_{t-1} \mathbf{X}_{t-1}^T)^{-1} \tag{5.1.77}$$

since

$$\mathbf{X}_t = \boldsymbol{\Phi}\mathbf{X}_{t-1} \tag{5.1.76a}$$

Note that the use of least squares implicitly assumes a (stochastic) AR model. For the special case $q = 2p$ using limited data, since $\mathbf{X}_{t-1}$ is a nonsingular square matrix, obviously

$$\boldsymbol{\Phi} = \mathbf{X}_t \mathbf{X}_{t-1}^{-1} \tag{5.1.77a}$$

as initially proposed in the ITD approach, which assumes the deterministic AR model (5.1.76). The eigenvalues of $\boldsymbol{\Phi}$ are λ_l of (5.1.74) from which poles μ_l and natural frequencies and damping ratios are found, whereas eigenvectors of $\boldsymbol{\Phi}$ provide mode shapes.

Eigensystem Realization Algorithm (ERA)

The eigensystem realization algorithm (ERA) described in Juang and Pappa (1984) attempts to exploit a combination of Ho–Kalman (1965) algorithm with the singular-value decomposition for curve fitting in both time and frequency domains. The Ho–Kalman algorithm is a system identification tool for determining the order of a deterministic system (dimension of the state space) which equals the rank of the pulse response matrix (or its Fourier transform, the FRF matrix). However, as explained in Section 3.5.3, when possibly noisy measurement data is used in the pulse-response matrix, its rank may exceed the system order. Therefore, the ERA uses the SVD of the data matrix to remove the smaller singular values and corresponding singular vectors, hopefully associated with noise, to obtain a smoothed version of the pulse response matrix that provides the modal parameters; some noise modes that may still remain are later eliminated by other criteria such as modal amplitude coherence. Hence we will review the basic elements of the ERA in this curve-fitting section. More details about the ERA and its relation to other curve-fitting methods such as the ITD and the polyreference may be found in Juang (1986).

For a deterministic system with state equation

$$\mathbf{x}(k+1) = \mathbf{A}\mathbf{x}(k) + \mathbf{B}\mathbf{u}(k), \quad k = 0, 1, 2, \ldots \quad (5.1.78)$$

and output $\mathbf{y}(k)$ given by

$$\mathbf{y}(k) = \mathbf{C}\mathbf{x}(k) \quad (5.1.79)$$

the impulse response matrix (see Chapter 4) is given by

$$\mathbf{Y}(k) = \mathbf{C}\mathbf{A}^k \mathbf{B} \quad (5.1.80)$$

If the data matrix of Section 3.2.1 is generalized by using these impulse response matrices in place of data values, we can form an $(r+1) \times (s+1)$ block Hankel (same elements along each diagonal perpendicular to the main diagonal) matrix

$$\mathbf{H}(k) = \begin{bmatrix} \mathbf{Y}(k) & \mathbf{Y}(k+1) & \cdots & \mathbf{Y}(k+s) \\ \mathbf{Y}(k+1) & \mathbf{Y}(k+2) & \cdots & \mathbf{Y}(k+s+1) \\ \vdots & \vdots & & \vdots \\ \mathbf{Y}(k+r) & \mathbf{Y}(k+r+1) & \cdots & \mathbf{Y}(k+r+s) \end{bmatrix} \quad (5.1.81)$$

$$= \mathbf{Q}_r \mathbf{A}^k \mathbf{P}_s \quad (5.1.82)$$

where the observability matrix

$$\mathbf{Q}_r = \begin{bmatrix} \mathbf{C} \\ \mathbf{C}\mathbf{A} \\ \vdots \\ \mathbf{C}\mathbf{A}^r \end{bmatrix} \quad (5.1.83)$$

and the controllability matrix

$$\mathbf{P}_s = [\mathbf{B}, \mathbf{A}\mathbf{B}, \ldots, \mathbf{A}^s \mathbf{B}] \quad (5.1.84)$$

If the system is controllable, observable, and has a state space of dimension n, then, recall from Section 4.6, that $\mathbf{P}_s$, $\mathbf{Q}_r$, and $\mathbf{A}$, have each rank n, when $(r+1) \geq n$ and $(s+1) \geq n$; it is then clear that $\mathbf{H}(k)$ has rank n. This was illustrated for the simple scalar case in Section 3.2.1.

It would therefore seem that to find the dimension of the state space all that one has to do is to form a block data matrix from impulse response (or its

Fourier transform FRF) or large enough r and s so that it is rank deficient, that is, rank $< (r + 1)$ and rank $< (s + 1)$; then the rank of the matrix gives the dimension of the state space. Unfortunately, even for deterministic systems, round-off errors in the data make the task of rank determination quite difficult; for stochastic systems there is no rank deficiency at all, as illustrated in Section 3.2.1.

Hence the ERA proposes the SVD, a natural tool for quantifying near rank deficiency. The characteristics of the noise are used to decide the dimension n such that the singular values $\sigma_{n+1}, \sigma_{n+2}, \ldots, \sigma_p$ of $\mathbf{H}(0)$ can be considered to be negligible and due to noise:

$$\mathbf{H}(0) = \mathbf{Q}_r \mathbf{P}_s = \mathbf{U} \, \text{diag}(\sigma_1, \sigma_2, \ldots, \sigma_n, \sigma_{n+1}, \ldots, \sigma_p) \mathbf{V}^T \qquad (5.1.85)$$

Then, letting the corresponding singular value and singular vector matrices to be Σ_n, $\mathbf{U}_n$, $\mathbf{V}_n$, that is,

$$\Sigma_n = \text{diag}(\sigma_1, \sigma_2, \ldots, \sigma_n) \qquad (5.1.86)$$

$$\mathbf{U} = [\mathbf{U}_n, \mathbf{u}_{n+1}, \ldots], \quad \mathbf{V} = [\mathbf{V}_n, \mathbf{v}_{n+1}, \ldots] \qquad (5.1.87)$$

the state matrix is taken as (see Section 3.5.3 for review)

$$\mathbf{A}_n = \Sigma_n^{-1/2} \mathbf{U}_n^T \mathbf{H}(1) \mathbf{V}_n \Sigma_n^{-1/2} \qquad (5.1.88)$$

The eigenvalues of $\mathbf{A}_n$ provide λ_i, from which the poles and then natural frequencies and damping ratios can be recovered; the eigenvectors of $\mathbf{A}_n$ provide the modal vectors. The procedure is the same in frequency domain, using FRF matrices instead of impulse matrices, except that $\mathbf{U}$ and $\mathbf{V}$ may now be complex and hence conjugate transpose needs to be used instead of ordinary transpose.

In spite of its control theory overtones and resultant matrix manipulations, the ERA is basically the same as the simple ITD method. Matrices $\mathbf{A}$, $\mathbf{B}$ are incorporated in the equations of motion and $\mathbf{C}$ can be accounted for in the measurement process and its calibration. The ITD uses the relation between lagged data vectors $\mathbf{x}_t$ and $\mathbf{x}_{t-1}$, just as the ERA uses the relation between the lagged data matrices $\mathbf{H}(0)$ and $\mathbf{H}(1)$ to obtain an estimate of the state matrix. Recall that the state space formulation of a given system is not unique. The ITD, ERA, and polyreference methods are the least squares, minimal, and canonical realizations of the same system. As pointed out by Juang (1986), it is true that conceptually, for ideal data from a deterministic system, ordinary least squares may have numerical difficulties. But this problem can also be overcome by using the SVD for least squares estimation; in fact, as discussed in Section 3.5.3, the use of the SVD is far more reliable in least squares estimation than in rank determination from noisy real data. Moreover, for real data, noise makes up the rank deficiency, there are no real singularities, and the decision about which singular values come from the noise becomes arbitrary and adhoc.

5.1.6 Limitations of the FFT

The Fourier transform and its generalization the Laplace transform are ideal mathematical tools for theoretical analysis of dynamic systems. Their ability of reducing the time domain operations of differentiation and integration to multiplication and division in the frequency domain simplify the analysis from calculus to algebra. The resultant graphical presentation of characteristics such as the FRF and the spectrum provide excellent visual aids to the analysis. Even the restriction of the Laplace transform $H(s)$ to the Fourier transform $H(\omega)$ depicted in Figures 5.1–5.3 is not a limitation of the Fourier transform; under most practical conditions when they are well behaved, $H(s)$ can be recovered from a theoretically known $H(\omega)$.

When this argument is used to recover $H(s)$ from a discrete distorted realization of $H(\omega)$ by the FFT with limited frequency resolution, severe limitations of the method arise and seriously affect the results when the response data contains noise. Overwhelmed by the mathematical elegance and graphical appeal of the Fourier transform together with the computational efficiency of FFT, many of these limitations escape the attention of most practitioners, although they have been discussed in the literature, for example, by Mitchell (1986). Commercial Fourier analyzers provide some remedies to circumvent these limitations, but these remedies merely transfer the problems somewhere else and at best yield good frequency estimates of nearly undamped modes, while distorting many other aspects. These distortions get compounded when curve-fitting methods described earlier are applied to FRF obtained by the FFT and even more so to impulse response obtained by inverse FFT of the already distorted FRF. Moreover, now these distortions get magnified when one attempts to recover three-dimensional $H(s)$ from two-dimensional distorted $H(\omega)$, like scaling the mountain peaks from the foothills mentioned earlier. Mathematical elegance and graphical clarity is already eroded in the raw FFT estimates of the FRF; sophisticated curve-fitting algorithms used to smooth them further erode the computational advantage of the FFT.

The basics of these limitations were discussed in Section 2.9 to show how they are either absent or minimized in the DDS approach. They spring from three key facts: (1) Fourier transformation is not an ergodic operation, (2) the FFT only gives samples of the Fourier transform, and (3) the FFT is a transform of only a finite time record. We will now discuss and illustrate the practical effects of these limitations, the remedies applied to overcome them in commercial Fourier analyzers, and the problems caused by these remedies themselves. This should help the reader to determine when to use raw FFT, with perhaps simpler curve-fitting for quick results, and when the DDS method would be more beneficial. As argued in Section 2.9.9, it turns out that the FFT with the remedies discussed below is enough to detect nearly undamped frequencies; in all other cases the DDS is more beneficial.

The fact that the data used in the FFT comes from a finite record length sampled uniformly at fixed number of points (between 256 and 4096 for most analyzers) leads to the limitations of (1) frequency resolution, (2) leakage, (3)

BRIEF REVIEW AND LIMITATIONS OF FFT-BASED APPROACHES 205

aliasing, and (4) noise. We will discuss each of these with their remedies available on commercial analyzers, without going into the details of the curve-fitting methods which were reviewed earlier.

Frequency Resolution and Zoom

The limited frequency resolution of the FFT is a direct consequence of the finite time record. If the record length in time is T, the lowest nonzero frequency obtained and the uniform frequency interval in the FFT is $1/T$. The simple remedy of taking arbitrarily large T to get arbitrarily fine resolution is not always effective in practice because (1) record over large T may not be available in many cases such as transients, (2) if available it may not stay "stationary" over large T, and (3) if it stays stationary, the finite fixed number of data points stored in commercial Fourier analyzers will cause other problems.

An immediate effect of improving the frequency resolution by increasing the record length with fixed number of data points is a limitation of the highest allowable frequency. If the record length is T and number of data points is N, the frequency resolution (and the lowest frequency estimated) is $1/T$, sampling interval is T/N, and the highest allowable (Nyquist) frequency is $N/2T$, which immediately shows the compromise between the frequency resolution or the lowest frequency and the highest frequency.

This compromise is graphically illustrated in Figure 5.5. The figure depicts two modes with damped frequencies 12 and 650 Hz. For $N = 1024$ points, if

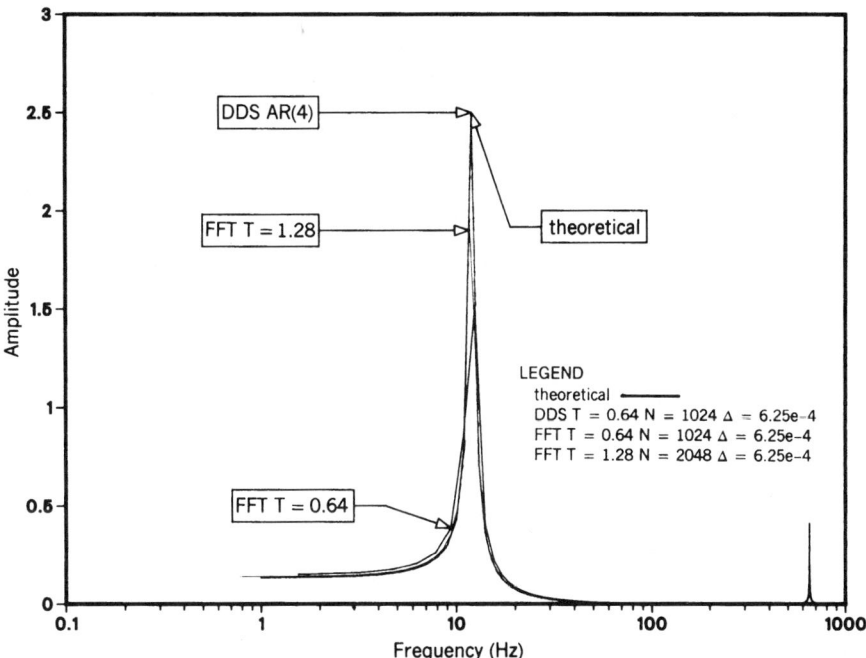

Figure 5.5 Frequency resolution compromise.

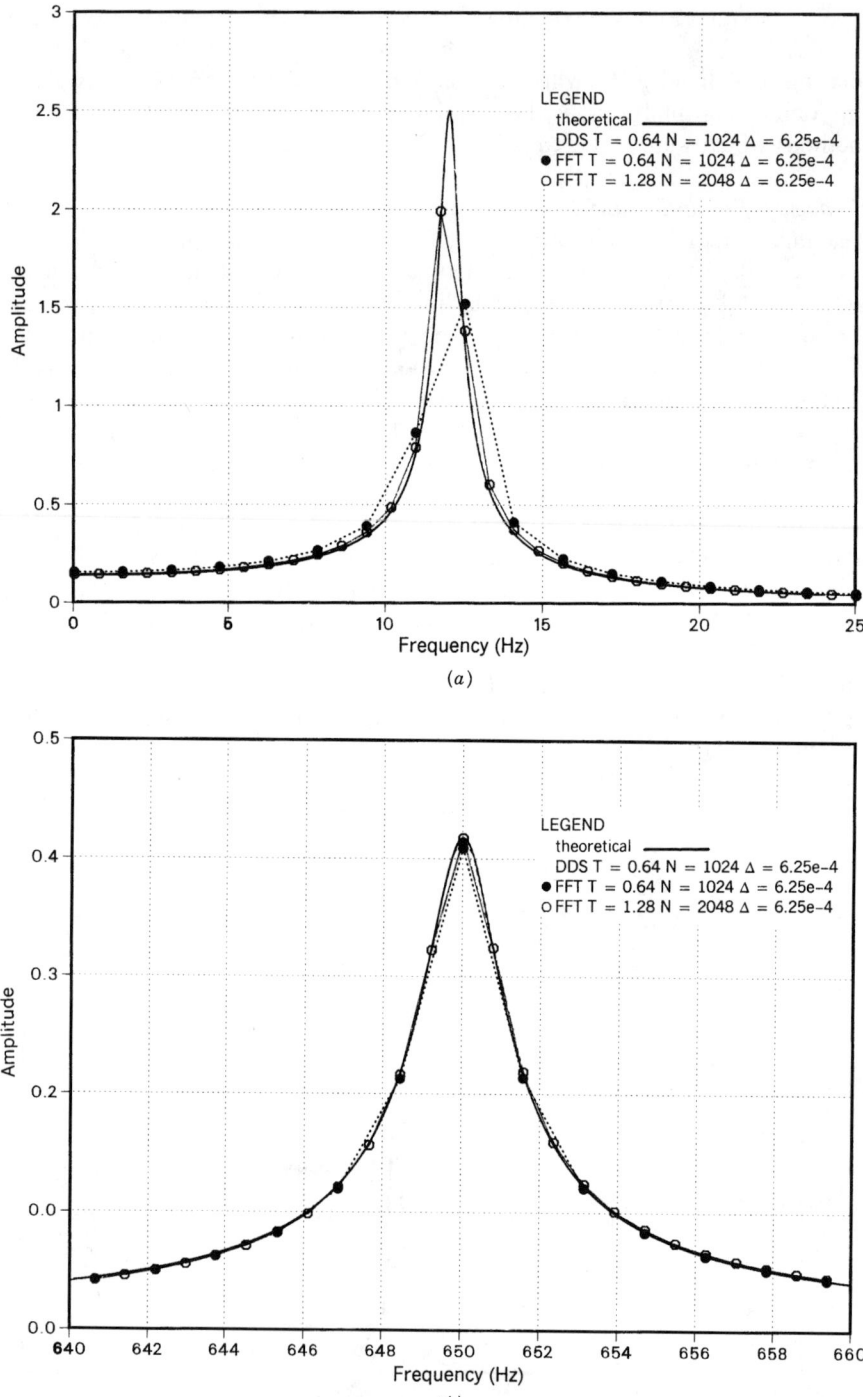

Figure 5.6 (*a*) Magnification of 12 Hz peak in Figure 5.5. (*b*) Magnification of 650 Hz peak in Figure 5.5.

$T = 0.64$ sec is chosen so that the Nyquist frequency $(1024/1.28 =)$ 800 Hz is well above the second mode, then simultaneous estimation of both modes by the FFT gives very poor results for the first mode compared to the DDS; this is further clarified in the magnified regions around the peaks shown in Figure 5.6. It seems that both the frequency and damping are overestimated; although the situation improves when N is increased to 2048, the frequency is still underestimated and damping overestimated. The second mode is well estimated by both techniques at $T = 0.64$ itself.

The magnification and separation of well-separated modes, similar to that depicted in Figures 5.5 and 5.6, can be accomplished on commercial Fourier analyzers by a technique popularly known as "zoom." Two commonly used methods of zoom consist of controlled aliasing, also called mixing or heterodyning, as generally used for tuning household radios and televisions, and translating the frequency origin. In the first method, if the frequency range of interest is between ω_1 and ω_2, a band-pass filter is applied over this range and the result aliased into the interval 0 to $(\omega_2 - \omega_1)$. In the second method the original time signal is multiplied by a cosine, say $\cos(\omega_1 t)$, so that

$$A \sin(\omega t) \cos(\omega_1 t) = \frac{A}{2} [\sin(\omega - \omega_1)t + \sin(\omega + \omega_1)t]$$

and the high-frequency components are filtered out, thus effectively translating the origin to ω_1. Although the zoom leaves the peak frequencies intact, it may affect other characteristics such as damping and mode shapes unless the modes have low damping and are well separated.

Leakage and Windowing

Leakage is another facet of frequency resolution and was already encountered in Figure 5.6(*a*). When the actual peak frequency does not fall on one of the multiples of $1/T$ at which the FFT is computed, the height of the peak falls short, as shown in Figure 5.6(*a*). But the total energy of vibration must be represented correctly; therefore, the reduced amplitude near the peak is made up by increased amplitude in neighboring frequencies on both sides. The energy "leaks" on both sides and the peak flattens and gives an impression of damping higher than it actually is.

Thus leakage is a direct consequence of the peak frequency not being a multiple of $1/T$, or, to put it equivalently, the time record of length T not containing an integral number of cycles of the peak frequency, as clarified in Figure 5.7. It arises from the fundamental limitation of the FFT, that is also the key to its computational efficiency: FFT assumes that the discrete set of N points comes from a trigonometric polynomial of frequencies that are multiples of $1/T$. Thus the FFT assumes the finite record of length T to be repeating on both sides with period T as depicted in Figure 5.8 and discussed in Section 2.9.6. Figure 5.8*c* and *e* provide a graphical representation of equation (2.9.8). If the sampled

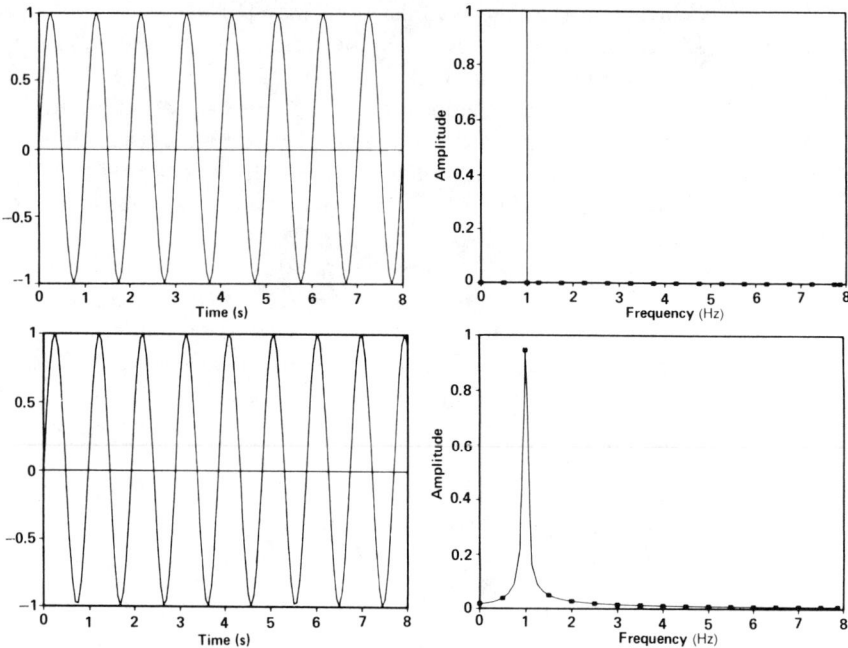

Figure 5.7 Integral number of cycles in the record length—no leakage (top); nonintegral number of cycles in the record length—leakage (bottom).

time record does contain an integral number of cycles, as shown in Figure 5.8a and b, the input assumed in FFT shown in Figure 5.8c matches with the original signal (Figure 5.8a) from which the time record has been sampled. Otherwise, the signal used in FFT is quite different from the actual signal, as shown in Figure 5.8d and e. The discontinuities at the ends of the periodic time record of length T seen in Figure 5.8e require many additional frequency components in the Fourier representation, which cause the leakage.

The effect of leakage can be so severe that it might mask neighboring small but distinct peaks. The frequencies of such nearly undamped peaks can be recovered by the remedy of windowing available on most Fourier analyzers. Such windowing attempts to eliminate the discontinuities causing the leakage by damping the signal at both ends of the time record. But such windowing adds its own damping, causing additional leakage! Good Fourier analyzer literature usually warns us about the dangers of leakage and its remedy, windowing, as illustrated in the photographs from Hewlett Packard (1989) reproduced in Figure 5.9. Note that the two neighboring peaks of Figure 5.9b are completely masked by leakage in d, and although the windowing has separated the peaks at correct frequencies in e, it has added its own leakage with increased damping in the first and third peak. Thus windowing reduces resolution.

BRIEF REVIEW AND LIMITATIONS OF FFT-BASED APPROACHES

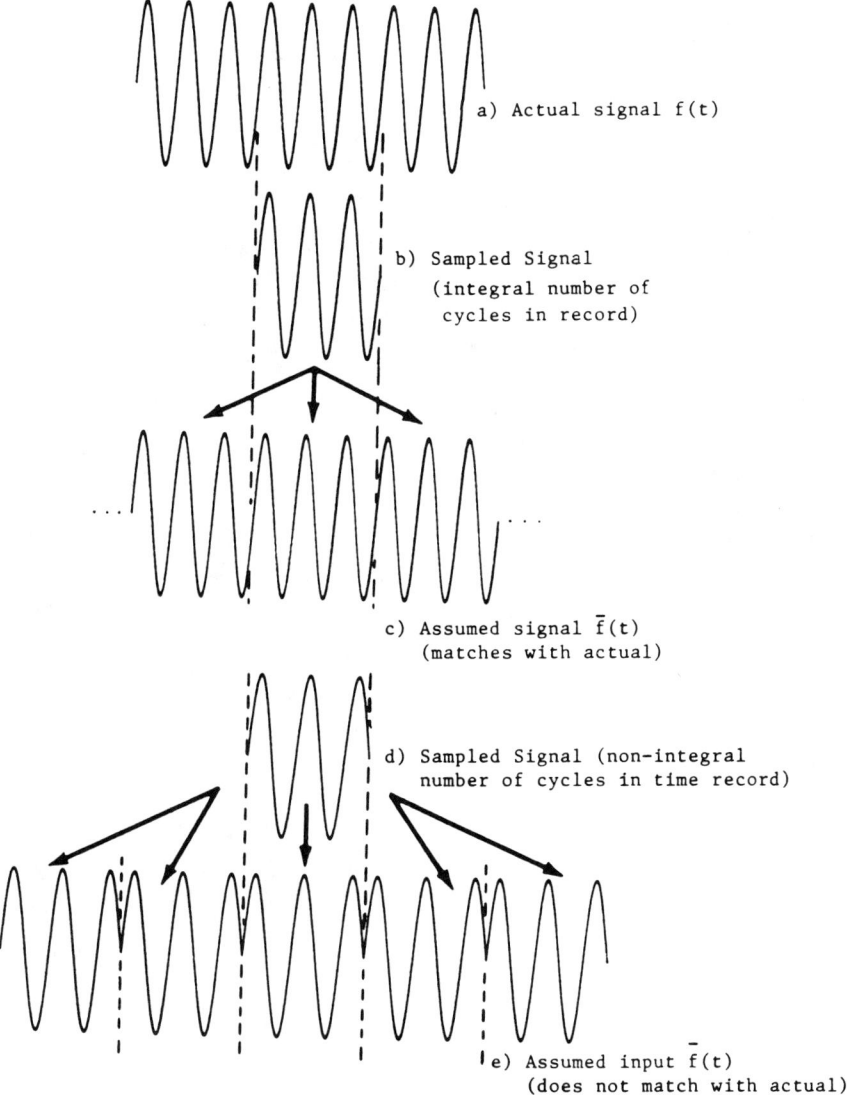

Figure 5.8 Assumed signal $\bar{f}(t)$ in FFT.

The effect of added damping due to the window can be analytically evaluated by using the known window function in the Fourier transform. This is particularly simple for an exponential window, say $e^{-\alpha_0 t}$, for which $H(\omega)$ can be obtained by

$$H(\omega) = H(s)|_{s=\alpha_0 + i\omega}$$

210 MODAL ANALYSIS

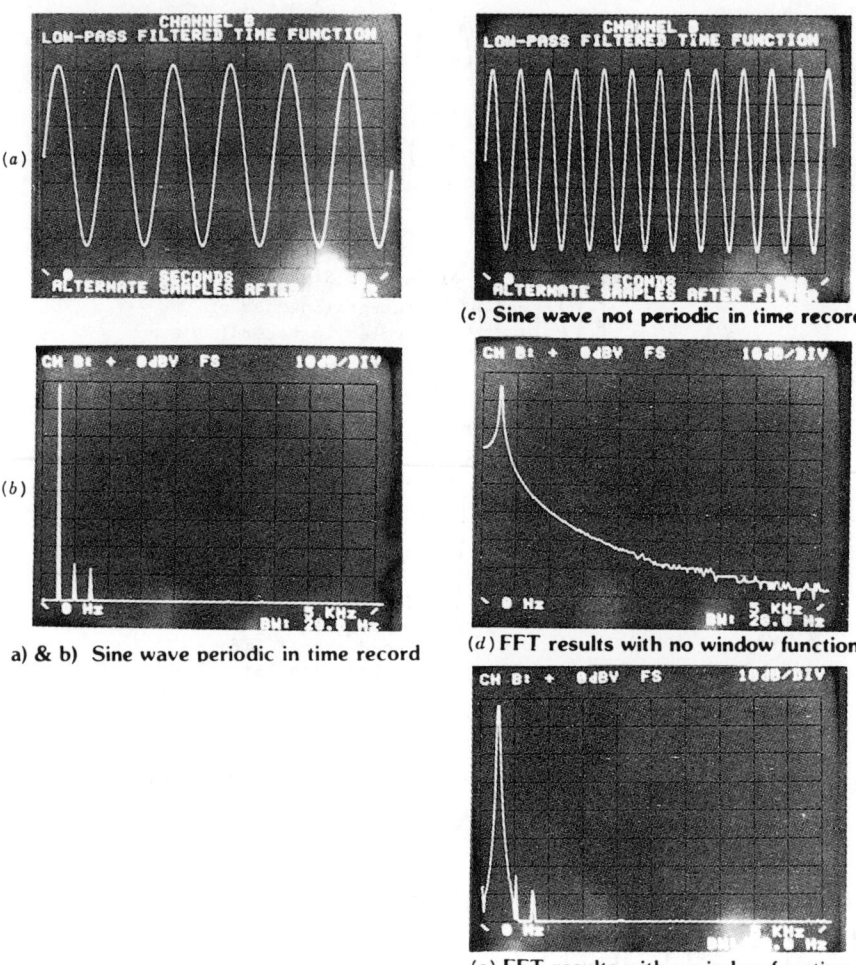

(a)

(b)

a) & b) Sine wave periodic in time record

(c) Sine wave not periodic in time record

(d) FFT results with no window function

(e) FFT results with a window function

Figure 5.9 Leakage reduction with windowing. (Reproduced from Hewlett Packard (1989) with permission, courtesy of Hewlett-Packard Company.)

instead of the usual formula

$$H(\omega) = H(s)|_{s=i\omega}$$

However, this requires a good knowledge of the actual $H(\omega)$ we are trying to estimate, and many approximations to implement it in practice. In fact, a proper choice of the type of window and its parameters itself depends upon knowledge of $H(\omega)$. Hence the "optimal" choice and implementation of a window has been called skilled "window carpentry" in spectral estimation literature with some justification, as it involves a similar trial-and-error approach.

Aliasing and Antialiasing Filters

The finite-time record length T used in the FFT, instead of the infinite length assumed in the Fourier transform, limits the low frequency or frequency resolution to $1/T$, and causes leakage when violated. On the other hand, the finite sampling interval $\Delta\ (=T/N)$ used in the FFT, instead of the zero interval assumed in the Fourier transform, limits the high frequency called Nyquist frequency to $1/2\Delta\ (=N/2T)$, and causes aliasing when violated. Thus inadequate frequency resolution or leakage and aliasing are the two sides of the same coin and arise from the FFT as a frequency domain image of N points sampled from a record of length T.

Aliasing superimposes the contribution of all frequencies beyond the range of 0 to $F_s/2$, where $F_s = 1/\Delta = N/T$ is the sampling frequency, by folding around 0 and $F_s/2$ as many times as necessary. This folding, mathematically represented by equation (2.9.12), is graphically illustrated in Figure 5.10. The long tail of the FRF beyond the Nyquist frequency $F_s/2$ first folds around $F_s/2$, a part of which between 0 and $F_s/2$ directly adds its aliased contribution to the original FRF. The other part falling below 0 folds a second time around 0 and also adds to the original FRF near zero frequencies. If this second fold had continued beyond $F_s/2$ it would fold again for a third time. The aliased version of the FRF is clearly quite different from the original one, and would give incorrect results when curve fitted by simple methods, such as the circle-fit method.

Since the FFT treats computation at all the frequencies independently, it is necessary to take the sampling frequency at least twice the frequency beyond

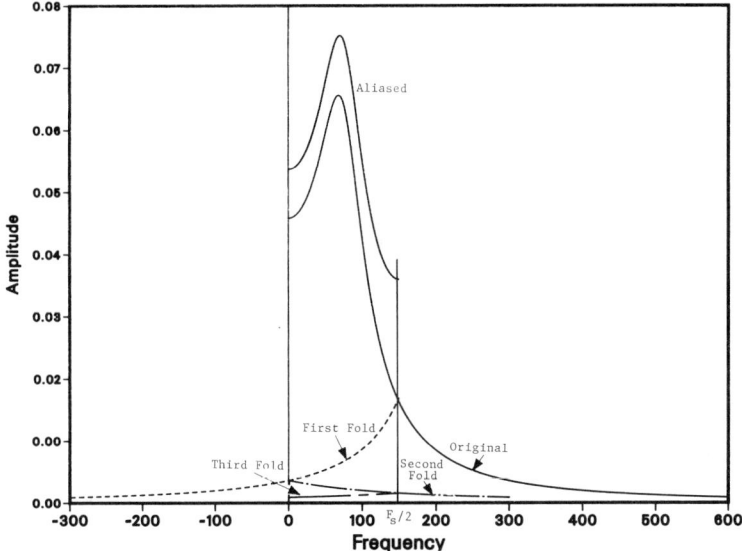

Figure 5.10 Aliasing—multiple folding around 0 and $F_s/2$.

which the FRF is nearly zero. Then the tail beyond $F_s/2$ and hence its folding around $F_s/2$ would be negligible and so also aliasing. But large $F_s = N/T$ with generally fixed N in commercial analyzers requires T to be small, thus reducing the resolution and possibly increasing leakage.

Aliasing does affect the DDS method also. However, since the DDS concentrates on the poles in the time domain, it only requires that the sampling frequency be at least twice the peak frequency. This is illustrated in Figure 5.11. The FRF has a peak at 5 Hz, but the FRF is close to zero only near 50 Hz, so the sampling frequency for the FFT must be beyond 100 Hz, whereas the DDS needs the sampling frequency only beyond 10 Hz. This is further clarified in Figure 5.12, which compares the theoretical FRF with those estimated by the DDS and FFT methods. Figures 5.11 and 5.12 illustrate graphically that aliasing affects the DDS less than the FFT. Figure 5.13 shows the effect of aliasing on the antiresonance, often ignored in vibration analysis.

The common remedy for aliasing provided in the commercial Fourier analyzers is the antialiasing filter. Since the effects of aliasing are inevitable if the FRF extends beyond $F_s/2$, and they are uncertain since the FRF is unknown, they must be avoided at all cost. Most analyzers automatically apply the antialiasing filters based on the specified frequency span so as to reject all the frequencies above the span. Moreover, the filter cannot have a sharp cut-off at the highest frequency of interest, but must have a gradual roll-off beyond it. Then the sampling frequency must be chosen twice the frequency at the end of the roll-off. Thus the sampling frequency usually turns out to be two and a half to four times the highest frequency of interest when antialiasing filters are used.

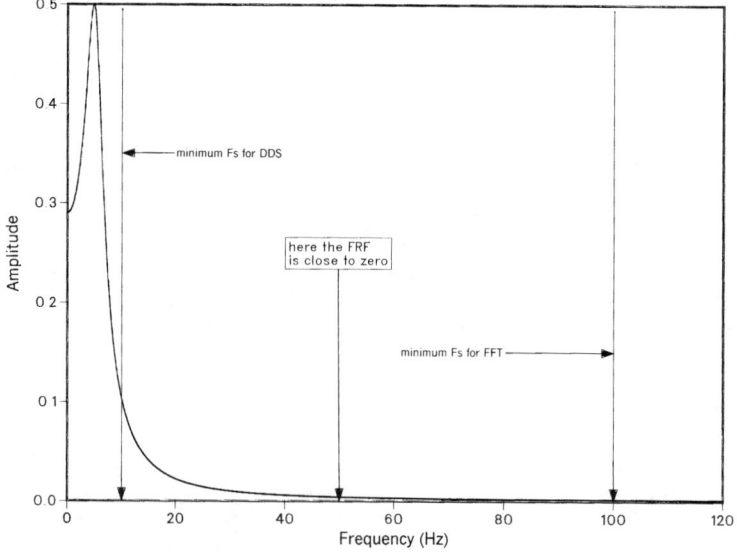

Figure 5.11 Minimum sampling frequencies for DDS and FFT.

BRIEF REVIEW AND LIMITATIONS OF FFT-BASED APPROACHES   213

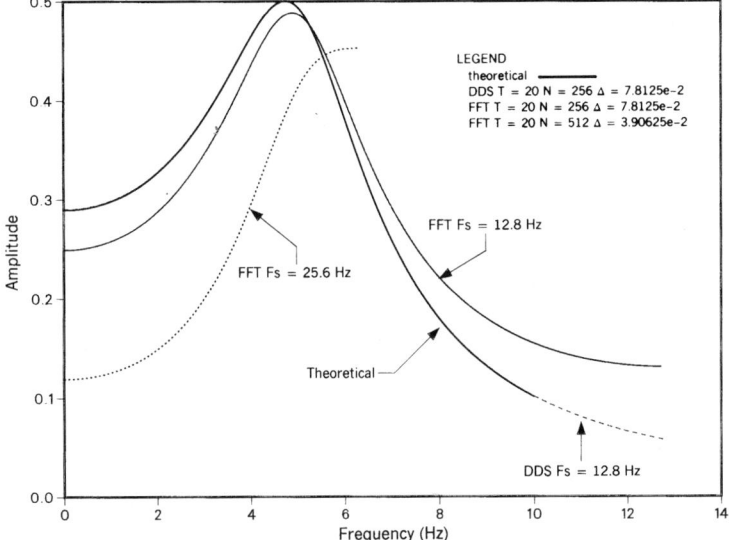

Figure 5.12   Comparison of FRF by DDS and FFT.

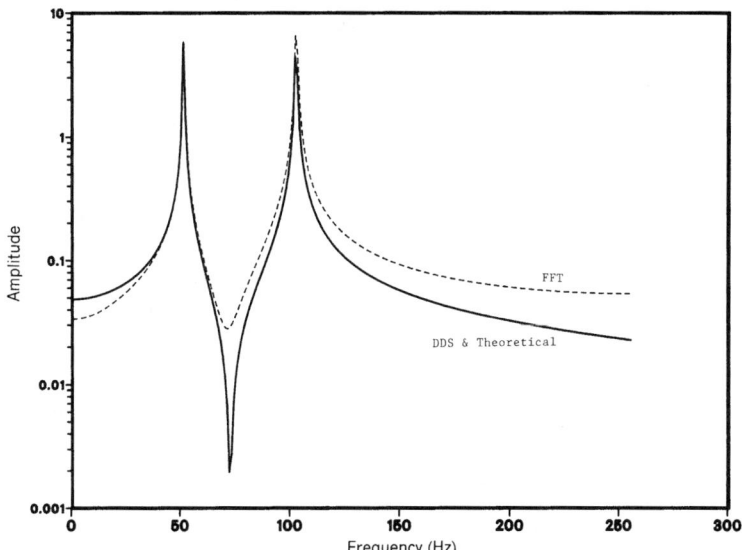

Figure 5.13   Underestimation of antiresonance by FFT.

This increased sampling frequency not only decreases T because of fixed N, providing a lower frequency resolution, but also correspondingly reduces the number of computed FFT points to $2N/5$ and $N/4$ instead of the full $N/2$ over the frequency span of interest.

Noise and Averaging

Real data processing in practice has to reckon with noise, be it from the actual system or measurement process, or at least from the round-off and the computation process. Theoretically, since the noise is a random process, its Fourier transform does not exist. However, an FFT of the given N data points can always be computed; so what this nonexistence of the Fourier transform means in practice is that the FFT's of two different samples from the same noisy system could be quite different. This is illustrated in Figure 5.14, which shows an FFT of $N = 1024$-point sample from a noise-added FRF against the DDS-estimated FRF from only 128 points.

Note the compounded effect of aliasing, leakage, and noise in Figure 5.14. Although for the deterministic signals the effect of aliasing and leakage can be predicted and remedied based on some knowledge of the system and sophisticated curve fitting, such as polynomial or MDOF algorithms, it is when they are compounded with significant noise that the remedies may fail in practice.

In the presence of noise in the FRF, it is tempting to "average" out the noise using several FFT's of noisy data. Ordinary averaging of FRF's does not help, as shown in Figure 5.15. If the original time-domain signal is stationary, the Fourier transform of its autocovariance function, called power spectrum, is well defined. Therefore, averaging the power spectra, commonly known as RMS (or power) averaging, is useful in obtaining more accurate estimates of the total power, including that of the signal and noise when both are stationary random processes. Hence spectral, RMS, or power averaging is not recommended for noisy measurements of deterministic response such as impulse or sinusoidal excitation responses, which are not stationary. Such averaging does not separate or reduce the noise as is often expected. This will be discussed more fully in Chapter 6.

In contrast, the DDS method precisely accomplishes the separation of the deterministic part and removal of noise in the time domain, as simply illustrated mathematically for the AR(1) case in Chapter 2. (Refer to equation (2.2.1), which readily generalizes to arbitrary order by the state space format.) The Fourier transform of this deterministic part or signal alone produces the clear and accurate DDS estimates shown in Figure 5.14, using only 128 points as against 1024 point records used in the FFT results.

Some Fourier analyzers provide a capability of improving the signal-to-noise ratio by a technique called linear averaging. This is essentially time-domain averaging and can reduce the noise effect if the deterministic part of the data can be precisely synchronized and remains exactly the same in all samples. Then the average of the deterministic part remains the same and the variance of the noise is reduced to σ^2/n from the original σ^2 if n sample records are used. (See

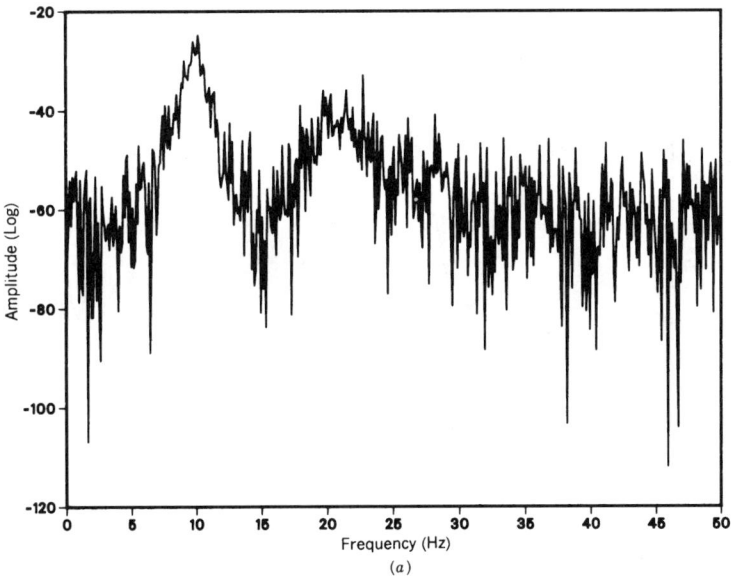

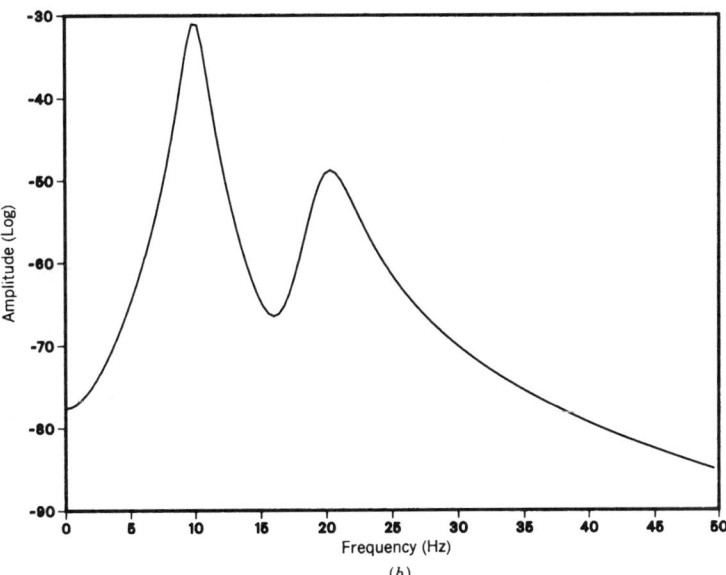

Figure 5.14 (a) FFT of noisy data 1024 Points. (b) DDS frequency response of 128 points from the same data. (Note the difference in scales.)

216 MODAL ANALYSIS

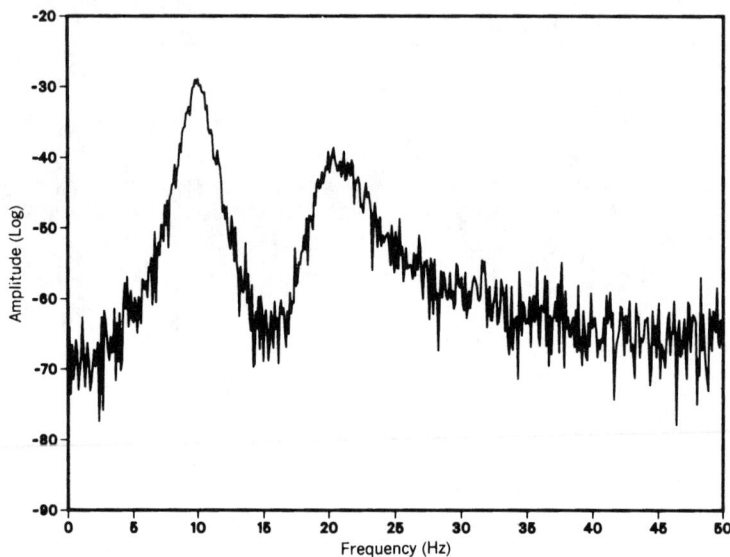

Figure 5.15 Ten sample averaged FFT of 1024 points each (one of these used in Fig. 5.14).

equation (2.9.17) if necessary.) The DDS results accomplish this in a much more effective way using a single record, thus eliminating the need for synchronization.

Other remedies common in modal analysis practice of FFT-based FRF estimation consist of choosing special inputs such as chirp or burst random, different windows for smoothing the FFT estimates, and various FRF estimators, depending upon the conjectured frequency content, nature, and source of noise, and so on. The systematic bias introduced in such FFT-based FRF estimation and the considerable improvement provided by DDS with much less data are illustrated in Pandit and Yao (1991). It is left as an exercise to show that the FFT of $c\lambda^t$ is biased (underestimated) by a factor $(1 - \lambda^N)$ when N data points are used. Such bias cannot be corrected in the prevalent methods of estimating the FRF by dividing the cross-spectrum and the auto-spectrum obtained by averaging the FFTs. (See Section 6.2.)

5.2 THE DDS MODELING FOR MODAL ANALYSIS

The DDS approach to modal analysis estimates the state matrix directly from the data. The eigenvalues of the state matrix then provide the poles which yield the natural frequencies and damping ratios, whereas the eigenvectors provide the modal vectors. The first task in synthesizing the state matrix is to obtain an adequate model. We will therefore outline the DDS modeling procedure, present the necessary estimation methods suitable for modal analysis, and discuss the application of some of the computational algorithms reviewed in Chapter 3 for implementing these methods.

5.2.1 Modeling

The DDS modeling strategy evolved in Chapter 2 is applicable to a $p \times 1$ data vector $\mathbf{X}_t$. In modal analysis, the elements of this vector consist of response as well as excitation measurements when available; in scalar modeling each will be modeled separately, in vector modeling all simultaneously measured (synchronized) data will be modeled together as a p vector. We will therefore formulate the vector case, the scalar case follows by setting $p = 1$.

The basic DDS modeling procedure developed in Pandit (1973) and illustrated for scalar and vector cases in Pandit and Wu (1977, 1983) consists of successively fitting, for $n = 1, 2, \ldots$, the ARMAV$(n, n-1)$ model:

$$\mathbf{X}_t = \boldsymbol{\phi}_1 \mathbf{X}_{t-1} + \boldsymbol{\phi}_2 \mathbf{X}_{t-2} + \cdots + \boldsymbol{\phi}_n \mathbf{X}_{t-n} + \mathbf{C}$$
$$- \boldsymbol{\theta}_1 \mathbf{a}_{t-1} - \boldsymbol{\theta}_2 \mathbf{a}_{t-2} - \cdots - \boldsymbol{\theta}_{n-1} \mathbf{a}_{t-n+1} + \mathbf{a}_t \quad (5.2.1)$$

$$E(\mathbf{a}_t \mathbf{a}_{t-k}^T) = \mathbf{0}, \quad k \neq 0, \quad E(\mathbf{a}_t \mathbf{a}_t^T) = \sigma_a^2, \quad E(\mathbf{a}_t) = \mathbf{0}$$

where $\mathbf{X}_t$, $\mathbf{a}_t$, and $\mathbf{C}$ are $p \times 1$ column vectors of data, residuals, and (DC) constants, respectively, the $p \times p$ matrices $\boldsymbol{\phi}_k$, $\boldsymbol{\theta}_k$ are the autoregressive and moving average parameters, and the $p \times p$ symmetric matrix σ_a^2 is the variance–covariance matrix of the residual vector $\mathbf{a}_t$ (see Appendix A for review of necessary statistics).

For a given n, the parameter elements ϕ_{ijk}, θ_{ijk}, and c_i are estimated from a set of N vector observations by minimizing the determinant of the conditional sum of squares and products (CSP) matrix

$$\mathbf{A}_0 = \left\{ \sum_{t=1}^{N} a_{it} a_{jt} \right\} \quad (5.2.2)$$

where a_{it}'s are the elements $\mathbf{a}_t$ computed successively from (5.2.1).

The order n is increased (generally by 1 in the vector case and 2 in the scalar case) until the reduction in the determinant of the CSP matrix is statistically insignificant. For a scalar case the determinant is just the residual sum of squares and the significance can be judged by the scalar F criterion (2.5.1) adopted from the linear least squares theory (e.g., see Rao, 1965). For the vector case it generalizes to

$$\Lambda = \frac{|A_0|}{|A_1|} \quad (5.2.3)$$

$$F = \frac{(1 - \Lambda^{1/l})/ps}{\Lambda^{1/l}/(kl - 2\lambda)} \sim F[ps, (kl - 2\lambda)] \quad (5.2.4)$$

with

$$k = N - r - \frac{(p+s+1)}{2}$$

$$l = \sqrt{\frac{p^2 s^2 - 4}{p^2 + s^2 - 5}}$$

$$\lambda = \frac{ps - 2}{4}$$

where

A_0 = CSP matrix of the fitted (higher-order) model

r = p times the number of matrix parameters in the fitted (higher-order) model plus 1 if **C** is estimated

s = p times the number of matrix parameters restricted to zero (to yield the lower-order model)

A_1 = CSP matrix of the (lower-order) model with less parameters

N = Number of (vector) observations

p = Number of data series

When the F value calculated by (5.2.4) is less than the table value at a predetermined significance level such as 5% in comparing the nth and $(n+1)$st-order models, the ARMAV($n,n-1$) model is taken as the adequate model. It should be emphasized that the F test is strictly applicable only when $\mathbf{a}_t$'s are uncorrelated at different t. However, its application at lower orders before the $\mathbf{a}_t$'s have become uncorrelated usually indicates significant improvement and hence correctly suggests that the order should be increased.

The minimization of the determinant is a computational burden that becomes intense when moving average parameters are present, since they require iterative methods. For scalar models the computational burden of such iterative methods is still reasonable, but not so for vector models. Fortunately, the need for the moving average parameters is much less for vector models with large p compared to the scalar models (see Section 2.4.3). In particular, for modal applications, if synchronized multichannel data acquisition equipment is available, then the number of channels and hence p is quite large. Thus both the reduced need for moving average parameters and the excessive computational burden caused by them work in favor of omitting them and using ARV models. The examples in this chapter will therefore be limited to ARV models in the vector case.

An immediate advantage of the ARV models is that the parameter estimates of the row vector $\{\phi_{ijk}\}$ for a given i can be obtained by (separately) minimizing the RSS Σa_{it}^2 using linear least squares:

$$X_{it} = \phi_{i11} X_{1t-1} + \phi_{i12} X_{1t-2} + \cdots + \phi_{i1n} X_{1t-n} + \phi_{i21} X_{2t-1} + \phi_{i22} X_{2t-2}$$
$$+ \cdots + X_{i2n} X_{2t-n} + \cdots + \phi_{ipn} X_{pt-n} + c_i + a_{it} \qquad (5.2.5)$$

The least squares set-up for N observations may be summarized by

$$\mathbf{Y}_i = {}_0^n\mathbf{X} \begin{bmatrix} c_i \\ \boldsymbol{\phi}_{i..} \end{bmatrix} + \mathbf{a}_{i.}, \qquad i = 1, 2, \ldots p \qquad (5.2.6)$$

where

$$\mathbf{Y}_i = \begin{bmatrix} X_{i(n+1)} \\ X_{i(n+2)} \\ \vdots \\ X_{iN} \end{bmatrix} \qquad \mathbf{a}_{i.} = \begin{bmatrix} a_{i(n+1)} \\ a_{i(n+2)} \\ \vdots \\ a_{iN} \end{bmatrix}$$

$${}_0^n\mathbf{X} = \begin{bmatrix} 1 & X_{11} & X_{21} & \cdots & X_{p1} & X_{12} & X_{22} & \cdots & X_{p2} & \cdots & X_{1n} & X_{2n} & \cdots & X_{pn} \\ 1 & X_{12} & X_{22} & \cdots & X_{p2} & X_{13} & X_{23} & \cdots & X_{p3} & \cdots & X_{1(n+1)} & X_{2(n+1)} & \cdots & X_{p(n+1)} \\ \vdots & \vdots & \vdots & & \vdots & \vdots & \vdots & & \vdots & & \vdots & \vdots & & \vdots \\ 1 & X_{1(N-n)} & X_{2(N-n)} & \cdots & X_{p(N-n)} & X_{1(N-n+1)} & X_{2(N-n+1)} & \cdots & X_{p(N-n+1)} & \cdots & X_{1(N-1)} & X_{2(N-1)} & \cdots & X_{p(N-1)} \end{bmatrix}$$

$$(5.2.7)$$

$$\boldsymbol{\phi}_{i..} = [\phi_{i1n}, \phi_{i2n}, \ldots, \phi_{ipn}, \phi_{i1(n-1)}, \phi_{i2(n-1)}, \ldots, \phi_{ip(n-1)}, \ldots, \phi_{i11},$$
$$\phi_{i21}, \ldots, \phi_{ip1}]^T \qquad (5.2.8)$$

For measurements with appreciable noise, either in excitation (even with random or pseudo random) and/or in the output response measurements, the system must be considered to be stochastic. Then the order n is increased until the residual a_{it} are uncorrelated and the F test computed by (5.2.3) and (5.2.4) indicates insignificance.

For systems with negligible noise, this approach can still be used. However, it may yield model orders so high that it is not useful in practice because one is more interested in the deterministic system characteristics than the noise. In such cases, lower-order models can be obtained by using other criteria such as the residuals falling below the noise floor, desired parameters converging with requisite accuracy, small enough deterministic response prediction error, and small enough modal amplitudes for the added modes. Some of these criteria will be illustrated with examples. A useful indicator of model adequacy also consists of the closeness of the actual "power," "energy," or data sum of squares with that predicted by the fitted model; the adequate model has the predicted data sum of squares that is lower and closer to the actual than that of an inadequate model. For stable/stationary data, the data variance and the model predicted variance can be used, expressions for which will be given in Chapter 6.

5.2.2 Estimation

Although separate minimization of the residual sum of squares (RSS) Σa_{it}^2 for each i, calculated for increasing order n, already reduces the computational burden of the ARV models, it can be further reduced by utilizing three key features of the ARV models seen from (5.2.6)–(5.2.8):

1. The $_0^n X$ matrix is the same for all series Y_i.
2. Column vectors Y_i form a part of the matrix $^{(n+1)}_{0}X$.
3. Many columns of $_0^n X$ are shifted versions of others.

Features 2 and 3 allow the use of computational efficiencies introduced by Toeplitz–Henkel matrix structures. We will only indicate these in the present context, the reader is referred to books on matrix computations such as Golub and Van Loan (1987) for details. The notation of this section is somewhat cumbersome, but it will become clear at the end that there is a method in the madness! A reader only interested in results may skip Sections 5.2.2 and 5.2.3 and proceed directly to Section 5.3 without affecting readability.

The ARV(n) model (5.2.5) can be succinctly written, for all i together, as

$$X_t - \phi_1 X_{t-1} - \cdots - \phi_n X_{t-n} - C = a_t$$

or

$$X_t^T - X_{t-1}^T \phi_1^T - \cdots - X_{t-n}^T \phi_n^T - C^T = a_t^T \tag{5.2.9}$$

which is (5.2.1) with all $\theta_i = 0$. In view of (5.2.9), the data matrix (5.2.7) may be succinctly written as

$$\underset{(N-n) \times (np+1)}{_0^n X} = \begin{bmatrix} 1 & X_1^T & X_2^T & \cdots & X_n^T \\ 1 & X_2^T & X_3^T & \cdots & X_{n+1}^T \\ \vdots & \vdots & \vdots & & \vdots \\ 1 & X_{(N-n)}^T & X_{(N-n+1)}^T & \cdots & X_{(N-1)}^T \end{bmatrix} \tag{5.2.10}$$

$$= [\, 1 \;\vdots\; _1^n X \,] \tag{5.2.10a}$$

$$= [\, ^{-n}_{0} X \;\vdots\; ^n X \,] \tag{5.2.10b}$$

Note that the scalar version of the data matrix $_1^n \mathbf{X}$, that may be denoted by $_1^n X$, is the same as the Henkel data matrix introduced in Section 3.2.1 in dealing with scalar deterministic AR models. In fact, the matrix $_1^n \mathbf{X}$ is block Henkel in that all its elements on the ith diagonal perpendicular to the main diagonal consist of the same row vectors X_i^T. Also, $^n \mathbf{X}$ denotes the last p columns of $_0^n \mathbf{X}$, $^{-n}_{0}\mathbf{X}$ denotes the remaining matrix after omitting $^n \mathbf{X}$ from $_0^n \mathbf{X}$, and $\mathbf{1}$ denotes a column vector of 1's. Preceding indexes $_0^n$ of $\mathbf{X}$ denote that the column $\mathbf{1}$ is included at the beginning and $_1^n$ denotes that it is not. The parameters can similarly be compressed by using the state matrix ϕ defined in Section 2.4.4. The autoregressive parameter matrices $\phi_1, \phi_2, \ldots, \phi_n$ form the top matrix row of the state matrix ϕ. If this row in reverse order is denoted by $^n\phi$, that is the $p \times np$

matrix

$$^n\phi = [\phi_n, \phi_{n-1}, \ldots, \phi_1] \tag{5.2.11}$$

$$p \times np$$

then

$$\underset{np \times p}{^n\phi^T} = \begin{bmatrix} \phi_n^T \\ \phi_{n-1}^T \\ \vdots \\ \phi_1^T \end{bmatrix} = [\phi_{1\cdot\cdot}, \phi_{2\cdot\cdot}, \ldots, \phi_{p\cdot\cdot}] \tag{5.2.11a}$$

in the notation of (5.2.8). Finally, similar to $^n\mathbf{X}$, let

$$\underset{(N-n) \times p}{^{(n+1)}\mathbf{a}} = \begin{bmatrix} \mathbf{a}_{(n+1)}^T \\ \mathbf{a}_{(n+2)}^T \\ \vdots \\ \mathbf{a}_N^T \end{bmatrix} = [\mathbf{a}_{1\cdot}, \mathbf{a}_{2\cdot}, \ldots, \mathbf{a}_{p\cdot}] \tag{5.2.12}$$

in the notation of (5.2.6).

Now, for $t = (n+1), (n+2), \ldots, N$, the equations of the ARV(n) model (5.2.9) can be written as a single matrix equation:

$$^{(n+1)}\mathbf{X} - {}_{0}^{-(n+1)}\mathbf{X} \begin{bmatrix} \mathbf{C}^T \\ {}^n\phi^T \end{bmatrix} = {}^{(n+1)}\mathbf{a} \tag{5.2.13a}$$

or, even more compactly

$$^{(n+1)}_{0}\mathbf{X} \begin{bmatrix} -\mathbf{C}^T \\ -{}^n\phi^T \\ \mathbf{I} \end{bmatrix} = {}^{(n+1)}\mathbf{a} \tag{5.2.13}$$

If $\mathbf{C} = \mathbf{0}$, either because the data are average (DC) subtracted, or because we choose to account for constant offsets in the data by characteristic roots of 1 (see Section 2.7.1), these equations can be simplified by omitting the column $\mathbf{1}$ as

$$^{(n+1)}\mathbf{X} - {}_{1}^{-(n+1)}\mathbf{X}{}^n\phi^T = {}^{(n+1)}\mathbf{a} \tag{5.2.13b}$$

or

$$^{(n+1)}_{1}\mathbf{X} \begin{bmatrix} -{}^n\phi^T \\ \mathbf{I} \end{bmatrix} = {}^{(n+1)}\mathbf{a} \tag{5.2.13c}$$

Since the columns of $^{(n+1)}\mathbf{a}$ are $\mathbf{a}_{i\cdot}$ of (5.2.6), they are minimized to zero by the estimates of the parameters $\mathbf{C}, \phi$ obtained from (5.2.13) as the solution of the homogeneous linear equation

$${}^{(n+1)}_{0}\mathbf{X}\begin{bmatrix} -\mathbf{C}^T \\ -{}_n\boldsymbol{\phi}^T \\ \mathbf{I} \end{bmatrix} = \mathbf{0} \qquad (5.2.14)$$

$$(N-n)\times(np+p+1) \quad (np+p+1)\times p$$

that is the same as the nonhomogeneous linear equation

$$-{}^{(n+1)}_{0}\mathbf{X}\begin{bmatrix} \mathbf{C}^T \\ {}_n\boldsymbol{\phi}^T \end{bmatrix} = {}^{(n+1)}_{0}\mathbf{X}$$

(5.2.14a)

$$(N-n)\times(np+1)\,(np+1)\times p \qquad (N-n)\times p$$

When $\mathbf{C} = \mathbf{0}$, the corresponding results follow by dropping $\mathbf{C}$ and changing the preceding subscript 0 to 1, as seen from (5.2.13b and c); this is the case for all the subsequent results of this section and hence we will not repeat them again. Recalling the discussion of the generalized inverse from Section 3.5.3, it is clear that the parameter estimates are given by the generalized or pseudo-inverse of the data matrix $-{}^{(n+1)}_{0}\mathbf{X}$ denoted by $-{}^{(n+1)}_{0}\mathbf{X}^\dagger$ as

$$\begin{bmatrix} \hat{\mathbf{C}}^T \\ {}_n\hat{\boldsymbol{\phi}}^T \end{bmatrix} = -{}^{(n+1)}_{0}\mathbf{X}^\dagger \, {}^{(n+1)}\mathbf{X} \qquad (5.2.15)$$

For deterministic systems, $\mathbf{a}_t = \mathbf{0}$ in (5.2.9), which now is the same as (5.2.14a) with $N = (np + n + 1)$; the square matrix $-{}^{(n+1)}_{0}\mathbf{X}$ is now of full rank $= (np + 1)$, the generalized inverse reduces to the usual inverse:

$$-{}^{(n+1)}_{0}\mathbf{X}^\dagger = -{}^{(n+1)}_{0}\mathbf{X}^{-1}, \qquad N = np + n + 1 \qquad (5.2.16)$$

and (5.2.15) with this gives the exact solution of $(np + 1)p$ equations (5.2.14a) in $(np + 1)p$ unknowns. It follows from the basic theory of linear equations reviewed in Section 3.1.2 that the rank of the data matrix ${}^{(n+1)}_{0}\mathbf{X}$ is also $(np + 1)$ so that (5.2.14 and 5.2.14a) are consistent, that is,

$$\text{Rank } ({}^{(n+1)}_{0}\mathbf{X}) = \text{Rank } (-{}^{(n+1)}_{0}\mathbf{X} \, \vdots \, {}^{(n+1)}\mathbf{X}) = \text{Rank } (-{}^{(n+1)}_{0}\mathbf{X}) = (np + 1)$$

(5.2.17)

so the columns of ${}^{(n+1)}_{0}\mathbf{X}$ are linearly dependent, whereas those of $-{}^{(n+1)}_{0}\mathbf{X}$ are not. In fact, for such a deterministic system, even with $N \geq (np + n + 1)$,

$$\text{Rank } ({}^{j}_{0}\mathbf{X}) = (jp + 1) < (np + 1), \qquad \text{for } j < n \qquad (5.2.18)$$

$$= (np + 1), \qquad \text{for } j \geq n$$

and with $\mathbf{C} = \mathbf{0}$,

$$\text{Rank } (_1^j \mathbf{X}) = jp < np, \quad \text{for } j < n \quad (5.2.18a)$$
$$= np, \quad \text{for } j \geq n$$

In other words, the rank of the data matrix $_0^j\mathbf{X}$ or $_1^j\mathbf{X}$ continues to increase as (the column size of the matrix determined by) j is increased so long as j does not exceed n, but does not increase any more when $j \geq n$. This is the crux of the Ho–Kalman algorithm which the ERA attempts to implement using the SVD to determine the rank and hence the "true" system order n, as discussed in Section 5.1.5. Unfortunately, this is true only for deterministic systems. For stochastic systems, there is no such rank limitation at the true order and hence the singular values may not show a clear break or jump to zero or small values after the true order is exceeded.

For stochastic systems, then, the linear equations (5.2.14 and 5.2.14a) are inconsistent, that is, for $N > (np + n + 1)$,

$$\text{Rank } (_0^{(n+1)}\mathbf{X}) = \text{Rank } (^{-(n+1)}_0\mathbf{X} \ \vdots \ ^{(n+1)}\mathbf{X}) > \text{Rank } (^{-(n+1)}_0\mathbf{X}) \quad (5.2.19)$$

so the columns of $^{(n+1)}_0\mathbf{X}$ are linearly independent, the system of equations is overdetermined and has no exact solution because the columns of $^{(n+1)}\mathbf{X}$ are not in $R(^{-(n+1)}_0\mathbf{X})$. Therefore, we seek to minimize a suitable norm of the columns of $^{(n+1)}\mathbf{a}$ which is the difference of the two sides of (5.2.14a). For our usual choice of the squared norm, this leads to the least squares estimates (5.2.15) with the generalized inverse defined by

$$^{-(n+1)}_0\mathbf{X}^\dagger = (^{-(n+1)}_0\mathbf{X}^T \ ^{-(n+1)}_0\mathbf{X})^{-1} \ ^{-(n+1)}_0\mathbf{X}^T \quad (5.2.20)$$

in which the usual inverse of the square matrix in parenthesis is always well defined for a stochastic system because it is nonsingular:

$$\text{Rank } (^{-(n+1)}_0\mathbf{X}) = \text{Rank } (^{-(n+1)}_0\mathbf{X}^T \ ^{-(n+1)}_0\mathbf{X}) = (np + 1) \quad (5.2.21)$$
$$(N - n) \times (np + 1) \qquad (np + 1) \times (np + 1)$$

Since a deterministic system is a special case of the stochastic one, the least squares estimates should also yield the exact parameters when the system is exactly deterministic. This is true in theory. But in practice, when the true system order n is unknown and the DDS modeling procedure is applied to such a hypothetical system with increasing order j, numerical difficulties may be encountered when j exceeds the "true" order. The reason is that, as shown by (5.2.18),

$$\text{Rank } (^{-(j+1)}_0\mathbf{X}) = \text{Rank } (^{-(j+1)}_0\mathbf{X}^T \ ^{-(j+1)}_0\mathbf{X}) = (np + 1) < (jp + 1),$$
$$(N - j) \times (jp + 1) \qquad (jp + 1) \times (jp + 1)$$
$$\text{for } j > n \quad (5.2.22)$$

so that the square matrix in (5.2.20) with j replacing n is singular and its inverse does not exist. Practical experience with simulated data has shown that this difficulty can be avoided by using computation more precise than the data, for example, single-precision data and double-precision computation. The round-off error in the data is enough to make the system stochastic. Some examples illustrating this will be given later. However, a computationally expensive but numerically reliable way of dealing with such rank-deficient matrices arising from perfectly deterministic systems of finite order is to use the SVD.

Recall from Section 3.5.3 that if the true order of a deterministic system is n so that by (5.2.18) the rank of the data matrix $^{-(j+1)}_{0}\mathbf{X}$ stays at $(np + 1)$ for $j \geq n$, and its SVD is given by

$$^{-(j+1)}_{0}\mathbf{X} = \mathbf{U} \text{ diag } (\sigma_1, \sigma_2, \ldots, \sigma_{(np+1)}, 0, 0, \ldots, 0)\mathbf{V}^T, \quad j \geq n \quad (5.2.23)$$

then the least squares estimates are obtained, with the generalized inverse defined by

$$^{-(j+1)}_{0}\mathbf{X}^\dagger = \mathbf{U} \text{ diag } (\sigma_1^{-1}, \sigma_2^{-1}, \ldots, \sigma_{(np+1)}^{-1}, 0, 0, \ldots, 0)\mathbf{V}^T, \quad j \geq n \quad (5.2.24)$$

using (5.2.15) with n replaced by j for $n \leq j \leq (N - np - 1)$. Although (5.2.24) can also be used for stochastic data, there is no actual rank deficiency, the singular values σ_i in (5.2.23) may reduce slowly, and the determination of rank $(np + 1)$ beyond which they may be considered to be zero becomes difficult, subjective, and ad hoc, as illustrated in Section 3.5.3. Even for stationary stochastic time series data, from which deterministic components have been removed, the singular values of the Henkel covariance matrix which are squares of the singular values of the corresponding data matrix do not jump to zero after the rank $(np + 1)$. Although the theoretical Henkel covariance matrix has a rank limitation of $(np + 1)$, its sampled version or the data version does not, as explained above. For this reason the DDS employs F test and other criteria mentioned in the last section to determine the order n, rather than estimating the rank of the Henkel data or covariance matrix.

The occurrence of rank deficiency for real data is somewhat rare in practice and hence the DDS approach does not employ rank estimates to determine the order for the following reasons:

1. A linear system of finite order n is usually an approximation to the real system which may be nonlinear and/or of infinite order. We only approximate it as well as permitted by the data; this is one of the basic tenets of the DDS approach.
2. Real data always contains at least measurement noise making the system stochastic. Singular values contributed by such noise need not necessarily be small.
3. Any noise, even round-off and computational noise, can make up for the rank deficiency even when it exists for the actual system in theory.

4. Unless the noise modes are correctly modeled and separated as done by the DDS approach, the accuracy of the actual system parameters will not improve.

For a given n, of course, the SVD is the most accurate method of least squares estimation, particularly when near rank deficiencies are likely to be encountered. At the expense of some extra computation, it frees us from the possibility of numerical instabilities.

5.2.3 Computation Methods, Their Speeds and Accuracies

The least squares estimates can be computed by many methods. We will consider three factorization methods with successively lower speed but higher accuracy and better numerical stability: Cholesky, QR, and SVD. These methods were reviewed in Sections 3.4.7 and 3.5.3 for a general $m \times n$ matrix $\mathbf{A}$ and the application of QR factorization was illustrated for scalar AR models in Sections 3.2.1 and 3.2.2. We will now present them in the context of ARV(n) models needed for modal analysis.

Although we present only the least squares methods, which make no assumptions about the data in the spirit of the DDS philosophy outlined in Chapter 1, estimation can often be improved in quality and/or computational efficiency by other methods when the data are known to satisfy some restrictive assumptions. For example, estimation for data satisfying the deterministic plus white noise assumption can be improved by total least squares; see Section 2.8.4 as well as Pandit and Yao (1991). Yule–Walker equations (see Section 6.3.1), essentially equivalent to the Cholesky method, can be used for stationary stochastic data; their least squares formulation using more sample covariances at higher lags is useful for deterministic plus stationary stochastic data, and so on.

The Cholesky Factorization
The Cholesky method of solving normal equations is the most computationally efficient procedure because it primarily works with a smaller dimensional square matrix in the solution process and the AR modeling leads to a near-Toeplitz structure, providing further efficiency comparable to the Levinson–Durbin algorithms. In its implementation we first form the square matrix of (5.2.20), say

$$\mathbf{F} = {}^{-(n+1)}_{0}\mathbf{X}^T \, {}^{-(n+1)}_{0}\mathbf{X} \tag{5.2.25}$$

$(np + 1) \times (np + 1)$

and the right-hand side of the normal equations

$$\mathbf{D} = {}^{-(n+1)}_{0}\mathbf{X}^T \, {}^{(n+1)}\mathbf{X} \tag{5.2.26}$$

$(np + 1) \times p$

Next the Cholesky factorization

$$\mathbf{F} = \mathbf{L}\mathbf{L}^T \tag{5.2.27}$$

is computed to obtain the lower triangular matrix $\mathbf{L}$. Forward elimination is used to solve for $(np + 1) \times p$ matrix $\mathbf{Y}$ by

$$\mathbf{L}\mathbf{Y} = \mathbf{D} \tag{5.2.28}$$

and finally the parameters are obtained by backward elimination from

$$\mathbf{L}^T \begin{bmatrix} \mathbf{C}^T \\ _n\boldsymbol{\phi}^T \end{bmatrix} = \mathbf{Y} \tag{5.2.29}$$

The operation count is roughly $(N - n)np^2 + (np)^3/6$, obtained from the value $(mn + n^3/6)$ for the scalar AR case (see Section 3.4.7). As shown in Table 5.1, it is the fastest among the three methods. It is also (therefore?) the least accurate, because its accuracy depends upon the *square* of the condition number of the data matrix. If ε^2 denotes the relative squared error in a parameter estimate vector, then

$$\varepsilon^2 \simeq u\kappa(^{-(n+1)}_0\mathbf{X}^T \, ^{-(n+1)}_0\mathbf{X}) = u\left(\frac{\sigma_1}{\sigma_{(np+1)}}\right)^2 \tag{5.2.30}$$

where u is the unit round-off error of the computer. Note that for a strictly deterministic system the error would be infinite when the true order is exceeded because (5.2.23) shows that $\sigma_{(jp+1)} = 0$ for $j > n$.

The QR Factorization
The computational advantage and also the loss of accuracy of the Cholesky method arises, in the general case of an $m \times n$ matrix $\mathbf{A}$ with $m > n$, from the use

TABLE 5.1 Computation in Solving the LS Problem for the ARV Model $(N-n \geq np)^a$

Algorithm	Approximate Flop Count
Normal equations using Cholesky (near Toeplitz ARV)	$(N - n)np^2 + (np)^3/6$
Normal equations using Cholesky	$(N - n)(np)^2/2 + (np)^3/6$
Householder orthogonalization	$(N - n)(np)^2 - (np)^3/3$
Modified Gram–Schmidt	$(N - n)(np)^2$
Golub–Reinsch SVD	$2(N - n)(np)^2 + 4(np)^3$
Chan SVD	$(N - n)(np)^2 + 17(np)^3/3$

N, number of observations for each series; n, model order; p, number of series.

of the smaller $n \times n$ matrix $\mathbf{A}^T\mathbf{A}$. The loss of accuracy can be recovered, at the expense of extra computation, if we could directly use the larger matrix $\mathbf{A}$ instead of $\mathbf{A}^T\mathbf{A}$. This is indeed possible by QR factorization because recall from Section 3.4.7 that the Cholesky method itself is derivable using the implicit QR factorization

$$\mathbf{A}^T\mathbf{A} = (\mathbf{QR})^T\mathbf{QR} = \mathbf{R}^T\mathbf{Q}^T\mathbf{QR} = \mathbf{R}^T\mathbf{R} = \mathbf{LL}^T$$

Therefore it is logical to use QR factorization of the data matrix $^{-(n+1)}_{0}\mathbf{X}$ in (5.2.14) to improve the accuracy of least squares estimates.

In the notation of (5.2.10) and (5.2.10b), let (see (3.2.1))

$$^{(n+1)}_{0}\mathbf{X} = [^{-(n+1)}_{0}\mathbf{X} \ \vdots \ ^{(n+1)}\mathbf{X}] = \begin{bmatrix} 1 & \mathbf{X}_1^T & \cdots & \mathbf{X}_n^T & \vdots & \mathbf{X}_{n+1}^T \\ 1 & \mathbf{X}_2^T & \cdots & \mathbf{X}_{n+1}^T & \vdots & \mathbf{X}_{n+2}^T \\ \vdots & \vdots & & \vdots & \vdots & \vdots \\ 1 & \mathbf{X}_{(N-n)}^T & \cdots & \mathbf{X}_{(N-1)} & \vdots & \mathbf{X}_N^T \end{bmatrix}$$

$$= \mathbf{Q}[\mathbf{R} \ \vdots \ \mathbf{T}_{n+1}]$$

$$= [\mathbf{1}, \mathbf{q}_1, \mathbf{q}_2, \ldots, \mathbf{q}_n] \begin{bmatrix} 1 & t_{01} & \cdots & t_{0n} & \vdots & t_{0(n+1)} \\ 0 & t_{11} & \cdots & t_{1n} & \vdots & t_{1(n+1)} \\ \vdots & \vdots & & \vdots & \vdots & \vdots \\ 0 & 0 & \cdots & t_{nn} & \vdots & t_{n(n+1)} \end{bmatrix} \quad (5.2.31)$$

where $^{-(n+1)}_{0}\mathbf{X} = \mathbf{QR}$ denotes the QR decomposition of the data matrix $^{-(n+1)}_{0}\mathbf{X}$, $\mathbf{t}_{0i}$ are $1 \times p$ row vectors, $\mathbf{t}_{ii}$ are $p \times p$ upper triangular matrices with 1's on their diagonals, $\mathbf{t}_{ij}$, $i < j$, are $p \times p$ matrices, and $\mathbf{q}_i$ are $(N-n) \times p$ matrices all of whose columns for $i = 1$ to n together with the $(N-n) \times 1$ column vector $\mathbf{1}$ form an orthogonal basis of the columns of $^{-(n+1)}_{0}\mathbf{X}$. Substituting this QR decomposition in the deterministic ARV(n) equation (5.2.14) we see that the orthogonality of $\mathbf{q}_i$ implies

$$[\mathbf{R} \ \vdots \ \mathbf{T}_{n+1}] \begin{bmatrix} -\hat{\mathbf{C}}^T \\ -{}_n\hat{\boldsymbol{\phi}}^T \\ \mathbf{I} \end{bmatrix} = \mathbf{0} \quad (5.2.32a)$$

that is,

$$\begin{bmatrix} 1 & t_{01} & t_{02} & \cdots & t_{0n} \\ 0 & t_{11} & t_{12} & \cdots & t_{1n} \\ \vdots & \vdots & \vdots & & \vdots \\ 0 & 0 & 0 & \cdots & t_{nn} \end{bmatrix} \begin{bmatrix} \hat{\mathbf{C}}^T \\ \hat{\boldsymbol{\phi}}_n^T \\ \vdots \\ \hat{\boldsymbol{\phi}}_1^T \end{bmatrix} = \begin{bmatrix} t_{0(n+1)} \\ t_{1(n+1)} \\ \vdots \\ t_{n(n+1)} \end{bmatrix} \quad (5.2.32)$$

For $\mathbf{C} = \mathbf{0}$ we need to omit the first column and row of this equation, which in scalar form is the same as (3.2.2). The parameter estimates are therefore obtained by the matrix version of the recursive computation (3.2.3):

$$\mathbf{t}_{ii}\,\hat{\boldsymbol{\phi}}_{n-i+1}^{T} = \mathbf{t}_{i(n+1)} - \sum_{j=i+1}^{n} \mathbf{t}_{ij}\,\hat{\boldsymbol{\phi}}_{n-j+1}^{T}, \qquad i = n, n-1, \ldots, 1 \qquad (5.2.33)$$

$$\hat{\mathbf{C}}^{T} = \mathbf{t}_{0(n+1)} - \sum_{j=1}^{n} \mathbf{t}_{0j}\,\hat{\boldsymbol{\phi}}_{n-j+1}^{T} \qquad (5.2.34)$$

For a given i, since $\mathbf{t}_{ii}$ is an upper triangular matrix with 1's on its diagonal, equations in (5.2.33) can be solved for columns of $\boldsymbol{\phi}_{n-i+1}^{T}$ simply by back substitution starting at the bottom. Although we have given an algebraic proof above, it would be an interesting exercise to prove (5.2.32) by the orthogonal basis argument used for proving (3.2.2).

The residuals can be recovered by going one step further in the QR decomposition and using $\mathbf{q}_{(n+1)}\mathbf{t}_{(n+1)(n+1)}$, as illustrated for the scalar case in Section 3.2.2. They can also be computed directly from the data by using the parameters from (5.2.33) and (5.2.34) in the model (5.2.9); this saves the additional step in the QR decomposition and hence it is computationally faster. The orthogonal basis argument provides an interesting interpretation of the relation between the residuals and $\mathbf{q}_{(n+1)}$, which is left as an exercise (see also Section 6.4.1).

We have given the entire formulation in this section for fixed N and n. In the practical implementation of the DDS modeling procedure, it is usually computationally more efficient to choose the highest possible order, say $n_{\max}$ and then use $(N - n_{\max})$ rows for all the data matrices. Then, as the order n is increased, only the last columns of $\mathbf{R}$ need to be updated, new parameters computed by (5.2.33) and (5.2.34), and residuals or prediction errors computed using (5.2.9) or (5.3.2a), respectively. This involves some loss of information at low order models.

The QR factorization may be computed either by Householder or (modified) Gram–Schmidt methods. The solution of the general least squares problem for an $m \times n$ matrix $\mathbf{A}$ has an operation count $n^2[m - (n/3)]$, and mn^2, respectively, which translates into $(np + 1)^2[(N - n) - (np/3)]$ and $(N - n)(np + 1)^2$ in the present case; $\mathbf{Q}$ is not required for the least squares solution, but is available in the Gram–Schmidt method. The computation of $\mathbf{Q}$ is not necessary for computing the parameters or the residuals. $\mathbf{Q}$ requires additional $[mn^2 - (n^3/3)]$ operations in the Householder method, thus needing more computation than the Gram–Schmidt method; however, the error in orthogonalization of the Gram–Schmidt method depends both on the round-off error of the computer as well as the condition number of $^{-(n+1)}_{0}\mathbf{X}$, whereas that of Householder method depends only on the round-off error.

For small residuals compared to the data, the relative error of the least squares problem by the QR factorization primarily depends upon the condition of $^{-(n+1)}_{0}\mathbf{X}$. As the residuals approach the data in magnitude, however, the relative error starts to depend more on the square of the condition. But in such a case the data is stochastic, the condition number is relatively small, particularly since n never exceeds the true order by a large amount when the DDS approach of successively increasing n is followed, and hence dependence on square of the condition is not harmful. Reference may be made to Golub and Van Loan (1987, Theorem 6.1–3) for more details of the sensitivity of least squares method.

Summarizing, the Householder method appears to be the best choice for DDS modeling of data with small noise when QR factorization is to be used for better accuracy. It allows better accuracy compared with the Cholesky method, with a moderate increase in computation. Its applications for the simple AR(1) model was illustrated in Section 3.4.6.

The Singular Value Decomposition

For a strictly deterministic system, the residuals are zero when the correct order is reached, stay zero beyond, and the condition number becomes infinite beyond the correct order. Therefore, one would expect numerical difficulties in the DDS modeling of data from nearly deterministic systems when the correct order is exceeded. These difficulties are more likely to occur in the Cholesky method whose errors depend on the square of the condition number, than in the QR factorization whose errors depend primarily on the condition number. The difficulties can be avoided by using computation precision better than that of the data used. (Note that the common remedy of column pivoting used for rank-deficient problems is not required in the present case because the zero q_i and zero diagonal elements of **R** always occur at the end.) However, a general purpose reliable method that will not cause numerical difficulties even in strictly deterministic systems is the SVD. This eliminates the numerical difficulties at the expense of much more computation and by shifting the difficulty to the determination of the rank beyond which the singular values may be considered to be zero.

One way to incorporate the SVD in the DDS modeling of strictly deterministic systems is to increase the order until the singular values in (5.2.23) start falling below a predetermined number, choose n such that $r = (np + 1)$ and then estimate the parameters using (5.2.24) and (5.2.15). The choice of σ_r below which the singular values would be considered zero becomes subjective, ad hoc, and difficult and hence we generally do not recommend it.

Recall from Section 3.5.3 that the singular values of **A** are the square roots of the eigenvalues of $\mathbf{A}^T\mathbf{A}$. However, the formation of $\mathbf{A}^T\mathbf{A}$ can lead to a loss of information. A preferable method of computing the SVD implemented in Golub and Reinsch (1970) is to first use Householder matrices on both sides of **A** to reduce it to a bidiagonal form **B** and then apply an iterative implicit symmetric QR algorithm for finding the eigenvalues of $\mathbf{B}^T\mathbf{B}$ by nearly diagonalizing **B**. In our case, since $m \gg n$, it is advisable to first triangularize $\binom{n+1}{0}\mathbf{X}$ before applying this procedure. The operation count for an $m \times n$ **A** is $[mn^2 + (17/3)n^3]$, much higher than the preceding methods; the count in our case is listed in Table 5.1 as the Chan SVD.

5.3 MODAL AND SYSTEM PARAMETER IDENTIFICATION

It is clear from the development in Chapters 2–4 that the DDS methodology provides us with statistically significant eigenvalues and eigenvectors from the recorded data, either univariate or multivariate. The application of the DDS methodology to modal analysis then involves selecting a proper set among these and getting the parameters of interest such as the natural frequency, damping ratio, scaled mode shapes, and finally mass, stiffness, and damping matrices. Although the impulse response matrix is also available from the DDS model, we will not consider it further because it is generally an intermediate step in the prevalent methods to get these parameters.

We will summarize the necessary results for such an application of the DDS methodology to modal analysis in this section. Two most commonly used excitations, impulse and sinusoidal, will be considered, although the method can be used for any known excitation for which a theoretical response of the given system can be derived. Random excitations are even easier to deal with by the DDS method, and will be considered but elaborated only later in the subsequent Chapter 6 that deals with transfer functions or FRF's. As emphasized earlier, all the parameters except system matrices can be obtained from response data when excitation measurement is not available, but its nature such as impulse, sinusoidal, or random is known.

In the conventional modal analysis, both the input excitation and output response are measured. The ratio of their FFT's or spectra provides the FRF's and curve fitting the FRF's provides the natural frequencies, damping ratios, and residues as reviewed in Section 5.1; this information in matrix form is often called the modal model. The information in the modal model of structures with real mode shapes can be displayed in a mode shape plot, depicting a grid of spatially located measurement points and the scaled amplitude plotted at each point in a positive or negative direction. The information can also be displayed by animated mode shapes depicting the motion of each grid point in time on a video screen; this display is possible for complex mode shapes as well.

The information in the modal model can be used to construct the full FRF matrix, often called the response model, provided enough FRF measurements are available. Since such an FRF matrix can be constructed only from a column of FRF measurements, some additional measurements not used in constructing the matrix can be used as a check on the accuracy of the whole process. Finally, the full FRF matrix yields the system matrices or the mass, spring, and damping matrices, often called the spatial model of the system.

In the DDS approach, the information in the modal model follows directly from the eigenvalues and eigenvectors of the state matrix based on the output data alone, no excitation measurement is needed. If the input excitation measurement is also included in the data used for DDS modeling, then the spatial model can also be obtained directly in addition to the modal model. The FRF matrix (Fourier transform of the impulse response matrix) is also directly available from the transfer function of the ARMAV models by substituting $B = e^{-i2\pi f \Delta}$ to yield the response model. However, it does not serve any purpose other than graphical display and hence will not be considered further in this chapter; it will be briefly discussed in Chapter 6.

5.3.1 Modal Model/Parameters: Frequency, Damping, and Mode Shapes

Recall from Sections 3.6.2, 4.3, and 4.5 that the ARV(n) model for the p-variate mean-subtracted vector $\mathbf{X}_t = (X_{1t}, X_{2t}, \ldots, X_{pt})^T$:

$$\mathbf{X}_t = \boldsymbol{\phi}_1 \mathbf{X}_{t-1} + \boldsymbol{\phi}_2 \mathbf{X}_{t-2} + \cdots + \boldsymbol{\phi}_n \mathbf{X}_{t-n} + \mathbf{a}_t \tag{5.3.1}$$

has the modal decomposition

$$\mathbf{X}_t = \mathbf{I}_p \boldsymbol{\phi}^{t-n+1} \mathbf{X}_0^{(n-1)} + \sum_{j=0}^{t-n} \mathbf{G}_j \mathbf{a}_{t-j} \tag{5.3.2}$$

where $\mathbf{X}_0^{(n-1)}$ is $np \times 1$ column vector of initial values with elements denoted by X_{0i}

$$\mathbf{X}_0^{(n-1)} = (\mathbf{X}_{n-1}^T, \mathbf{X}_{n-2}^T, \ldots, \mathbf{X}_0^T)^T = (X_{01}, X_{02}, \ldots, X_{0(np)})^T \tag{5.3.3}$$

$\mathbf{I}_p$ is a $p \times np$ matrix with a $p \times p$ identity followed by zeros:

$$\mathbf{I}_p = [\mathbf{I}, \mathbf{0}, \ldots, \mathbf{0}] \tag{5.3.4}$$
$$p \times np \quad p \times p$$

and the Green's function $\mathbf{G}_j$ is the $p \times p$ matrix in the top left corner of $\boldsymbol{\phi}^j$, that is,

$$\mathbf{G}_j = \mathbf{I}_p \boldsymbol{\phi}^j \mathbf{I}_p^T \tag{5.3.5}$$

with the discrete state matrix $\boldsymbol{\phi}$ defined by

$$\underset{np \times np}{\boldsymbol{\phi}} = \begin{bmatrix} \boldsymbol{\phi}_1 & \boldsymbol{\phi}_2 & \cdots & \boldsymbol{\phi}_{n-1} & \boldsymbol{\phi}_n \\ \mathbf{I} & \mathbf{0} & \cdots & \mathbf{0} & \mathbf{0} \\ \mathbf{0} & \mathbf{I} & \cdots & \mathbf{0} & \mathbf{0} \\ \vdots & \vdots & & \vdots & \vdots \\ \mathbf{0} & \mathbf{0} & \cdots & \mathbf{I} & \mathbf{0} \end{bmatrix} \tag{5.3.6}$$

The $p \times np$ matrix $\mathbf{I}_p$ picks the first p rows when applied from left and picks the first p columns when applied from right as $\mathbf{I}_p^T$.

232   MODAL ANALYSIS

The response prediction by the deterministic part $\hat{\mathbf{X}}_t$, which is the first term in the right-hand side of (5.3.2), may also be recursively computed as

$$\hat{\mathbf{X}}_t = \mathbf{X}_t, \qquad 0 \leq t \leq (n-1)$$

$$= \sum_{i=1}^{n} \hat{\phi}_i \hat{\mathbf{X}}_{t-i}, \qquad t \geq n \qquad (5.3.2a)$$

so that the prediction error $(\mathbf{X}_t - \hat{\mathbf{X}}_t)$ is the second term on the right-hand side of (5.3.2). This estimate of the deterministic part is approximate and suffices for many modal analysis applications. Better estimates are found by nonlinear least squares method, see, for example, Pandit and Wu (1983, Chapter 10). Such nonlinear least squares routines take the initial data vector $\mathbf{X}_0^{(n-1)}$ and the ϕ_i obtained from the linear least squares fitted ARV model as initial values, and iteratively refine either one of them or both $\mathbf{X}_0^{(n-1)}$ and ϕ_i to minimize the residual or prediction error sum of squares.

Assuming that ϕ is diagonalizable, that is, it has a diagonal spectral matrix λ for convenience, its spectral decomposition may be written as

$$\phi = \mathbf{L}\lambda\mathbf{L}^{-1} \quad \text{or} \quad \phi^j = \mathbf{L}\lambda^j\mathbf{L}^{-1} \qquad (5.3.7)$$

where

$$\lambda = \text{diag}\{\lambda_i\}$$

is the matrix of eigenvalues (that will be a Jordan form when ϕ is *not* diagonalizable) and $\mathbf{L}$ is the matrix of eigenvectors of ϕ with a special form

$$\mathbf{L} = \begin{bmatrix} \ell_1 \lambda_1^{n-1} & \ell_2 \lambda_2^{n-1} & \cdots & \ell_{np} \lambda_{np}^{n-1} \\ \ell_1 \lambda_1^{n-2} & \ell_2 \lambda_2^{n-2} & \cdots & \ell_{np} \lambda_{np}^{n-2} \\ \vdots & \vdots & & \vdots \\ \ell_1 & \ell_2 & \cdots & \ell_{np} \end{bmatrix} \qquad (5.3.8)$$

due to the special form of the state vector $[\mathbf{X}_t^T, \mathbf{X}_{t-1}^T, \ldots, \mathbf{X}_{t-n+1}^T]^T$. Here ℓ_i are $p \times 1$ column vectors that become 1 for the scalar case and then the matrix (5.3.8) reduces to the form in Section 3.6.2. They are replaced by generalized eigenvectors when ϕ is not diagonalizable.

The ℓ_i are the modal vectors called scaled mode shapes, because, as eigenvectors, they may be arbitrarily scaled without affecting the state matrix. Note that they are the modal vectors of both the discrete time representation $\mathbf{X}_t$ governed by an ARV model as well as the continuous time representation $\mathbf{X}(t)$ governed by the vector differential equation; they are the columns of the bottom $p \times np$ submatrix of the eigenvector matrix $\mathbf{L}$, and can also be alternatively expressed by using the autoregressive matrix polynomial:

$$[\lambda_i^n \mathbf{I} - \lambda_i^{n-1}\phi_1 - \cdots - \phi_n]\ell_i = \mathbf{0}, \, i = 1, 2, \ldots, np \qquad (5.3.7a)$$

For an ideal system with p degrees of freedom, noise-free data $\mathbf{X}_t$ from properly chosen p locations will lead to an ARV(2) model so that $n = 2$. Then $\ell_1, \ell_2, \ldots, \ell_{2p}$ give the modal vectors or scaled mode shapes and the corresponding system poles or the eigenvalues are given by

$$\mu_i = \frac{1}{\Delta} \ln(\lambda_i), \quad i = 1, 2, \ldots, 2p \tag{5.3.9}$$

where Δ is the sampling interval. The natural frequency ω_i and damping ratio ζ_i can be found from a complex conjugate pair of poles say $\mu_i, \bar{\mu}_i$ by

$$\mu_i, \bar{\mu}_i = -\zeta_i \omega_i \pm j \omega_i \sqrt{1 - \zeta_i^2} = \sigma_i \pm j \Omega_i \tag{5.3.10a}$$

or directly from the discrete eigenvalues $\lambda_i, \bar{\lambda}_i$ as

$$\sigma_i = \frac{1}{2\Delta} \ln(\lambda_i \bar{\lambda}_i) \tag{5.3.10b}$$

$$\Omega_i = \frac{1}{\Delta} \arctan\left[\frac{\operatorname{Im}(\lambda_i)}{\operatorname{Re}(\lambda_i)}\right] \tag{5.3.10c}$$

where Ω_i is the damped frequency $\omega_i \sqrt{1 - \zeta_i^2}$, and $j = \sqrt{-1}$.

Since the state vector in continuous time is of the form $[\mathbf{x}(t)^T, \dot{\mathbf{x}}(t)^T]^T$, the eigenvector of the state matrix $\mathbf{A}$ corresponding to the eigenvalue μ_i is $[\ell_i^T, \mu_i \ell_i^T]^T$. Hence the continuous-time state matrix $\mathbf{A}$ is given by

$$\mathbf{A} = \mathbf{M} \boldsymbol{\mu} \mathbf{M}^{-1} = \begin{bmatrix} \mathbf{0} & \mathbf{I} \\ -\mathbf{m}^{-1}\mathbf{k} & -\mathbf{m}^{-1}\mathbf{c} \end{bmatrix} \tag{5.3.11}$$

$2p \times 2p$

where $\boldsymbol{\mu}$ is the matrix of characteristic roots or poles

$$\boldsymbol{\mu} = \operatorname{diag}\{\mu_i\} \tag{5.3.12}$$

$2p \times 2p$

and the modal matrix $\mathbf{M}$ may be denoted by

$$\mathbf{M} = \begin{bmatrix} \ell_1 & \ell_2 & \cdots & \ell_{2p} \\ \mu_1 \ell_1 & \mu_2 \ell_2 & \cdots & \mu_{2p} \ell_{2p} \end{bmatrix} \text{ or } \begin{bmatrix} \ell_1 & \bar{\ell}_1 & \cdots & \bar{\ell}_p \\ \mu_1 \ell_1 & \bar{\mu}_1 \bar{\ell}_1 & \cdots & \bar{\mu}_p \bar{\ell}_p \end{bmatrix} \tag{5.3.13}$$

where the second notation is applicable when all the eigenvalues are complex, as is often the case in practical modal analysis; we will use the first, more general form. The diagonal matrix $\boldsymbol{\mu}$ gives the frequencies and damping ratios, whereas

M gives the scaled mode shapes $\ell_1, \ell_2, \ldots, \ell_{2p}$; the two together constitute the modal model. Since μ_i as well as ℓ_i occur in complex conjugate pairs, the modal vectors can be expressed in terms of p amplitudes and p phases; for real modes the phases are restricted to 0 or 180° and then the p amplitude vectors with respective plus and minus signs give the $p \times p$ modal matrix used in the normal mode theory of classical vibrations. The continuous-time state matrix **A** really summarizes the modal model since μ and **M** follow from its spectral decomposition.

If desired for graphical presentation, the estimated FRF matrix **H** can be obtained using (5.1.69a) with **M**, $\mathbf{M}^{-1}$ replacing **S**, $\mathbf{S}^T$ and with $n = p$, whereas the impulse response matrix can be obtained by the methods of Chapter 4. They are not needed either for estimating the modal model or for predicting the response, both of which are obtained more directly from the state matrix **A**. This response prediction of course provides only relative magnitudes at the measured points because the modal/eigenvectors ℓ_i are only scaled mode shapes, they can be multiplied by an arbitrary constant without altering **A** or $\mathbf{m}^{-1}\mathbf{c}$ and $\mathbf{m}^{-1}\mathbf{k}$. The determination of the scaling factor needed in predicting the actual response by

$$\ddot{\mathbf{x}} + \mathbf{m}^{-1}\mathbf{c}\dot{\mathbf{x}} + \mathbf{m}^{-1}\mathbf{k}\mathbf{x} = \mathbf{m}^{-1}\mathbf{f}$$

requires modeling forced response data from a system with measured or estimated force. This will be discussed for impulse, sinusoidal, and random forcing function later. The knowledge of the forcing function can be used to recover **m** and then **c** and **k** using **A**. This is logical since **A** or $\mathbf{m}^{-1}\mathbf{c}$ and $\mathbf{m}^{-1}\mathbf{k}$ summarize the inherent dynamic nature of the system and can even be recovered from transient response to initial conditions without requiring any external excitation force.

The consequent nonuniqueness of modal vectors ℓ_i up to a complex scaling factor is therefore not only logical but also useful in practice. For example, what can we do if we find that $n > 2$ and it seems that, say $(p + 1)$ frequencies may belong to the system and the rest to noise? Do we have to repeat the whole modal test again using $(p + 1)$ measurements? The answer is that although such a test may provide more information, it is not essential. All that is essential is that we measure two points: one from the old set of p measurements called the reference point and the additional point. We then scale the modal vectors in the two sets of p and 2 measurements such that the reference point has the same value, say 1, and thus recover the $(p + 1) \times 1$ modal vectors as well as $(p + 1) \times (p + 1)$ matrices μ, **M**, and **A**. The argument extends to any number of sets of measurements with a common reference point, as illustrated in Section 5.6.3.

5.3.2 Selection of Modes

This brings us to the question: How do we select the modes to be included in the modal model and the subsequent spatial model yielding **m**, **c**, and **k**? How do we separate the system modes of interest from those due to noise, measurement instrumentation, computation, and so on? This question cannot be sidetracked by intuitively selecting the number of measurements p by prior knowledge of the system and then artificially restricting the model order $n = 2$. Such a short-cut might not give the DDS procedure enough freedom to model the extraneous modes which will then ruin the accuracy of parameters of interest.

The initial number of measurements or degrees-of-freedom (DOF) p can be selected on the basis of following:

1. Prior knowledge of the system.
2. The degree of approximation by a p-DOF lumped mass system to a possibly infinite-DOF continuous system demanded by the application.
3. Number of simultaneous (synchronized) measurements possible on the available hardware.

This initial choice can be subsequently augmented as discussed above, or, in rare special cases, curtailed as discussed later.

Once the number of measurements is selected, the order n of the ARV(n) model should be selected by objective criteria such as the F test, modal amplitudes, stability of estimates as the order is increased, and the prediction error that is the difference between the data and the deterministic (homogeneous) part of the model. The selection of modes should be undertaken only after the adequate order n has been independently determined by such criteria. We will now provide some guidelines for such selection.

The frequencies and damping ratios provided by the eigenvalues μ_i yield the simplest mode selection criterion. Knowledge of the frequency range of the system or application of interest may be used to reject the low-frequency and high-frequency modes. Similarly, modes with high damping may be rejected because they damp out quickly. If the response to a sinusoidal force is being modeled, the known frequencies of the excitation should also be rejected. For a more general forced response, including that due to random input excitation, the modes with frequencies dominant in the input must be rejected as not belonging to the system; such dominance can be judged by low damping ratio and relatively high amplitudes. Global nature of eigenvalues and the consistency of resultant mode shapes with prior knowledge, such as a rigid body nature or finite element results, may also be used to select system modes.

The second category of criteria relate to the amplitudes obtained from the possibly complex modal vectors ℓ_i whose conjugates $\bar{\ell}_i$ are also modal vectors. Computational modes usually have very small amplitudes and can be spotted and rejected readily. These amplitudes may be considered from two aspects: the transient and the forced response. Since ℓ_i are indeterminate up to a complex scaling factor, their lengths or magnitudes are not helpful in ascertaining their importance until the scaling factor is resolved by a proper normalization. As we will now show, elements of the column vector $\mathbf{L}^{-1}\mathbf{X}_0^{(n-1)}$ provide the natural scaling factors from the transient or deterministic part and the corresponding scaling factor from the stochastic part of the decomposition (5.3.2) allows us to compare the properly scaled magnitudes of ℓ_i with the randomness or noise in the ith mode.

236   MODAL ANALYSIS

Let us denote the elements, $1 \times np$ rows, and $np \times p$ column submatrices of $\mathbf{L}^{-1}$ by l^{ij}, ℓ^i, and $\mathbf{L}^i$, respectively, so that

$$\underset{np \times np}{\mathbf{L}^{-1} = \{l^{ij}\}} = \begin{bmatrix} \ell^1 \\ \ell^2 \\ \vdots \\ \ell^{np} \end{bmatrix} = \underset{np \times p}{[\mathbf{L}^1, \mathbf{L}^2, \ldots, \mathbf{L}^n]} \tag{5.3.14}$$

whereas $\mathbf{L}$ given by (5.3.8) has $[\ell_1, \ell_2, \ldots, \ell_{np}]$ as its bottom p rows. Note further that if the ith row of $\mathbf{L}^1$ is denoted by $\mathbf{L}^{i1}$, then

$$\underset{np \times p}{\mathbf{L}^1} = \begin{bmatrix} \mathbf{L}^{11} \\ \mathbf{L}^{21} \\ \vdots \\ \mathbf{L}^{np1} \end{bmatrix} \quad \text{and} \quad \underset{p \times p}{\ell_i \lambda_i^{n-1} \mathbf{L}^{i1} = \mathbf{g}_i} \tag{5.3.15}$$

in the notation of Sections 4.4 and 4.5. Then, using (5.3.2)–(5.3.8) it can be shown that the modal decomposition takes the form

$$\mathbf{X}_t = \sum_{i=1}^{np} \ell_i \left[\ell^i \mathbf{X}_0^{(n-1)} \lambda_i^t + \sum_{j=0}^{t-n} \mathbf{L}^{i1} \mathbf{a}_{t-j} \lambda_i^{j+n-1} \right] \tag{5.3.16}$$

$$= \sum_{i=1}^{np} \left[\ell_i s_i \lambda_i^t + \sum_{j=0}^{t-n} \mathbf{g}_i \lambda_i^j \mathbf{a}_{t-j} \right] \tag{5.3.16a}$$

$$= \sum_{i=1}^{np} \left[\ell_i \left(\sum_{k=1}^{np} l^{ik} X_{0k} \right) \lambda_i^t + \sum_{j=0}^{t-n} \ell_i \left(\sum_{k=1}^{p} l^{ik} a_{k(t-j)} \right) \lambda_i^{j+n-1} \right] \tag{5.3.16b}$$

where the scale factors s_i of the modal vectors ℓ_i are the elements of the column vector $\mathbf{L}^{-1} \mathbf{X}_0^{(n-1)}$

$$\mathbf{L}^{-1} \mathbf{X}_0^{(n-1)} = \begin{bmatrix} s_1 \\ s_2 \\ \vdots \\ s_{np} \end{bmatrix} = \begin{bmatrix} \sum_{k=1}^{np} l^{1k} X_{0k} \\ \sum_{k=1}^{np} l^{2k} X_{0k} \\ \vdots \\ \sum_{k=1}^{np} l^{(np)k} X_{0k} \end{bmatrix} \tag{5.3.17}$$

so that $\ell_i s_i$ are uniquely determined by the initial data values denoted by (5.3.3):

$$\mathbf{X}_0^{(n-1)} = (\mathbf{X}_{n-1}^T, \mathbf{X}_{n-2}^T, \ldots, \mathbf{X}_0^T)^T = (X_{01}, X_{02}, \ldots, X_{0(np)})^T$$

although the ℓ_i may be arbitrarily scaled, and $\mathbf{g}_i$ are the rank one matrices decomposing the Green's function $\mathbf{G}_j$:

$$\mathbf{G}_j = \sum_{i=1}^{np} \mathbf{g}_i \lambda_i^j, \qquad \mathbf{g}_i = \begin{bmatrix} l_{1i} \\ l_{2i} \\ \vdots \\ l_{pi} \end{bmatrix} [l^{i1}, l^{i2}, \ldots, l^{ip}] \lambda_i^{n-1} \qquad (5.3.18)$$

Note the similarity and differences of this decomposition with the continuous- and discrete-time versions of (4.1.1) and (4.3.1) and particularly the remarks made regarding (4.1.1). Such a comparison will show that the present state space description combines the multivariate structure of Chapter 4 with the Vandermonde structure of scalar models in Chapter 3. Hence knowing λ_i, ℓ_i can be found using (5.3.7) or (5.3.7a), and $\ell_i s_i$, $\mathbf{g}_i$ found alternatively without inverting $\mathbf{L}$, as illustrated in Section 4.5.

Since the modal vector $\ell_i s_i$ of the mode λ_i is complex as seen from (5.3.16a), the first criterion that emerges is the average modal amplitude

$$\mathbf{AM}_i = \sum_{t=n}^{N} (\bar{\ell}_i^T \ell_i)^{1/2} |s_i| |\lambda_i|^{(t-n)} = \frac{(\bar{\ell}_i^T \ell_i)^{1/2} |s_i| (1 - |\lambda_i|^{(N-n+1)})}{(1 - |\lambda_i|)} \qquad (5.3.19)$$

The average modal amplitude $\mathbf{AM}_i$ characterizes the strength of the ith mode in the deterministic part of $\mathbf{X}_t$ over the recorded data. Although its discrete form (5.3.19) is more appropriate because it is obtained from discrete data, its continuous version, which may be physically more meaningful, can be readily obtained by integrating the exponential envelope from $\lambda_i = \exp(\sigma_i + j\Omega_i)\varDelta$ as

$$\mathbf{CM}_i = \frac{-(\bar{\ell}_i^T \ell_i)^{1/2} |s_i|}{\sigma_i} \qquad (5.3.20)$$

which is the area under the envelope $\ell_i s_i \exp(\sigma_i t)$. Another advantage of the continuous time average modal amplitude (5.3.20) is that it does not depend upon the number of observations. Note that since ℓ_i can be normalized to be of unit length, the average amplitude is primarily characterized by the scale factor $|s_i|$ from the deterministic part and the damping behavior characterized by $|\lambda_i|$ and σ_i in the respective cases. Both these criteria can be utilized by any modal analysis method and not by DDS alone.

Since $\sigma_i = -\zeta_i \omega_i$, the two criteria $\mathbf{AM}_i$ and $\mathbf{CM}_i$ penalize high-frequency modes for the same damping ratio. Such penalty can be removed either by using ζ_i alone or by multiplying these criteria by ω_i. It is also possible to use "power" instead of amplitude by using the squared quantities in (5.3.19) and (5.3.20). A comparison of modal power at the given frequencies obtained from the deterministic part is useful in validating finite element models in particular and simulation models in general. If the natural frequencies, damping ratios, and modal powers match for the DDS models fitted to the simulation and real data, the simulation model is valid, otherwise it is not valid, as illustrated in Pandit and Lin (1991).

Referring to the fundamental decomposition (5.3.16), it is clear that the corresponding scaling factor $\mathbf{L}^{i1}\mathbf{a}_{t-j}$ from the other, stochastic part, is random. Since it has a zero mean, it is characterized by its expected squared value or the variance (see Appendix A at the end of the book for review).

$$\text{Var}\left(\sum_{k=1}^{p} l^{ik} a_{k(t-j)}\right) = \text{Var}(\mathbf{L}^{i1}\mathbf{a}_{t-j}) = \mathbf{L}^{i1}\sigma_a^2 \bar{\mathbf{L}}^{i1T} \quad (5.3.21)$$

Hence the modal variability at a given time t is

$$\text{MV}_i^t = \boldsymbol{\ell}_i^T \boldsymbol{\ell}_i \mathbf{L}^{i1} \sigma_a^2 \bar{\mathbf{L}}^{i1T} \left[\frac{1 - |\lambda_i|^{2(t-n+1)}}{1 - |\lambda_i|^2}\right] \quad (5.3.22)$$

So the total modal variance over the number of observations that characterizes the stochastic part, just as AM_i given by (5.3.19) does for the deterministic part, is

$$\text{MV}_i = \sum_{t=n}^{N} \text{MV}_i^t = \frac{\bar{\boldsymbol{\ell}}_i^T \boldsymbol{\ell}_i \mathbf{L}^{i1} \sigma_a^2 \bar{\mathbf{L}}^{i1T} \left[(N - n + 1) - \dfrac{|\lambda_i|^2(1 - |\lambda_i|^{2(N-n+1)})}{(1 - |\lambda_i|^2)}\right]}{(1 - |\lambda_i|^2)}$$

$$(5.3.23)$$

Therefore, an overall index of the modal signal-to-noise ratio over the observed data for the ith mode may be summarized as

$$\text{MSN}_i = \frac{\text{AM}_i}{\sqrt{\text{MV}_i}}$$

$$= \frac{|s_i|(1 + |\lambda_i|)(1 - |\lambda_i|^{(N-n+1)})}{[\mathbf{L}^{i1}\sigma_a^2 \bar{\mathbf{L}}^{i1T}((N-n)(1 - |\lambda_i|^2) - 2|\lambda_i|^2 + |\lambda_i|^{2(N-n+2)} + 1)]^{1/2}}$$

$$(5.3.24)$$

On the other hand, for a given time t, a similar index may be defined as the ratio of an individual term in (5.3.19) (before summing over t) with $\sqrt{\text{MV}_i^t}$,

$$\text{MSN}_i^t = \frac{|s_i||\lambda_i|^{(t-n)}(1 - |\lambda_i|^2)^{1/2}}{\{\mathbf{L}^{i1}\sigma_a^2 \bar{\mathbf{L}}^{i1T}[1 - |\lambda_i|^{2(t-n+1)}]\}^{1/2}} \quad (5.3.25)$$

This index MSN_i^t may be particularly useful for highly damped modes which will be quickly buried in noise when MSN_i is used.

Thus, in the absence of any a priori knowledge of the amplitudes of modal vectors, they may be first ranked on the basis of the average modal amplitude

AM_i or CM_i. A convenient method would then be to add the amplitudes of all modes not already rejected on the basis of a priori frequency and damping information, and then reject the last modes which comprise say 1 or 5% of the amplitude; the selected modes would thus have 99 or 95% of the total amplitude. The modes thus selected may now be ranked according to the signal-to-noise ratio MSN_i and the highest ranking ones among them be selected so that, together with the ones definitely preferable on the basis of a priori frequency and damping information, they comprise the p modes needed to construct the state matrix. Such a procedure would be rational and, at the same time, allow for a modal analyst's intuitive a priori knowledge, if available.

In practice, one may often have a priori modal vectors available from earlier work such as finite element analysis or even earlier modal tests. A correlation coefficient [similar to the estimate of ϕ_1 of an AR(1) model] may then be defined using the earlier and current vectors. Such a criterion is called the modal assurance criterion, MAC, see Ewins (1985). The correlation coefficient may be obtained by taking the inner (dot) product of the two vectors and normalizing it by dividing with the square root of the product of their norms. However, the inner product of two different complex vectors would be complex and even though its phase information may be useful in practice, the MAC is defined by taking the absolute value of this correlation coefficient. If $\mathbf{b}_i$ is the earlier modal vector, including the scaling factor with normalization consistent with the one used in obtaining the present vector $\ell_i s_i$, then the criterion is defined as

$$\text{MAC} = \frac{|\bar{\mathbf{b}}_i^T \ell_i|}{\sqrt{\bar{\mathbf{b}}_i^T \mathbf{b}_i} \sqrt{\bar{\ell}_i^T \ell_i}} \quad (5.3.26)$$

A similar criterion has been utilized in Ibrahim and Mikulcik (1978) to distinguish system modes from (computational) noise modes by comparing the two modal vectors obtained from the same data by shifting the origin. In our notation, if the two modal vectors are obtained with scaling factors s_i^n and s_i^{n+m} by shifting the origin m sampling intervals, say using $\mathbf{X}_0^{(n-1)}$ and $\mathbf{X}_0^{(n+m-1)}$, then the coefficient, called the modal confidence factor, is simply

$$\text{MCF}_i = \frac{|s_i^{n+m}|}{|s_i^n||\lambda_i|^m} \quad (5.3.27)$$

For a mode with no noise $|s_i^{n+m}| = |s_i^n||\lambda_i|^m$ and hence the criterion will give the value 1; otherwise, the value will be different from 1. In practice, a value larger than 1 is replaced by its reciprocal.

The criterion MCF_i only compares the reduction in the modal vector caused by damping over one arbitrarily selected time interval $m\Delta$ with the theoretical reduction given by $|\lambda_i|^m = |e^{m\mu_i \Delta}|$. Therefore, when the significant noise in the measurement data causes even slight error in the damping estimate, it may give unnecessarily low value. On the other hand, transients caused by noise modes in the deterministic part may give artificially close values to 1. Hence an overall criterion such as MSN_i that takes into account all the shifts would be more consistent for noisy data, as has been empirically confirmed.

In the rare case when the number of measurements is excessively large, it may happen that the adequate order of the ARV model $n \leq 2$. After some of the modes have been rejected by any of the criteria given above, the question arises: How do we curtail the initial choice of the number of measurements and choose p less than the number of measurements? This question is easier to answer by using the same logic as the one used in formulating the indexes for selecting the modes. The answer is easier because we do not have to worry about the behavior of the modes in time. The selected number of modes now becomes our new p, say p' and we simply form the $p \times p'$ rectangular matrix

$$[\ell_1, \ell_2, \ldots, \ell_{p'}]$$

and select $p' < p$ rows of this matrix with the largest length/norm.

5.4 SPATIAL MODEL: MASS, STIFFNESS, AND DAMPING MATRICES

We have seen in Section 5.3 how to obtain the state matrix **A** that essentially contains $\mathbf{m}^{-1}\mathbf{k}$ and $\mathbf{m}^{-1}\mathbf{c}$ from the transient response data. This transient response may be from a response to unknown initial conditions, unknown impulse input, or even unknown forcing function in theory. However, in practice, the first two responses provide better results. In terms of parameters, this provides frequencies, damping ratios, and scaled mode shapes. We will now show how to resolve the scaling and get the mass, damping, and stiffness matrices, **m**, **c**, and **k**.

The reason why the transient response cannot resolve the scaling is that it is a solution to the homogeneous equation

$$\mathbf{m}\ddot{\mathbf{x}} + \mathbf{c}\dot{\mathbf{x}} + \mathbf{k}\mathbf{x} = 0$$

which for a nonsingular mass matrix **m** is equivalent to

$$\ddot{\mathbf{x}} + \mathbf{m}^{-1}\mathbf{c}\dot{\mathbf{x}} + \mathbf{m}^{-1}\mathbf{k}\mathbf{x} = 0$$

and hence can give only $\mathbf{m}^{-1}\mathbf{c}$ and $\mathbf{m}^{-1}\mathbf{k}$. To resolve the scaling and recover **m**, **c**, and **k** separately, it is necessary to use the forced response from

$$\mathbf{m}\ddot{\mathbf{x}} + \mathbf{c}\dot{\mathbf{x}} + \mathbf{k}\mathbf{x} = \mathbf{F} \quad (5.4.1)$$

or

$$\ddot{\mathbf{x}} + \mathbf{m}^{-1}\mathbf{c}\dot{\mathbf{x}} + \mathbf{m}^{-1}\mathbf{k}\mathbf{x} = \mathbf{m}^{-1}\mathbf{F} \quad (5.4.1a)$$

SPATIAL MODEL: MASS, STIFFNESS, AND DAMPING MATRICES

The general approach is therefore to get $\mathbf{m}^{-1}\mathbf{k}$ and $\mathbf{m}^{-1}\mathbf{c}$ from the transient response modeling and analysis, and then get enough equations by equating the theoretical solution of (5.4.1) with the corresponding values obtained by modeling the forced response to solve for $\mathbf{m}$, $\mathbf{c}$, and $\mathbf{k}$. We will first indicate the format of this general approach, which is applicable to any sustained input or excitation extending over time, although in practice it works best when the excitation is capable of producing a steady-state response. The explicit formulas for the equations differ with the nature of excitation and some typical examples of these will be subsequently presented to illustrate the general approach.

5.4.1 The General Approach

The transient data modeling and the choice of modal vectors in Section 5.3 provides us with the state matrix

$$\begin{bmatrix} \mathbf{0} & \mathbf{I} \\ -\mathbf{m}^{-1}\mathbf{k} & -\mathbf{m}^{-1}\mathbf{c} \end{bmatrix} = \begin{bmatrix} \mathbf{0} & \mathbf{I} \\ \mathbf{A}_{21} & \mathbf{A}_{22} \end{bmatrix} = \mathbf{A} = \mathbf{M}\mu\mathbf{M}^{-1} \quad (5.4.2a)$$

defining the $p \times p$ submatrices

$$\mathbf{A}_{21} = -\mathbf{m}^{-1}\mathbf{k}, \qquad \mathbf{A}_{22} = -\mathbf{m}^{-1}\mathbf{c} \quad (5.4.2b)$$

For a diagonalizable $\mathbf{A}$ the modal matrix is, see (5.3.13),

$$\mathbf{M} = \begin{bmatrix} \ell_1 & \ell_2 & \cdots & \ell_{2p} \\ \mu_1\ell_1 & \mu_2\ell_2 & \cdots & \mu_{2p}\ell_{2p} \end{bmatrix} \text{ or } \begin{bmatrix} \ell_1 & \bar{\ell}_1 & \cdots & \bar{\ell}_p \\ \mu_1\ell_1 & \bar{\mu}_1\bar{\ell}_1 & \cdots & \bar{\mu}_p\bar{\ell}_p \end{bmatrix}$$

and denoting its inverse by

$$\mathbf{M}^{-1} = \begin{bmatrix} \mathbf{L}^{11} & \mathbf{L}^{12} \\ \mathbf{L}^{21} & \mathbf{L}^{22} \\ \vdots & \vdots \\ \mathbf{L}^{2p1} & \mathbf{L}^{2p2} \end{bmatrix} \quad (5.4.2)$$

where $\mathbf{L}^{i1}$ and $\mathbf{L}^{i2}$ and $1 \times p$ row vectors; it can be verified that the equations for $\mathbf{m}^{-1}\mathbf{k}$ and $\mathbf{m}^{-1}\mathbf{c}$ are given by

$$-\mathbf{k} = \mathbf{m} \sum_{i=1}^{2p} \mu_i^2 \ell_i \mathbf{L}^{i1} = \mathbf{m}\mathbf{A}_{21} \quad (5.4.3)$$

$$-\mathbf{c} = \mathbf{m} \sum_{i=1}^{2p} \mu_i^2 \ell_i \mathbf{L}^{i2} = \mathbf{m}\mathbf{A}_{22} \quad (5.4.4)$$

Note that these $\mathbf{L}^{i1}$ and $\mathbf{L}^{i2}$ are *not* the same as the ones from $\mathbf{L}^{-1}$ in (5.3.15). Each of the matrix equations (5.4.3 and 5.4.4) contains p^2 linear equations in elements of $\mathbf{m}$, $\mathbf{k}$, and $\mathbf{c}$. If $\mathbf{m}$ and either $\mathbf{k}$ or $\mathbf{c}$ are symmetric, then we need p equations from the forced response to solve for $\mathbf{m}$ and either $\mathbf{k}$ or $\mathbf{c}$ in conjunction with (5.4.3) or (5.4.4) and then the remaining $\mathbf{c}$ or $\mathbf{k}$ can be solved using the remaining (5.4.4) or (5.4.3). In fact, such p equations from the forced response together with (5.4.3) and (5.4.4) can be used to solve for $\mathbf{m}$, $\mathbf{c}$, and $\mathbf{k}$ if any two of them are symmetric so that they contain $(2p^2 + p)$ unknowns. Recall that a symmetric $p \times p$ matrix contains $p(p + 1)/2$ unknowns. Such p equations can be obtained using responses from p locations to a single force at one location.

On the other hand, if $\mathbf{m}$, $\mathbf{c}$, and $\mathbf{k}$ are not symmetric, then we need p^2 equations from the forced response to solve for them, now containing $3p^2$ unknowns. These p^2 equations may be obtained in many different ways, for example, p sets of p responses collected by exciting each of the p locations separately, p responses to excitation of different frequency applied simultaneously at every location, or p responses to an excitation containing p distinct frequencies at one location.

To get the p equations in the symmetric case from the forced response, all we need is a modal vector of responses, say $\mathbf{r}_i = (r_{1i}, r_{2i}, \ldots, r_{pi})^T$ to an excitation vector, say $\mathbf{f}_i = (f_{1i}, f_{2i}, \ldots, f_{pi})^T$. These vectors may be obtained directly by instrumentation or by modeling the data by the methods of Section 5.3 and taking $\mathbf{r}_i$ and $\mathbf{f}_i$ as the modal vectors of appropriate modes from the decomposition of fitted models. In the unsymmetric case, we need p pairs of such vectors $\mathbf{r}_i, \mathbf{f}_i, i = 1, 2, \ldots, p$, to be obtained in a similar fashion. These vectors are then related using (5.4.1) as we now show. Additional forced-response measurements may be used for confirmation purposes; a much larger set of measurements may be used via the least squares method.

In developing the general approach we can assume the excitation vector to be of the form of a constant column vector times a *scalar* time function:

$$\mathbf{F}(t) = [f_1, f_2, \ldots, f_p]^T u(t) = \mathbf{f} u(t) \tag{5.4.5}$$

because a general excitation $\mathbf{F}(t) = [f_1 u_1(t), f_2 u_2(t), \ldots, f_p u_p(t)]^T$ can be expressed as

$$\mathbf{F}(t) = [f_1, 0, \ldots, 0]^T u_1(t) + [0, f_2, \ldots, 0]^T u_2(t) + \cdots$$
$$+ [0, 0, \ldots, f_p]^T u_p(t)$$

and the response obtained as the sum of responses to the individual terms. Hence the state space solution of (5.4.1) is given by

$$\begin{bmatrix} \mathbf{x} \\ \dot{\mathbf{x}} \end{bmatrix} = e^{\mathbf{A}t} \begin{bmatrix} \mathbf{x}(0) \\ \dot{\mathbf{x}}(0) \end{bmatrix} + \int_0^t e^{\mathbf{A}(t-v)} \begin{bmatrix} \mathbf{0} \\ \mathbf{m}^{-1} \mathbf{F}(v) \end{bmatrix} dv \tag{5.4.6}$$

and the response is found by setting $\mathbf{A} = \mathbf{M}\boldsymbol{\mu}\mathbf{M}^{-1}$ to be

$$\begin{bmatrix} \mathbf{x}(t) \\ \dot{\mathbf{x}}(t) \end{bmatrix} = e^{\mathbf{A}t} \begin{bmatrix} \mathbf{x}(0) \\ \dot{\mathbf{x}}(0) \end{bmatrix} + \mathbf{M} \operatorname{diag}\{e^{\mu_i t} b_i(t)\} \mathbf{M}^{-1} \begin{bmatrix} \mathbf{0} \\ \mathbf{m}^{-1}\mathbf{f} \end{bmatrix} \quad (5.4.7)$$

where

$$b_i(t) = \int_0^t e^{-\mu_i v} u(v)\, dv \quad (5.4.8)$$

In theory, it is possible to use either the transient or the steady-state part of the forced response to get the needed additional equations. In practice, it is better to let the transients die out and then make the measurement of steady-state data for which a simpler model is adequate. Alternatively, the steady-state part may be chosen by omitting the transient part. The steady-state displacement and velocity response can be predicted from (5.4.7) and equated with the same estimated from the fitted model, say $\mathbf{x}_s(t)$ and $\dot{\mathbf{x}}_s(t)$. Equating these gives us

$$\begin{bmatrix} \mathbf{x}_s(t) \\ \dot{\mathbf{x}}_s(t) \end{bmatrix} = \mathbf{M} \operatorname{diag}\{e^{\mu_i t} b_i(t)\} \mathbf{M}^{-1} \begin{bmatrix} \mathbf{0} \\ \mathbf{m}^{-1}\mathbf{f} \end{bmatrix} \quad (5.4.9)$$

This equation provides us with the remaining equations involving $\mathbf{m}^{-1}$ to complete the estimation of $\mathbf{m}$, $\mathbf{c}$, and $\mathbf{k}$ in conjunction with (5.4.3) and (5.4.4). In the symmetric case a single force $\mathbf{f}u(t)$ is enough to give the additional p equations. In the unsymmetric case p sets of equations (5.4.9) with p forcing functions $\mathbf{f}_k u_k(t)$ with distinctly identifiable $u_k(t)$ are needed. [Mathematically, $u_k(t)$ must be linearly independent.] Although in theory the needed p^2 equations can be obtained by using the excitation $\mathbf{f}_1 u_1(t) + \cdots + \mathbf{f}_p u_p(t)$ at one location, in practice it is better to use multiple-point excitation data with each point excited by $\mathbf{f}_k u_k(t)$. The specific form of equations arising out of (5.4.9) to complement (5.4.3) and (5.4.4) would depend upon the form of $u(t)$. We now give some examples. Although we have discussed the theoretically minimum number of tests, in practice more tests can always be used for reconfirmation or for finding better estimates by the use of linear least squares.

5.4.2 Impulse and Random Excitation

The impulse excitation is the simplest type of excitation that can be applied by a hammer, and if carefully applied, can provide both the spatial model and the modal model in one test. If impulses of strength $f_1, f_2, \ldots, f_p$ are applied at p locations, with at least one of them nonzero, then

$$\mathbf{F}(t) = [f_1, f_2, \ldots, f_p]^T \delta(t) = \mathbf{f}\delta(t) \quad (5.4.10)$$

and

$$b_i(t) = 1 \quad (5.4.11)$$

In implementing (5.4.9), its left-hand side may be obtained from the model for the response data, using the methods of Section 5.3.2 to select the modes. Then, for impulse response, (5.4.9) takes the form

$$\begin{bmatrix} \mathbf{x}_s(t) \\ \dot{\mathbf{x}}_s(t) \end{bmatrix} = e^{\mathbf{A}t} \begin{bmatrix} \mathbf{0} \\ \mathbf{m}^{-1}\mathbf{f} \end{bmatrix} \quad (5.4.12)$$

Therefore, at $t = 0$

$$\dot{\mathbf{x}}_s(0) = \mathbf{m}^{-1}\mathbf{f} \quad (5.4.13)$$

and the desired p equations can be written as

$$\mathbf{m}\dot{\mathbf{x}}_s(0) = \mathbf{m} \sum_{i=1}^{2p} \ell_i s_i \mu_i = \mathbf{f} \quad (5.4.14)$$

Thus equations (5.4.3), (5.4.4), and (5.4.14) provide the necessary $2p^2 + p$ equations to solve for $\mathbf{m}$, $\mathbf{c}$, and $\mathbf{k}$ when two of them are symmetric. In (5.4.14) μ_i, ℓ_i, and s_i are found from the modal model, whereas $\mathbf{f}$ is obtained from the strength of the impulse excitation.

For the unsymmetric case, the impulse test needs to be repeated p times using excitations $\mathbf{f}_k \delta(t), k = 1, 2, \ldots, p$ with linearly independent vectors $\mathbf{f}_k$. The simplest way to achieve this is to impulse one location at a time so that $\mathbf{f}_1 = [f_1\ 0 \ldots 0]^T, \mathbf{f}_2 = [0\ f_2 \ldots 0]^T, \ldots, \mathbf{f}_p = [0 \ldots f_p]^T$ which are clearly linearly independent. Then equation (5.4.13) shows that a test impulsing kth location solves for the kth column of $\mathbf{m}^{-1}$ and repeating for $k = 1, 2, \ldots, p$ yields all columns of a possibly unsymmetric $\mathbf{m}^{-1}$. Using this $\mathbf{m}^{-1}$ in (5.4.3) and (5.4.4) yields the possibly unsymmetric $\mathbf{c}$ and $\mathbf{k}$. Note that each of these tests should theoretically yield the same modal model from which μ_i, ℓ_i and $\mathbf{L}^{i1}$ will be used in (5.4.3) and (5.4.4). In practice of course this is usually not the case and some form of averaging of the parameters or alternatively least squares fitting must be used.

Such averaging is automatically accomplished when the impulse response is obtained via the transfer function or the FRF from the ARMAV model fitted to the random input excitation data; this will be discussed in Section 6.2.3. Since the impulse response so obtained is with respect to a unit impulse, $f_k = 1$ when the random input is used at the kth location. The preceding remarks for the symmetric and unsymmetric cases apply to random input excitation as well. Models from the response to such random excitation provide a comprehensive characterization and analysis of random vibrations, particularly when the excitation originates from natural sources.

The impulse test by a hammer has the advantages that it allows both the modal model and the scaling determination from the forced response to be accomplished by the same test, it is conceptually easy to understand, and it is relatively easy to carry out. Its practical limitations stem from the difficulty of experimentally implementing an impulse and obtaining an estimate of its strength f_k needed in (5.4.13). Theoretically, an impulse $\delta(t)$ is a signal of infinite height concentrated over an infinitesimal interval at zero, with a finite area over that interval equal to its strength. Practically, therefore, an impulse must be concentrated on a small interval at the beginning to well approximate its theoretical form. Moreover, it must be sampled at a very high sampling rate, much higher than the response signals, to get a good representation of the true signal. This is necessary to get a good estimate of the strength f_k, either from the area or by taking the FFT of the time signal and taking the height of the best-fitting step function as an estimate of the area. [Recall that the Fourier transform of an impulse $f_k \delta(t)$ is the step $f_k s(t)$.]

Even more troublesome, in practice, is the determination of "zero" time. Since the measuring instrument must be triggered to begin the measurement at rest sufficiently before the start of the impulse to capture it fully, determining the $t = 0$ initial point from the inevitably noisy measurements, and synchronizing it with the response measurements, is not easy. Yet the point $t = 0$ is extremely crucial in estimation, as seen from the results given above, and a small error in determining this point could cause large errors in the resolution of scaling; the modal model, however, is not affected by this point.

5.4.3 Sinusoidal Excitation

These problems with the impulse excitation can be resolved by sinusoidal excitation, which provides a sustained input. The steady-state output from such excitation will not depend so crucially upon the starting point $t = 0$ as in the case of impulse excitation. A reasonable procedure then would be to first get the modal model from the free response data, either using suitable (unknown) initial conditions that will excite all modes, or using impulse response data where the forcing impulse need not be measured. This will give $\mathbf{m}^{-1}\mathbf{c}$ and $\mathbf{m}^{-1}\mathbf{k}$ by (5.4.3) and (5.4.4) and the additional equations needed must be obtained from (5.4.9) which we will now specialize for the sinusoidal input. It is convenient to express the real $\mathbf{F}(t)$ in the complex form:

$$\mathbf{F}(t) = \mathbf{f}e^{\alpha t} + \bar{\mathbf{f}}e^{-\alpha t} \qquad (5.4.15)$$

where $\mathbf{f}$ is a complex vector to express a phase lag between the excitation force and the response and α is imaginary to represent the forcing frequency. Then, considering only the part $\mathbf{f}e^{\alpha t}$

$$b_i(t) = \int_0^t e^{-\mu_i v} e^{\alpha v} \, dv$$
$$= \frac{[e^{(\alpha - \mu_i)t} - 1]}{(\alpha - \mu_i)} \qquad (5.4.16)$$

and, in the steady state,

$$e^{\mu_i t} b_i(t) \to \frac{e^{\alpha t}}{(\alpha - \mu_i)} \qquad (5.4.17)$$

The left-hand side of equation (5.4.9) and $\mathbf{f}$ on its right-hand side must be obtained from the forced measurement model, the remaining parameters are obtained from the free response model. Let the steady-state forced response data be modeled and analyzed by the methods of Sections 5.2 and 5.3. Suppose $\ell_\alpha s_\alpha$ and $\mathbf{f}$ are the modal vectors *with* scaling for the forced response and the force measurement respectively. If they are modeled separately, the modal vectors are obtained from (5.3.16a) as the amplitudes of λ_α containing the forcing frequency α. If they are modeled together as a vector, say $(x_1, x_2, \ldots, x_p, F_1, F_2, \ldots, F_k)^T$, where k can vary from 1 to p, then the corresponding modal vectors are $(\ell_\alpha^T s_\alpha, f_1, f_2, \ldots, f_p)^T = (\ell_\alpha^T s_\alpha, \mathbf{f}^T)^T$ where some elements of $\mathbf{f}$ would be zero when $k < p$.

Thus, using $\ell_\alpha s_\alpha$, $\mathbf{f}$ from the steady-state forced response model with forcing frequency α, and $\mathbf{M}$, μ_i from the free-response model, equation (5.4.9) specializes to

$$\begin{bmatrix} \ell_\alpha s_\alpha/\alpha^2 \\ \ell_\alpha s_\alpha/\alpha \end{bmatrix} = \mathbf{M} \, \text{diag} \left\{ \frac{1}{\alpha - \mu_i} \right\} \mathbf{M}^{-1} \begin{bmatrix} 0 \\ \mathbf{m}^{-1}\mathbf{f} \end{bmatrix} \quad (5.4.18a)$$

that is,

$$[\alpha \mathbf{I} - \mathbf{A}] \begin{bmatrix} \ell_\alpha s_\alpha/\alpha^2 \\ \ell_\alpha s_\alpha/\alpha \end{bmatrix} = \begin{bmatrix} 0 \\ \mathbf{m}^{-1}\mathbf{f} \end{bmatrix} \quad (5.4.18)$$

where we have assumed that the forced response is measured by the acceleration data, since accelerometers are in fact the most common instruments used in practice. Hence the acceleration $\ell_\alpha s_\alpha$ has been converted to velocity by dividing by α. If, on the other hand, displacement data is used for both free- and forced-response modeling, then the displacement version of (5.4.9) specializes to

$$[\alpha \mathbf{I} - \mathbf{A}] \begin{bmatrix} \ell_\alpha s_\alpha \\ \alpha \ell_\alpha s_\alpha \end{bmatrix} = \begin{bmatrix} 0 \\ \mathbf{m}^{-1}\mathbf{f} \end{bmatrix} \quad (5.4.19)$$

Now, substituting $\mathbf{A}$ from (5.4.2a), the acceleration and displacement measurement of forced response yield the following versions of (5.4.9):

$$\mathbf{m}[\alpha^2 \mathbf{I} - \alpha \mathbf{A}_{22} - \mathbf{A}_{21}]\ell_\alpha s_\alpha/\alpha^2 = \mathbf{f} \quad (5.4.20)$$

$$\mathbf{m}[\alpha^2 \mathbf{I} - \alpha \mathbf{A}_{22} - \mathbf{A}_{21}]\ell_\alpha s_\alpha = \mathbf{f} \quad (5.4.21)$$

The $p \times p$ submatrices $\mathbf{A}_{21} = -\mathbf{m}^{-1}\mathbf{k}$ and $\mathbf{A}_{22} = -\mathbf{m}^{-1}\mathbf{c}$ are already known from the free response by (5.4.3) and (5.4.4); hence (5.4.20) and (5.4.21) also follow directly from (5.4.1). Equations (5.4.20) and (5.4.21) are analogous to (5.4.14) and give us the needed p equations in the symmetric case and p^2 equations by repeating them with linearly independent $\mathbf{f}_1, \mathbf{f}_2, \ldots, \mathbf{f}_p$ in the unsymmetric case. Again, the simplest set of linearly independent $\mathbf{f}_k$ are when only the kth location is forced and then (5.4.18) and (5.4.19) show that such a test yields the kth column of $\mathbf{m}^{-1}$ and when all the p columns are recovered by (5.4.20) or (5.4.21), they can be used to solve for $\mathbf{c}$ and $\mathbf{k}$ using (5.4.3) and (5.4.4) from the free response data.

5.4.4 Step Excitation

Although step excitation is difficult to implement for most modal tests of mechanical structures, it is often used as a step-relaxation technique. It is necessary to use displacement measurements for both free and forced response, and it is left as an exercise to show the corresponding results by setting $\alpha = 0$:

$$-\mathbf{m}\mathbf{A}_{21}\ell_0 s_0 = \mathbf{f} \qquad (5.4.22)$$

Here elements of $\mathbf{f}$ contain the magnitudes of steps at different locations and $\ell_0 s_0$ are the modal vectors of steady-state forced-response displacements; these may be available directly from the measurements or can be obtained as coefficients of the discrete roots equal to 1 from the ARMAV models fitted to the entire response to a step input. All the remarks made regarding symmetric and unsymmetric cases made earlier apply here and delineate the generality of the approach. Equations (5.4.3), (5.4.4), and (5.4.22), using free and forced respectively, solve for $\mathbf{m}$, $\mathbf{c}$, and $\mathbf{k}$.

5.5 ILLUSTRATIVE EXAMPLES—SCALAR MODELS

We will now illustrate the preceding theory by results obtained from both simulated and experimental data. Simulated data results clarify the procedure and verify the accuracy of estimates; they are primarily based on Pandit and Mehta (1984, 1985, and 1988). Experimental results also present a comparison with FFT and are taken from Pandit and Jacobson (1988). Scalar models are used to illustrate ARMA modeling results; vector models favor AR for reasons discussed in Sections 2.4.3 and 5.2.1 and will be illustrated in Section 5.6.

5.5.1 Nonclassically Damped System

The displacement responses X_{1t}, X_{2t}, and X_{3t} to the initial condition $x_1(0) = 1$ were generated for a 3-DOF system using ACSL (advanced continuous simulation language). The spatial model consists of the mass damping and stiffness matrices

$$\mathbf{m} = \begin{bmatrix} 3 & 0 & 0 \\ 0 & 2 & 0 \\ 0 & 0 & 1 \end{bmatrix}, \quad \mathbf{c} = \begin{bmatrix} 0.1 & -0.1 & 0.0 \\ -0.1 & 0.3 & -0.2 \\ 0.0 & -0.2 & 0.2 \end{bmatrix},$$

$$\mathbf{k} = \begin{bmatrix} 6 & -2 & 0 \\ -2 & 3 & -1 \\ 0 & -1 & 1 \end{bmatrix}$$

The ACSL generates its own computational noise, together with that due to the round-off error of the single precision data and the nonlinear least squares routine used for estimation. Hence the data could be considered to be stochastic and the ARMA$(2n, 2n - 1)$ modeling strategy with increasing n was applied to it for illustration; these results are given in Pandit and Mehta (1984). Here we reproduce the closely spaced mode results after giving pure AR results for exact simulation first.

The exact simulation used the state space solution to generate the data with precision the same as that of the computer. Since this 3-DOF system has six characteristic roots, the AR(6) model should be adequate, which is verified in Table 5.2. The theoretical frequencies and damping ratios are obtained from the eigenvalues of the state matrix $\mathbf{A}$ formed from the above $\mathbf{m}$, $\mathbf{c}$, and $\mathbf{k}$. The amplitudes and phase angles follow from the corresponding complex conjugate pairs in $e^{\mathbf{A}t}\mathbf{x}(0)$ with $\mathbf{x}(0) = (1,0,0,0,0,0)^T$ in the present case.

The AR(6) model fitted to this exact simulated data has six characteristic roots λ_i which yield the natural frequencies and damping ratios by relations (5.3.9) and (5.3.10a). The amplitudes and phase angles are obtained from the modal "scalars" s_i using (5.3.17) with $n = 6$ and $p = 1$. The match with the theoretical values is excellent, as expected. Phase angles different from zero and 180° confirm the nonclassical damping that should yield complex mode shapes.

TABLE 5.2 Comparison of 3-DOF Theoretical and Simulated Results

	Natural Frequency ω_i(rad/s)	Damping Ratio ζ_i	Amplitude $2\lvert s_i\rvert$	Phase $\angle s_i$ (Degrees)
Theoretical				
$x1$	0.583987	0.0158689	0.1029186	10.571
	1.207716	0.0788843	0.3825884	18.764
	1.637197	0.0532114	0.5565169	− 15.385
$x2$	0.583987	0.0158689	0.2560983	9.197
	1.207716	0.0788843	0.3382794	− 3.487
	1.637197	0.0532114	0.5905598	178.944
$x3$	0.583987	0.0158689	0.3865747	6.712
	1.207716	0.0788843	0.7434070	− 168.279
	1.637197	0.0532114	0.3633321	18.784
AR(6) Model				
$x1$	0.583971	0.0158585	0.1029165	10.570
	1.207714	0.0788985	0.3825901	18.764
	1.637197	0.0532118	0.5565197	− 15.386
$x2$	0.583997	0.0158585	0.2560951	9.195
	1.207714	0.0788985	0.3382836	− 3.486
	1.637197	0.0532118	0.5905622	178.945
$x3$	0.583997	0.0158585	0.3865813	6.710
	1.207714	0.0788985	0.7434129	− 168.279
	1.637197	0.0532118	0.3633322	18.784

Phase angles w.r.t. cosine.

The modal vectors can be synthesized by stacking s_i for each scalar AR model in a column and arbitrarily normalizing them. Such modal vectors or mode shapes ℓ_i scaled to give 1 for x_1 measurement are shown in Table 5.3. The table presents a complete modal model for this example and also lists vector modeling results (discussed later) for comparison. Since the vector model uses the information from all the data sets simultaneously for estimating each parameter, its results are even more accurate than those from scalar models.

The ACSL simulation, which is not as exact, requires an ARMA(12,11) model, whereas added noise (17% signal-to-noise ratio) requires an ARMA(26,25) model to get results of comparable accuracy. The modal decomposition of a typical such model is illustrated in Table 5.5 of the next section on closely spaced modes.

5.5.2 Closely Spaced Modes

To demonstrate the ability of the general procedure to resolve systems with high modal density, the ACSL simulation was used to generate data for the classically (proportionally) damped system with matrices

$$\mathbf{m} = \begin{bmatrix} 1 & 0 & 0 \\ 0 & 1.2 & 0 \\ 0 & 0 & 1 \end{bmatrix}, \quad \mathbf{c} = \begin{bmatrix} 0.3 & 0 & 0 \\ 0 & 0.36 & 0 \\ 0 & 0 & 0.3 \end{bmatrix}, \quad \mathbf{k} = \begin{bmatrix} 3 & -1 & -1 \\ -1 & 3 & -1 \\ -1 & -1 & 3 \end{bmatrix}$$

using the initial conditions $x_2(0) = 2$, $\dot{x}_2(0) = 0.5$, and the rest zero. Although they could be modeled by AR models to get results similar to Tables 5.2 and 5.3, since the ACSL simulation is not exact they were again modeled by the ARMA $(2n, 2n - 1)$ strategy. The effect of model order on the predicted modal parameters and RSS is shown in Table 5.4.

The lower-order models are unable to separate the closely spaced second and third modes. The relatively high value of the RSS, however, reflects this incomplete resolution and shows that the dynamic content of the data has not been entirely accounted for. A sharp decrease in the RSS, to a level corresponding to single-precision digital accuracy at order 28 accurately indicates when all the three modes have been identified. As an example, the computer output of the modal decomposition of the adequate ARMA(28,27) model for X_{3t} has been reproduced in Table 5.5. Since $N = 500$, for the ARMA(28,27) model $\hat{\sigma}_a^2 = 0.8882 \times 10^{-11}/500 = 1.776 \times 10^{-14}$, $\hat{\sigma}_a = 1.3328 \times 10^{-7}$, similar to the AR(1) modeling in Chapter 2 (see Section 2.5.5).

Comparing σ_a with the amplitudes of s_i ($=c_i$) in Table 5.5 clearly shows that only the first three are the system modes and the rest either noise or computational modes. Thus modal selection can be accomplished by the simple criterion (2.8.20) conjectured in Chapter 2 for separating the deterministic modes from the stochastic ones. Note that s_i ($=c_i$ of equation (2.8.20)) and g_i are

TABLE 5.3 Comparison of Theoretical and Simulated Results for Scalar and Vector Modeling

	Theoretical	AR(6)	ARV(2)
$\dfrac{\omega_1}{\zeta_1}$	$\dfrac{0.583987}{0.015869}$	$\dfrac{0.583988}{0.015859}$	$\dfrac{0.583987}{0.015869}$
ℓ_1[Amp, Ph°]	$\begin{bmatrix} 1.0, & 0.0 \\ 2.48836, & -1.37 \\ 3.75612, & -3.86 \end{bmatrix}$	$\begin{bmatrix} 1.0, & 0.0 \\ 2.48838, & -1.38 \\ 3.75626, & -3.86 \end{bmatrix}$	$\begin{bmatrix} 1.0, & 0.0 \\ 2.48836, & -1.37 \\ 3.75612, & -3.86 \end{bmatrix}$
$\dfrac{\omega_3}{\zeta_3}$	$\dfrac{1.207717}{0.078894}$	$\dfrac{1.207714}{0.078899}$	$\dfrac{1.207717}{0.078894}$
ℓ_3[Amp, Ph°]	$\begin{bmatrix} 1.0, & 0.0 \\ 0.88419, & -22.25 \\ 1.94310, & 172.96 \end{bmatrix}$	$\begin{bmatrix} 1.0, & 0.0 \\ 0.88419, & -22.23 \\ 1.94311, & 172.96 \end{bmatrix}$	$\begin{bmatrix} 1.0, & 0.0 \\ 0.88419, & -22.25 \\ 1.94310, & 172.96 \end{bmatrix}$
$\dfrac{\omega_5}{\zeta_5}$	$\dfrac{1.637197}{0.053211}$	$\dfrac{1.637197}{0.053212}$	$\dfrac{1.637197}{0.053211}$
ℓ_5[Amp, Ph°]	$\begin{bmatrix} 1.0, & 0.0 \\ 1.06117, & -165.67 \\ 0.65287, & 34.17 \end{bmatrix}$	$\begin{bmatrix} 1.0, & 0.0 \\ 1.06117, & -165.67 \\ 0.65286, & 34.17 \end{bmatrix}$	$\begin{bmatrix} 1.0, & 0.0 \\ 1.06117, & -165.67 \\ 0.65287, & 34.17 \end{bmatrix}$

TABLE 5.4 Effect of Model Order—Closely Spaced Modes (X_{1t})

Autoregressive Order	Residual Sum of Squares	ω_1	ζ_1	ω_3	ζ_3	ω_5	ζ_5
6	0.4966×10^{-2}	1.6951	0.10150	4.8819	0.10142		
10	0.4696×10^{-7}	0.9567	0.16383	1.9746	0.08037		
16	0.3242×10^{-7}	0.9578	0.15408	1.9763	0.08112		
24	0.6427×10^{-8}	0.9670	0.15487	1.8821	0.08003		
28	0.8615×10^{-11}	0.9669	0.15503	1.8888	0.07965	2.0009	0.07471
36	0.7931×10^{-11}	0.9669	0.15513	1.8886	0.07941	2.0007	0.07488

TABLE 5.5 Modal Decomposition of an ARMA(28,27) Model (X_{3t})

λ_i DISCRETE COMPLEX ROOTS REAL	IMAG	ω_i (rad/s) NATURAL FREQUENCY	ζ_i DAMPING RATIO	$s_i(c_i)$ AMP/PHASE°	g_i MAG/PHASE°
0.9879	0.0636	0.96694 + 000	0.15521 + 000	0.31163 + 000 89.9915	0.1612 + 003 −70.7688
0.9821	0.1249	0.18865 + 001	0.79345 − 001	0.10468 + 000 89.5389	0.3052 + 003 −67.8494
0.9811	0.1324	0.20005 + 001	0.74792 − 001	0.24790 + 000 −90.1920	0.2865 + 003 70.9459
0.5684	0.6361	0.12733 + 002	0.18563 + 000	0.31664 − 006 26.8257	0.5209 − 001 28.3065
0.5121	0.7878	0.14812 + 002	0.62511 − 001	0.19545 − 006 −79.5163	0.5707 − 001 −36.3884
0.2544	0.9367	0.19414 + 002	0.22825 − 001	0.13162 − 006 −71.6944	0.3048 − 001 −17.3325
−0.0174	0.9683	0.23090 + 002	0.20638 − 001	0.20162 − 006 −98.4527	0.1658 − 001 31.1961

−0.1354	0.9494	0.25467 + 002	0.24449 − 001	0.18845 − 006 177.5400	0.8502 − 001 82.3259
−0.4323	0.8368	0.30454 + 002	0.29244 − 001	0.12896 − 006 −168.2382	0.3332 − 001 −5.8817
−0.6236	0.7149	0.34025 + 002	0.23016 − 001	0.20787 − 006 160.4468	0.5193 − 001 55.8081
−0.7289	0.6067	0.36394 + 002	0.21666 − 001	0.19633 − 006 48.9158	0.6988 − 001 38.9458
−0.8989	0.3998	0.40484 + 002	0.59964 − 002	0.96510 − 007 −59.8373	0.1469 − 001 79.8239
−0.9567	0.1376	0.44584 + 002	0.11361 − 001	0.20466 − 006 −143.1680	0.7001 − 002 18.3289
−0.4142				−0.199443 − 006	0.12833 + 000
1.0000				−0.18828 − 006	0.16504 + 003

Phase angles w.r.t. cosine. ± three digits indicate multiplication by power of 10.
$\Delta = 0.67265 - 001$.
RSS = 0.88822 − 011.

the coefficients of the root λ_i in the deterministic and stochastic parts respectively. As discussed at the end of Section 2.8.4, c_1, c_2 and c_3 are much larger than σ_a confirming the deterministic nature of these modes. However, because of ARMA modeling and the simulation as well as computational noise, $c_4, c_5, \ldots$, corresponding to the coefficients of the noise modes, are *not* smaller than σ_a. Also, contrary to the expectation, g_1, g_2, g_3 of the stochastic part corresponding to the deterministic modes are *not* small, because small estimation errors of these modes enter the stochastic part as additional noise.

The information about the selected 3 modes from tables such as Table 5.5 for X_{1t}, X_{2t} and X_{3t} has been summarized in Table 5.6. The frequencies and the damping ratios have been averaged over the values from the models for each set of data; ARMA(28,27) has been adequate for X_{1t} and X_{3t}, whereas ARMA(26,25) was adequate for X_{2t}. The mode shapes ℓ_i have been scaled so that the first element for X_{1t} is 1. The table essentially provides the modal model for this example. The theoretical values have also been listed in Table 5.6 for comparison. The agreement is seen to be excellent, showing that even closely spaced modes can be recovered using high enough model order.

TABLE 5.6 Comparison of Results—Closely Spaced Modes

	Theoretical	ARMA
ω_1	0.96697	0.96698
ζ_1	0.15512	0.15503
ℓ_1	$\begin{bmatrix} 1 \\ 1.06498 \\ 1 \end{bmatrix}$	$\begin{bmatrix} 1 \\ 1.06670 \\ 1.00090 \end{bmatrix}$
ω_3	1.88810	1.8888
ζ_3	0.07944	0.07965
ℓ_3	$\begin{bmatrix} 1 \\ -1.56498 \\ 1 \end{bmatrix}$	$\begin{bmatrix} 1 \\ -1.4359 \\ 1.0071 \end{bmatrix}$
ω_5	2.0000	2.0009
ζ_5	0.075	0.07471
ℓ_5	$\begin{bmatrix} 1 \\ 0 \\ -1 \end{bmatrix}$	$\begin{bmatrix} 1 \\ 0 \\ -1.001 \end{bmatrix}$

Even more significant aspect of this example is that the modal density is higher than the upper limit for effective modal resolution using single-exciter frequency domain methods. Lang (1983) has suggested that there is an intrinsic limit on the degree of damping and frequency proximity beyond which FFT-based FRF methods cannot adequately separate individual modes. The average modal density of n clustered or closely spaced modes is defined as

$$\text{Average Modal Density} = \frac{\text{Average Modal Bandwidth}}{\text{Average Frequency Spacing}}$$

$$= \frac{1}{n}\sum_{i=1}^{n} 2\zeta_i \omega_i \bigg/ \frac{(\omega_n - \omega_1)}{(n-1)}$$

It has been reported that in regions where the modal density exceeds 1, modal parameters cannot be successfully identified by FFT-based FRF techniques, regardless of the resolution and sophistication of the analysis system; spatial decomposition techniques such as the polyreference method must be used.

It is seen that the modal density in the present example is 2.5, much higher than the limit 1 for the frequency domain methods. Yet the modes have been resolved by the DDS method as shown above. This suggests that the DDS method using a relevant adequacy criterion is largely unaffected by strong modal overlap and that multi-point excitation is not necessarily required. This will be confirmed for vector modeling of real data in Section 5.6.2.

5.5.3 Spatial Model—Mass, Stiffness and Damping Matrices

To illustrate the method for obtaining the spatial model as simply as possible using scalar ARMA models, a 2-DOF system with the following matrices was simulated with the help of ACSL.

$$\mathbf{m} = \begin{bmatrix} 5 & 0 \\ 0 & 10 \end{bmatrix}, \quad \mathbf{c} = \begin{bmatrix} 1 & -0.5 \\ -0.5 & 1.5 \end{bmatrix}, \quad \mathbf{k} = \begin{bmatrix} 4 & -2 \\ -2 & 6 \end{bmatrix}$$

Since $\mathbf{c} = 0.25\mathbf{k}$, the system is classically (proportionally) damped and therefore has real mode shapes. Although the DDS method can readily handle complex mode shapes as illustrated in Section 5.5.1, this example will illustrate the classical vibration theory that allows the use of the transpose of the modal matrix instead of its inverse.

Following the method of Section 5.4.1, the free response of displacements X_{1t} and X_{2t} was generated using the initial conditions $x_2(0) = 1$. The modal decompositions of ARMA(10,9) models, considered adequate for both these displacements, are shown in Table 5.7. It is clear from their global nature and the amplitudes of $s_i (=|s_i|)$ that only the first two modes are the system modes. The modal vectors can be assembled from the two scalar modeling results as

TABLE 5.7 Free-Response Decompositions—ARMA(10,9)

	λ_i Discrete Roots	ω_i (rad/s) Natural Frequency	ζ_i Damping Ratio	s_i Amplitude	*Phase⁰
$x_1(t)$	$0.9899 \pm j0.0831$	0.6324	0.07905	0.33424	4.4822
	$0.9750 \pm j0.1293$	1.0000	0.12502	0.67195	7.1979
	$-0.1865 \pm j0.8561$	13.473	0.073824	0.81419×10^{-6}	-71.1124
	$-0.5246 \pm j0.5440$	17.721	0.11892	0.13912×10^{-5}	-167.8332
	-0.8934			0.26356×10^{-6}	
	1.000			0.77035×10^{-6}	
$x_2(t)$	$0.9899 \pm j0.0831$	0.6324	0.07909	0.33437	4.5512
	$0.9750 \pm j0.1293$	1.000	0.12504	0.33596	-172.8073
	$-0.0125 \pm j0.7726$	12.100	0.16802	0.53523×10^{-7}	-23.9172
	$-0.6211 \pm j0.6739$	17.438	0.03765	0.19805×10^{-7}	-80.0652
	-0.9312			0.48927×10^{-7}	
	0.4711			0.21699×10^{-7}	

* Phases w.r.t. cosine.

ILLUSTRATIVE EXAMPLES—SCALAR MODELS

$(j = \sqrt{-1})$

$$\ell_1 s_1 = \begin{bmatrix} 0.33424 \ e^{j4.4822°} \\ 0.33437 \ e^{j4.5512°} \end{bmatrix}, \ell_2 s_2 = \bar{\ell}_1 \bar{s}_1$$

$$\ell_3 s_3 = \begin{bmatrix} 0.67195 \ e^{j7.1979°} \\ 0.33596 \ e^{-j172.8073°} \end{bmatrix}, \ell_4 s_4 = \bar{\ell}_3 \bar{s}_3$$

Scaling them so that $x_1(t)$ has the element 1 gives

$$\ell_1 = \begin{bmatrix} 1 \\ 1.0004 \ e^{j0.07°} \end{bmatrix} \simeq \begin{bmatrix} 1 \\ 1.0004 \end{bmatrix} = \ell_2$$

$$\ell_3 = \begin{bmatrix} 1 \\ 0.49998 \ e^{-j180.005°} \end{bmatrix} \simeq \begin{bmatrix} 1 \\ -0.49998 \end{bmatrix} = \ell_4$$

showing that the mode shapes are real as expected. Together with the roots μ_i given by the natural frequencies and damping ratios in Table 5.7, this provides us with the modal matrix

$$\mathbf{M} = \begin{bmatrix} \ell_1 & \ell_2 & \ell_3 & \ell_4 \\ \mu_1 \ell_1 & \mu_2 \ell_2 & \mu_3 \ell_3 & \mu_4 \ell_4 \end{bmatrix}, \quad \begin{aligned} \mu_1, \mu_2 &= -0.05 \pm j0.63046 \\ \mu_3, \mu_4 &= -0.125 \pm j0.99216 \end{aligned}$$

Therefore

$$\mathbf{M}^{-1} = \begin{bmatrix} 0.16662 - j0.013214 & 0.33324 - j0.026429 & j0.079237 - j1.5848 \\ 0.16662 + j0.013214 & 0.33324 + j0.026429 & -j0.079237 & j1.5848 \\ 0.33338 - j0.042002 & -0.33325 + j0.041985 & -j1.007460 & j1.0071 \\ 0.33338 + j0.042002 & -0.33325 - j0.041985 & j1.007460 & -j1.0071 \end{bmatrix}$$

$$= \begin{bmatrix} L^{11} & L^{12} \\ L^{21} & L^{22} \\ L^{31} & L^{32} \\ L^{41} & L^{42} \end{bmatrix}$$

and the bottom 2 × 2 submatrices $\mathbf{A}_{21} = -\mathbf{m}^{-1}\mathbf{k}$ and $\mathbf{A}_{22} = -\mathbf{m}^{-1}\mathbf{c}$ of the 4 × 4 continuous-time state matrix $\mathbf{A}$ are given by equations (5.4.3) and (5.4.4) as

$$-\mathbf{k} = \mathbf{m}\mathbf{A}_{21} = \mathbf{m} \sum_{i=1}^{4} \mu_i^2 \ell_i \mathbf{L}^{i1} = \mathbf{m} \begin{bmatrix} -0.80006 & 0.39917 \\ 0.20003 & -0.59993 \end{bmatrix}$$

$$-\mathbf{c} = \mathbf{m}\mathbf{A}_{22} = \mathbf{m} \sum_{i=1}^{4} \mu_i^2 \ell_i \mathbf{L}^{i2} = \mathbf{m} \begin{bmatrix} -0.20001 & 0.099747 \\ 0.05005 & -0.149986 \end{bmatrix}$$

so that the state matrix **A** is given by

$$\mathbf{A} = \begin{bmatrix} 0 & 0 & 1 & 0 \\ 0 & 0 & 0 & 1 \\ -0.80006 & 0.39917 & -0.20001 & 0.099747 \\ 0.20003 & -0.59993 & 0.05005 & -0.149985 \end{bmatrix}$$

This completes the modal model and allows us to write $2p^2 = 8$ equations in the elements of **m** and **c** and **k**. It is clear that $\mathbf{m}^{-1}\mathbf{c}$ and $\mathbf{m}^{-1}\mathbf{k}$ obtained from these equations match very well with the true ones.

The next task in determining **m**, **c**, and **k** is to resolve the scaling of the mode shapes with the help of forced response. Assuming **m** and **k** to be symmetric, only one mass needs to be forced to get the remaining $p = 2$ required equations. Hence the displacement responses to a sinusoidal function applied at mass 1 were generated by ACSL. The steady-state data was modeled by ARMA$(2n, 2n-1)$ strategy and a relatively lower-order ARMA$(4,3)$ model turns out to be adequate. The modal decomposition of the models for the two responses and the forcing function is shown in Table 5.8. The sinusoidal frequency 1.2609 rad/sec is the only significant mode in all the models as expected. Hence $\alpha = j1.2609$ and we can use the displacement relation (5.4.21) with

$$\mathbf{f} = \begin{bmatrix} -j \\ 0 \end{bmatrix}, \quad \ell_\alpha s_\alpha = \begin{bmatrix} 0.50365 \; e^{-j113.3481} \\ 0.10479 \; e^{j38.3389} \end{bmatrix}$$

so that

$$\mathbf{m}[\alpha^2 \mathbf{I} - \alpha \mathbf{A}_{22} - \mathbf{A}_{21}]\ell_\alpha s_\alpha = \mathbf{m}\begin{bmatrix} -j0.19995 \\ 0 \end{bmatrix} = \begin{bmatrix} -j \\ 0 \end{bmatrix}$$

to get the necessary two equations. Solving these together with the eight obtained from the free response gives

$$\mathbf{m} = \begin{bmatrix} 5.00125 & 0 \\ 0 & 9.9785 \end{bmatrix} \quad \mathbf{c} = \begin{bmatrix} 1.0003 & -0.49886 \\ -0.49897 & 1.49676 \end{bmatrix}$$

$$\mathbf{k} = \begin{bmatrix} 4.0013 & -1.9960 \\ -1.9960 & 5.9864 \end{bmatrix}$$

where we have assumed **m** and **k** to be symmetric.

Since the mode shapes are real, it is not necessary to use the 4×4 state matrix; it is enough to use a 2×2 modal matrix of the classical vibration theory and its transpose as discussed in the review of Section 5.1.4. It is left as an exercise to show that such a modal matrix is given by

$$\mathbf{S} = \begin{bmatrix} 0.25822 & -0.36511 \\ 0.25832 & 0.18255 \end{bmatrix}$$

TABLE 5.8 Forced-Response Decompositions—ARMA(4,3)

	λ_i Discrete Roots	ω_i (rad/s) Natural Frequency	ζ_i Damping Ratio	s_i Amplitude	*Phase⁰
$x_1(t)$	$0.9860 \pm j0.1688$	1.2609	0.57669×10^{-7}	0.50365	-113.3481
	$0.4789 \pm j0.8042$	15.871	0.313561	0.15711×10^{-7}	-6.8705
$x_2(t)$	$-0.9860 \pm j0.1688$	1.2609	0.13896×10^{-7}	0.10479	38.3389
	$-0.4569 \pm j0.7807$	15.824	0.47693×10^{-1}	0.60120×10^{-8}	-179.7739
$F(t)$	$0.9965 \pm j0.0834$	1.2609	0.81759×10^{-7}	2.0000	90.0001
	-0.3101			0.34621×10^{-6}	
	-0.9997				

*Phases w.r.t. cosine.

TABLE 5.9 Comparison of Results

	ω_1	ζ_1	s_1	ω_2	ζ_2	s_2
Theoretical	0.63246	0.079059	0.25819	1.0	0.125	-0.36515
			0.25819			0.18257
DDS	0.63244	0.079069	0.25822	1.0	0.12503	-0.36511
			0.25832			0.18255

and now using (5.1.44) with ω_i and ζ_i from Table 5.7 yields the same **m**, **c**, and **k** as above. Comparison of the theoretical and DDS results is given in Table 5.9.

5.5.4 Modal Parameters of a Disc–Brake Rotor—Structural Modification

To illustrate the application of the procedure to real structures, we summarize the results and discussion from Pandit and Jacobson (1988). The disc-brake rotor, due to its symmetry, causes the vibration-induced brake squeal. If any structural modifications are made to reduce the squeal, it is necessary to determine all the dynamic characteristics of both the original and modified rotor to assess their effectiveness.

In order to compare the DDS method with a typical Fourier Analyzer-based modal analysis package, a parallel data collection scheme was developed. This consisted of a Hewlett-Packard 5420A Fourier Analyzer (HP), and a Nicolet Fourier Analyzer and modal analysis package. The reason why two Fourier Analyzers were chosen stems from the fact that a data transfer link already existed between the HP and the Univac mainframe computer via the Commodore 8032 personal computer (PC).

The experimental procedure involved testing the disc-brake rotor under various boundary conditions. The first experimental endeavor involved testing a fixed-free rotor, the disc being bolted at its center and free at its periphery. The second case tested was that of a free-free rotor configuration, which consisted of the rotor resting on packing foam (the stiffness of the packing foam is much less than that of the rotor itself, and thereby a free-free boundary condition is simulated). The third case was also a free-free condition, but the rotor was structurally modified. The modified rotor will be discussed later.

For all three test conditions the rotor was discretized into three annular rings: two rings, each consisting of 24 test points located 15 degrees apart near the free edge boundary, and the remaining 12 points on an annular ring located on the center hub. Figure 5.16 shows the model used to represent the disc-brake rotor.

Fixed-Free Rotor
Detailed tables of modal parameters, similar to Table 5.7, may be found in Pandit, Jacobson, and Shapton (1985); main results will be graphically illustrated here. The fixed-free study showed that DDS required an ARMA(54,53) model to characterize adequately all the dynamic characteristics of the rotor. This ARMA(54,53) model resulted in 24 second-order modes and 6 first-order modes, that is, 24 sets of complex conjugate roots and 6 real roots.

In the DDS modal analysis output, the natural frequencies vary from point to point on the structure, so only the modal data that correspond to the FFT analyzer was used if it appeared at all locations analyzed on the rotor. In the event that closely spaced frequencies from the DDS method appeared for all locations on the rotor, two mode shapes were calculated. These sets of closely spaced frequencies suggest repeated roots due to the geometric symmetry. These

ILLUSTRATIVE EXAMPLES—SCALAR MODELS 261

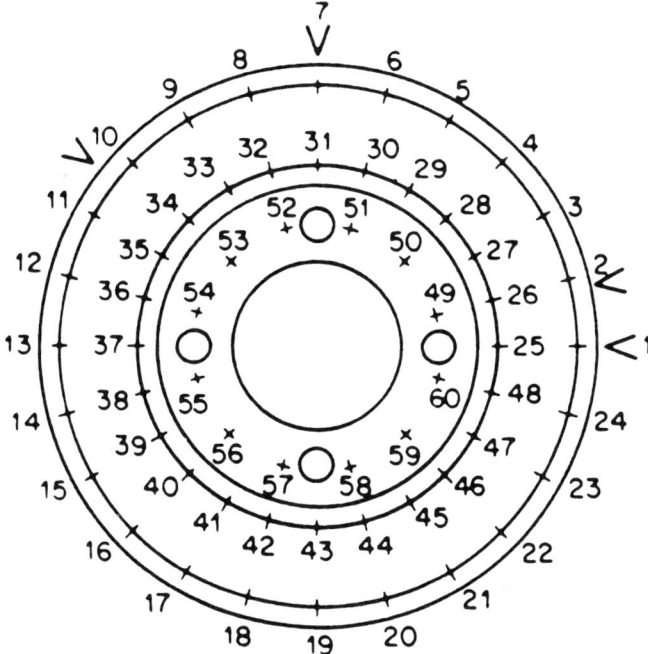

Figure 5.16 Schematic location of measuring points. (Location of four bolts for fixing is shown by small circles. Arrowheads indicate locations of rib removal.)

do not actually appear at the same frequency; some nonhomogeneity separates them into two distinct but very closely spaced frequencies. Although these frequencies are slightly separated, when the mode shapes are plotted, they appear to be the same mode, which shows that the mode shapes are repeated, as illustrated later.

Owing to the large computing time, only every third point on the structure was modeled for this first experimental case, and only the first four modes at the lowest frequencies were plotted. To improve spatial resolution at the higher frequencies, the mode shapes require modeling of additional points since a nodal line may fall between the two points on the rotor that were modeled.

The scaled mode shapes ℓ_i were then broken into real and imaginary components because complex mode shapes are difficult to visualize. Only the real parts of the mode shapes are plotted. In identifying the frequency of each mode, a global average of the frequencies within a small bandwidth at each point location was used. For example, for a given mode the frequencies may range between 763 and 774 Hz and average to 769 Hz, which is the first mode plotted in Figure 5.17. The corresponding frequencies and damping ratios are listed. Regions of positive and negative real parts of the mode shapes have been indicated by plus and minus signs.

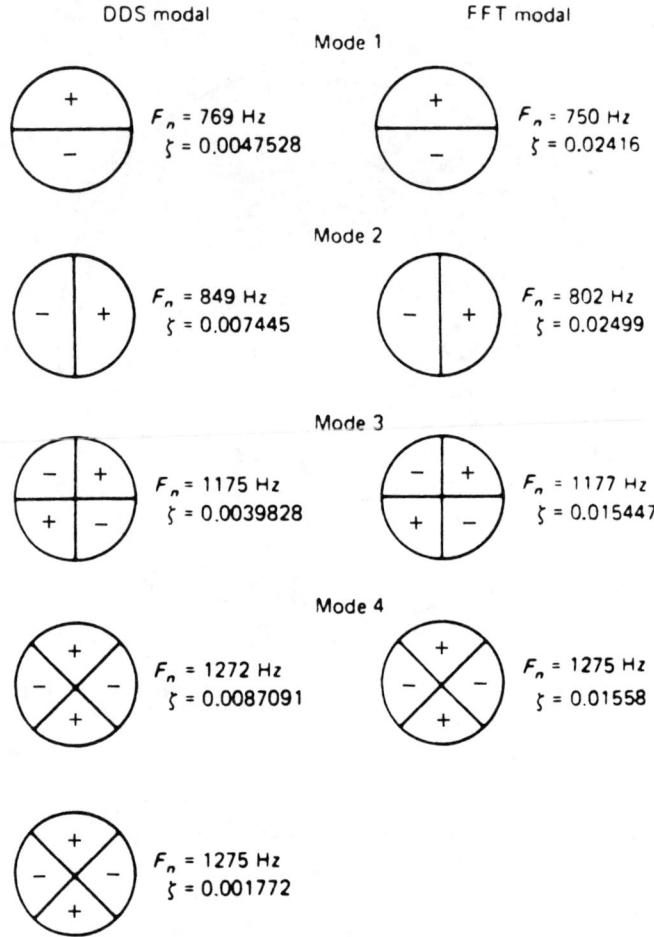

Figure 5.17 Mode shapes for the unmodified fixed-free disc-brake rotor.

Considering both the real and imaginary components of the mode shape one can relate this to two types of vibration. The real part is referred to as a real mode, and the imaginary component is actually a traveling wave propagating around the rotor. This can be seen on the Nicolet analyzer, if instead of extracting the modal parameters for only real modes one extracts the complex modes. The mode shape appears to be rotating as well as vibrating in a real mode fashion. This concept will be utilized later when structural modification is applied to the rotor.

Free-Free Unmodified Rotor

In this second experimental investigation the rotor was modeled in a free-free boundary condition by placing the rotor on packing foam. Testing the rotor in a free-free state allows generation of a modal model of the rotor itself. In the first investigation the mode shapes of the rotor are affected by what the rotor is bolted to, the torque that is applied, and so on, which is probably closer to reality. In this second test case an accurate model of the rotor itself is needed so that effective structural modification can be applied.

To reduce the effort in collecting data, only one-half-of the rotor was modeled. Owing to the symmetry of the rotor, only half of the rotor need be modeled; the mode shapes can be reflected across a nodal diameter to obtain the complete rotor characteristics.

The DDS results indicate that an ARMA(60,59) model fully describes the dynamic characteristics of the free-free rotor. Because of the free-free boundary condition all of the frequencies appear to be higher than in the fixed-free test, as expected. Later, the modes obtained for the fixed-free rotor and the free-free rotor are compared to examine the similarities.

In this second case attention is concentrated on only six modes. These modes are the ones with the most power or largest mode participation factors. These modes also appear to be in the frequency range where human hearing is the most sensitive. In this range the disc-brake squeal propensity is the largest. The mode shapes for the DDS method were extracted and are presented in Figure 5.18 along with the mode shapes obtained by the Nicolet system. The mode shapes are similar for both methods but the damping and frequencies are marginally different in some cases.

Free-Free Modified Rotor

The third test case involved testing the modified disc-brake rotor with a free-free boundary condition. The rotor was structurally modified by selectively removing ribs from the structure. Figures 5.16 and 5.19 show the locations by arrowheads at points 1, 2, 7, and 10.

The DDS method indicates that an ARMA(76,75) model is required to fully describe the modal parameters of the modified disc-brake rotor. For this investigation the entire rotor was modeled, to predict accurately the mode shapes, since the structural modification applied to the rotor now made it geometrically unsymmetric.

Figure 5.19 shows the mode shapes obtained via the DDS method and the mode shapes obtained by the Nicolet system. Note that the mode shapes are similar for both methods, but the damping terms and the natural frequencies are different; damping is overestimated by FFT. At the natural frequency of 1110 Hz the damping ratio is 0.003372 for the unmodified, while for the structurally modified rotor the corresponding natural frequency of 1086 Hz is associated with a damping ratio an order of magnitude larger, at 0.013910. Experimentally one can notice the different damping ratio when the rotor is impacted by the impulse hammer. The unmodified rotor *rings* when struck, while the structurally modified rotor produces a *thudding* noise.

Discussion

This investigation indicates that removing the ribs unsymmetrically causes a destruction of the symmetry of the rotor, therefore eliminating a constant

264 MODAL ANALYSIS

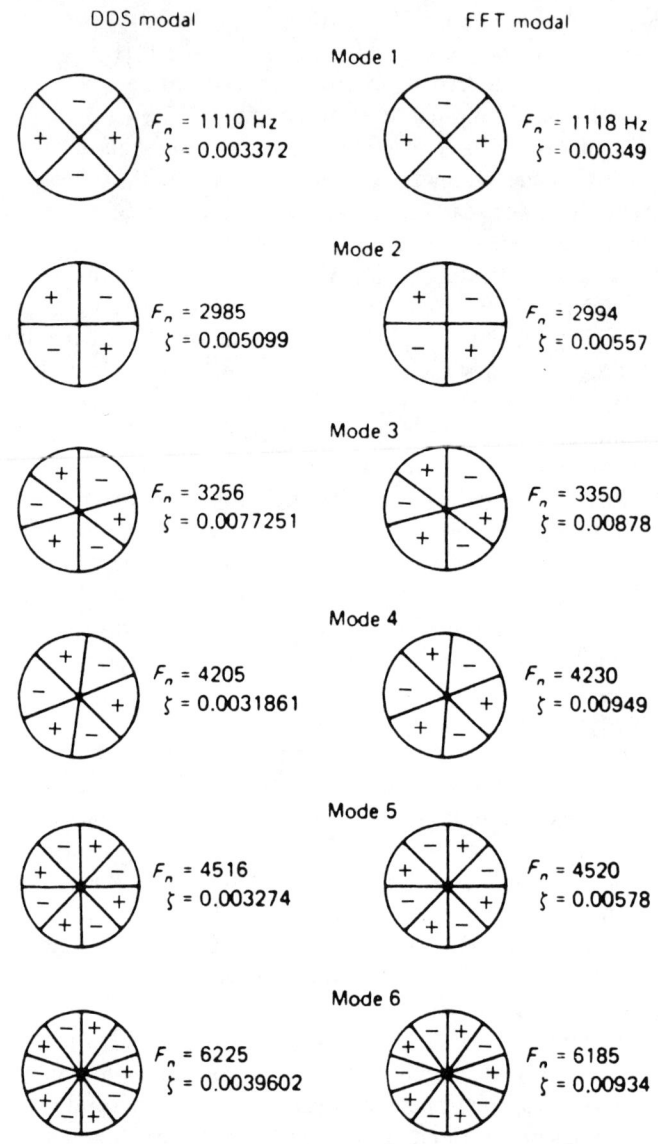

Figure 5.18 Mode shapes for the unmodified free-free disc-brake rotor.

frequency forcing function. The unsymmetric rib spacing caused the travelling wave, induced by the impulse force, to have a component which *reflected* off of the ribs, thus leading to a cancellation effect. This can be seen by noting that for the modified rotor the apparent damping of the mode corresponding to a frequency of 1100 Hz increased by an order of magnitude.

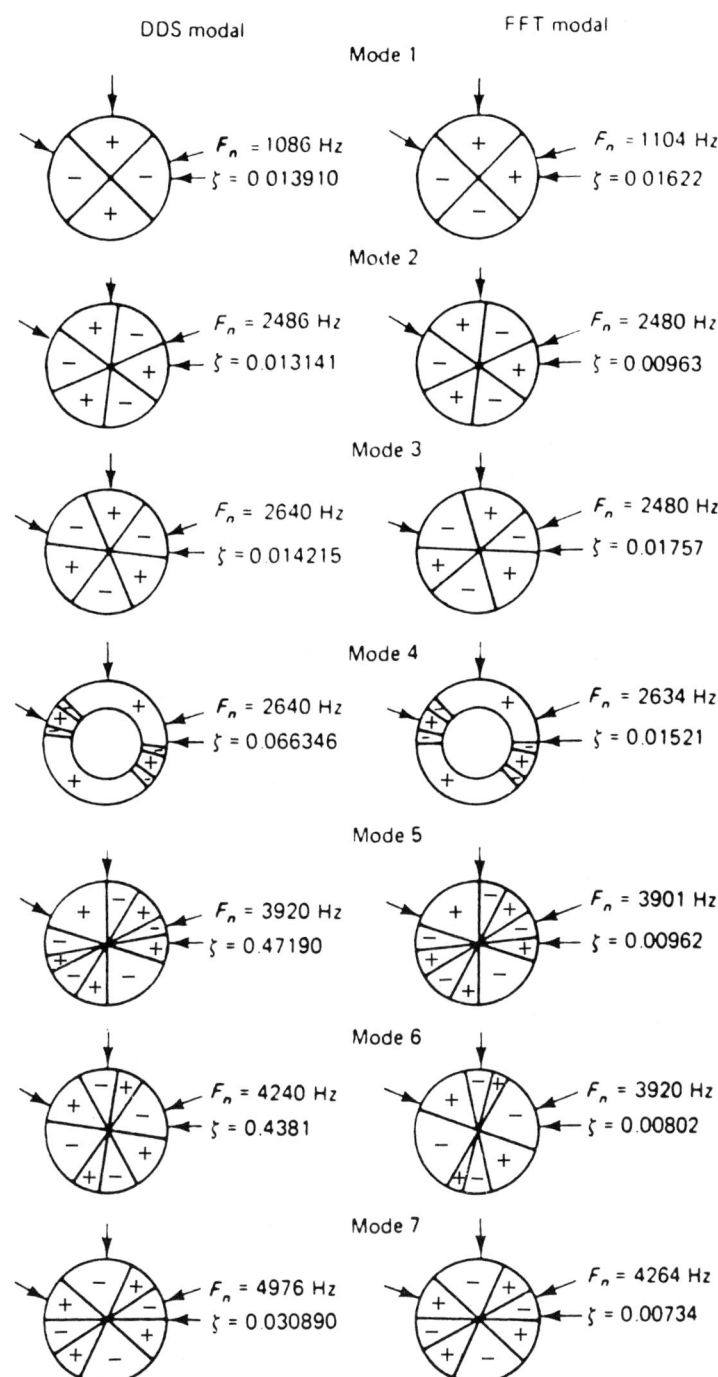

Figure 5.19 Mode shapes for the modified free-free disc-brake rotor.

The experimental results for the free-free unmodified rotor show that the DDS method fits a 60th-order difference equation model that translates into 29 second-order modes of vibration. This fact indicates that a 29th-order, lumped-parameter model can be used to fully characterize the disc-brake rotor. The subsequent elimination of ribs from the structure increases the flexibility at various points on the structure and necessitates increasing this lumped parameter model to 36th order to account for the increased flexibility.

This investigation showed that the DDS method can extract the nearly repeated roots and detect more frequency content in a given bandwidth than the FFT counterpart. The damping of each mode of vibration of the rotor can be extracted more accurately by the DDS method as already illustrated by simulated examples earlier. This is possible because in DDS each pole in the complex plane is extracted, whereas in the FFT method only the projection of the pole onto the imaginary axis is considered, as discussed in Sections 2.9.3 and 5.1.1. For systems with low damping both methods should extract similar values because the poles are relatively close to the imaginary axis. When a structure has high inherent damping and the system's poles are not close to the imaginary axis, the DDS method becomes a superior method for obtaining the damping values. This becomes evident when structural modification is applied to the rotor. The estimates of the damping values from the two methods differ significantly. For example, for the first mode of vibration the DDS method shows the damping ratio to be 0.01391 while the Nicolet system shows the damping ratio to be 0.01622. With the latter method, when the poles do not appear on the imaginary axis, the projection onto this axis makes the peak in the frequency response appear *more* rounded in appearance.

Although the modification increases the damping and resists the propensity to squeal, it may also induce imbalance in the rotor. The effect of this imbalance needs to be carefully examined *in situ* before implementing the modification in practice.

5.6 ILLUSTRATIVE EXAMPLES—VECTOR MODELS

The procedure for vector models is essentially the same. ARV models rather than ARMAV are fitted for reasons discussed in Sections 2.4.3 and 5.2.1. The adequate ARV model yields the discrete-time state matrix ϕ. The spectral decomposition of the state matrix $\phi = \mathbf{L}\lambda\mathbf{L}^{-1}$ yields the mode shapes ℓ_i as the columns of the bottom $p \times np$ submatrix, of $\mathbf{L}$, and λ yields the characteristic roots μ_i from which the natural frequencies and damping ratios are obtained. The elements of ℓ_i are then arbitrarily scaled, generally one location has unit value. This procedure, applied to free-response data, gives the modal model.

If, further, spatial model consisting of $\mathbf{m}$, $\mathbf{c}$, and $\mathbf{k}$ is desired, the same procedure is applied to the forced-response data as well as the force measurement data. If two of these matrices are assumed to be symmetric, one forcing frequency is enough to get ℓ_α and $\mathbf{f}$ from the modal vectors at the frequency

ω, $\alpha = j\omega$. If the responses and forces are modeled together, as one vector, then these ℓ_α and $\mathbf{f}$ are obtained from the overall modal vector, for example, from $[\ell_\alpha^T, \mathbf{f}^T]^T$ if the response data forms the first and the force data forms the last elements of the data vector used in the ARV models. If they are separately modeled, the procedure is similar to the scalar models illustrated earlier except that $\ell_\alpha s_\alpha$ and $\mathbf{f}$ are available directly from their respective vector models and need not be assembled element by element as in the scalar case. If symmetry is not assumed, p different tests with each location forced separately or fewer tests with p distinct frequencies are necessary; any configuration that provides p linearly independent $\mathbf{f}_k$, $k = 1, 2, \ldots, p$ and their response modal vectors suffices. These $\ell_\alpha s_\alpha$ and $\mathbf{f}$ are used to resolve the scaling using (5.4.20/21) and the modal model parameters are used in (5.4.3) and (5.4.4) to solve for $\mathbf{m}$, $\mathbf{c}$, and $\mathbf{k}$.

Since the procedure is essentially the same as in the scalar cases outlined earlier in detail, we will only give the final results. The numerical steps leading to these results are straightforward and should be clear from Sections 5.3 and 5.4; they are too numerous to repeat here and hence will be omitted.

5.6.1 Modal Model—Simulated and Real Systems

Results from a 3-DOF simulated system modeled by ARV(2) model were given in Table 5.3 and it was emphasized in Section 5.5.1 that they are more accurate than the scalar results because more information is used in determining any individual parameter. Similar accuracy has been obtained using an ARV(2) model for the closely spaced modes example of Section 5.5.2. More simulated examples to show that the same fifth decimal place accuracy is obtainable with light as well as heavy damping are given in Pandit, Helsel, and Evensen (1986); we reproduce the following results for a real system from this reference.

A simple lumped mass system was constructed for the testing of a real system. The whole system was put in tension to help eliminate many of the rotation modes of the system. An accelerometer was mounted to the top of each mass. The system was excited into free vibration by simply displacing either mass downward and releasing it. A Hewlett Packard 5420 signal analyzer was used to digitize the accelerometer data into 512 point records.

The masses of this system were weighed on a balance beam. The springs were calibrated by hanging known weights on them and measuring the deflections. By using the "proper" initial displacements, each mode was individually excited and the damping ratio of each mode was calculated using the log-decrement method. For example, to excite the first mode, both masses were given the same initial displacement and released.

An ARV(2) model was adequate to fit this real data. This indicates that the data was relatively free of noise. Table 5.10 compares the measured parameters to those obtained by the DDS method. Since it was not possible to measure the individual damping parameters of the system, the frequencies and mode shapes were estimated assuming no damping. This should cause little error since the damping ratio of each mode was less than 2%. All in all, however, the results

TABLE 5.10 Comparison of DDS and Theoretical Mode Shapes and Natural Frequencies for a Real 2-DOF System

Parameter	Theoretical		DDS	
	Mag.	Phase (deg.)	Mag.	Phase (Deg.)
First mode shape	1.0	0.0	1.0	0.0
	0.893	0.0	0.9697	0.8
	(no damping)			
Second mode shape	1.0	0.0	1.0	0.0
	1.15	180	1.075	178.5
	(no damping)			
First natural frequency (Hz)	12.02		12.21	
Second natural frequency (Hz)	20.92		21.04	
	(no damping)			
First mode damping ratio	0.018		0.013	
Second mode damping ratio	0.0001		0.0002	
$m^{-1}k$	$\begin{bmatrix} -10770 & 5666 \\ 5817 & -12220 \end{bmatrix}$		$\begin{bmatrix} -11377 & 5666 \\ 5911 & -11975 \end{bmatrix}$	
$m^{-1}c$	NA NA	NA NA	$\begin{bmatrix} -1.008 & -2.031 \\ 0.1266 & -1.062 \end{bmatrix}$	

from the DDS method matched the measured parameters within the experimental error in the measured parameters. The ACSL simulations illustrate the validity of the DDS method and show that the frequencies and damping ratio can be estimated accurately to within 5 decimal places and mass, damping, and stiffness matrices with only 0.04% error. The real data, however, represents the true test of the method's usefulness. The results presented here show that the method works well within experimental error (considering the difficulty in measuring the stiffness of coil springs).

5.6.2 Modal and Spatial Model—An experimental Torsional System

An experimental verification of the procedure will now be described from Helsel, Evensen, and Pandit (1988). A torsional system consisting of a straight circular shaft with three cylindrical flywheels was built and tested; this system was selected because the accuracy of the model obtained using the DDS method could be reasonably evaluated by theoretical means and by conventional FFT-based modal analysis. Torsional fixed-fixed, fixed-free, and free-free conditions could be obtained and it was possible to excite the system so that its torsional measurements would be contaminated by bending modes having similar natural frequencies. Results for only the fixed-fixed condition will be listed for brevity.

Experimental Setup

Three steel disks (1/2 in. thick × 4, 5, and 6 in. diameter) were each fastened to a steel circular shaft (1/2 in. diameter × 13.5 in. long) by means of two setscrews. Half-inch aluminium cubes screwed to flats on the edge of each disk served as accelerometer bases. These cubes and the attached accelerometers were counterbalanced by appropriately sized aluminium cubes screwed to flats on the opposite sides of the disks. Similarly-balanced 3/4-in. cubes, screwed at 90° offsets to the accelerometer bases, served as loading bases for applying external torques. The ends of the shafts were bolted to heavy steel tubes which in turn were mounted on a massive isolated bed.

To obtain either fixed or free end conditions, a steel spacer which could be replaced with a pillow block bearing was positioned between the steel tube and the end of the shaft. Fixed-fixed, fixed-free, and free-free shaft end conditions were tested using this arrangement. Viscous damping was applied to one of the disks by forming a clay trough around the disk and pouring in heavy silicone oil to a depth of about one-third disk diameter.

Rotational acceleration was sensed through piezoelectric accelerometers glued to the accelerometer bases. Forces were sensed through piezoelectric force transducers screwed to the loading bases. Dynamic external torques were applied through the loading bases by means of an horizontally oriented electromagnetic shaker, attached to the force transducer through a flexible nylon sting. The shaker was driven by a sine generator through a power amplifier. The free response of the structure was generated by impulsing the loading base of a disk with a nylon-tipped hammer, striking along a direction perpendicular to the sensing axis of the accelerometer to reduce the sensing of any lateral bending modes.

Acceleration and force signals were passed through signal conditioners and low-pass antialiasing filters, then simultaneously digitized on a four-channel analyzer (12-bit, 1024 points per record per channel). The digitized signals were downloaded to a mainframe computer via telephone modem for DDS modeling and computing the mass, stiffness, and damping matrices. The autoregressive model order was established by successively increasing it until the modal parameter estimates stabilized. The signals were monitored at each stage of the conversion process using digital and analog oscilloscopes. A digital signal analyzer was also used to obtain FFT-based spectra of the data.

Basic Procedure

To test the consistency of the DDS approach, three separate analyses were performed for each of the three support conditions. In the first phase of testing, a modal analysis was performed using a small, nylon-tipped hammer to excite the system, while simultaneously measuring the free response at each disk. This test was performed three times, applying the excitation to a different rotational mass each time. In the second phase of testing, the structure was driven sinusoidally at each acceleration transducer location to produce three separate estimates of the mass, stiffness, and damping matrices. This approach tested the variance between the different estimates and helped to identify the best method for identifying these properties.

The DDS-based modal parameters were verified by comparing them with values obtained from conventional, FFT-based modal analysis and from theoretical analysis. The FFT-based analysis was performed employing impact excitation to all three disks and referring results to a single reference accelerometer. Two independent modal tests were conducted: one by relatively inexperienced students using the ZONIC 6080 multichannel analyzer equipped with MODAL-PLUS software by Structural Dynamics Research Corporation (SDRC), and the other by an experienced investigator using a Nicolet 6601 modal analysis system with a Nicolet 660B dual-channel analyzer. A MDOF curve-fitting routine (polynomial least squares) was used to estimate the mode shapes. Since the DDS-based modeling technique has an inherent ability to model complex mode shapes, the complex-mode curve-fitting option (enabling mode shapes with amplitude and phase) was used. In the FFT-based modal analysis, an exponential window was applied to all data. The theoretical evaluation was performed assuming zero damping, with the rotational inertias and torsional stiffnesses calculated using simple handbook formulas.

Modal Parameters

Modal parameters were estimated for the fixed-fixed, fixed-free, and free-free conditions. Table 5.11, which summarizes results for the fixed-fixed condition, shows that the third disk rotates very little compared to the other two disks in the third mode. The third disk is evidently situated near a node of the third vibration mode, so that impulsing this disk does little to excite the third mode. This is corroborated in Table 5.11 by the DDS analysis, which shows that the motion of the third disk in the third mode is so small that even an ARV(18) model could not resolve it. However, all the mode shapes were resolved when disk 1 or disk 2 were impulsed.

On the other hand, the FFT-based analysis was unable to resolve the third mode because whenever the third disk was impulsed the third mode was poorly excited. The DDS method produced very small variations between the estimated modal parameters; the parameters produced by the experienced FFT operator displayed more variation, while the inexperienced operator produced unacceptable variations. Comparing the DDS-based modal parameters against both theoretical and FFT-based parameters illustrates the precision and accuracy of the DDS method. For the fixed-free condition (one fixed support replacing by a bearing), the third disk was also found to be near a node of the third mode; in fact, the DDS and FFT methods displayed practically the same performances displayed in Table 5.11.

For the free-free condition, the analysis confirmed that the first torsional natural frequency is zero and that only two modes are oscillatory. Again, the third mode was weakly excited when the third disk was impulsed, but was strongly excited when either disk 1 or disk 2 were excited. Agreement between all

TABLE 5.11 Comparison of Modal Analysis Techniques Applied to the Fixed-Fixed Torsional System. No Added Damping

Mode	Modal Parameters		Theory	DDS, Impulse Applied To			FFT, Ref Disk 3
				Disk 1	Disk 2	Disk 3	
1	Nat'l Frequency, Hz		92.8	91.3	91.0	91.2	91.5
	Damping Ratio		0.0	0.0014	0.0015	0.0009	0.0104
	Mode Shape	Disk 1	1.40/0.0	1.31/0.0	1.30/−1.	1.30/−1.	1.30/7.0
	(Mag/Phase°)	Disk 2	1.94/0.0	1.69/1.0	1.70/1.0	1.70/1.0	1.75/4.0
		Disk 3	1.00/0.0	1.00/0.0	1.00/0.0	1.00/0.0	1.00/0.0
2	Nat'l Frequency, Hz		204.4	200.3	200.3	199.9	201.4
	Damping Ratio		0.0	0.0018	0.0021	0.0017	0.0075
	Mode Shape	Disk 1	0.232/180	0.234/179	0.237/180	0.237/180	0.248/178
	(Mag/Phase°)	Disk 2	0.223/180	0.230/179	0.231/178	0.231/178	0.245/179
		Disk 3	1.00/0.0	1.00/0.0	1.00/0.0	1.00/0.0	1.00/0.0
3	Nat'l Frequency, Hz		360.1	352.5	353.0	—	356.5
	Damping Ratio		0.0	0.0030	0.0033	—	0.0089
	Mode Shape	Disk 1	35.3/0.0	30.4/6.0	28.1/−3.	—	256/122
	(Mag/Phase°)	Disk 2	5.96/180	5.15/175	4.80/177	—	−51.2/−91
		Disk 3	1.00/0.0	1.00/0.0	1.00/0.0	—	1.00/0.0
	ARV Model order			10	10	18	

three modal analysis techniques was reasonable, though the DDS method was more precise.

Mass, Stiffness, and Damping Matrices

The lumped rotational inertia, torsional damping, and torsional stiffness matrices were estimated under the three support conditions. Theoretically, these matrices are obtainable by subjecting any single disk to a sinusoidal force at any frequency. However, there are practical considerations in choosing the forcing frequency. If the structure is lightly damped, the phase between the force and response is very erratic near a response frequency. Therefore, it is advisable to choose a forcing frequency that falls between the resonance frequencies, where the phase is more stable. Table 5.12 gives results for the fixed-fixed conditions, where the symmetric mass and stiffness matrices obtained by forcing each disk separately are compared against the general unsymmetric matrices obtained by combining the data from all three single-disk tests.

Each of the symmetric mass and stiffness matrices compare reasonably well with the theoretical matrices, though the columns corresponding to the driven disk are in best agreement with the theoretical values. The general mass and stiffness matrices were also in reasonable agreement with theory: the off-diagonal terms of the mass matrices never exceeded eight percent of the smallest mass; the near-zero off-diagonal terms of the stiffness matrix were an order of magnitude smaller than the other terms. No theoretical damping estimates were available for comparison with the experimentally derived values; the actual damping values were so small that they would be hard to resolve by any method. The fact that some of the off-diagonal terms are positive and some of the diagonal terms are negative suggests that this matrix may be a good indicator of the effects of experimental and computational error.

In the fixed-free case, the mass matrices were again found to match the theoretical matrix, with the general mass matrix being the best match. The stiffness matrices all showed general agreement with theory; all identified the removal of the torsional spring between the third disk and the end support, as evidenced by the fact that the row two, column three term is approximately equal to the row three, column three term. This illustrates that the physical properties of the system can be uniquely identified from the data itself; no assumptions need to be imposed during the modeling process. The damping matrix exhibited the same behavior observed for the fixed-fixed system.

These patterns were also observed in the free-free comparisons. The stiffness matrices properly identified the removal of the torsional stiffnesses from each end of the structure while maintaining the proper value of torsional stiffness between the disks. Values of the damping matrix were found to be larger than those of previous damping matrices, perhaps because of the added damping from the greased bearings at the two free ends of the structure. Evidence is also seen in the increased modal damping ratios, and the fact that the diagonal terms of the damping matrix were strongly positive while the off-diagonal terms were negative.

TABLE 5.12 Comparison of Theoretical and Estimated Mass, Stiffness, and Damping Matrices from the Fixed-Fixed Torsional System. No Added Damping

Condition	m lb-sec^2-in./(rad)	k lb-in./(rad)	c lb-sec-in./(rad)
Theory	$\begin{bmatrix} 0.0115 & 0.0 & 0.0 \\ 0.0 & 0.0513 & 0.0 \\ 0.0 & 0.0 & 0.0259 \end{bmatrix}$	$\begin{bmatrix} 52923 & -35282 & 0 \\ -35282 & 50963 & -15681 \\ 0 & -15681 & 39202 \end{bmatrix}$	$\begin{bmatrix} — & — & — \\ — & — & — \\ — & — & — \end{bmatrix}$
Disk 2 impulsed Disk 1 forced	$\begin{bmatrix} 0.0119 & -0.0009 & -0.0004 \\ -0.0009 & 0.0505 & -0.0010 \\ -0.0004 & 0.0010 & 0.0248 \end{bmatrix}$	$\begin{bmatrix} 52779 & -37311 & -881 \\ -37311 & 54108 & -16089 \\ -811 & -16089 & 35897 \end{bmatrix}$	$\begin{bmatrix} 0.361 & 1.274 & 0.173 \\ -0.386 & -0.792 & -0.505 \\ 0.115 & 0.224 & 0.132 \end{bmatrix}$
Disk 2 impulsed Disk 2 forced	$\begin{bmatrix} 0.0109 & 0.0001 & -0.0002 \\ 0.0001 & 0.0516 & 0.0005 \\ -0.0002 & 0.0005 & 0.0232 \end{bmatrix}$	$\begin{bmatrix} 47668 & -33307 & -787 \\ -33307 & 50892 & -14309 \\ -787 & -14309 & 33156 \end{bmatrix}$	$\begin{bmatrix} 0.325 & 1.155 & 0.150 \\ -0.354 & -0.677 & -0.493 \\ 0.104 & 0.214 & 0.113 \end{bmatrix}$
Disk 2 impulsed Disk 3 forced	$\begin{bmatrix} 0.0131 & -0.0001 & 0.0000 \\ -0.0001 & 0.0608 & 0.0011 \\ 0.0000 & 0.0011 & 0.0263 \end{bmatrix}$	$\begin{bmatrix} 57442 & -40373 & -654 \\ -40373 & 60424 & -16064 \\ -654 & -16064 & 37384 \end{bmatrix}$	$\begin{bmatrix} 0.393 & 1.393 & 0.184 \\ -0.423 & -0.820 & -0.582 \\ 0.118 & 0.250 & 0.124 \end{bmatrix}$
General result: Disk 2 impulsed All disks forced one at a time	$\begin{bmatrix} 0.0119 & 0.0002 & -0.0001 \\ -0.0009 & 0.0515 & 0.0009 \\ -0.0005 & 0.0005 & 0.0264 \end{bmatrix}$	$\begin{bmatrix} 52166 & -38092 & -2230 \\ -36499 & 54087 & -15454 \\ -807 & -13647 & 37869 \end{bmatrix}$	$\begin{bmatrix} 0.356 & 1.265 & 0.165 \\ -0.385 & -0.789 & -0.505 \\ 0.109 & 0.210 & 0.124 \end{bmatrix}$

Fixed-Fixed Torsional System with Added Damping

Viscous damping was added to the center disk of the fixed-fixed torsional system by surrounding that disk with a clay trough and filling the 0.016-in. clearance with highly viscous silicone oil. If the internal damping of the steel is zero in comparison with the damping of the oil, the theoretical damping matrix would show a nonzero value only in the second diagonal term. Although it was not possible to confirm the accuracy of the derived damping matrix, the DDS analysis produced a damping matrix whose second diagonal term was an order of magnitude larger than the rest of the matrix terms.

The mass, stiffness, and damping matrices were estimated using 10 different schemes. First, the free response to impulsing disk 1 was combined with the forced response to forcing disk 1 to generate a set of symmetric matrices. This same free response was also used with the forced responses to forcing disks 2 and 3 to produce two more sets of symmetric matrices. Similarly, the free responses to impulsing disks 2 and 3 were combined with the forced responses to forcing each disk to produce six more sets of symmetric matrices. Finally, the free response to impulsing disk 2 was combined with all three forced responses to generate a set of general, unsymmetric matrices.

In general, the matrix column corresponding to a particular mass was found to be best estimated when the forcing function was applied to that mass; this was especially true in the case of the third disk, which was located at a node of the third mode. The estimated value of a particular lumped mass improved as its distance from the forced mass was reduced, suggesting that the forced mass should be centrally located in the structure for the best results. Obviously, forcing every mass would produce the best result, but the results indicate that useful results can nevertheless be obtained by forcing only one mass.

Although the damping values could not be confirmed, Table 5.13 compares some of the damping matrices that were calculated from this scheme. Without exception, the row two, column two term is about 10 times larger than most of the other terms; these other terms are of the same order of magnitude as the damping terms computed for the fixed-fixed system with no additional damping.

Fixed-Fixed Torisonal System with Measurements Contaminated by Bending Modes

In this test, the fixed-fixed system was struck in a direction parallel to the axes of the accelerometers, so that both the torsional and vertical bending modes were strongly excited and detected by the accelerometers. The bending modes were excited to illustrate the capability of the DDS technique to handle closely spaced modes. An FFT Zoom transform showed that the first vertical bending mode (90.3 Hz) and the first torsional mode (91.5 Hz) produced equal contributions to the accelerometer signals, while in previous tests the former had been maintained at 10–15 dB below the latter. In this test, only the free response from impulsing disk 1 was used, and only the DDS-based modal analysis was performed. The record length was increased to capture enough cycles and system time constants so that only the signal-to-noise ratio (S/N) and frequency

TABLE 5.13 Comparison of Damping Matrices c from the Fixed-Fixed Torsional System Before and After Adding Damping to Disk 2[a]

Condition	No Added Damping	Damping Added to Disk 2
Disk 2 Impulsed Disk 1 Forced	$\begin{bmatrix} 0.361 & 1.274 & 0.173 \\ -0.386 & -0.792 & -0.505 \\ 0.115 & 0.224 & 0.132 \end{bmatrix}$	$\begin{bmatrix} 0.201 & -0.365 & -0.286 \\ 0.351 & 5.161 & 0.239 \\ -0.332 & 0.763 & 0.251 \end{bmatrix}$
Disk 2 Impulsed Disk 2 Forced	$\begin{bmatrix} 0.325 & 1.155 & 0.150 \\ -0.354 & -0.677 & -0.493 \\ 0.104 & 0.214 & 0.113 \end{bmatrix}$	$\begin{bmatrix} 0.209 & -0.517 & -0.312 \\ 0.329 & 4.938 & 0.248 \\ -0.346 & 0.654 & 0.252 \end{bmatrix}$
Disk 2 Impulsed Disk 3 Forced	$\begin{bmatrix} 0.393 & 1.393 & 0.184 \\ -0.423 & -0.820 & -0.582 \\ 0.118 & 0.250 & 0.124 \end{bmatrix}$	$\begin{bmatrix} 0.248 & -0.864 & -0.406 \\ 0.310 & 5.435 & 0.332 \\ -0.345 & 0.685 & 0.252 \end{bmatrix}$

[a] Units: lb-sec-in./(rad).

TABLE 5.14 Data Characteristics and Adequate ARV Model Orders Required to Obtain Estimates of the Modal Parameters for the Fixed-Fixed Torsional System with Strong Bending Modes

Characteristics	Torsional Modes			Bending Modes		
	Mode 1	Mode 2	Mode 3	Mode 1	Mode 2	Mode 3
Natural frequency (Hz)	91.5	201.	355.	90.3	224.	384.
Damping ratio	0.0018	0.0033	0.0044	0.0025	0.0051	0.015
Captured record length						
Time constants	1.1	4.3	10.	1.5	7.4	37.
Cycles of free vibration	94	206	363	92	229	393
S/N Ratio (dB)						
Disk 1	32	24	59	30	34	18
Disk 2	38	27	47	34	20	20
Disk 3	32	38	35	29	34	32
Adequate model order for:						
Frequency estimation	2	4	2	8	6	5
Mode shape estimation	8	7	2	11	7	—
Mode phase estimation	2	2	2	11	6	—
Damping ratio estimation	9	7	2	14	8	8

spacing could influence the autoregressive (AR) model order. Table 5.14 shows the characteristics of this data, and summarizes the ARV model orders needed to obtain stable estimates.

The three torsional modes and the three identified bending modes were readily identified after the ARV model orders had been increased to the point where stable estimates were indicated. For example, the 355 Hz torsional mode was well isolated from the other modes and was represented by adequate S/N, so an ARV(2) model order was adequate to establish all its modal parameters. On the other hand, the 201 and 224 Hz modes were well separated, but because of their low S/N ratios ARV(8) models were required to adequately define their mode shapes and damping ratios. The 91.5 and 90.3 Hz modes had reasonable S/N ratios, but because these frequencies were so close, ARV(9) and ARV(14) models, respectively, were required to define the damping ratios. Lastly, because the antialiasing filter was set at 390 Hz, it required an ARV(8) model to establish a stable estimate of the damping ratio for the 384 Hz mode, while the low S/N ratio made it difficult to establish the mode shape even with an ARV(21) model.

The significant result of this study was that the DDS technique can be used to simultaneously identify several modes, even those with closely spaced frequencies, by simply increasing the model order until the modal frequency and damping ratio estimates stabilize. This confirms that the scalar modeling results of Section 5.5.2 for simulated data readily extend to vector modeling of real data. No attempt was made to verify the bending mode shapes, but the modal parameters associated with the three torsional modes were nearly identical to those obtained from the previous analyses.

Remarks
Without exception, the DDS method displayed better correspondence between tests than the FFT-based method did. This was true of all the modal parameters, but particularly the phase of the mode shapes, which were always estimated correctly by this technique. The FFT-based estimates exhibited large variations attributable to the choice of curve-fitting algorithms and to the degree of experience of the user. In some instances, good FFT-based fits could not be achieved. On the other hand, once an adequate model order was established in the DDS method, the modal parameters were achieved almost automatically. Even an inexperienced user should be able to apply the DDS method and get accurate, consistent results by following good data acquisition principles.

5.6.3 Modal Analysis of a V-6 Engine

To illustrate the method for in situ measurements we give some results from Pandit, Hu, and Schilke (1990). Impulse test data was taken at 14 points on a V-6 engine without removing it from a truck. Three-directional acceleration data was collected using SD 380A 4-Channel digital signal analyzer. Model adequacy was determined by the F test applied to the prediction error of the deterministic part calculated using (5.3.2a). Here we give results of Z-directional measurements at selected points.

Table 5.15 shows the test configuration of four sets of measurements, each containing the common pt20. Using the F test on the prediction error, the adequate ARV model for $p = 4$ series turns out to be ARV(20) for all four sets of measurements. The undamped natural frequencies and damping ratios given by the adequate models are listed in Tables 5.16–5.19. For a real eigenvalue μ_i, its

TABLE 5.15 V-6 Engine Test Measurements

Test	Channel 1	Channel 2	Channel 3	Channel 4
sp2z1	pt20	pt1	pt2	pt5
sp3z1	pt20	pt9	pt3	pt5
sp4z1	pt20	pt9	pt3	pt7
sp6z1	pt20	pt11	pt25	pt7

TABLE 5.16 First 20 Modes of Test sp2z1[a]

Mode	Frequency (Hz)	Damping (%)	Area (%)	MCF (%)
1	76.28	1.607	34.65	99.97
2	13.28	10.07	17.11	96.45
3	68.61	89.21	5.132	0.122
4	72.41	17.68	4.195	92.77
5	15.10	84.92	3.701	55.70
6	82.83	2.385	3.608	97.04
7	32.64	11.86	3.000	87.91
8	38.87	29.64	2.794	69.50
9	74.64	100.0	2.375	8.103
10	96.70	4.105	2.062	95.48
11	15.84	25.47	1.686	77.98
12	176.2	100.0	1.648	0.0
13	59.75	1.466	1.602	49.75
14	87.28	9.291	1.507	91.14
15	27.90	40.32	1.408	70.41
16	55.05	9.617	1.389	98.62
17	2.648	80.47	1.373	86.62
18	242.2	7.452	1.134	56.92
19	60.18	24.09	0.831	78.85
20	79.72	10.43	0.804	77.19

[a] This table shows the first 20 modes (ranked by area) of sp2z1 given by the ARV(20) model with the area ratio and MCF.

absolute value has been listed under frequency and damping listed as 100%. The modal confidence factors (MCF) and the area ratio percentage of the envelope of each mode to the sum of all are calculated by (5.3.20), to help distinguish the physical modes from the computational modes. The area ratio by (5.3.20) was necessary because the MCF could not distinguish between the strong and weak modes. Since the ℓ_i's were normalized to unit length, the area criterion percentage reduces to

$$C_i = \frac{\dfrac{|s_i|}{\sigma_i}}{\sum\limits_{i=1}^{np} \dfrac{|s_i|}{\sigma_i}} 100\%, \qquad i = 1, 2, \ldots, np$$

where s_i is scaling factor of the ith mode, when the modal vectors are normalized to a unit length; σ_i is the real part of the ith characteristic root or eigenvalue; and np is the number of eigenvalues for p series of the ARV(n) model. For a complex conjugate pair of eigenvalues, the two ratios are added together.

TABLE 5.17 First 20 Modes of Test sp3z1[a]

Mode	Frequency (Hz)	Damping (%)	Area (%)	MCF (%)
1	76.29	1.527	20.70	96.91
2	13.10	9.817	15.33	97.77
3	75.29	33.20	8.531	35.95
4	10.40	21.10	5.136	93.30
5	84.15	16.81	4.821	90.99
6	17.62	22.90	4.782	99.45
7	31.45	9.504	4.684	96.98
8	2.779	100.0	3.055	92.98
9	82.15	1.859	2.566	81.57
10	33.60	100.0	2.558	90.90
11	60.14	2.175	2.490	53.27
12	52.66	24.55	2.166	58.04
13	181.0	14.86	1.777	22.97
14	53.57	4.210	1.516	85.99
15	10.42	30.15	1.509	1.959
16	96.60	3.916	1.496	98.82
17	210.7	4.452	1.423	94.24
18	32.23	27.67	1.279	79.22
19	30.60	100.0	1.258	12.40
20	218.2	5.993	1.147	81.61

[a] This table shows the first 20 modes (ranked by area) of sp3z1 given by the ARV(20) model with the area ratio and MCF.

Compared to the system mode, the pseudo mode has three distinguishing characteristics. (1) The amplitude of the pseudo mode is relatively small. (2) If the amplitude is not small, the damping ratio is usually very high. If we look at the time response of each mode, the area under the envelope of the time response of the pseudo mode is small. (3) The amplitudes calculated using different initial values are not consistent—the MCF will be much less than 1 in this case.

The area ratio of each mode can be used to cut off a lot of modes, because when the area ratio is small, the mode is weak compared to other modes. In addition, the MCF may be used to separate the physical modes from the remaining modes.

Out of all the modes estimated, two are common for all the measurement points, their time responses are nearly 50% of the total measurement. The complex mode shapes of these two modes are synthesized through the common reference. Table 5.20 shows how the modes are synthesized. For those points that appear in several sets of data, their mode shapes are averaged.

The idle speed of the tested truck is about 800 rpm, which is about 13.13 Hz; 6 times the idle speed is about 80 Hz. The two significant modes happen to be

TABLE 5.18 First 20 Modes of Test sp4z1[a]

Mode	Frequency (Hz)	Damping (%)	Area (%)	MCF (%)
1	76.89	1.528	23.38	94.03
2	13.13	10.64	14.13	98.54
3	54.46	59.35	6.706	38.06
4	17.22	27.81	5.983	90.18
5	11.92	41.04	4.581	78.75
6	31.41	10.13	4.228	93.61
7	82.20	2.087	3.143	83.40
8	206.7	9.107	2.594	80.25
9	155.5	21.31	2.488	31.69
10	203.8	5.443	2.466	90.75
11	50.22	28.97	2.349	87.22
12	213.2	4.299	2.267	88.12
13	59.00	59.00	3.829	65.95
14	3.772	100.0	1.935	91.79
15	68.98	18.00	1.778	70.84
16	53.36	6.866	1.632	86.28
17	97.81	3.120	1.594	83.73
18	148.2	2.344	1.501	99.36
19	173.3	5.171	1.372	93.26
20	2.848	100.0	1.285	94.82

[a] This table shows the first 20 modes (ranked by area) of sp4z1 given by the ARV(20) model with the area ratio and MCF.

ILLUSTRATIVE EXAMPLES—VECTOR MODELS   281

TABLE 5.19 First 20 Modes of Test sp6z1[a]

Mode	Frequency (Hz)	Damping (%)	Area (%)	MCF (%)
1	76.04	1.586	28.89	99.09
2	53.46	100.0	14.19	0.090
3	48.79	100.0	14.17	0.192
4	13.87	11.70	9.187	92.35
5	66.23	54.19	5.407	9.959
6	82.46	1.762	2.388	90.39
7	182.4	14.69	2.254	33.66
8	96.81	5.084	2.000	88.82
9	31.85	8.511	1.950	76.13
10	9.548	28.88	1.613	94.29
11	77.71	7.878	1.579	94.45
12	79.66	31.11	1.524	16.14
13	61.62	18.61	1.321	60.61
14	145.2	24.98	1.209	2.179
15	181.8	8.022	1.185	78.78
16	24.77	59.62	0.993	51.26
17	39.57	28.71	0.850	93.26
18	146.2	7.596	0.844	71.50
19	201.9	6.387	0.680	95.50
20	198.8	2.692	0.620	97.42

[a] This table shows the first 20 modes (ranked by area) of sp6zl given by the ARV(20) model with the area ratio and MCF.

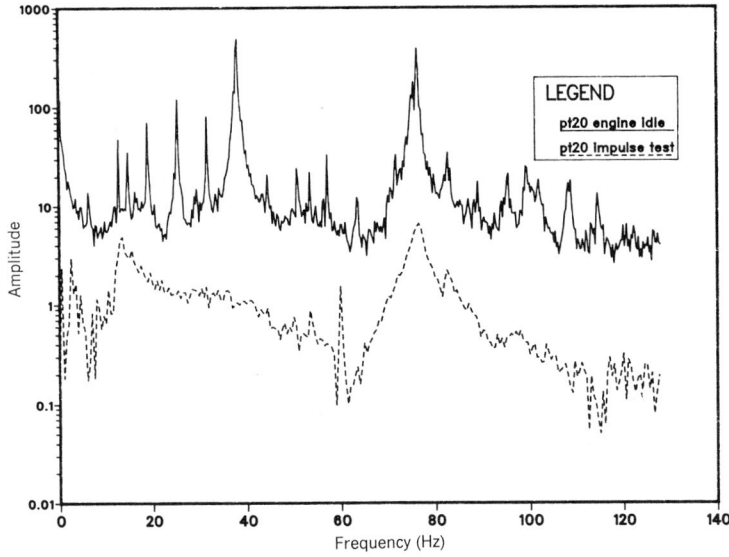

Figure 5.20 Measured (FFT) frequency response for impulse and idle tests.

TABLE 5.20 The Synthesized Mode Shapes for the First Two Modes[a]

	First Mode					Second Mode				
	sp2z1	sp3z1	sp4z1	sp6z1	Synthesized	sp2z1	sp3z1	sp4z1	sp6z1	Synthesized
Frequency (Hz)	13.29	13.10	13.14	13.87	13.35	76.28	76.29	76.19	76.04	76.20
Damping (%)	10.07	9.82	10.64	11.71	10.56	1.61	1.53	1.53	1.59	1.57
			Mode Shape: Amplitude (Phase°)							
pt20	1 (0)	1 (0)	1 (0)	1 (0)	1 (0)	1 (0)	1 (0)	1 (0)	1 (0)	1 (0)
pt1	1.42 (−4.3)				1.42 (−4.3)	1.70 (0.3)				1.70 (0.3)
pt2	1.04 (6.63)				1.04 (6.63)	0.72 (2.6)				0.72 (2.6)
pt3		0.22 (87.3)	0.35 (67.4)		0.29 (77.4)		0.18 (3.2)	0.18 (3.18)		0.18 (3.19)
pt5	0.66 (−7.5)	0.70 (0.9)			0.68 (−3.3)	1.21 (−2.86)	1.19 (−4.15)			1.20 (−3.51)
pt7			0.95 (−4.37)	0.99 (−3.32)	0.97 (−3.85)			0.95 (−4.37)	0.99 (−3.32)	0.97 (−3.85)
pt9		0.64 (160.8)	0.59 (146.3)		0.62 (153.6)		0.05 (296)	0.05 (−67.9)		0.05 (115.0)
pt11				0.95 (11.6)	0.95 (11.6)				0.57 (135.0)	0.57 (135.0)
pt25				0.74 (7.13)	0.74 (7.13)				3.05 (180.2)	3.05 (180.2)

[a] The table shows how the overall mode shapes are synthesized from the four sets of modeling result. The numbers in parentheses are the phase difference (degree) relative to the reference.

very close to the idle RPM and 6 times the RPM. For a V-6 engine, this could cause serious problems. The vibration of the engine at idle was also tested, and the FFT amplitudes of acceleration response of pt20 from the hammer test and the operating test shown in Figure 5.20 proved this point.

6

SPECTRUM ANALYSIS

Although the modeling procedure outlined in Chapter 5 was quite general, its application in that chapter was limited to situations where the deterministic part is of primary interest. Transient response, impulse response, step response, and sinusoidal response are of this variety; the stochastic or random parts of these data are only a nuisance. Fourier analysis, which is based on the assumption of *fixed* amplitudes, frequencies, and phases useful for deterministic signals, breaks down when applied to such noisy data, because the noise or stochastic part is characterized by *random* changes of amplitudes, frequencies, and phases. An adequate model obtained by the DDS approach that does not require such an assumption was shown to be capable of modeling both the deterministic and the stochastic part and of providing a good estimate of the former by naturally separating the latter.

There are, however, applications where the stochastic part, or noise itself is of interest. Forging impact noise is an example illustrating such an application that will be used later; characterization of such noise is sought not only to determine its frequency content but also to locate the source of the various frequency components, so that they can be suppressed to reduce noise levels. Modal testing of structures by random or pseudo-random excitation is another example; such excitation is useful because the structure is thereby subjected to a wide range of frequencies to "map" the frequency range.

Since Fourier analysis based on amplitudes breaks down for such noise or random process (see also Section 2.9.8), it is essential to find a *deterministic* function that completely characterizes the random process and is also amenable to Fourier analysis. Autocovariance function, which exists only for a stationary stochastic process, provides such a characterization; its Fourier transform is called autospectrum. For multiple series, cross-covariance and cross-spectrum may be defined in addition, to provide a relation between the series. Autocovariance at lag zero is the variance, and the autospectrum provides a frequency decomposition of the variance or expected absolute-value-squared amplitude, also called power by analogy with electrical signals. Hence the autospectrum is also known as power spectral density (PSD). This is called spectrum analysis and will be the subject of this chapter.

Note that random amplitudes are averaged over infinite record length and the phases of individual components are canceled when expectation of the absolute value squared is used in defining the autospectrum or power spectrum over a continuous range of frequencies. Since this is an analytical operation on a random process, just as Fourier series and transform are analytical operations on deterministic functions, it is best done on a mathematical model or function adequately fitted to the data, rather than on the data itself. As the DDS approach provides such models from the data, it should be a relatively simple matter to conduct spectrum analysis using these models. This is indeed true, as will be shown in this chapter. In fact all the information generally gleaned from spectrum plots (with considerable difficulty) and much more not directly available from such plots, can be readily obtained from the models, just as the information generally obtained from FRF plots was shown to be available in a concise and precise form from the models in the last chapter. Nevertheless, graphical presentation of the spectrum and FRF plots does have a visual appeal and therefore analytical expressions for these functions to provide clear, smooth plots will be given in this chapter. However, it is the modal decomposition of the model based on the spectral decomposition of the state matrix that provides the most comprehensive spectrum analysis in this chapter.

An alternative approach, which may be called empirical spectral analysis, is to obtain spectra at frequencies predetermined by the number of data points, using a computationally efficient algorithm such as FFT to transform the data in frequency domain, and to average their conjugate product, [e.g., Bendat and Piersol 1980]. For large record lengths, this is equivalent to using the FFT of the sample auto- or cross-covariance. Since sample autocovariance is itself a random process, and, moreover, is a very poor estimator of the theoretical autocovariance, such a procedure has all the problems of the deterministic signal, such as leakage and aliasing, together with those of random signals such as variance and bias, as discussed in Section 2.9. A further variant of this approach is to assume a model with fixed orders to have generated the data, guess the order by some ad hoc criteria, and fit the model to the data by trial and error. (It is to avoid such trial and error that the DDS approach does not assume a fixed order, but chooses the best possible representation allowed by the data from a predetermined sequence of models.) If the model is adequate, less-biased spectral estimators of higher resolution will result. We will not discuss these trial-and-error procedures, which work very well with low-order simulated data, but not so well with high-order real data. There is a vast literature on such empirical spectral estimation; the interested reader is referred to the recent surveys by Kay (1988) and Marple (1987).

286 SPECTRUM ANALYSIS

In Section 6.1 we use a simple first-order model to explain why a superior estimate of spectrum is obtained from the model compared to that from the data, and illustrate the explanation by a second-order model. Modal decomposition of the spectrum of a higher-order scalar ARMA model is introduced. Analytical expressions for the frequency response function or transfer function directly obtained from an ARMAV model are given in Section 6.2 together with their modal decomposition. This section shows that not only feedforward but also feedback paths can be identified by the DDS methodology. Spectral expressions for multiple-input–multiple-output systems are presented in Section 6.3 with two methods of their interpretation: one based on residual spectrum and the other on the variance–covariance of the white noise source. Sections 6.4 and 6.5 present illustrative applications to machine tool dynamics and forgehammer noise source identification together with illustrative examples of modeling and spectrum analysis based on experimentally generated data.

6.1 THE UNIVARIATE SPECTRUM

Recall from Chapter 2 that a zero mean stationary stochastic process X_t has a finite variance and its autocovariance is defined as (see Appendix A for review)

$$\gamma_k^{xx} = E(X_t X_{t-k}) \tag{6.1.1}$$

so that the variance, which is the covariance at zero lag, is given by

$$\gamma_0^{xx} = E(X_t^2) \tag{6.1.2}$$

Since the expectation operator E can be thought of as an average over infinite record, the variance can be considered to be the "power" of the varying signal X_t. The spectrum, or autospectrum $S_{xx}(\omega)$ forms a Fourier transform pair with autocovariance, so that if Δ denotes the sampling interval ($j = \sqrt{-1}$)

$$S_{xx}(\omega) = \frac{\Delta}{2\pi} \sum_{k=-\infty}^{\infty} \gamma_k^{xx} e^{-j\omega k \Delta}, \quad -\frac{\pi}{\Delta} \leq \omega \leq \frac{\pi}{\Delta} \tag{6.1.3}$$

$$\gamma_k^{xx} = \int_{-\pi/\Delta}^{\pi/\Delta} S_{xx}(\omega) e^{j\omega k \Delta} d\omega, \quad k = 1, \pm 1, \pm 2, \ldots \tag{6.1.4}$$

and in particular, for $k = 0$ we have the variance

$$\gamma_0^{xx} = \int_{-\pi/\Delta}^{\pi/\Delta} S_{xx}(\omega) d\omega \tag{6.1.5}$$

showing that the area under the autospectrum equals the power or variance. Hence the name power spectral density (PSD) for $S_{xx}(\omega)$, which shows how the variance or power is distributed over frequencies. Although we will use ω in radians per unit time, it can be converted to cycles per unit time (in particular Hz, which is cycles per second) by using $\omega = 2\pi f$ to get *formally*

$$S(f) = 2\pi S(\omega), \quad \omega = 2\pi f, \quad -\frac{1}{2\Delta} \leq f \leq \frac{1}{2\Delta} \qquad (6.1.3a)$$

It is sometimes useful (particularly in plotting) to normalize the autocovariance and autospectrum by dividing by the variance γ_0; they are then called the autocorrelation function ρ_k and the spectral density function and form a Fourier transform pair

$$\rho_k = \frac{\gamma_k^{xx}}{\gamma_0^{xx}} = \int_{-\pi/\Delta}^{\pi/\Delta} \frac{S_{xx}(\omega)}{\gamma_0^{xx}} e^{j\omega k \Delta} d\omega \qquad (6.1.6)$$

In particular,

$$\rho_0 = 1 = \int_{-\pi/\Delta}^{\pi/\Delta} \frac{S_{xx}(\omega)}{\gamma_0^{xx}} d\omega \qquad (6.1.7)$$

showing that the spectral density function $S_{xx}(\omega)/\gamma_0^{xx}$ is indeed a density function.

6.1.1 White Noise

The simplest stochastic process is the uncorrelated white noise

$$X_t = a_t, \qquad E(a_t) = 0 \qquad (6.1.8)$$

For the white noise process

$$\begin{aligned}\gamma_k^{aa} = E a_t a_{t-k} &= \delta_k \sigma_a^2 \qquad (6.1.9)\\ &= \sigma_a^2, \quad k = 0 \\ &= 0, \quad k \neq 0\end{aligned}$$

Therefore, its autospectrum is, by (6.1.3),

$$S_{aa}(\omega) = \frac{\Delta \sigma_a^2}{2\pi}, \quad -\frac{\pi}{\Delta} \leq \omega \leq \frac{\pi}{\Delta} \qquad (6.1.10)$$

which is flat over the frequency range, hence the name white noise, in analogy with the optical spectrum of white light, in which no particular frequency coresponding to a specific color dominates.

6.1.2 Spectrum Versus Sample Spectrum

One of the reasons for the popularity of empirical spectrum analysis is that an estimate of the spectrum $S(\omega)$, called sample spectrum $\hat{S}(\omega)$, can be readily computed without the trouble of fitting a model. This can be done by first defining a sample autocovariance function which replaces the infinite averaging operation of expectation in (6.1.1) by a finite averaging operation:

$$\hat{\gamma}_k^{xx} = \frac{1}{N} \sum_{t=k+1}^{N} (X_t - \bar{X})(X_{t-k} - \bar{X}), \qquad \bar{X} = \frac{1}{N} \sum_{t=1}^{N} X_t \qquad (6.1.11)$$

where the average $\bar{X}$ is subtracted because the data may not have a zero mean. A Fourier transform of this sample autocovariance then gives the sample spectrum. Unfortunately, there are three basic problems associated with this approach, and that is why we do not adopt it in this book. The finite number of observations used in computing the sample autocovariance (6.1.11) allows us to use $\hat{\gamma}_k$ for only $k = 0, 1, \ldots, M \ll N$. Thus the sample spectrum calculated using this truncated version $\hat{\gamma}_k$ in (6.1.3) can be highly biased, particularly when γ_k is significantly large for $k > M$. Secondly, $\hat{\gamma}_k$ is a very poor estimator of γ_k, as noted early in the time series literature by Kendall (1945). Finally, Fourier transformation is not an ergodic operation and destroys the consistency of $\hat{\gamma}_k$ when transformed to $\hat{S}(\omega)$; smoothing windows used to alleviate this problem introduce their bias, as discussed in Section 2.9.8. Moroever, deterministic parts causing trends in the data need to be removed before applying such a procedure; since these trends are not known a priori, their removal by trial and error may itself introdue other distortions.

To illustrate the root cause of these problems we reproduce Figures 6.1 and 6.2 from Jenkins and Watts (1968), showing the sample spectra for $N = 50$ and 100 of normal (discrete) white noise and of the data from an AR(2) process to be discussed shortly. The sample spectrum is so erratic that it is difficult to conjecture the underlying (theoretical) spectrum from it. Just like the AR(2) case illustrated in Figure 6.2, it is not difficult to see that sample spectrum of data from any ARMA model would behave much the same way, as these models can be expressed as linear combinations of white noise (a_t).

6.1.3 The AR(1) Model

For a stable/stationary AR(1) model with $|\phi_1| < 1$:

$$X_t = \phi_1 X_{t-1} + a_t \qquad (6.1.12)$$

we know from Chapter 3 that the Green's function is $G_j = \phi_1^j$ and hence X_t can be expressed as an infinite linear combination of white noise sequence a_t:

$$X_t = \sum_{j=0}^{\infty} G_j a_{t-j} = \sum_{j=0}^{\infty} \phi_1^j a_{t-j} \qquad (6.1.13)$$

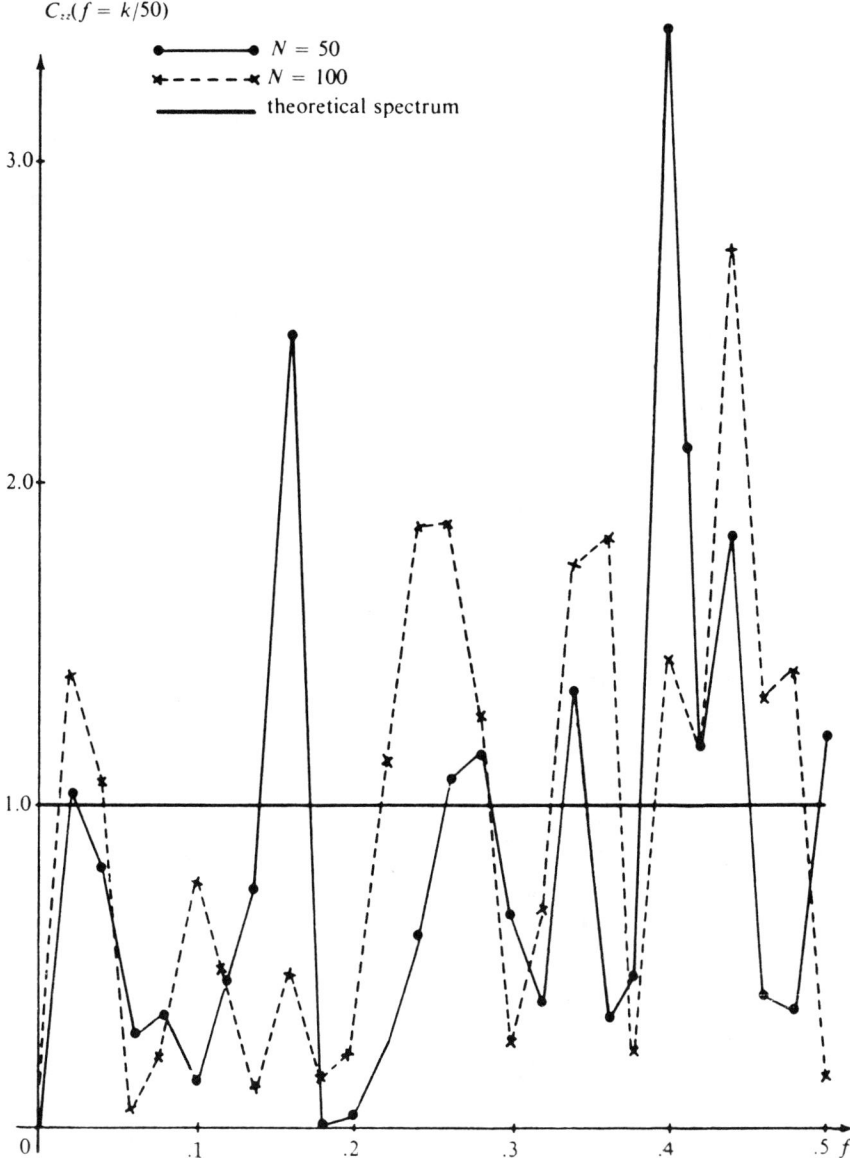

Figure 6.1 Sample spectra for the first half ($N = 50$) and the whole ($N = 100$) of a realization of discrete normal white noise. [Reproduced from Jenkins and Watts (1968) with permission.]

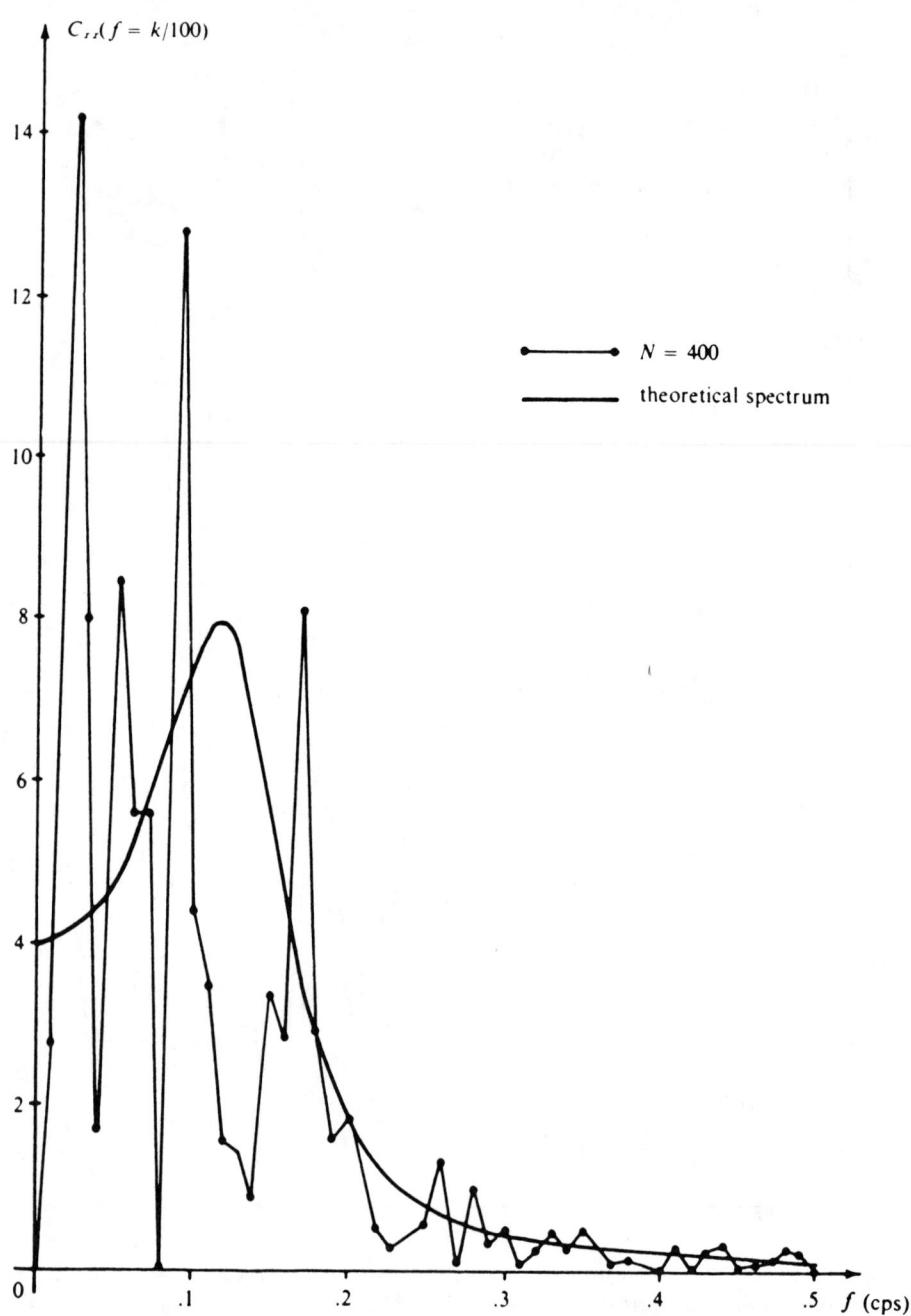

Figure 6.2 Sample spectrum for a realization of a second-order autoregressive process. [Reproduced from Jenkins and Watts (1968) with permission.]

Its autocovariance function is, using (6.1.1) and (6.1.9),

$$\gamma_k^{xx} = \gamma_{(-k)}^{xx} = \sigma_a^2 \sum_{j=0}^{\infty} G_j G_{j+k} = \frac{\sigma_a^2}{(1-\phi_1^2)} \phi_1^k \qquad (6.1.14)$$

Taking the Fourier transform defined by (6.1.3) gives the spectrum

$$S_{xx}(\omega) = \frac{\sigma_a^2 \Delta}{2\pi(1-\phi_1^2)} \left\{ -1 + \sum_{k=0}^{\infty} \phi_1^k [(e^{-j\omega\Delta})^k + (e^{j\omega\Delta})^k] \right\}$$

$$= \frac{\sigma_a^2 \Delta}{2\pi} \cdot \frac{1}{(1-\phi_1 e^{-j\omega\Delta})(1-\phi_1 e^{j\omega\Delta})} \qquad (6.1.15a)$$

$$= \frac{\Delta}{2\pi}(1-\phi_1 e^{-j\omega\Delta})^{-1} \sigma_a^2 (1-\phi_1 e^{j\omega\Delta})^{-1} \qquad (6.1.15b)$$

$$= \frac{\sigma_a^2 \Delta}{2\pi |1-\phi_1 e^{-j\omega\Delta}|^2} \qquad (6.1.15)$$

where we have given the form (6.1.15b) because it readily generalizes to other scalar as well as vector cases. The proof given above emphasizes why the stationarity condition $|\phi_1| < 1$ is essential to defining γ_k and $S(\omega)$.

These results can be expressed formally and more coveniently by the concept of a transfer function, which is essentially a steady-state concept that now applies because of the stationarity restriction. Using the backshift operator B to denote $BX_t = X_{t-1}$, or $B^j X_t = X_{t-j}$, the AR(1) model can be written as

$$X_t - \phi_1 X_{t-1} = a_t$$

$$(1-\phi_1 B)X_t = \phi(B)X_t = a_t$$

$$X_t = \frac{1}{(1-\phi_1 B)} a_t = \phi^{-1}(B) a_t \qquad (6.1.16)$$

Therefore, if a_t is considered as the input and X_t as the output, then $\phi^{-1}(B) = (1-\phi_1 B)^{-1}$ can be considered as the transfer function. Note that this transfer function is fictitious in that a_t is our own creation from modeling X_t, which is the only data available. Moreover, being stochastic processes, neither X_t nor a_t have Fourier transforms.

However, if a_t were periodic, say, $ae^{j\omega\Delta t}$, then X_t would be periodic, say $Xe^{j\omega\Delta t}$, and now the transfer function between input a_t and output X_t can be derived as

$$Xe^{j\omega\Delta t} - \phi_1 Xe^{j\omega\Delta(t-1)} = ae^{j\omega\Delta t}$$

or the transfer function

$$\frac{X}{a} = (1-\phi_1 e^{-j\omega\Delta})^{-1} \qquad (6.1.17)$$

$$= \phi^{-1}(B)|_{B=e^{-j\omega\Delta}} \qquad (6.1.18)$$

292 SPECTRUM ANALYSIS

This is also clear because for a periodic,

$$X_t = Xe^{j\omega\Delta t}, \quad BX_t = X_{t-1} = e^{-j\omega\Delta}X_t$$

Using this formal transfer function and (6.1.15) (or using the z transform of linear system theory, $B = z^{-1}$) the spectrum $S_{xx}(\omega)$ can be written as

$$S_{xx}(\omega) = \frac{\Delta}{2\pi}\phi^{-1}(\omega)\sigma_a^2\bar{\phi}^{-1}(\omega) \tag{6.1.19}$$

where

$$\phi(\omega) = \phi(B)|_{B = e^{-j\omega\Delta}}$$

Expressions (6.1.15) and (6.1.19) for the autospectrum of an AR(1) model show how the flat or constant spectrum $\sigma_a^2\Delta/2\pi$ of the patternless white noise a_t is altered by the transfer function into a spectrum of colored noise X_t, with the pattern of maximum at $\omega = 0$ and decaying at higher frequencies. These expressions also explain why a plot of the spectrum from a fitted model would be smooth, whereas that of the sample spectrum would not. The plot of the spectrum is essentially that of the analytic function $|\phi^{-1}(\omega)|^2$ scaled by $\hat{\sigma}_a^2\Delta/2\pi$, in which the erratic variation of the sampled white noise has been eliminated. On the other hand, a sample spectrum would be a plot of the sampled white noise spectrum in Figure 6.1 modified by an unknown transfer function. Thus the erratic variations of the sampled white noise spectrum are superimposed on the sample spectrum, but eliminated in the spectrum from a model. This is true for any ARMA model, as we now show.

6.1.4 The ARMA(2,1) Model

The ARMA(2,1) model

$$X_t = \phi_1 X_{t-1} + \phi_2 X_{t-2} + a_t - \theta_1 a_{t-1} \tag{6.1.20a}$$

may be written in the transfer function form

$$X_t = \frac{(1 - \theta_1 B)}{(1 - \phi_1 B - \phi_2 B^2)}a_t = \phi^{-1}(B)\theta(B)a_t \tag{6.1.20}$$

where

$$\phi(B) = (1 - \phi_1 B - \phi_2 B^2) = (1 - \lambda_1 B)(1 - \lambda_2 B), \quad \theta(B) = (1 - \theta_1 B)$$
$$\tag{6.1.20b}$$

Since its Green's function was shown in Chapter 3 to be of the form

$$G_j = g_1\lambda_1^j + g_2\lambda_2^j, \quad g_1 = \frac{(\lambda_1 - \theta_1)}{(\lambda_1 - \lambda_2)}, \quad g_2 = \frac{(\lambda_2 - \theta_1)}{(\lambda_2 - \lambda_1)} \tag{6.1.21}$$

for distinct roots, the autocovariance function is

$$\gamma_k^{xx} = \sigma_a^2 \sum_{j=0}^{\infty} G_j G_{j+k} = d_1 \lambda_1^k + d_2 \lambda_2^k \qquad (6.1.22)$$

where

$$d_1 = \left[\frac{g_1^2}{1-\lambda_1^2} + \frac{g_1 g_2}{1-\lambda_1 \lambda_2}\right] \sigma_a^2$$

$$d_2 = \left[\frac{g_2^2}{1-\lambda_2^2} + \frac{g_2 g_1}{1-\lambda_2 \lambda_1}\right] \sigma_a^2$$

and the variance

$$\gamma_0 = d_1 + d_2 \qquad (6.1.22a)$$

provided the stationarity conditions $|\lambda_i| < 1$, $i = 1, 2$ are satisfied.

Since the autocovariance function has the form of a sum of two first-order autocovariance functions, we can use the method of explicit Fourier transformation of γ_k used earlier and also the method via transfer function to write two alternate expressions for the spectrum

$$S_{xx}(\omega) = \frac{\Delta}{2\pi} \phi^{-1}(\omega)\theta(\omega)\sigma_a^2 \bar{\phi}^{-1}(\omega)\bar{\theta}(\omega) \qquad (6.1.23a)$$

$$= \frac{\sigma_a^2 \Delta}{2\pi} |\phi^{-1}(\omega)\theta(\omega)|^2 \qquad (6.1.23b)$$

$$= \frac{\Delta}{2\pi}[d_1(1-\lambda_1^2)|1-\lambda_1 e^{-j\omega\Delta}|^{-2} + d_2(1-\lambda_2^2)|1-\lambda_2 e^{-j\omega\Delta}|^{-2}] \quad (6.1.24)$$

As before (6.1.23a and b) provide smooth plots of the spectrum, as illustrated by the theoretical spectrum in Figure 6.2 in which the AR(2) model has $\theta_1 = 0$. The sample spectrum provides only a distorted picture of this spectrum.

What is more important is that (6.1.24) provides a modal decomposition of the spectrum. This is not available from the sample spectrum at all. For the ARMA(2,1) model, this modal decomposition is not very useful when the roots are complex (conjugate) because each component spectrum in (6.1.24) is also complex. But for real roots, this modal decomposition expresses the spectrum of a second-order system as a sum of two first-order systems, each characterized by the dynamic mode λ_i^t, $i = 1, 2$. Moreover d_1 is the component of the variance or power from mode 1 and d_2 from mode 2. As the reader might have surmised, such modal decomposition of the spectrum would be even more useful for higher-order scalar or vector models, as it would decompose a complex spectrum into simple ones owing to a first-order system with real roots or a second-order system with a complex conjugate pair of roots. This is indeed true, as we will now show.

Note that the autocovariance function is oscillatory whenever λ_1, λ_2 are complex, that is, when

$$\phi_1^2 + 4\phi_2 < 0 \qquad (6.1.25)$$

But this oscillatory behavior does not always show up as a peak in the spectrum. For example, (6.1.20) and (6.1.23a) show that for an AR(2) model with $\theta_1 = 0$, the spectrum has a maximum or minimum (i.e., a peak or a trough) only when

$$|\phi_1(1 - \phi_2)| < |4\phi_2| \qquad (6.1.26)$$

and this happens at a frequency ω_m given by

$$\cos(\omega_m \Delta) = -\frac{\phi_1(1 - \phi_2)}{4\phi_2} \qquad (6.1.27)$$

6.1.5 The ARMA(n,m) Model

The transfer function form of an ARMA(n,m) model

$$X_t - \phi_1 X_{t-1} - \cdots - \phi_n X_{t-n} = a_t - \theta_1 a_{t-1} - \cdots - \theta_m a_{t-m} \qquad (6.1.28a)$$

is

$$X_t = \phi^{-1}(B)\theta(B)a_t \qquad (6.1.28)$$

where

$$\phi(B) = (1 - \phi_1 B - \phi_2 B^2 - \cdots - \phi_n B^n)$$
$$= (1 - \lambda_1 B)(1 - \lambda_2 B) \cdots (1 - \lambda_n B) \qquad (6.1.28b)$$
$$\theta(B) = (1 - \theta_1 B - \cdots - \theta_m B^m) \qquad (6.1.28c)$$

For $m < n$, the Green's function has the form, for distinct roots,

$$G_j = g_1 \lambda_1^j + g_2 \lambda_2^j + \cdots + g_n \lambda_n^j \qquad (6.1.29)$$

$$g_i = (\lambda_i^{n-1} - \theta_1 \lambda_i^{n-2} - \cdots - \theta_m \lambda_i^{n-m-1}) \bigg/ \prod_{\substack{k=1 \\ k \neq i}}^{n} (\lambda_i - \lambda_k)$$

with the autocovariance function

$$\gamma_k^{xx} = \sigma_a^2 \sum_{j=0}^{\infty} G_j G_{j+k} = d_1 \lambda_1^k + d_2 \lambda_2^k + \cdots + d_n \lambda_n^k \qquad (6.1.30)$$

where

$$d_i = \left(\frac{g_i g_1}{1 - \lambda_i \lambda_1} + \frac{g_i g_2}{1 - \lambda_i \lambda_2} + \cdots + \frac{g_i g_n}{1 - \lambda_i \lambda_n} \right) \sigma_a^2, \quad i = 1, 2, \ldots, n$$

$$(6.1.30a)$$

and the variance

$$\gamma_0 = d_1 + d_2 + \cdots + d_n \tag{6.1.30b}$$

provided the stationarity conditions $|\lambda_i| < 1, i = 1, 2, \ldots, n$ are satisfied. The autospectrum has now the alternate transfer function and the explicit or modal decomposition forms

$$S_{xx}(\omega) = \frac{\Delta}{2\pi} \phi^{-1}(\omega)\theta(\omega)\sigma_a^2 \bar{\phi}^{-1}(\omega)\bar{\theta}(\omega) \tag{6.1.31a}$$

$$= \frac{\sigma_a^2 \Delta}{2\pi} |\phi^{-1}(\omega)\theta(\omega)|^2 \tag{6.1.31b}$$

$$= \frac{\Delta}{2\pi} [d_1(1 - \lambda_1^2)|1 - \lambda_1 e^{-j\omega\Delta}|^{-2} + d_2(1 - \lambda_2^2)|1 - \lambda_2 e^{-j\omega\Delta}|^{-2}$$
$$+ \cdots + d_n(1 - \lambda_n^2)|1 - \lambda_n e^{-j\omega\Delta}|^{-2}] \tag{6.1.32}$$

If λ_i is real, so are g_i and d_i; if λ_i, λ_{i+1} form a complex conjugate pair, so do g_i, g_{i+1} and d_i, d_{i+1}. Thus the basic modal decomposition of the Green's function carries into the autocovariance function and its Fourier transform the autospectrum. A higher-order scalar system and all its key characteristics are decomposed into first- or second-order components. In particular, the variance or power of the signal is split into components d_i, owing to a real root λ_i representing a first-order system, and $(d_i + d_{i+1})$, resulting from a complex conjugate pair λ_i, λ_{i+1} representing a second-order system. Note that unlike the Parseval relation of Fourier transform which restricts a similar decomposition to positive powers only, the d_i's or $(d_i + d_{i+1})$ are allowed to be negative and hence can represent a variance- or power-enhancing as well as suppressing effect of the modes.

In addition to these quantitative indicators, the analytic function (6.1.31b) for the autospectrum provides clear and smooth plots showing a peak at zero frequency corresponding to first-order modes and other peaks at frequencies corresponding to second-order modes whenever their damping is small enough. Moreover, a component by component plot of (6.1.32), using a real λ_i or complex conjugate pairs of λ_i, λ_{i+1} together at a time, clearly shows how that overall spectrum is built by combing the first- and second-order component system spectra. Each component has a clear and smooth appearance, similar to the *theoretical* spectrum in Figure 6.2.

The other advantage of the analytic expressions (6.1.31a and b) as the ratio polynomials in ω is that they can be integrated precisely by a computationally efficient method to yield the variance γ_0 of the fitted model. Unlike (6.1.30a and b), this procedure does not require the computation of λ_i and g_i, but follows directly from σ_a^2, ϕ_i, and θ_i of the fitted model. The procedure readily generalizes, and becomes even more computationally advantageous, for vector models discussed later in Section 6.3. Recall from Section 5.2.1 on modeling that the closeness of the data variance with that predicted by the model provides a good indicator of model adequacy. The variance γ_0 of the fitted model needed for this comparison can be rapidly computed by integrating (6.1.31a and b) before reaching the adequate model; the modal decomposition (6.1.30b) of the variance is usually of interest only for the adequate model. Although we have presented λ_i's as the roots of the scalar autoregressive polynomial $\phi(B)$ in (6.1.28b), recall from Chapters 2 and 3 that they represent the eigenvalues of the (companion) state matrix ϕ. Thus the modal decomposition of the autocovariance, variance, and spectrum is a simple consequence of the spectral decomposition of the state matrix. It would therefore readily extend to the vector case, as we will show in Section 6.3.

It should also be emphasized, as clearly shown in the derivation given above, that although the basic modal decomposition of the Green's function is applicable to general, possibly unstable/nonstationary systems with the presence of deterministic components, it extends to the autocovariance function and to the spectrum only under the stationarity restriction, when such deterministic components have been removed. For this reason the former has been emphasized in this book and the latter only briefly discussed in this last chapter for the sake of completeness. The development given above also demonstrates how easily the latter follows from the former.

The plot of the spectrum given by (6.1.31a) with the plots of its first- and second-order components in (6.1.32) adding up to the whole spectrum does have its graphical appeal. However, in this modern era of computer-aided graphics, it is not difficult to show a plot of the original data with the plots of its first- and second-order components, constructed using the modal decomposition of the Green's function and the residuals, adding up to the original data. Such a graphical version of the modal decomposition has been illustrated in Pandit and Weber (1990) for machine vision applications, where the modal decomposition can literally be "visualized."

6.2 FREQUENCY RESPONSE FUNCTION (FRF) OR TRANSFER FUNCTION

Before presenting the multivariate generalization of the univariate spectrum analysis discussed in the last section, we will present a characterization of a stationary system which is often the aim of spectrum analysis, namely the frequency transfer function or the frequency response function. The name frequency transfer function is common in control literature to connote the ratio of output to input, whereas the name frequency response function (FRF) is common in vibration and modal testing practice to connote the ratio of response to excitation in the frequency domain.

For a system with deterministic data, therefore, it can be obtained simply by taking the ratio of the Fourier transform of the output response with that of the input excitation. For a system with random or stochastic data, often obtained by testing the system with random input excitation, the Fourier transform does not exist. However, for stable stationary systems, the spectra exist. In addition to the autospectrum for each signal defined earlier, now, since there are two different signals, their cross-covariance function and its Fourier transform cross-spectrum exist. (These will be formally defined in the next section, because we do not need them to obtain the FRF in this section.) Denoting the autospectra of the output response X and input excitation F by S_{XX}, S_{FF}, respectively, their cross-spectra by S_{XF}, S_{FX}, and the FRF or transfer function by H, it is easy to see from the preceding development that

$$S_{XX}(\omega) = |H(\omega)|^2 S_{FF}(\omega)$$
$$S_{XF}(\omega) = H(\omega) S_{FF}(\omega)$$
$$S_{XX}(\omega) = H(\omega) S_{FX}(\omega)$$

Hence, theoretically, two alternate expressions for the transfer function are given by

$$H_1(\omega) = S_{XF}(\omega)/S_{FF}(\omega)$$
$$H_2(\omega) = S_{XX}(\omega)/S_{FX}(\omega)$$

It is therefore a common practice to estimate the transfer function from the ratios of *sample* auto- and cross-spectra with the help of these expressions and to test how good these estimates are by the squared coherence function

$$\kappa^2(\omega) = \frac{H_1(\omega)}{H_2(\omega)}$$

For a good estimate it would be close to its theoretical value 1, for poor estimates it would be much less than one. Unfortunately, the sample cross-spectra inherit all the problems of sample auto-spectra discussed earlier, and hence this approach requires the same amount of ad hoc methods and trial-and-error procedures.

Although the transfer function between X_t and a_t obtained in the preceding section was fictitious, it is clear that a similar transfer function between output X_{2t} and input X_{1t} obtained from a bivariate ARMAV model would be a genuine transfer function or FRF. Such a transfer function in the B operator form ($B = z^{-1}$ for those familiar with z transform) was already demonstrated in Section 4.4 for the bivariate ARV(1) model. Changing B to $e^{-j\omega\Delta}$, demonstrated in Section 6.1, should yield the FRF or transfer function directly from the model; there is no need to use auto- and cross-spectra.

Expressions for obtaining the FRF directly from the fitted models are developed in this section for the ARMAV models. To keep the development simple, diagonal moving average matrices θ_i will be used so that an a_{it} and its lagged term a_{it-k} are present only in the model for the series X_{it} and not in others. Actually, as illustrated in Sections 6.4.2 and 6.4.4, upper triangular θ_i can be readily justified and merely adds additional noise terms in the input model(s), with which we are not concerned when considering the input-to-output transfer functions. For vibration systems, since ARV models suffice, this restriction has no effect. We will first discuss the single-input–single-output case and then generalize to multiple-input–single-output and multiple-input–multiple-output cases.

6.2.1 Single-Input–Single-Output (SISO) System

Recall from Chapters 2 and 4 that a bivariate ARMAV(n,m) model with diagonal θ_i matrices may be written as

$$X_{1t} - \phi_{111}X_{1t-1} - \cdots - \phi_{11n}X_{1t-n} = \phi_{121}X_{2t-1} + \cdots + \phi_{12n}X_{2t-n}$$
$$+ a_{1t} - \theta_{111}a_{1t-1}$$
$$- \cdots - \theta_{11m}a_{1t-m} \quad (6.2.1a)$$

$$X_{2t} - \phi_{221}X_{2t-1} - \cdots - \phi_{22n}X_{2t-n} = \phi_{211}X_{1t-1} + \cdots + \phi_{21n}X_{1t-n}$$
$$+ a_{2t} - \theta_{221}a_{2t-1} - \cdots - \theta_{22m}a_{2t-m}$$
$$(6.2.2a)$$

The transfer-function forms of these models are

$$X_{1t} = -\phi_{11}^{-1}(B)\phi_{12}(B)X_{2t} + \phi_{11}^{-1}(B)\theta_{11}(B)a_{1t} \quad (6.2.1)$$

where

$$\phi_{11}(B) = (1 - \phi_{111}B - \phi_{112}B^2 - \cdots - \phi_{11n}B^n)$$
$$\phi_{12}(B) = -(\phi_{121} + \phi_{122}B + \phi_{123}B^2 + \cdots + \phi_{12n}B^{n-1})B$$
$$\theta_{11}(B) = (1 - \theta_{111}B - \theta_{112}B^2 - \cdots - \theta_{11m}B^m)$$

and

$$X_{2t} = -\phi_{22}^{-1}(B)\phi_{21}(B)X_{1t} + \phi_{22}^{-1}(B)\theta_{22}(B)a_{2t} \quad (6.2.2)$$

where

$$\phi_{22}(B) = (1 - \phi_{221}B - \phi_{222}B^2 - \cdots - \phi_{22n}B^n)$$
$$\phi_{21}(B) = -(\phi_{211} + \phi_{212}B + \phi_{213}B^2 + \cdots + \phi_{21n}B^{n-1})B$$
$$\theta_{22}(B) = (1 - \theta_{221}B - \theta_{222}B^2 - \cdots - \theta_{22m}B^m)$$

No distinction is made at the modeling stage between the input and the output. But now the transfer function will require distinguishing between the input and the output. If X_{1t} is the input and X_{2t} is the output, the main (feedforward) transfer function or the FRF is given by

$$H_{21}(\omega) = -\phi_{22}^{-1}(\omega)\phi_{21}(\omega) = -\phi_{22}^{-1}(B)\phi_{21}(B)|_{B=e^{-j\omega\Delta}} \quad (6.2.3)$$

The feedback transfer function is given by

$$H_{12}(\omega) = -\phi_{11}^{-1}(\omega)\phi_{12}(\omega) = -\phi_{11}^{-1}(B)\phi_{12}(B)|_{B=e^{-j\omega\Delta}} \quad (6.2.4)$$

which would be zero if there were no feedback, so that $\phi_{12i} = 0, i = 1, 2, \ldots, n$.

Since the FRF (6.2.3) has the same form as that of an ARMA($n, n-1$) model transfer function given earlier with $\theta_i = \phi_{21(i+1)}, i = 1, 2, \ldots, (n-1)$ and $\theta_0 = -1$ replaced by ϕ_{211}, the modal decomposition of the FRF can be written as, for $m < n$,

$$H_{21}(\omega) = \frac{(\phi_{211} + \phi_{212}B + \cdots + \phi_{21n}B^{n-1})B}{(1 - \phi_{221}B - \phi_{222}B^2 - \cdots - \phi_{22n}B^n)}\bigg|_{B=e^{-j\omega\Delta}}$$

$$= \sum_{i=1}^{n} \frac{c_i}{(1 - \lambda_i B)}\bigg|_{B=e^{-j\omega\Delta}} \quad (6.2.5)$$

where

$$(1 - \phi_{221}B - \phi_{222}B^2 - \cdots - \phi_{22n}B^n) = (1 - \lambda_1 B)(1 - \lambda_2 B) \cdots (1 - \lambda_n B) \quad (6.2.5a)$$

$$c_i = (\phi_{211}\lambda_i^{n-1} + \phi_{212}\lambda_i^{n-2} + \cdots + \phi_{21n})/\lambda_i \prod_{\substack{k=1 \\ k \neq i}}^{n} (\lambda_i - \lambda_k), \, i = 1, 2, \ldots, n \quad (6.2.5b)$$

The FRF or transfer function is complex and so are its modal components; therefore, they must be plotted as pairs of real and imaginary parts or magnitude and phase angle, each of which now has a decomposition by modes. Again the real λ_i or a complex conjugate pair of λ_i's provide a first-order or second-order system modal contribution. This modal decomposition is better represented by the discrete impulse response h_j given by

$$h_j = c_1\lambda_1^j + c_2\lambda_2^j + \cdots + c_n\lambda_n^j, \quad j \geq 1$$
$$= 0, \quad j \leq 0 \quad (6.2.6)$$

This impulse response can also be plotted as a whole with individual components adding up to it, real roots giving exponentially decaying first-order system components and complex conjugate pairs of roots giving exponentially decaying sinusoidal components.

Note that this impulse response has been obtained after the noise has been correctly modeled and naturally separated by the second terms on the right-hand sides of (6.2.1) and (6.2.2). In the current practice, such impulse response is often obtained by applying inverse Fourier transform to the FRF or transfer function synthesized in the frequency domain, which suffers from problems such as lack of resolution, leakage, and aliasing characteristic of deterministic signals and also bias and variance (random) errors characteristic of stochastic signals. The modal selection criteria developed in Section 5.3.2 can be usefully employed to select the necessary modes to be retained in the impulse response function obtained by the DDS approach. Note that the *gain* of an asymptotically stable system, which is the eventual step response as $t \to \infty$, is simply obtained by substituting $B = 1$ in $H(B)$, and $D = 0$ or $s = 0$ in $H(D)$ or $H(s)$, thus generalizing the first- and second-order system results of Sections 3.6.1 and 3.6.2.

Moreover, the impulse response function of the feedback path, when it exists, can also be obtained and decomposed by the approach given above. This is not possible by the standard FFT-based approach that uses sample spectra, which assumes no feedback.

6.2.2 Multiple-Input–Single-Output (MISO) System

The SISO system procedure of the last section readily extends to the MISO system. To avoid repetition, we will restrict ourselves to the main feedforward FRF or transfer function. The transfer functions for the feedback loops follow analogously, as seen for the SISO system in the last section.

If $X_{1t}, X_{2t}, \ldots, X_{(p-1)t}$ are the inputs and X_{pt} is the output, the main feedforward transfer function form of the model relating the output to the inputs is given by

$$X_{pt} = \sum_{k=1}^{p-1} -\phi_{pp}^{-1}(B)\phi_{pk}(B)X_{kt} + \phi_{pp}^{-1}(B)\theta_{pp}(B)a_{pt} \qquad (6.2.7)$$

where

$$\phi_{pp}(B) = (1 - \phi_{pp1}B - \phi_{pp2}B^2 - \cdots - \phi_{ppn}B^n)$$

$$\phi_{pk}(B) = -(\phi_{pk1} + \phi_{pk2}B + \phi_{pk3}B^2 + \cdots + \phi_{pkn}B^{n-1})B,$$

$$k = 1, 2, \ldots, p-1$$

$$\theta_{pp}(B) = (1 - \theta_{pp1}B - \theta_{pp2}B^2 - \cdots - \theta_{ppm}B^m)$$

The $(p-1)$ FRFs can now be written as

$$H_{pk}(\omega) = -\phi_{pp}^{-1}(\omega)\phi_{pk}(\omega) = -\phi_{pp}^{-1}(B)\phi_{pk}(B)|_{B=e^{-j\omega\Delta}} \qquad (6.2.8)$$

Their explicit form and modal decomposition are given by

$$H_{pk}(\omega) = \frac{(\phi_{pk1} + \phi_{pk2}B + \cdots + \phi_{pkn}B^{n-1})B}{(1 - \phi_{pp1}B - \phi_{pp2}B^2 - \cdots - \phi_{ppn}B^n)}\bigg|_{B=e^{-j\omega\Delta}}$$

$$= \sum_{i=1}^{n} c_{pki}/(1 - \lambda_i B)\bigg|_{B=e^{-j\omega\Delta}}, \quad k = 1, 2, \ldots, p-1 \qquad (6.2.9)$$

where

$$(1 - \phi_{pp1}B - \phi_{pp2}B^2 - \cdots - \phi_{ppn}B^n) = (1 - \lambda_1 B)(1 - \lambda_2 B) \cdots (1 - \lambda_n B) \quad (6.2.9a)$$

$$c_{pki} = (\phi_{pk1}\lambda_i^{n-1} + \phi_{pk2}\lambda_i^{n-2} + \cdots + \phi_{pkn})/\lambda_i \prod_{\substack{l=1 \\ l \neq i}}^{n} (\lambda_i - \lambda_l), \quad i = 1, 2, \ldots, n \quad (6.2.9b)$$

Finally, the impulse response functions for an impulse in the kth input are given by

$$h_{pkj} = c_{pk1}\lambda_1^j + c_{pk2}\lambda_2^j + \cdots + c_{pkn}\lambda_n^j, \quad j \geq 1 \quad (6.2.10)$$
$$= 0, \quad j \leq 0$$

6.2.3 The Multiple-Input–Multiple-Output (MIMO) System

The SISO system procedure of Section 6.2.1 also readily extends the MIMO system, but it now requires matrices and some additional notation to keep track of the various inputs and outputs. Let there be p inputs, q outputs, and corresponding residuals, forming the elements of column vectors $\mathbf{X}_t = [\mathbf{X}_t^{FT}, \mathbf{X}_t^{RT}]^T$, $\mathbf{a}_t = [\mathbf{a}_t^{FT}, \mathbf{a}_t^{RT}]^T$:

Input forcing function: $\mathbf{X}_t^F = [X_{1t}, X_{2t}, \ldots, X_{pt}]^T$

Output response: $\mathbf{X}_t^R = [X_{(p+1)t}, X_{(p+2)t}, \ldots, X_{(p+q)t}]^T$ (6.2.11)

with the residual vectors $\mathbf{a}_t^F$ and $\mathbf{a}_t^R$ defined similarly. Although ARV models suffice for most MIMO systems, we will use the more general ARMAV formulation for completeness. Splitting the parameter matrices ϕ_i and θ_i of the ARMAV (n, m) model relating the inputs and the outputs as

$$\phi_i = \begin{bmatrix} \phi_i^{FF} & \phi_i^{FR} \\ \phi_i^{RF} & \phi_i^{RR} \\ {\scriptstyle q \times p} & {\scriptstyle q \times q} \end{bmatrix} \begin{matrix} {\scriptstyle p \times p} & {\scriptstyle p \times q} \\ & \end{matrix}, \quad i = 1, 2, \ldots, n \quad (6.2.12a)$$

and similarly, using block diagonal θ_i for the sake of simplicity,

$$\theta_i = \begin{bmatrix} \theta_i^{FF} & 0 \\ 0 & \theta_i^{RR} \end{bmatrix}, \quad i = 1, 2, \ldots, m \quad (6.2.12b)$$

the transfer function models (6.2.1) and (6.2.2) can be readily generalized to

$$\mathbf{X}_t^F = -\phi_{FF}^{-1}(B)\phi_{FR}(B)\mathbf{X}_t^R + \phi_{FF}^{-1}(B)\theta_{FF}(B)\mathbf{a}_t^F \quad (6.2.13)$$

where

$$\phi_{FF}(B) = (\mathbf{I} - \phi_1^{FF}B - \phi_2^{FF}B^2 - \cdots - \phi_n^{FF}B^n)$$
$$\phi_{FR}(B) = -(\phi_1^{FR} + \phi_2^{FR}B + \cdots + \phi_n^{FR}B^{n-1})B$$
$$\theta_{FF}(B) = (\mathbf{I} - \theta_1^{FF}B - \theta_2^{FF}B^2 - \cdots - \theta_m^{FF}B^m)$$

and

$$\mathbf{X}_t^R = -\phi_{RR}^{-1}(B)\phi_{RF}(B)\mathbf{X}_t^F + \phi_{RR}^{-1}(B)\theta_{RR}(B)\mathbf{a}_t^R \quad (6.2.14)$$

where

$$\phi_{RR}(B) = (\mathbf{I} - \phi_1^{RR}B - \phi_2^{RR}B^2 - \cdots - \phi_n^{RR}B^n)$$
$$\phi_{RF}(B) = -(\phi_1^{RF} + \phi_2^{RF}B + \cdots + \phi_n^{RF}B^{n-1})B$$
$$\theta_{RR}(B) = (\mathbf{I} - \theta_1^{RR}B - \theta_2^{RR}B^2 - \cdots - \theta_m^{RR}B^m)$$

Therefore, the main feedforward transfer function or the FRF matrix is given by

$$\underset{q \times p}{\mathbf{H}^{RF}(\omega)} = -\phi_{RR}^{-1}(\omega)\phi_{RF}(\omega) = -\phi_{RR}^{-1}(B)\phi_{RF}(B)|_{B=e^{-j\omega\Delta}} \quad (6.2.15)$$

and the feedback transfer function is given by

$$\underset{p \times q}{\mathbf{H}^{FR}(\omega)} = -\phi_{FF}^{-1}(\omega)\phi_{FR}(\omega) = -\phi_{FF}^{-1}(B)\phi_{FR}(B)|_{B=e^{-j\omega\Delta}} \quad (6.2.16)$$

which would be zero if there is no feedback, so that $\phi_i^{FR} = \mathbf{0}, i = 1, 2, \ldots, n$.

Since the FRF matrix $\mathbf{H}^{RF}$ contains a matrix polynomial $\phi_{RR}^{-1}(B)$ of the ARV(n) model, its modal decomposition follows straightforwardly from the theory in Chapters 3–5 as

$$\mathbf{H}^{RF}(\omega) = (\mathbf{I} - \phi_1^{RR}B - \phi_2^{RR}B^2 - \cdots - \phi_n^{RR}B^n)^{-1}(\phi_1^{RF} + \phi_2^{RF}B$$
$$+ \cdots + \phi_n^{RF}B^{n-1})B|_{B=e^{-j\omega\Delta}}$$
$$= \sum_{i=1}^{nq} \frac{\mathbf{c}_i}{(1-\lambda_i B)}\bigg|_{B=e^{-j\omega\Delta}} \quad (6.2.17)$$

where λ_i (assumed distinct) are the eigenvalues of the state matrix,

$$\underset{nq \times nq}{\boldsymbol{\phi}_R} = \begin{bmatrix} \phi_1^{RR} & \phi_2^{RR} & \cdots & \phi_{n-1}^{RR} & \phi_n^{RR} \\ \mathbf{I} & \mathbf{0} & \cdots & \mathbf{0} & \mathbf{0} \\ \vdots & \vdots & & \vdots & \vdots \\ \mathbf{0} & \mathbf{0} & \cdots & \mathbf{I} & \mathbf{0} \end{bmatrix} \quad (6.2.18)$$

$$= \mathbf{L}_R \operatorname{diag}\{\lambda_i\}\mathbf{L}_R^{-1}$$

the eigenvector matrix

$$\mathbf{L}_R = \begin{bmatrix} \ell_1^R \lambda_1^{n-1} & \ell_2^R \lambda_2^{n-1} & \cdots & \ell_{nq}^R \lambda_{nq}^{n-1} \\ \ell_1^R \lambda_1^{n-2} & \ell_2^R \lambda_2^{n-2} & \cdots & \ell_{nq}^R \lambda_{nq}^{n-2} \\ \vdots & \vdots & & \vdots \\ \ell_1^R & \ell_2^R & \cdots & \ell_{nq}^R \end{bmatrix} \quad (6.2.19)$$

and denoting the rows of the left $nq \times q$ submatrix of $\mathbf{L}_R^{-1}$ [see (5.3.14) and (5.3.15)] by $\mathbf{L}_R^{i1}$:

$$\mathbf{L}_R^{-1} = \begin{bmatrix} \mathbf{L}_R^{11} & \cdots \\ \mathbf{L}_R^{21} & \cdots \\ \vdots & \vdots \\ \mathbf{L}_R^{nq1} & \cdots \end{bmatrix} \quad (6.2.20)$$

$$\mathbf{c}_i = \ell_i^R \mathbf{L}_R^{i1} (\boldsymbol{\phi}_1^{RF} \lambda_i^{n-1} + \boldsymbol{\phi}_2^{RF} \lambda_i^{n-2} + \cdots + \boldsymbol{\phi}_n^{RF})/\lambda_i \quad (6.2.21)$$

Finally the $q \times p$ impulse response matrix is given by

$$\mathbf{h}_j^{RF} = \mathbf{c}_1 \lambda_1^j + \mathbf{c}_2 \lambda_2^j + \cdots + \mathbf{c}_{nq} \lambda_{nq}^j, \quad j \geq 1 \quad (6.2.22)$$
$$= \mathbf{0}, \, j \leq 0$$

which is the inverse Fourier transform of $\mathbf{H}^{RF}(\omega)$. The modal decomposition of the feedback FRF $\mathbf{H}^{FR}(\omega)$ follows similarly with the corresponding $\mathbf{h}_j^{FR}$. Selection of modes can now be accomplished using Section 5.3.2 and the resultant FRF or transfer function used for modal analysis or control system design.

The FRF matrix $\mathbf{H}^{RF}(\omega)$ has the proper scaling of eigenvectors or mode shapes. Therefore, similar to impulse testing with measured impulse force, such an FRF, obtained by a single random input excitation providing the impulse response matrix, not only yields the natural frequencies, damping ratios, and scaled mode shapes giving the modal model, but also yields the necessary scaling to obtain $\mathbf{m}$, $\mathbf{c}$, and $\mathbf{k}$ of the spatial model. The modal model is obtained as in Section 5.3.1, mode selection is accomplished as in Section 5.3.2, and, finally, $\mathbf{m}$, $\mathbf{c}$, and $\mathbf{k}$ are computed by the formulas of Section 5.4.2. Such an approach to random vibrations is far more satisfactory than that based on correlation/spectrum, although these characteristics, if desired, can be readily obtained by the expressions given in Section 6.3. Moreover, nonlinear systems can also be modeled by this approach, which provides a stochastic linearization of nonlinear systems; see Pandit (1977b).

This MIMO case is general enough to specialize not only to SISO and MISO system, dealt with earlier, but also to the single-input–multiple-output (SIMO) system, which we therefore will not elaborate. A combination of these cases can be used to get a modal or spatial model with minimum testing effort. For

6.3 THE MULTIVARIATE SPECTRUM

It was pointed out in the last section that the multivariate spectrum matrix, which includes auto- and cross-spectrum as its elements, is not necessary for finding the FRF matrix or the transfer function by the DDS approach. However, there are other applications, such as acoustics and noise source analysis, where the auto- and cross-spectrum, and the coherence function derived from these, are themselves of interest. Once an adequate ARMAV model is fitted to the data, the multivariate spectrum matrix can be found as easily as the univariate spectrum from the ARMA models in Section 6.1. The notation becomes cumbersome and the computation, of course, is much more extensive.

The basic definitions of Section 6.1 for the univariate autocovariance readily generalize to the multivariate case. The lagged covariance matrix is defined as (see Appendix A for review)

$$\pmb{\gamma}_k = \{\gamma_{ijk}\} = E(\mathbf{X}_t \bar{\mathbf{X}}^T_{t-k}) = \bar{\pmb{\gamma}}^T_{(-k)} = \{\bar{\gamma}_{ji(-k)}\} \tag{6.3.1}$$

where $\mathbf{X}_t = (X_{1t}, X_{2t}, \ldots, X_{pt})^T$ is the zero mean vector of all series, including all the inputs and outputs. Although the definition is given for complex stochastic processes for generality, $\pmb{\gamma}_k$ is real when it is mostly used for real data in practice. The diagonal of the matrix $\pmb{\gamma}_k$ contains the autocovariances and the off-diagonal elements contain the cross-covariances. Note that for nonzero lag $k \neq 0$, this matrix is not Hermitian, $\pmb{\gamma}_k \neq \bar{\pmb{\gamma}}^T_k$. In particular, the variance–covariance matrix at (zero lag)

$$\pmb{\gamma}_0 = E(\mathbf{X}_t \bar{\mathbf{X}}^T_t) \tag{6.3.2}$$

is Hermitian, and symmetric when real. The Fourier transform of the lagged covariance matrix $\pmb{\gamma}_k$ is the multivariate spectrum (matrix) $\mathbf{S}(\omega)$, and the two form a Fourier transform pair:

$$\mathbf{S}(\omega) = \frac{\Delta}{2\pi} \sum_{k=-\infty}^{\infty} \pmb{\gamma}_k e^{-j\omega k \Delta}, \qquad -\frac{\pi}{\Delta} \leq \omega \leq \frac{\pi}{\Delta} \tag{6.3.3}$$

$$\pmb{\gamma}_k = \int_{-\pi/\Delta}^{\pi/\Delta} \mathbf{S}(\omega) e^{j\omega k \Delta} d\omega, \qquad k = 0, \pm 1, \pm 2, \ldots \tag{6.3.4}$$

and in particular,

$$\pmb{\gamma}_0 = \int_{-\pi/\Delta}^{\pi/\Delta} \mathbf{S}(\omega) d\omega \tag{6.3.5}$$

As in the univariate case, we can change the units from ω radians to f cycles per unit time by $\mathbf{S}(f) = 2\pi \mathbf{S}(\omega)$, $\omega = 2\pi f$. Note that the matrix $\mathbf{S}(\omega)$ is Hermitian:

$$\mathbf{S}(\omega) = \{S_{ij}(\omega)\} = \bar{\mathbf{S}}^T(\omega) = \{\bar{S}_{ji}(\omega)\} \qquad (6.3.3a)$$

Its diagonal element $S_{ii}(\omega)$ is the autospectrum of X_{it}, which is real, and the off-diagonal element $S_{ij}(\omega)$ is the cross-spectrum of X_{it} and X_{jt}, $i \neq j$, which is complex. It can be shown that a variance–covariance matrix and $\mathbf{S}(\omega)$ are nonnegative definite, so that the variance–covariances and auto- and cross-spectra satisfy

$$|\gamma_{ijk}|^2 \leq \gamma_{ii0}\gamma_{jj0} \qquad (6.3.6)$$

$$|S_{ij}|^2 \leq S_{ii}(\omega)S_{jj}(\omega) \qquad (6.3.7)$$

This leads to the definitions of the cross-correlation function

$$\rho_{ijk} = \frac{\gamma_{ijk}}{\sqrt{\gamma_{ii0}\gamma_{jj0}}}, \qquad 0 \leq |\rho_{ijk}| \leq 1 \qquad (6.3.8)$$

and the squared coherence function

$$\kappa_{ij}^2(\omega) = \frac{|S_{ij}(\omega)|^2}{S_{ii}(\omega)S_{jj}(\omega)}, \qquad 0 \leq \kappa_{ij}^2(\omega) \leq 1 \qquad (6.3.9)$$

It is left as an exercise to check that this definition of the coherence function agrees with the one given at the beginning of Section 6.2.

The multivariate white noise

$$\mathbf{X}_t = \mathbf{a}_t, \qquad E(\mathbf{a}_t) = \mathbf{0} \qquad (6.3.10)$$

with the covariance function

$$\gamma_k^a = E(\mathbf{a}_t \mathbf{a}_{t-k}^T) = \delta_k \sigma_a^2 \qquad (6.3.11)$$

$$= \sigma_a^2, \qquad k = 0$$

$$= \mathbf{0}, \qquad k \neq 0$$

has the multivariate spectrum, using (6.3.3):

$$\mathbf{S}^a(\omega) = \frac{\Delta}{2\pi}\sigma_a^2, \qquad -\frac{\pi}{\Delta} \leq \omega \leq \frac{\pi}{\Delta} \qquad (6.3.12)$$

which is real for real $\mathbf{a}_t$ and constant over the frequency range $-\pi/\Delta$ to π/Δ radians per unit time.

6.3.1 The ARMAV(n,m) Model and Its Modal Decomposition

The transfer function form of an ARMAV(n,m) model

$$\mathbf{X}_t - \boldsymbol{\phi}_1 \mathbf{X}_{t-1} - \cdots - \boldsymbol{\phi}_n \mathbf{X}_{t-n} = \mathbf{a}_t - \boldsymbol{\theta}_1 \mathbf{a}_{t-1} - \cdots - \boldsymbol{\theta}_m \mathbf{a}_{t-m} \qquad (6.3.13a)$$

is

$$\mathbf{X}_t = \boldsymbol{\phi}^{-1}(B)\boldsymbol{\theta}(B)\mathbf{a}_t \qquad (6.3.13)$$

where

$$\boldsymbol{\phi}(B) = (\mathbf{I} - \boldsymbol{\phi}_1 B - \cdots - \boldsymbol{\phi}_n B^n) \qquad (6.3.13b)$$

$$\boldsymbol{\theta}(B) = (\mathbf{I} - \boldsymbol{\theta}_1 B - \cdots - \boldsymbol{\theta}_m B^m) \qquad (6.3.13c)$$

For $m < n$, and assuming distinct eigenvalues λ_i for the state matrix for convenience, the Green's function matrix $\mathbf{G}_j$ has the explicit form

$$\mathbf{G}_j = \mathbf{g}_1 \lambda_1^j + \mathbf{g}_2 \lambda_2^j + \cdots + \mathbf{g}_{np} \lambda_{np}^j, \quad j \geq 0 \qquad (6.3.14)$$
$$= \mathbf{0}, \, j < 0$$

where the state matrix

$$\underset{np \times np}{\boldsymbol{\phi}} = \begin{bmatrix} \boldsymbol{\phi}_1 & \boldsymbol{\phi}_2 & \cdots & \boldsymbol{\phi}_{n-1} & \boldsymbol{\phi}_n \\ \mathbf{I} & \mathbf{0} & \cdots & \mathbf{0} & \mathbf{0} \\ \vdots & \vdots & & \vdots & \vdots \\ \mathbf{0} & \mathbf{0} & \cdots & \mathbf{I} & \mathbf{0} \end{bmatrix} \qquad (6.3.15)$$

$$= \mathbf{L} \, \text{diag}\{\lambda_i\} \mathbf{L}^{-1}$$

the eigenvector matrix

$$\mathbf{L} = \begin{bmatrix} \boldsymbol{\ell}_1 \lambda_1^{n-1} & \boldsymbol{\ell}_2 \lambda_2^{n-1} & \cdots & \boldsymbol{\ell}_{np} \lambda_{np}^{n-1} \\ \boldsymbol{\ell}_1 \lambda_1^{n-2} & \boldsymbol{\ell}_2 \lambda_2^{n-2} & \cdots & \boldsymbol{\ell}_{np} \lambda_{np}^{n-2} \\ \vdots & \vdots & & \vdots \\ \boldsymbol{\ell}_1 & \boldsymbol{\ell}_2 & \cdots & \boldsymbol{\ell}_{np} \end{bmatrix} \qquad (6.3.16)$$

and denoting the *rows* of the left $np \times p$ submatrix of $\mathbf{L}^{-1}$ by $\mathbf{L}^{i1}$ (see (5.3.14)

THE MULTIVARIATE SPECTRUM 307

and (5.3.15)),

$$\mathbf{L}^{-1} = \begin{bmatrix} \mathbf{L}^{11} & \cdots \\ \mathbf{L}^{21} & \cdots \\ \vdots & \vdots \\ \mathbf{L}^{np1} & \cdots \end{bmatrix} \qquad (6.3.17)$$

$$\mathbf{g}_i = \ell_i \mathbf{L}^{i1}(\mathbf{I}\lambda_i^{n-1} - \boldsymbol{\theta}_1 \lambda_i^{n-2} - \cdots - \boldsymbol{\theta}_m \lambda_i^{n-m-1}) \qquad (6.3.18)$$

Under the stationarity conditions $|\lambda_i| < 1, i = 1, 2, \ldots, np$, the lagged covariance matrix is given by, assuming real $\mathbf{X}_t$,

$$\boldsymbol{\gamma}_k = \sum_{j=0}^{\infty} \mathbf{G}_{j+k} \sigma_a^2 \mathbf{G}_j^T = \mathbf{d}_1 \lambda_1^k + \mathbf{d}_2 \lambda_2^k + \cdots + \mathbf{d}_{np} \lambda_{np}^k \qquad (6.3.19)$$

where

$$\mathbf{d}_i = \left[\frac{\mathbf{g}_i \sigma_a^2 \mathbf{g}_1^T}{(1 - \lambda_i \lambda_1)} + \frac{\mathbf{g}_i \sigma_a^2 \mathbf{g}_2^T}{(1 - \lambda_i \lambda_2)} + \cdots + \frac{\mathbf{g}_i \sigma_a^2 \mathbf{g}_{np}^T}{(1 - \lambda_i \lambda_{np})} \right] \qquad (6.3.19a)$$

and the (zero lag) variance–covariance matrix

$$\boldsymbol{\gamma}_0 = \mathbf{d}_1 + \mathbf{d}_2 + \cdots + \mathbf{d}_{np} \qquad (6.3.19b)$$

Its implicit form (often used for parameter estimation by using $\hat{\boldsymbol{\gamma}}_k$ obtained from data in place of $\boldsymbol{\gamma}_k$) is

$$(\mathbf{I} - \boldsymbol{\phi}_1 B - \cdots - \boldsymbol{\phi}_n B^n) \boldsymbol{\gamma}_k = (\sigma_a^2 \mathbf{G}_{-k}^T - \boldsymbol{\theta}_1 \sigma_a^2 \mathbf{G}_{1-k}^T - \cdots - \boldsymbol{\theta}_m \sigma_a^2 \mathbf{G}_{m-k}^T),$$

$$0 \le k \le m \qquad (6.3.19c)$$

$$= \mathbf{0}, \quad k > m \qquad (6.3.19d)$$

Equations (6.3.19d) show that $\boldsymbol{\gamma}_k$ satisfies the *deterministic* ARV(n) model for large enough k; the scalar AR form of these equations for $p = 1, m = 0$ are called Yule–Walker equations. Since $\boldsymbol{\gamma}_k \ne \boldsymbol{\gamma}_{(-k)}$ for the vector case, they generalize only for $k \ge \max(n, m + 1)$.

Therefore, the multivariate spectrum $\mathbf{S}(\omega)$ can now be written in the transfer function form as well as the modal decomposition form:

$$\mathbf{S}(\omega) = \frac{\Delta}{2\pi} \boldsymbol{\phi}^{-1}(\omega) \boldsymbol{\theta}(\omega) \sigma_a^2 \bar{\boldsymbol{\theta}}^T(\omega) \bar{\boldsymbol{\phi}}^{-T}(\omega) \qquad (6.3.20a)$$

$$= \frac{\Delta}{2\pi} \Big[\mathbf{d}_1(1 - \lambda_1^2)|1 - \lambda_1 e^{-j\omega\Delta}|^{-2} + \mathbf{d}_2(1 - \lambda_2^2)|1 - \lambda_2 e^{-j\omega\Delta}|^{-2}$$

$$+ \cdots + \mathbf{d}_{np}(1 - \lambda_{np}^2)|1 - \lambda_{np} e^{-j\omega\Delta}|^{-2} \Big] \qquad (6.3.20b)$$

308   SPECTRUM ANALYSIS

Although this is formally similar to the modal decomposition of the ARMA spectrum, it is much more extensive. It decomposes every element of $\mathbf{S}(\omega)$, the diagonal autospectra $S_{ii}(\omega)$ as well as the off-diagonal cross-spectra $S_{ij}(\omega)$, into modal components with first-order system terms for real λ_i and the addition of a pair of terms for the second-order system having complex conjugate λ_i, λ_{i+1}. The elements of the matrix (6.3.20a) are analytic functions of ω and provide smooth plots of autospectra, plots of real and imaginary parts or magnitude and phase of cross-spectra, and the corresponding coherence function defined by (6.3.9). Each of these plots can be shown to be built up of mode by mode plots of the terms in (6.3.20b). Other remarks at the end of Section 6.1.5 apply here also. The element by element derivations of the special cases of (6.3.20b), for the bivariate ARV(1) model with $p = 2$ illustrated in Section 4.4, are left as an interesting exercise; the interested reader may find some of them in Pandit (1973).

For large p, such modal decomposition might be too extensive to be of practical use. There are two other ways of summarizing the information in $\mathbf{S}(\omega)$ in a less extensive manner. These are presented in the next two sections.

6.3.2 Decomposition by White Noise Variance–Covariance

Using the basic Fourier relation (6.3.5) and the modal decomposition of γ_0 by (6.3.19a and b) together:

$$\gamma_0 = \int_{-\pi/\Delta}^{\pi/\Delta} \mathbf{S}(\omega) d\omega$$

$$= \mathbf{d}_1 + \mathbf{d}_2 + \cdots + \mathbf{d}_{np}$$

$$= \sum_{i,j=1}^{np} \frac{\mathbf{g}_i \sigma_a^2 \mathbf{g}_j^T}{(1 - \lambda_i \lambda_j)} \tag{6.3.21}$$

Thus, while $\mathbf{S}(\omega)$ gives a qualitative picture of how the variance–covariance or power is distributed over a continuous band of frequencies, the modal decomposition shows how it is quantitatively distributed over the modal frequencies contained in the complex conjugate pairs λ_i, λ_{i+1} contributing peaks in the auto- and cross-spectra and real λ_i contributing a peak at zero frequency.

However, in many studies, such as noise source identification illustrated later by example, this modal decomposition may be of only secondary interest. One may be primarily interested in finding out how various noise sources contribute to the noise at a given location. If we consider a_{it} as the essence of noise in X_{it} which unfortunately also affects all other $X_{jt}, j \neq i$, we can use the decomposition (6.3.21) to show how the variances σ_{aii} and covariances σ_{aij} of the multivariate white noise $\mathbf{a}_t$ affect all the variances (power) and covariances.

Denoting

$$\boldsymbol{\sigma}_a^2 = \{\sigma_{aij}\}, \quad \mathbf{g}_i = \{g_{kli}\} \qquad (6.3.22)$$

the decomposition (6.3.21) may be written as

$$\boldsymbol{\gamma}_0 = \{\gamma_{kl0}\} = \left\{ \sum_{r,s=1}^{p} \sigma_{ars} \sum_{i,j=1}^{np} g_{kri}g_{lsj}/(1 - \lambda_i \lambda_j) \right\} \qquad (6.3.23)$$

Thus σ_{ars} times the second summation on the right-hand side of (6.3.23) is the contribution of σ_{ars} to the variance ($k = l$) or covariance ($k \neq l$) γ_{kl0}. Similar manipulation of the operators in (6.3.20a) provides the decomposition of the auto- and cross-spectra by σ_{ars} at each frequency ω, which can be plotted to get a visual picture of the decomposition and integrated to get (6.3.23), see the remarks about (6.1.31a and b). The second summation term can thus be alternatively found directly from the model parameters, without requiring the computation of λ and g, as illustrated later in Section 6.4.2. For uncrosscorrelated white noises $\sigma_{ars} = 0$ for $r \neq s$ and the decomposition given above simplifies to

$$\boldsymbol{\gamma}_0 = \{\gamma_{kl0}\} = \left\{ \sum_{r=1}^{p} \sigma_{ar}^2 \sum_{i,j=1}^{np} g_{kri}g_{lrj}/(1 - \lambda_i \lambda_j) \right\} \qquad (6.3.24)$$

where we have set $\sigma_{arr} = \sigma_{ar}^2$ for the variance of a_{rt}.

For example, if $X_{1t}, X_{2t}, \ldots, X_{(p-1)t}$ are acceleration measurements of vibrating structures and X_{pt} is the noise measurement, assuming uncrosscorrelated a_{it}'s (which can be checked from modeling results) the decomposition of the noise variance γ_{ppo} is given by

$$\gamma_{ppo} = \sum_{r=1}^{p} \sigma_{ar}^2 \sum_{i,j=1}^{np} g_{pri}g_{prj}/(1 - \lambda_i \lambda_j) \qquad (6.3.25)$$

6.3.3 Partial Correlation/Coherence and Conditioned Spectra

One of the reasons why both the standard transfer functions and multiple spectra forming the elements of the multivariate spectrum $\mathbf{S}(\omega)$ are difficult to interpret in practice is that the simple idea of a relation between two variables is overwhelmed by the presence of too many other variables. The same problem arises in the time domain in considering the covariance/correlation. If we have p measurements $X_{1t}, X_{2t} \ldots X_{pt}$ on a system, how much of the correlation/coherence between say X_{1t} and X_{pt} is due to their direct dependence, and how much of it is due to the indirect dependence via other variables $X_{2t}, X_{3t}, \ldots, X_{(p-1)t}$?

The answer is in the corresponding partial correlation/coherence functions. Recall that the entities representing a relation between two variables—covariance and cross-spectrum—form a Fourier transform pair, whereas squared correlation and coherence represent their normalized magnitude squares. Partial correlation/coherence represent the correlation/coherence between two variables after the linear effects of other variables have been removed. The variance/covariance and auto/cross-spectrum of the residuals left after removing such linear effects of other variables are called residual or conditioned variance/covariance and auto/cross-spectrum, respectively. The correlation/coherence defined from these by standard definitions are called partial correlation/coherence, to distinguish them from the ordinary correlation/coherence in which the linear effects of other variables are mixed up.

Building upon the work of Dodds and Robson (1975), Bendat and Piersol (1980) have summarized the results and physical interpretation of these partial functions, which may be referred to for more details and recursive relations derived without explicitly using the matrix decomposition. We will present here a simpler version based on LU decomposition for which computationally efficient algorithms are available; see for example, Golub and VanLoan (1987, algorithm 5.1-2).

Recall from Chapter 3 that the QR decomposition based on the Gram–Schmidt procedure precisely provides the needed mechanism for removing linear dependences. Applying it formally to the vector $\mathbf{X}_t$, the positive definite matrix $\boldsymbol{\gamma}_k = E(\mathbf{X}_t \bar{\mathbf{X}}_{t-k}^T)$ and its Fourier transform $\mathbf{S}(\omega)$, that is a Hermitian matrix, admit LU decomposition (similar to the one used for the Cholesky method in Section 3.4.7) where $\mathbf{L}$ is lower triangular and $\mathbf{U}$ is upper triangular. In fact, such positive definite matrices can be diagonalized by triangular matrices with 1's on their diagonals and the diagonal matrix having all positive elements [see (5.1.37)]. Absorbing this diagonal matrix in $\mathbf{U}$ provides the following decomposition

$$\boldsymbol{\gamma}_k = \begin{bmatrix} \gamma_{11k} & \gamma_{12k} & \cdots & \gamma_{1pk} \\ \gamma_{21k} & \gamma_{22k} & \cdots & \gamma_{2pk} \\ \gamma_{31k} & \gamma_{32k} & \cdots & \gamma_{3pk} \\ \vdots & \vdots & \vdots & \vdots \\ \gamma_{p1k} & \gamma_{p2k} & \cdots & \gamma_{ppk} \end{bmatrix} = \begin{bmatrix} 1 & 0 & 0 & \cdots & 0 \\ l_{21} & 1 & 0 & \cdots & 0 \\ l_{31} & l_{32} & 1 & \cdots & 0 \\ \vdots & \vdots & \vdots & \vdots & \vdots \\ l_{p1} & l_{p2} & l_{p3} & \cdots & 1 \end{bmatrix}$$

$$\times \begin{bmatrix} \gamma_{11} & \gamma_{12} & \gamma_{13} & \cdots & \gamma_{1p} \\ 0 & \gamma_{22.1} & \gamma_{23.1} & \cdots & \gamma_{2p.1} \\ 0 & 0 & \gamma_{33.12} & \cdots & \gamma_{3p.12} \\ \vdots & \vdots & \vdots & \vdots & \vdots \\ 0 & 0 & 0 & \cdots & \gamma_{pp.(p-1)!} \end{bmatrix} \quad (6.3.26)$$

and since $\mathbf{S}(\omega)$ is Hermitian, $\mathbf{\bar{S}}(\omega) = \mathbf{S}^T(\omega)$, it also admits a further decomposition $\mathbf{L D \bar{L}}^T$:

$$\mathbf{S}(\omega) = \begin{bmatrix} S_{11}(\omega) & S_{12}(\omega) & \cdots & S_{1p}(\omega) \\ S_{21}(\omega) & S_{22}(\omega) & \cdots & S_{2p}(\omega) \\ S_{31}(\omega) & S_{32}(\omega) & \cdots & S_{3p}(\omega) \\ \vdots & \vdots & \vdots & \vdots \\ S_{p1}(\omega) & S_{p2}(\omega) & \cdots & S_{pp}(\omega) \end{bmatrix}$$

$$= \begin{bmatrix} 1 & 0 & 0 & \cdots & 0 \\ L_{21} & 1 & 0 & \cdots & 0 \\ L_{31} & L_{32} & 1 & \cdots & 0 \\ \vdots & \vdots & \vdots & \vdots & \vdots \\ L_{p1} & L_{p2} & L_{p3} & \cdots & 1 \end{bmatrix} \begin{bmatrix} S_{11} & S_{12} & S_{13} & \cdots & S_{1p} \\ 0 & S_{22.1} & S_{23.1} & \cdots & S_{2p.1} \\ 0 & 0 & S_{33.12} & \cdots & S_{3p.12} \\ \cdots & \cdots & \cdots & \vdots & \cdots \\ 0 & 0 & 0 & \cdots & S_{pp.(p-1)!} \end{bmatrix} \quad (6.3.27a)$$

$$= \begin{bmatrix} 1 & 0 & 0 & \cdots & 0 \\ L_{21} & 1 & 0 & \cdots & 0 \\ L_{31} & L_{32} & 1 & \cdots & 0 \\ \vdots & \vdots & \vdots & \vdots & \vdots \\ L_{p1} & L_{p2} & L_{p3} & \cdots & 1 \end{bmatrix} \begin{bmatrix} S_{11} & 0 & 0 & \cdots & 0 \\ 0 & S_{22.1} & 0 & \cdots & 0 \\ 0 & 0 & S_{33.12} & \cdots & 0 \\ \vdots & \vdots & \vdots & \vdots & \vdots \\ 0 & 0 & 0 & \cdots & S_{pp.(p-1)!} \end{bmatrix}$$

$$\times \begin{bmatrix} 1 & \bar{L}_{21} & \bar{L}_{31} & \cdots & \bar{L}_{p1} \\ 0 & 1 & \bar{L}_{32} & \cdots & \bar{L}_{p2} \\ 0 & 0 & 1 & \cdots & \bar{L}_{p3} \\ \vdots & \vdots & \vdots & \vdots & \vdots \\ 0 & 0 & 0 & \cdots & 1 \end{bmatrix} \quad (6.3.27b)$$

where $(p-1)! = 123 \ldots (p-1)$ as usual. The arguments k and ω have been omitted on the right-hand side for convenience.

Here $\gamma_{22.1}$ is the partial variance of X_{2t} after removing the effect of X_{1t}, $\gamma_{2p.1}$ is the partial covariance of X_{2t} with X_{pt} after removing the effect of X_{1t}, $\gamma_{3p.12}$ is the partial covariance of X_{3t} with X_{pt} after removing the effects of X_{1t} and X_{2t}, and so on. Similarly $S_{22.1}$ is the partial, conditioned, or residual autospectrum of X_{2t} after the effect of X_{1t} has been removed; $S_{2p.1}$ is the partial, conditioned, or residual cross-spectrum of X_{2t} and X_{pt} after the effect of X_{1t} has been removed; $S_{3p.12}$ is the partial, conditioned, or residual cross-spectrum of X_{3t} and X_{pt} after the effects of X_{1t} and X_{2t} have been removed. The partial correlation can now be defined as usual by dividing the partial covariance by the square root of the product of the partial variances, and the partial squared coherence is defined by the partial cross-spectrum absolute value squared divided by the product of the partial autospectra; see (6.3.8) and (6.3.9).

Note that such a decomposition requires a natural ordering of inputs. X_{1t} does not depend upon any other inputs. X_{2t} depends only on X_{1t}, X_{3t} on X_{1t} and X_{2t}, and so on. X_{pt}, the last variable, is usually the output of interest. Such ordering is extremely crucial.

The goal of the method is the prediction of the output spectrum from the multiple inputs. This follows from the multiplication of the last row and column (6.3.27a) as

$$S_{pp}(\omega) = L_{p1}(\omega)S_{1p}(\omega) + L_{p2}(\omega)S_{2p.1}(\omega) + \cdots + S_{pp.(p-1)!}(\omega) \quad (6.3.28)$$

where $S_{pp.(p-1)!}$ is the residual or noise auto-spectrum after all the effects of the inputs have been removed from X_{pt}.

In the present methods of implementing this technique, sample spectra are used on the left-hand side of (6.3.27) and the partial or the conditioned spectra on the right-hand side are recursively computed by the following relations from these. Since the sample spectra have large bias and variance errors, the end results for large number of inputs can be quite unreliable; this and other difficulties have been pointed out by Gersch, Brotherton, and Braun (1980). Using the smooth spectra obtained by the analytic functions from the fitted models considerably improves the appearance and numerical contribution of the conditioned spectra; this will be illustrated by examples in Section 6.4. In particular, Section 6.4.3 gives the simpler case of the results for $p = 3$, which may be consulted first if the derivations of the general results in this section appear to be involved.

The recursive expressions to compute the linear relations L_{ij} and the conditioned spectra $S_{ij.(i-1)!}$, $i \geq j$, follow from (6.3.27a and b) by row column multiplications. Comparing the diagonal, say (k, k), entries gives

$$S_{kk} = \sum_{j=1}^{k-1} |L_{kj}|^2 S_{jj.(j-1)!} + S_{kk.(k-1)!} \quad (6.3.29)$$

whereas comparing the off-diagonal (i, k) entries, $i > k$, gives

$$S_{ik} = \sum_{j=1}^{k-1} L_{ij} S_{jk.(j-1)!} + L_{ik} S_{kk.(k-1)!}, \quad i > k \quad (6.3.30a)$$

$$= \sum_{j=1}^{k-1} L_{ij} \bar{L}_{kj} S_{jj.(j-1)!} + L_{ik} S_{kk.(k-1)!}, \quad i > k \quad (6.3.30b)$$

$$S_{ki} = \sum_{j=1}^{k-1} \bar{L}_{ij} L_{kj} S_{jj.(j-1)!} + \bar{L}_{ik} S_{kk.(k-1)!}, \quad i > k \quad (6.3.30c)$$

using $\bar{S}_{ik} = S_{ki}$.

Since $\mathbf{S}(\omega)$ is nonnegative-definite, it is possible that $S_{kk.(k-1)!}$ is zero for some k; hence we have assumed $\mathbf{S}(\omega)$ to be positive definite to avoid this case. This assumption can be implemented by requiring the program to quit when $S_{kk.(k-1)} = 0$. Since the recursive calculation needs to be repeated for every ω, a considerable saving in memory is achieved by overwriting S_{ik} with L_{ik} if $i > k$. A computationally and memorywise efficient algorithm can then be succinctly written as

For $k = 1, 2, \ldots, p$
   For $j = 1, 2, \ldots, k-1$

$$S_{kk.(k-1)!} = S_{kk} - \sum_{j=1}^{k-1} |L_{kj}|^2 S_{jj.(j-1)!}$$

If $S_{kk.(k-1)!} = 0$
  then
      quit
  else
    For $i = k+1, k+2, \ldots, p$

$$L_{ik} = (S_{ik} - \sum_{j=1}^{k-1} \bar{L}_{ij} L_{kj} S_{jj.(j-1)!})/S_{kk.(k-1)!} \qquad (6.3.31a)$$

The physical interpretation of $L_{ik}(\omega)$ can be obtained from the following relation found by comparing (6.3.27a and b):

$$L_{ik}(\omega) = \frac{S_{ik.(k-1)!}}{S_{kk.(k-1)!}}, \qquad i > k, \qquad L_{ii} = 1 \qquad (6.3.31)$$

which was used in deriving (6.3.30b) from (6.3.30a). Thus $L_{ik}(\omega)$ is the (partial or conditioned) transfer function relating the output X_{it} with the input X_{kt} after the linear effects of inputs $X_{1t}, X_{2t}, \ldots, X_{(k-1)t}$ have been removed. Indeed, for $k = 1$, (6.3.31) specializes to

$$L_{i1}(\omega) = \frac{S_{i1}(\omega)}{S_{11}(\omega)} \qquad (6.3.32)$$

which would have been the usual FRF or transfer function from input excitation X_{1t} to output response X_{it}, if the inputs $X_{1t}, X_{2t}, \ldots, X_{(i-1)t}$ were uncorrelated. When these inputs are correlated, X_{1t} affects X_{it} not only directly by H_{i1}, but also via X_{2t} by H_{i2}, via X_{3t} by $H_{i3}, \ldots$, via $X_{(i-1)t}$ by $H_{i(i-1)}$; L_{i1} represents this total effect as we will show below in (6.3.35a).

Thus $L_{ik}(\omega)$ is the transfer function between the output X_{it} and the conditioned input, $X_{k.(k-1)!t}$ having the autospectrum $S_{kk.(k-1)!}$, whereas $H_{ik}(\omega)$ is the usual transfer function between X_{it} and X_{kt}. If we choose $i = p$ for definiteness, so that X_{pt} is the output and $X_{kt}, k = 1, 2, \ldots, (p-1)$ are the inputs, then the ordered inputs X_{kt} are linearly dependent and statistically correlated whereas the conditioned inputs $X_{k.(k-1)!t}$, from which the linear dependences have been successively removed, are linearly independent and statistically uncorrelated. Hence the LU decomposition essentially provides us with the Gram–Schmidt orthogonalization:

$$X_{pt} = H_{p1}(B)X_{1t} + H_{p2}(B)X_{2t} + \cdots + H_{p(p-1)}(B)X_{(p-1)t} + N_t \qquad (6.3.33)$$

$$= L_{p1}(B)X_{1t} + L_{p2}(B)X_{2.1t} + \cdots + L_{p(p-1)}(B)X_{(p-1).(p-2)!t}$$
$$+ X_{p.(p-1)!t} \qquad (6.3.34)$$

where $N_t \equiv X_{p.(p-1)!t}$ is the noise term, after the linear effects of all the inputs have been removed, having autospectrum $S_{pp.(p-1)!}$; (6.3.34) is simply a restatement of (6.3.28).

Thus H_{pk} express X_{pt} in the nonorthogonal basis, whereas L_{pk} express X_{pt} in the orthogonal basis. Hence it follows from Chapter 3 (see Section 3.2.1) that

$$\begin{bmatrix} L_{p1} \\ L_{p2} \\ \vdots \\ L_{p(p-1)} \end{bmatrix} = \begin{bmatrix} 1 & L_{21} & L_{31} & \cdots & L_{(p-1)1} \\ 0 & 1 & L_{32} & \cdots & L_{(p-1)2} \\ \vdots & \vdots & \vdots & & \vdots \\ 0 & 0 & 0 & \cdots & 1 \end{bmatrix} \begin{bmatrix} H_{p1} \\ H_{p2} \\ \vdots \\ H_{p(p-1)} \end{bmatrix} \qquad (6.3.35a)$$

and by backward substitution

$$H_{pi} = L_{pi} - \sum_{j=i+1}^{(p-1)} L_{ji} H_{pj}, \qquad i = (p-1), (p-2), \ldots, 1 \qquad (6.3.35b)$$

which completes the relations between ordinary and conditioned (partial) spectra and transfer functions. In fact, it is clear from the Gram–Schmidt procedure that (6.3.34) can be generalized for any $j > r$ to

$$X_{jt} = \sum_{i=1}^{r} L_{ji} X_{i.(i-1)!t} + X_{j.r!t} \qquad (6.3.36)$$

which yields the recursion for the conditioned variables and for the conditioned spectra

$$X_{j.r!t} = X_{j.(r-1)!t} - L_{jr} X_{r.(r-1)!t} \qquad (6.3.37)$$

$$S_{jj.r!} = S_{jj.(r-1)!} - |L_{jr}|^2 S_{rr.(r-1)!} \qquad (6.3.38)$$

It is left as an interesting exercise to derive all the preceding relations (6.3.28)–(6.3.32) starting from the fundamental Gram–Schmidt relation (6.3.36).

Similarly, the relation between the partial squared coherence, for $j > r$,

$$\kappa^2_{jr.(r-1)!} = \frac{|S_{jr.(r-1)!}|^2}{S_{jj.(r-1)!}S_{rr.(r-1)!}} \qquad (6.3.39)$$

and the conditioned or noise spectrum now follows from (6.3.31) and (6.3.38) as

$$S_{jj.r!} = S_{jj.(r-1)!}(1 - \kappa^2_{jr.(r-1)!}) \qquad (6.3.40)$$

$$= S_{jj}(1 - \kappa^2_{j1})(1 - \kappa^2_{j2.1}) \cdots (1 - \kappa^2_{jr.(r-1)!}) \qquad (6.3.40a)$$

showing how the noise spectrum is obtained from the output spectrum by successively removing the parts coherent with inputs. If X_{pt} is the output and $X_{1t}, \ldots, X_{(p-1)t}$ are the inputs, then the ratio of the spectrum coherent to all the inputs with the original spectrum is called the multiple squared coherence function:

$$\kappa^2_{p:(p-1)!} = \frac{(S_{pp} - S_{pp.(p-1)!})}{S_{pp}} \qquad (6.3.41)$$

and follows from (6.3.40a) by substituting $j = p, r = (p-1)$, as

$$\kappa^2_{p:(p-1)!} = 1 - (1 - \kappa^2_{p1})(1 - \kappa^2_{p2.1}) \cdots (1 - \kappa^2_{p(p-1).(p-2)!}) \qquad (6.3.42)$$

6.4 TWO- AND THREE-CHANNEL SISO AND MISO MODELING WITH EXAMPLES

We will now illustrate some of the results by examples of both experimentally generated and real-life data. The first two experimentally generated data results in Sections 6.4.2 and 6.4.3 are from Dabrowski (1985), whereas the real-life study of machine-tool dynamics in Section 6.4.4 is from Okafor and Pandit (1987). The interested reader is referred to these references for more details and additional references, although our notation in this book is slightly different. It should also be emphasized that better algorithms and software as well as hardware computer facilities become available with the progress of time and hence these results should be examined for their illustrative ability rather than their numerical/algorithmic and graphical accuracy.

Since we will be dealing with dominantly stochastic data, the moving average parameters must be included. Therefore the full-fledged ARMAV$(n, n-1)$ modeling strategy needs to be implemented; the ARV(n) short-cut justified on the basis of Section 2.4.3 in Section 5.2.1 for modal applications is no longer viable. The estimation of ARMAV model parameters is computationally intensive, whereas that of scalar ARMA models is manageable. Under a mild assumption that the a_{it} for different i are uncorrelated, the ARMA model can be extended to accommodate more than one series and then the parameters estimated by minimizing the sum of squares of a_{it} for each i separately, using a simple extension of the ARMA program, [e.g., see Pandit and Wu (1983)]. We will describe these extended models and their transformation to ARMAV models from this reference in Section 6.4.1; these models will be used in the examples of this section as well as in Section 6.5.

6.4.1 Extended ARMA Models and Their Transformation to ARMAV Models

For multiple series, say $X_{1t}, X_{2t}, \ldots, X_{pt}$ with X_{pt} as the output and the rest as inputs, a simple extension of the univariate ARMA($n, n-1$) model can be written as

$$X_{it} + \sum_{k=i+1}^{p} \phi_{ik0} X_{kt} = \sum_{k=1}^{p} \sum_{m=1}^{n_i} \phi_{ikm} X_{it-m} - \sum_{k=1}^{n_i-1} \theta_{iik} a_{it-k} + a_{it} \quad (6.4.1)$$

$$i = 1, 2, \ldots, p$$

which specializes to (6.2.2) and (6.2.7) in the transfer function form for SISO and MISO systems respectively when $i = p$. The inclusion of the terms ϕ_{ik0} removes the cross-correlation between a_{it} and a_{jt}, $i \neq j$, and the absence of the terms θ_{ijk} guarantees that the noise in the X_{it} model does not include a_{jt-k}. Thus the model allows feedback, but assumes that the input or feedback noise is uncorrelated with the output noise, which is not very restrictive in practice.

If $n_i = n$ for all i then the parameters of an ARMAV($n, n-1$) model are given by

$$\boldsymbol{\phi}_m = \boldsymbol{\phi}_0^{-1} \{\phi_{ikm}\}, \quad m = 1, 2, \ldots, n \quad (6.4.2a)$$

$$\boldsymbol{\theta}_k = \boldsymbol{\phi}_0^{-1} \operatorname{diag} \{\theta_{iik}\} \boldsymbol{\phi}_0, \quad k = 1, 2, \ldots, n-1 \quad (6.4.2b)$$

where $\boldsymbol{\phi}_0 = \{\phi_{ik0}\}$ and any parameters not available from (6.4.1) are set to zero for the corresponding elements in the matrix parameters. The variance–covariance (zero lag) matrix of the residual vector (multivariate white noise $\mathbf{a}_t$) is given by

$$\boldsymbol{\sigma}_a^2 = \boldsymbol{\phi}_0^{-1} \operatorname{diag} \{\sigma_{ai}^2\} \boldsymbol{\phi}_0^{-T} \quad (6.4.2c)$$

where σ_{ai}^2 is the residual variance of a_{it} in the model (6.4.1).

Thus the adequate models are found by the ($n, n-1$) strategy applied to (6.4.1) for each i. The transfer functions can be found by the expressions in Sections 6.2.2 and 6.2.3 for SISO and MISO cases respectively. The multivariate spectrum analysis can be conducted by first transforming these models to the ARMAV models by (6.4.2) and then using the desired expressions from Section 6.3. Experimentally generated and real-life illustrative examples follow in the next three sections.

6.4.2 Two-Channel SISO System

A low-pass analog filter with peak frequency 1000 Hz and damping ratio $\zeta = 0.08$ was used to generate the data. The input was 20 kHz white noise and the output was band-limited over 0–1600 Hz. An HP5420A digital signal analyzer with 12-bit digitization at either 102.4 or 1.024 kHz provided both time and frequency measurements. Analog filters were set to 30 or 3 kHz to avoid aliasing. The maximum bandwidth was 25.6 and 51.2 kHz for frequency and time measurements, respectively, with usable base band range of 0–25.6 kHz.

A 6400 Hz bandwidth was used for the two-channel FFT measurements. A 10-sample ensemble average provided the auto-, cross-spectrum, and the FRF or transfer function. A 3200 Hz time record bandwidth was used to collect a 512-point time series for DDS modeling, the actual bandwidth of HP5420A being twice the menu selection. The effect of the HP5420A antialiasing filter roll-off in the 3200–6400 Hz range shows up in the DDS models, but the FFT results are reliable only up to 3200 Hz; hence a comparison between the DDS and FFT results was restricted to the 0–3200 Hz range.

The time records collected on the HP5420A were sent to the UNIVAC mainframe computer using a CBM computer terminal linkage. After finding the adequate models (6.4.1) on the mainframe, they were transformed via (6.4.2) to ARMAV models, which were used to compute the DDS frequency domain functions. These functions in numerical form were then transformed back to the HP5420A for comparing the single-sample DDS results with 10 sample-averaged FFT results. The adequate model (6.4.1) turned out to have $n_i = 12$ for both the input and the output. After transforming to ARMAV, the two models can be written in the operator form using the notation of Section 6.2.1 as

$$\text{Input:} \quad \phi_{11}(B)X_{1t} = -\phi_{12}(B)X_{2t} + \theta_{11}(B)a_{1t} + \theta_{12}(B)a_{2t}$$
$$\text{Output:} \quad \phi_{22}(B)X_{2t} = -\phi_{21}(B)X_{1t} + \theta_{22}(B)a_{2t} \tag{6.4.3}$$

Note that the θ_i matrices are upper triangular.

The Transfer Function or FRF

Since there was no real feedback or feedback noise in this case, only the feedforward transfer function is of interest. Following (6.2.3) it is given by

$$H_{21}(f) = \left. \frac{-\phi_{21}(B)}{\phi_{22}(B)} \right|_{B = e^{-j2\pi f\Delta}}$$

and is shown in Figure 6.3. The magnitude in decibels with 1 volt^2 reference is shown by the solid line, whereas the phase between $\pm 180°$ is shown by the dotted line; this convention will be used throughout the rest of the section. Although the feedback transfer function and even the individual operators $\phi_{11}(B)$, $\phi_{12}(B)$, $\phi_{21}(B)$, $\phi_{22}(B)$ as well as $\theta_{11}(B)$, $\theta_{12}(B)$, and $\theta_{22}(B)$ can be

318 SPECTRUM ANALYSIS

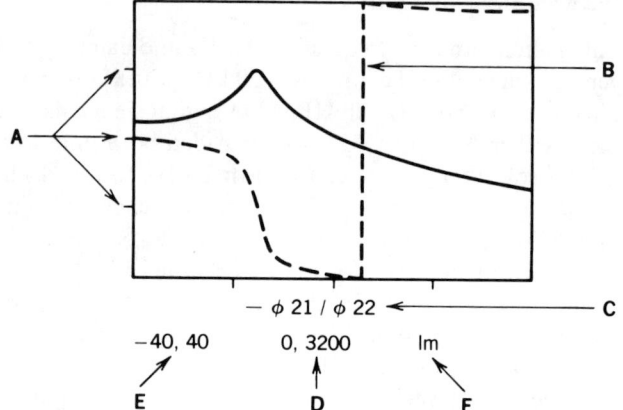

Figure 6.3 The transfer function of the SISO system and the scaling/labeling scheme for subsequent figures. (A) These are the axis half- and quarter-scale tick marks. (B) Dashed lines are usually phase plots ($\pm 180°$ on Y axis). (C) Characteristic. (D) X axis linear-scale frequency. (E) Y axis scale and format code letters (F): lm, log magnitude spectrum (re 1 volt^2); mg, linear magnitude spectrum; re, real component of spectrum.

similarly plotted, they are of no interest in this example. Note that the transfer function can be obtained without the computation of spectra. In Section 6.4.3 it will be computed from the spectra and then compared with that obtained from FFT which must use the spectra.

Decomposition by White Noise Variance–Covariance

We will now illustrate the decomposition of the auto- and cross-spectra by the white noise variance–covariance σ_{ars} given by (6.3.23), but computed by integrating the operators rather than using λ_i and g_i as remarked following (6.3.23). For two series, it is easy to manipulate the operators in (6.3.20a) illustrating the method.

The matrix form of the bivariate model (6.4.3) is given by

$$\phi(B)\mathbf{X}_t = \theta(B)\mathbf{a}_t$$

$$\phi(B) = \begin{bmatrix} \phi_{11}(B) & \phi_{12}(B) \\ \phi_{21}(B) & \phi_{22}(B) \end{bmatrix}, \quad \theta(B) = \begin{bmatrix} \theta_{11}(B) & \theta_{12}(B) \\ 0 & \theta_{22}(B) \end{bmatrix}$$

Let

$$\phi^{-1}(f)\theta(f)$$

$$= \phi^{-1}(B)\theta(B)|_{B = e^{-j2\pi f\Delta}}$$

$$= (\phi_{11}\phi_{22} - \phi_{12}\phi_{21})^{-1} \begin{bmatrix} \phi_{22}\theta_{11} & \phi_{22}\theta_{12} - \phi_{12}\theta_{22} \\ -\phi_{21}\theta_{11} & \phi_{11}\theta_{22} - \phi_{21}\theta_{12} \end{bmatrix} = \begin{bmatrix} b_{11} & b_{12} \\ b_{21} & b_{22} \end{bmatrix}$$

(6.4.4)

where we have omitted the argument (f) for brevity and summarized the operators in the matrix elements by b_{ij}, for example, $b_{11} = \phi_{22}\theta_{11}/(\phi_{11}\phi_{22} - \phi_{12}\phi_{21})$. Now the Hermitian multivariate spectrum matrix (6.3.20a) given by

$$\mathbf{S}(f) = \Delta \boldsymbol{\phi}^{-1}(f)\boldsymbol{\theta}(f)\sigma_a^2 \bar{\boldsymbol{\theta}}^T(f)\bar{\boldsymbol{\phi}}^{-T}(f)$$

takes the form

$$\begin{bmatrix} S_{11} & S_{12} \\ S_{21} & S_{22} \end{bmatrix} = \Delta \begin{bmatrix} b_{11} & b_{12} \\ b_{21} & b_{22} \end{bmatrix} \begin{bmatrix} \sigma_{a11} & \sigma_{a12} \\ \sigma_{a12} & \sigma_{a22} \end{bmatrix} \begin{bmatrix} \bar{b}_{11} & \bar{b}_{21} \\ \bar{b}_{12} & \bar{b}_{22} \end{bmatrix} \quad (6.4.5)$$

and straightforward matrix multiplication yields

$$S_{11} = \Delta\{\sigma_{a11}|b_{11}|^2 + \sigma_{a12}[2\,\text{Real}(b_{11}\bar{b}_{12})] + \sigma_{a22}|b_{12}|^2\} \quad (6.4.5a)$$

$$S_{12} = \bar{S}_{21}$$
$$= \Delta\{\sigma_{a11}b_{11}\bar{b}_{21} + \sigma_{a12}(b_{12}\bar{b}_{21} + b_{11}\bar{b}_{22}) + \sigma_{a22}b_{12}\bar{b}_{22}\} \quad (6.4.5b)$$

$$S_{22} = \Delta\{\sigma_{a11}|b_{21}|^2 + \sigma_{a12}[2\,\text{Real}(b_{22}\bar{b}_{21})] + \sigma_{a22}|b_{22}|^2\} \quad (6.4.5c)$$

Integrating these over the frequency range $-1/2\Delta$ to $1/2\Delta$ gives us the variances in (6.4.5a and c) and the auto-covariance in (6.4.6b) on the left-hand side; each term on the right-hand side provides the contribution due to σ_{aij}, in an alternative form of (6.3.23), without requiring the computation of λ and g; see equations (6.4.7a and b) below. Note that the contribution to the variances from the residual variances σ_{a11} and σ_{a22} given by the first and the last terms are always positive, but that from the residual covariance σ_{a12} given by the middle term can be negative.

Although the two-sided spectra $S_{ij}(f)$ are useful for analytical purposes, since $S_{ii}(f) = S_{ii}(-f)$ and $S_{ij}(-f) = \bar{S}_{ij}(f) = S_{ji}(f)$, the one-sided spectra defined by

$$G_{ij}(f) = 2S_{ij}(f), \quad f \geq 0 \quad (6.4.6)$$
$$= 0, \quad f < 0$$

are more useful for plotting and will be used.

Figure 6.4 graphically illustrates the decomposition of auto-spectra (6.4.5a and c); it also gives the plots of autospectra by the usual FFT methods for comparison. Figures 6.4a–c give the three terms on the right-hand side of (6.4.5a) for the input autospectrum S_{11}, which is plotted in Figure 6.4d. Figures 6.4e–f give the estimated S_{11} by FFT using 10 samples (average) and 1 sample respectively; the smoothing effect of averaging is clear. Similarly, Figures 6.4g–i give the three terms on the right-hand side of (6.4.5c) for the output autospectrum S_{22}, which is plotted in Figure 6.4j. Note the negative contribution in Figure 6.4h due to the middle term of (6.4.6c) resulting from the residual covariance σ_{a12}. Figures 6.4k and l give the estimated S_{22} by FFT using 10 samples (average) and 1 sample, respectively. Since the input is nearly the white

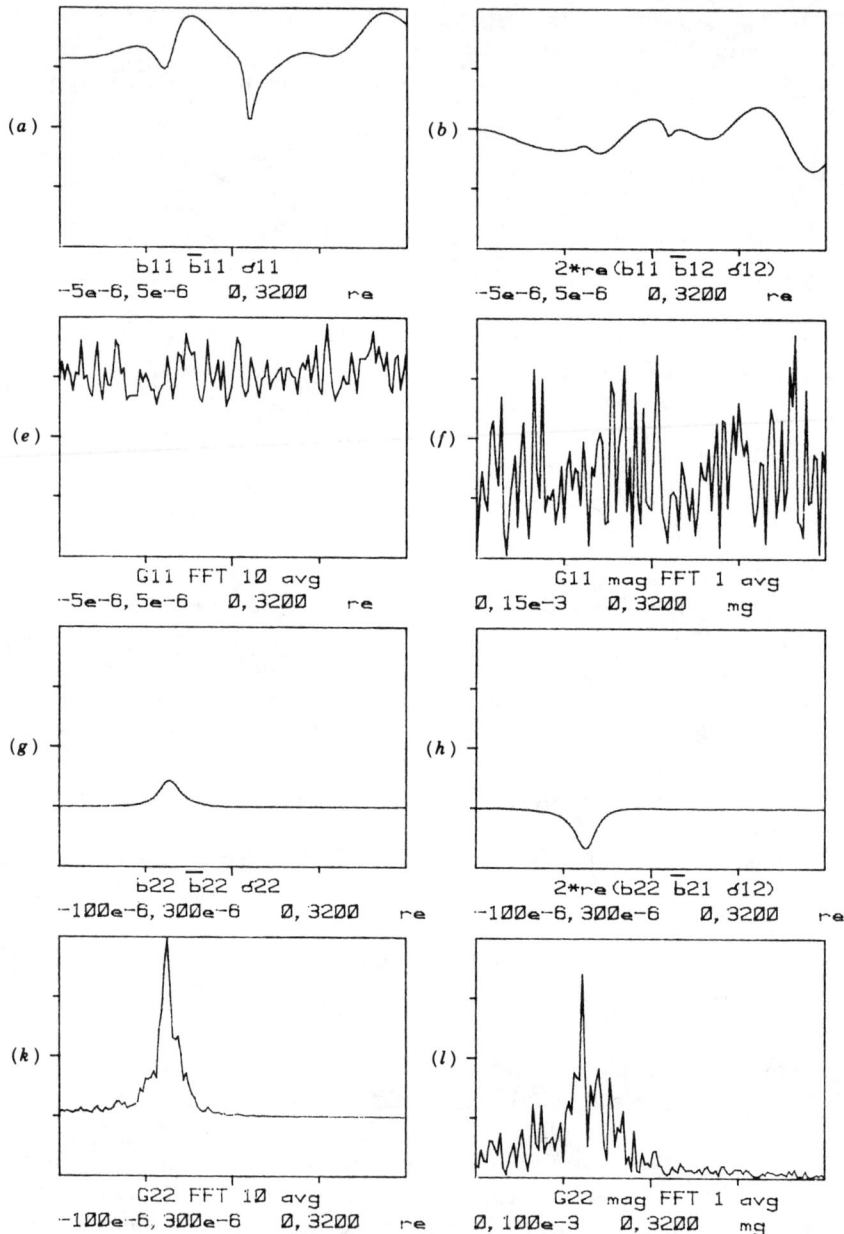

noise a_t, the FFT plots Figures 6.4k and l are practically the FFT spectra in Figures 6.4e and f superimposed on the DDS smooth spectrum S_{22} in Figure 6.4j. Thus the smoothness of the DDS plots results from replacing the white noise spectrum by a constant, while the FFT retains the choppy sample spectrum of the residual a_t's, as clarified in Section 6.1.2 and illustrated in Figures 6.1 and 6.2 for the scalar case.

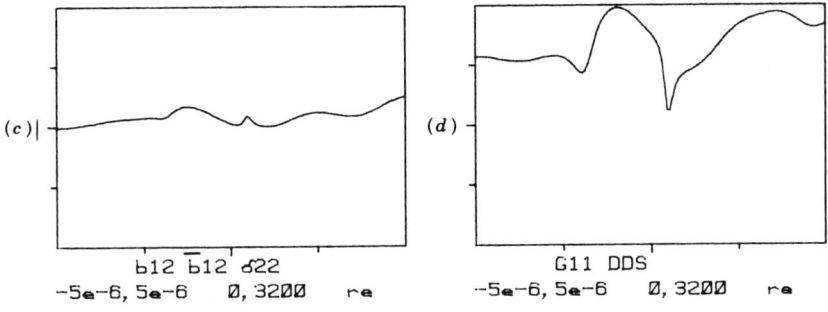

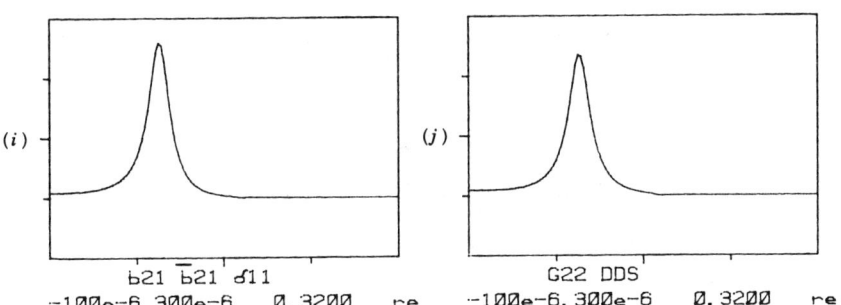

Figure 6.4 Decomposition of the autospectra by white noise variance–covariance (refer to Fig. 6.3 for scaling/labeling).

Integrating the smooth DDS autospectra provides the following variance decomposition of the input and the output, respectively, which is equivalent to (6.3.23):

$$\gamma_{110} = \sigma_{a11} \sum_{i,j=1}^{12} g_{11i}g_{11j}/(1 - \lambda_i\lambda_j) + \sigma_{a12} \sum_{i,j=1}^{12} (g_{11i}g_{12j} + g_{12i}g_{11j})/(1 - \lambda_i\lambda_j)$$

$$+ \sigma_{a22} \sum_{i,j=1}^{12} g_{12i}g_{12j}/(1 - \lambda_i\lambda_j)$$

$$\int S_{11} = \sigma_{a11}\Delta\int |b_{11}|^2 + \sigma_{a12}2\Delta\int \text{Real}(b_{11}\bar{b}_{12}) + \sigma_{a22}\Delta\int |b_{12}|^2 \quad (6.4.7a)$$

$$494.61 = 457.6 \ (92.52\%) - 36.93 \ (7.47\%) \quad\quad + 73.94 \ (14.95\%)$$

$$\gamma_{220} = \sigma_{a11} \sum_{i,j=1}^{12} g_{21i}g_{21j}/(1-\lambda_i\lambda_j)$$
$$+ \sigma_{a12} \sum_{i,j=1}^{12} (g_{21i}g_{22j} + g_{22i}g_{21j})/(1-\lambda_i\lambda_j)$$
$$+ \sigma_{a22} \sum_{i,j=1}^{12} (g_{22i}g_{22j}/(1-\lambda_i\lambda_j)$$
$$\int S_{22} = \sigma_{a11}\Delta \int |b_{21}|^2 + \sigma_{a12} 2\Delta \int \text{Real}(b_{22}\bar{b}_{21}) + \sigma_{a22}\Delta \int |b_{22}|^2 \quad (6.4.7b)$$
$$3.2733 = 3.3890\,(103.53\%) - 0.6779\,(20.71\%) + 0.5622\,(17.18\%)$$

where the integrals are with respect to frequency in Hertz over the range $-1/2\Delta$ to $1/2\Delta$.

Although the partial coherence and the conditioned spectra can be formally calculated, (6.3.28) shows that the only conditioned spectrum $S_{22.1}$ is just the noise spectrum, and $L_{21} = H_{21}$, which is the usual transfer function, as also trivially confirmed by (6.3.35a and b). Hence we will illustrate these by the more interesting MISO system in the next section.

6.4.3 Three-Channel MISO System Illustrating Conditioned Spectra

Figure 6.5 shows the operation amplifier and summing element configuration for generating the MISO system data. The inputs were driven by two independ-

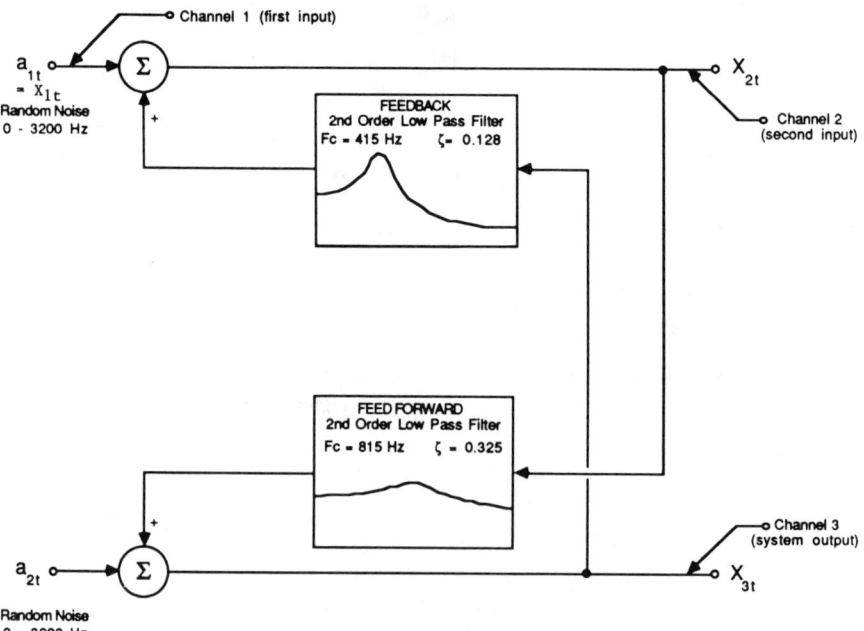

Figure 6.5 Configuration of the MISO System

ent random noise generators. Both the inputs and the outputs were recorded using a B and K 7003 FM tape recorder, band-limited by a Wavetek System 716 low-pass filter to the band 0–3200 Hz and then digitized on HP5420A to mimic a real life analog signal environment. We will give the results of a two-input $(a_{1t} = X_{1t}, X_{2t})$, one-output (X_{3t}) analysis. The spectral estimates from a 25-sample-averaged FFT were compared with a single sample ARMAV(6, 5) model.

Recursive Relations

The general recursive relations of Section 6.3.3 will now be specialized to $p = 3$, thus illustrating them, and then used for computing the conditioned spectra, transfer functions, and coherences plotted later.

The basic **LU** and **LDL̄**^{T} decompositions (6.3.27a and b) simplify to

$$\begin{bmatrix} S_{11} & S_{12} & S_{13} \\ S_{21} & S_{22} & S_{23} \\ S_{31} & S_{32} & S_{33} \end{bmatrix} = \begin{bmatrix} 1 & 0 & 0 \\ L_{21} & 1 & 0 \\ L_{31} & L_{32} & 1 \end{bmatrix} \begin{bmatrix} S_{11} & S_{12} & S_{13} \\ 0 & S_{22.1} & S_{23.1} \\ 0 & 0 & S_{33.12} \end{bmatrix}$$

$$= \begin{bmatrix} 1 & 0 & 0 \\ L_{21} & 1 & 0 \\ L_{31} & L_{32} & 1 \end{bmatrix} \begin{bmatrix} S_{11} & 0 & 0 \\ 0 & S_{22.1} & 0 \\ 0 & 0 & S_{33.12} \end{bmatrix} \begin{bmatrix} 1 & \bar{L}_{21} & \bar{L}_{31} \\ 0 & 1 & \bar{L}_{32} \\ 0 & 0 & 1 \end{bmatrix}$$

(6.4.8)

The output autospectrum can be written in alternative forms involving conditioned auto- and cross-spectra by the last row and column multiplication as

$$S_{33} = L_{31} S_{13} + L_{32} S_{23.1} + S_{33.12} \qquad (6.4.9)$$

$$= |L_{31}|^2 S_{11} + |L_{32}|^2 S_{22.1} + S_{33.12} \qquad (6.4.10)$$

thus specializing (6.3.28) and (6.3.29). For a positive definite $\mathbf{S}(\omega)$, the algorithm based on (6.3.30a–c) specializes to

$$\begin{aligned} L_{21} &= S_{21}/S_{11} \\ L_{31} &= S_{31}/S_{11} \\ S_{22.1} &= S_{22} - |L_{21}|^2 S_{11} \\ L_{32} &= (S_{23} - \bar{L}_{31} L_{21} S_{11})/S_{22.1} \\ S_{33.12} &= S_{33} - |L_{31}|^2 S_{11} - |L_{32}|^2 S_{22.1} \end{aligned} \qquad (6.4.11)$$

which also illustrates (6.3.31) and (6.3.32), since, in this case $S_{23.1} = S_{23} - \bar{L}_{31} L_{21} S_{11}$.

The ordinary and conditioned transfer functions relating the original and conditioned variables are defined by

$$X_{3t} = H_{31}(B)X_{1t} + H_{32}(B)X_{2t} + N_t \qquad (6.4.12)$$
$$= L_{31}(B)X_{1t} + L_{32}(B)X_{2.1t} + X_{3.12t} \qquad (6.4.13)$$

and they are related by

$$\begin{bmatrix} L_{31} \\ L_{32} \end{bmatrix} = \begin{bmatrix} 1 & L_{21} \\ 0 & 1 \end{bmatrix} \begin{bmatrix} H_{31} \\ H_{32} \end{bmatrix} \qquad (6.4.14a)$$

that is,

$$H_{32} = L_{32}$$
$$H_{31} = L_{31} - L_{21}H_{32} \qquad (6.4.14b)$$

which specialize (6.3.33)–(6.3.35). Equation (6.3.36) has only one more equation in addition to (6.4.13) above:

$$X_{2t} = L_{21}X_{1t} + X_{2.1t} \qquad (6.4.15)$$

whereas (6.3.37) and (6.3.38) become

$$X_{3.12t} = X_{3.1t} - L_{32}X_{2.1t}$$
$$X_{3.1t} = X_{3t} - L_{31}X_{1t}$$
$$X_{3.2t} = X_{3t} - L_{32}X_{2t} \qquad (6.4.16)$$
$$X_{2.1t} = X_{2t} - L_{21}X_{1t}$$

and

$$S_{33.12} = S_{33.1} - |L_{32}|^2 S_{22.1}$$
$$S_{33.1} = S_{33} - |L_{31}|^2 S_{11}$$
$$S_{22.1} = S_{22} - |L_{21}|^2 S_{11}$$

The ordinary and partial squared coherences

$$\kappa_{31}^2 = \frac{|S_{31}|^2}{S_{33}S_{11}}, \quad \kappa_{32}^2 = \frac{|S_{32}|^2}{S_{33}S_{22}}, \quad \kappa_{21}^2 = \frac{|S_{21}|^2}{S_{22}S_{11}}$$

$$\kappa_{32.1}^2 = \frac{|S_{32.1}|^2}{S_{33.1}S_{22.1}} \qquad (6.4.18)$$

are related to the ordinary and conditioned autospectra by

$$S_{33.12} = S_{33.1}(1 - \kappa_{32.1}^2) \qquad (6.4.19)$$
$$= S_{33}(1 - \kappa_{31}^2)(1 - \kappa_{32.1}^2) \qquad (6.4.19a)$$

$$S_{22.1} = S_{22}(1 - \kappa_{21}^2) \qquad (6.4.19b)$$

which illustrate (6.3.39) and (6.3.40) and show how the conditioned autospectra are obtained by successively removing the parts coherent with preceding inputs. Finally, the multiple squared coherence

$$\kappa_{3:12}^2 = \frac{S_{33} - S_{33.12}}{S_{33}} \qquad (6.4.20)$$

is in this case given by

$$\kappa_{3:12}^2 = 1 - (1 - \kappa_{31}^2)(1 - \kappa_{32.1}^2) \qquad (6.4.21)$$

which specializes (6.3.41) and (6.3.42), also follows from (6.4.19a) directly, and provides a measure of the proportion of the output X_{3t} coherent with both the inputs X_{1t} and X_{2t} at a given frequency.

Experimental Results

The plots of original and conditioned spectra, transfer functions, and (squared) coherences obtained by the preceding recursive relations applied to the FFT estimates are shown in Figure 6.6, whereas those based on the DDS estimates are shown in Figure 6.7. For ease of comparison both figures have the same layout. In both figures, the upper left-hand 3×3 set of plots show the six auto- and cross-spectrum estimates in the upper triangular form, whereas the three conditioned transfer functions L_{ij} are shown below the diagonal. The key to scaling abbreviation at the bottom of each plot is given in Figure 6.3. Note that for $p = 3$, $L_{32} = H_{32}$, and hence the only other ordinary transfer function H_{31} obtained by (6.4.14b) is given at the left bottom. As before, although we discuss S_{ij}, the plots of $G_{ij} = 2S_{ij}$ are given following (6.4.6). All coherences are κ^2 values.

Ordinary and conditioned coherences of the output X_{3t} with X_{1t}, and X_{3t} with X_{2t} after removing the linear effects of X_{1t}, are shown in the first two rows of the last column on the right, whereas the third row gives the multiple coherence computed by (6.4.21). The last three column elements in the bottom row show the three components of the output spectrum given by equation (6.4.10): the part predicted by the input X_{1t}, $S_{33:1} = |L_{31}|^2 S_{11}$, that predicted by input X_{2t} conditioned on X_{1t}, $S_{33:2} = |L_{32}|^2 S_{22.1}$, and finally the noise spectrum $S_{33.12}$; the corresponding power decomposition obtained by integrating these is presented in Table 6.1 using both millivolts squared units and percent of output power.

The signal-to-noise ratio in the last row is the ratio of the coherent output power with that of the residual noise power obtained by integrating $S_{33:12}$ and $S_{33.12}$, respectively.

The comparison of Figures 6.6 and 6.7 shows that the DDS gives smoother plots than FFT for all characteristics as expected. In particular, the FFT plots show strong spurious fluctuations beyond the 815 Hz roll-off, where there is no signal path between the inputs and the outputs. These fluctuations are most

326 SPECTRUM ANALYSIS

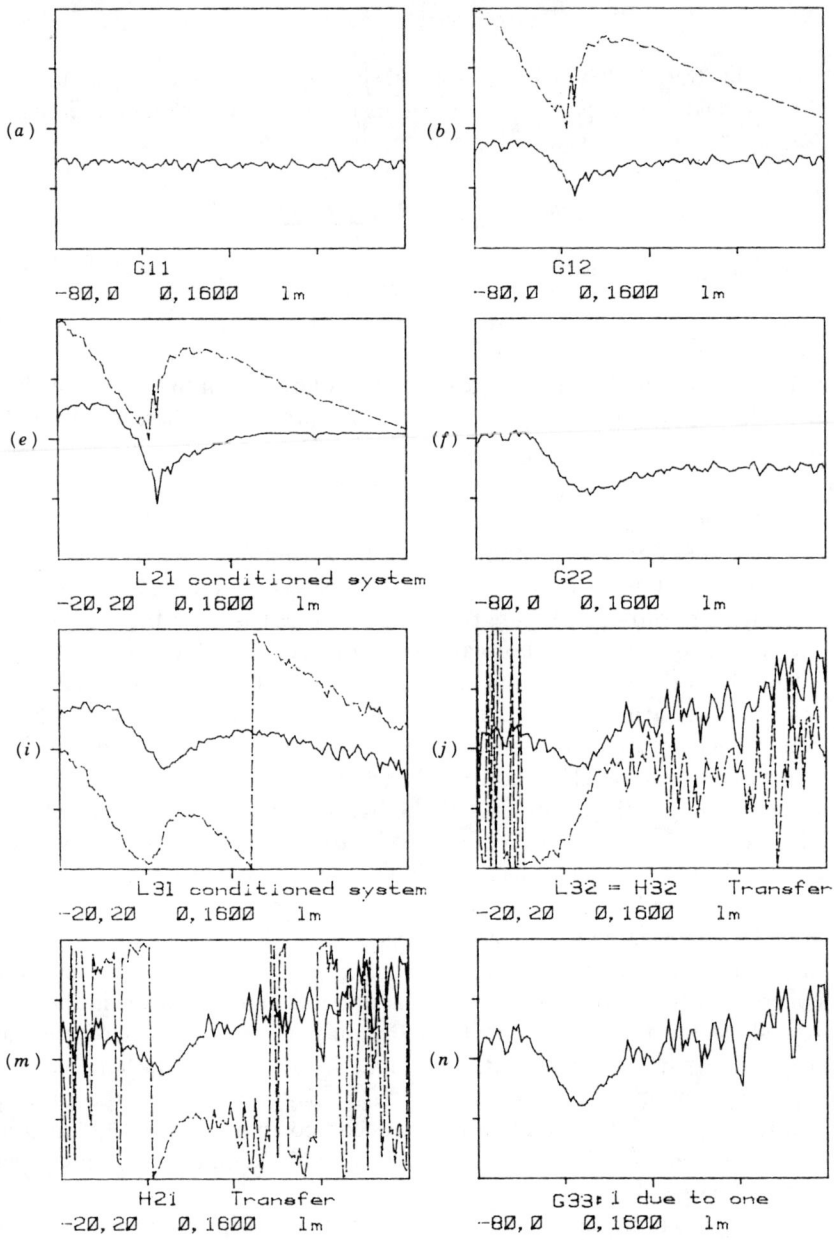

erratic for the partial coherence in Figure 6.6h, whereas the same for the DDS in Figure 6.7h shows the near-perfect coherence over the 0–800 Hz range of interest. The DDS multiple coherence in Figure 6.7l also shows finer resolution in the 800–1600 Hz range. The higher resolution of the DDS method is due to

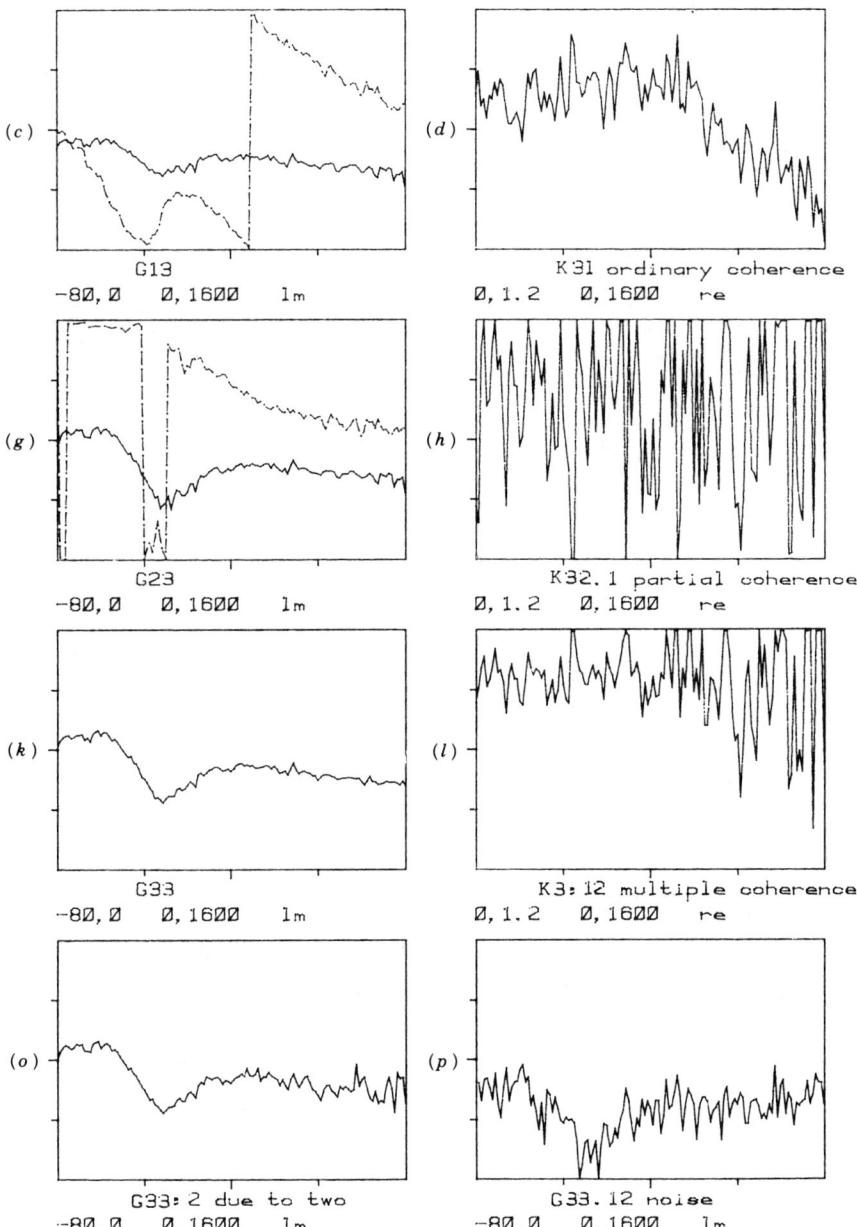

Figure 6.6 Spectra, transfer functions, and coherences of a two-input–one-output system by FFT (refer to Fig. 6.3 for scaling/labeling).

the large dynamic range of the fitted model. The smooth nature and higher resolution of the DDS characteristics is also reflected quantitatively in Table 6.1 and results in the signal-to-noise ratio for DDS being nearly seven times that for FFT.

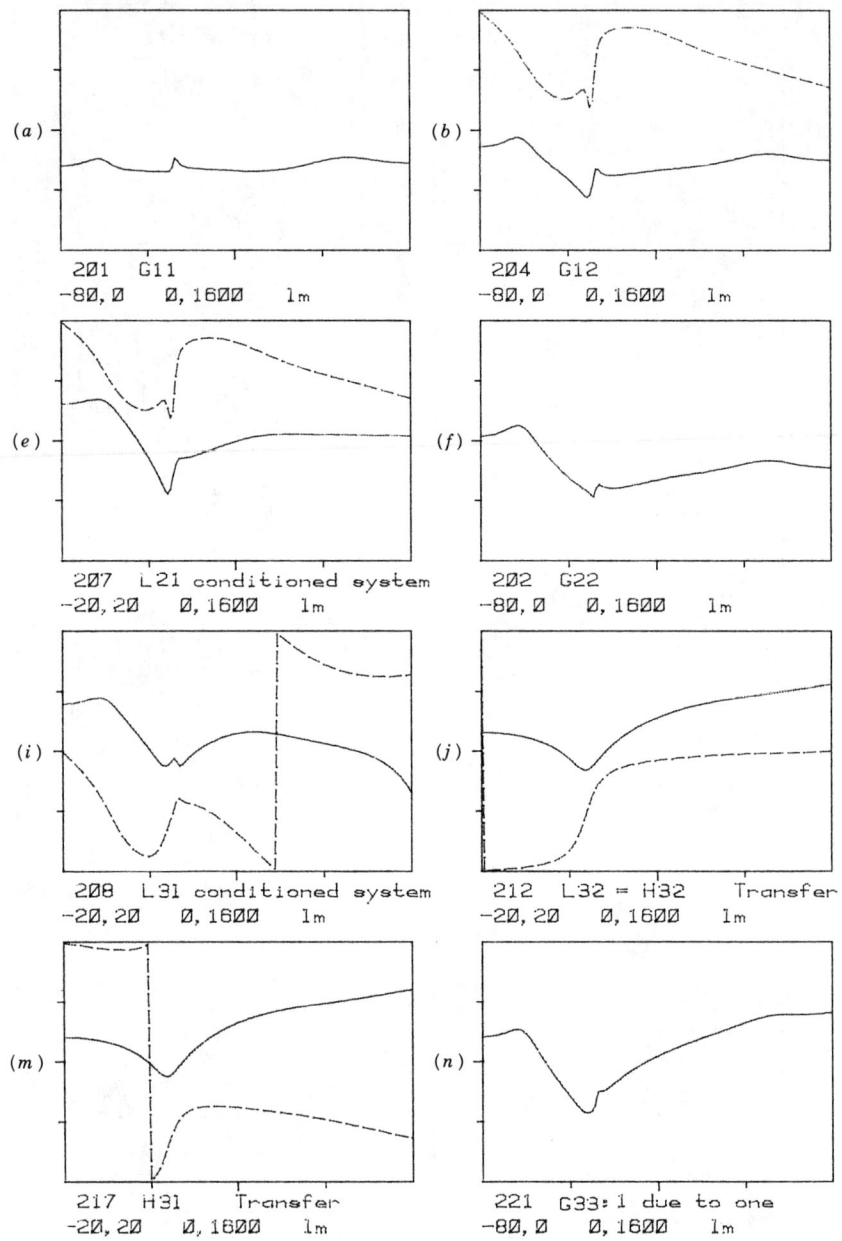

Table 6.1 shows that the FFT predicts 99.4% of the output with 5.8% noise, whereas the DDS predicts 99.1% of the output with about 0.9% noise. Note that the noise in the FFT prediction is larger than what would be expected (1–0.994 = 0.006 or 0.6%). The FFT noise spectrum shows that the statistical estimation errors are also included in the noise term of the decomposition.

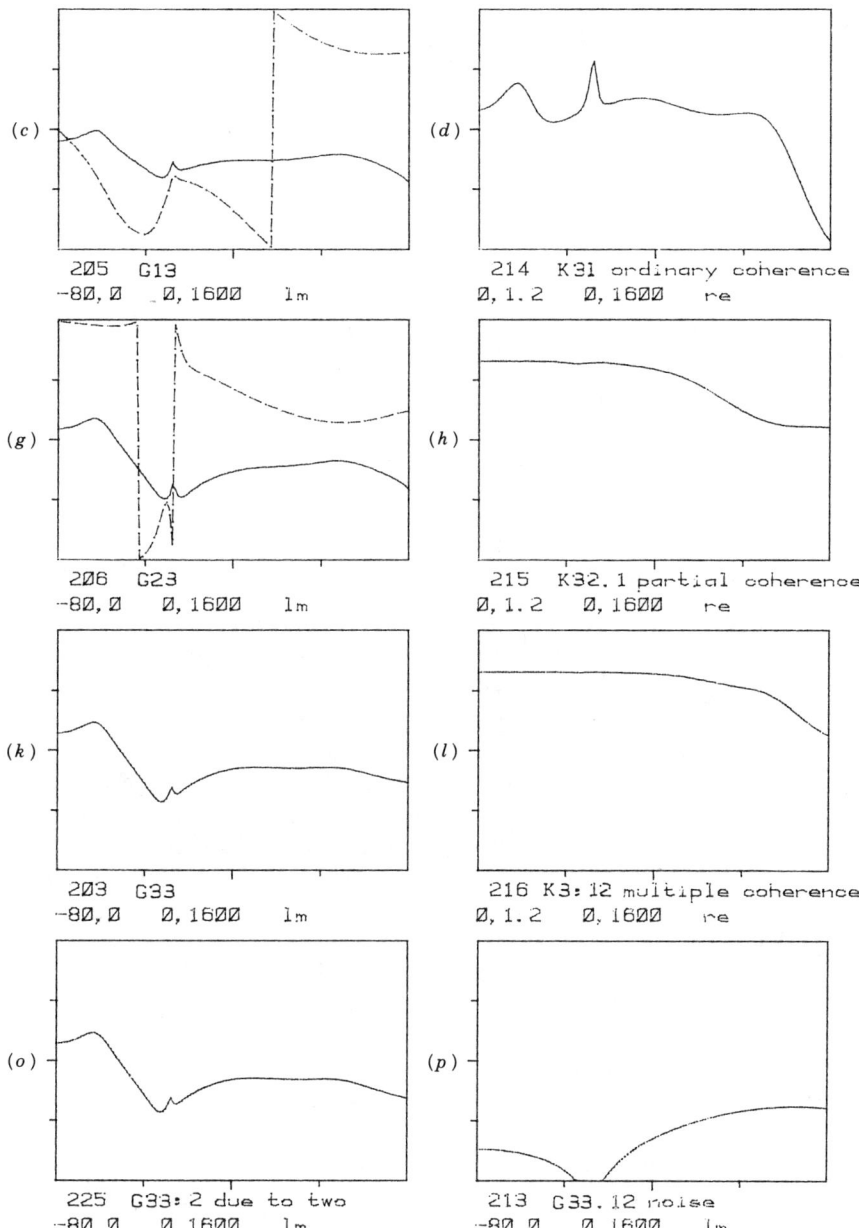

Figure 6.7 Spectra, transfer functions, and coherences of a two-input–one-output system by DDS (refer to Fig. 6.3 for scaling/labeling).

The conditioned and ordinary transfer functions are unable to resolve the feedforward and feedback paths and hence they cannot be resolved by FFT. The DDS method has resolved these paths and also provided the white noise variance–covariance decomposition. We will omit these since these paths will be

330 SPECTRUM ANALYSIS

Table 6.1 Output Power Decomposition for DDS and FFT

	FFT Power		DDS Power	
Integrating	Millivolt² (×0.001)	%	Millivolt² (×0.001)	%
S_{33}	8.7699	100.00	7.2990	100.00
$S_{33:12}$	8.7138	99.36	7.2349	99.12
$S_{33:1}$	6.4279	73.29	5.5284	75.74
$S_{33:2}$	2.2859	26.06	1.7065	23.38
$S_{33.12}$	0.5081	5.79	0.0641	0.88
S/N Ratio	17.1		112.9	

illustrated for a more realistic machine-tool system in Section 6.4.4 and the white noise variance–covariance decomposition was already illustrated in Section 6.4.2; the interested reader is referred to Dabrowski (1985) for these plots.

6.4.4 Feedback Transfer Function Identification of a Machine Tool

The prevalent methods of machine-tool dynamics identification generally use external exciting forces to simulate the cutting conditions. Such simulation is usually incomplete because of design and/or measurement difficulties and may result in inaccurate dynamic characteristics, in contrast to the one obtained under actual working conditions. Moreover, a machine tool in actual cutting is a closed-loop system, with the machine-tool structure in the main (feedforward) loop and the cutting process in the feedback loop, as shown in Figure 6.8. Separate identification of these loops necessitated by the prevalent methods based on FFT may not provide a realistic identification, whereas these paths cannot be resolved by the usual FFT methods based on input–output data alone. Hence, the DDS methodology was used to identify the major modes of a lathe in Okafor and Pandit (1987), which may be referred to for more details and earlier references relating some of the illustrative results reproduced in this section.

As shown in the schematic diagram of Figure 6.8a, the random cutting force generated in the cutting process acts on the machine tool and sets it into vibratory motion with displacement Y normal to the machined surface. Denoting the displacement response Y by X_{1t} and the input force F by X_{2t}, the scalar models (6.4.1) were fitted to the measured radial cutting force and simultaneously measured displacements of the headstock, toolpost, and tailstock units. These models were transformed to the ARMAV form by (6.4.2), which yields upper triangular θ_i matrices. Using the notation of Section 6.2.1, the transfer function form of these models becomes

$$X_{1t} = \frac{-\phi_{12}(B)}{\phi_{11}(B)} X_{2t} + \frac{\theta_{11}(B)}{\phi_{11}(B)} a_{1t} + \frac{\theta_{12}(B)}{\phi_{11}(B)} a_{2t} \qquad (6.4.23)$$

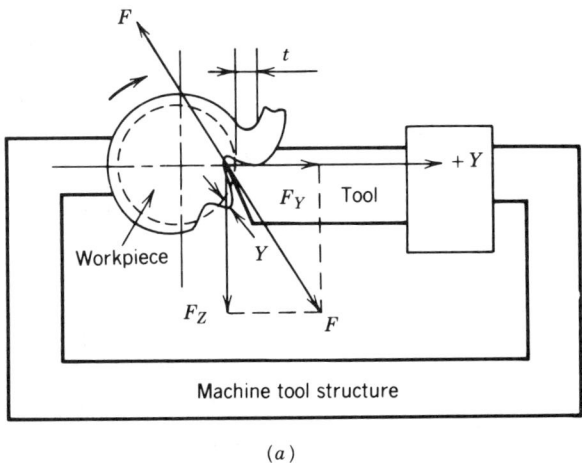

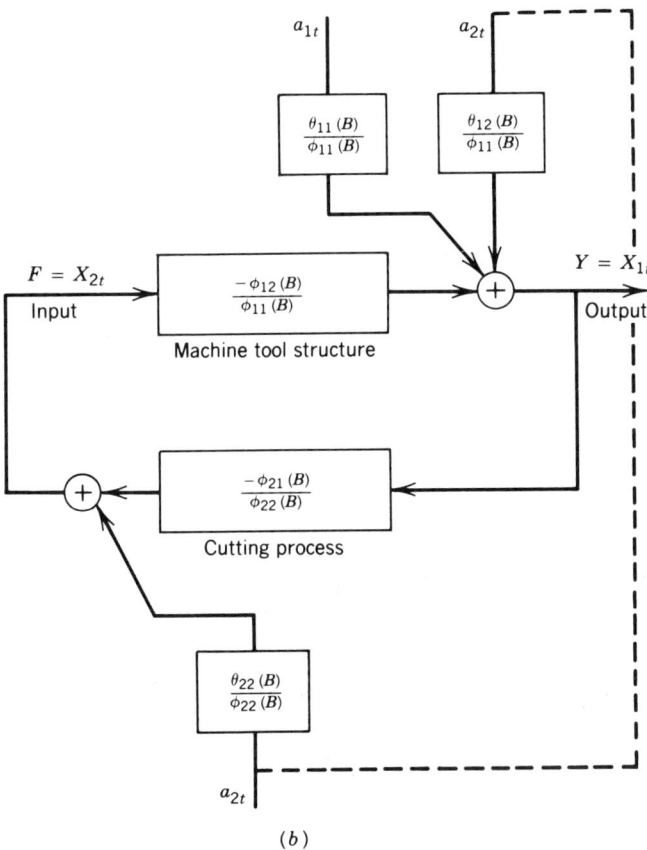

Figure 6.8 Machine tool as a closed-loop system: (*a*) Schematic; (*b*) block diagram. (Reproduced from Okafor and Pandit (1987), Courtesy of the Society of Manufacturing Engineers, Copyright 1987.)

332 SPECTRUM ANALYSIS

$$X_{2t} = \frac{-\phi_{21}(B)}{\phi_{22}(B)} X_{1t} + \frac{\theta_{22}(B)}{\phi_{22}(B)} a_{1t} \qquad (6.4.24)$$

and their block diagram is shown in Figure 6.8b.

Recall from Section 5.1, equation (5.1.3a), that the receptance of the machine-tool unit can now be obtained as

$$H(f) = \frac{Y(f)}{F(f)} = \frac{-\phi_{12}(B)}{\phi_{11}(B)} \bigg|_{B = e^{-j2\pi f \Delta}} \qquad (6.4.25)$$

and the cutting stiffness is similarly obtained from the operator $-\phi_{21}(B)/\phi_{22}(B)$, which will not be discussed further. The usual amplitude versus frequency and phase versus frequency plots of receptance will be given, although

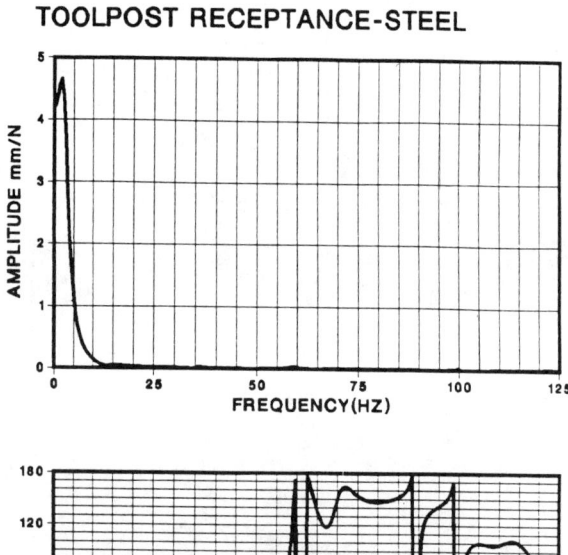

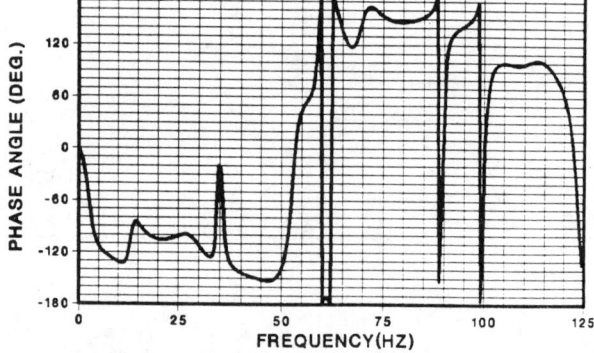

Figure 6.9 Toolpost receptance—lathe idling. (Reproduced from Okafor and Pandit (1987), Courtesy of the Society of Manufacturing Engineers, Copyright 1987.)

the alterative polar plot common in the machine-tool literature can also be readily obtained from (6.4.25).

Figure 6.9 gives typical plots of the toolpost receptance in amplitude and phase form for the cutting speed set at 730 feet/min and the feed rate of 0.0052 in./rev. when the machine tool is idling, that is, the tool is not contacting the workpiece. Figure 6.10 gives the plots of the same characteristics under the same speed and feed when the tool is cutting with a depth of cut 0.2 in. The natural frequencies and damping ratios of the identified major modes (present in other conditions not reproduced here) are listed in Table 6.2; these are calculated directly from the characteristic roots of the operators in the denominators of the respective transfer functions.

It is clear that the dynamics when actual cutting is in progress is quite different from that under idle condition. Comparison of Figures 6.9 and 6.10 shows that there is an overall shift to higher frequencies. The frequency of

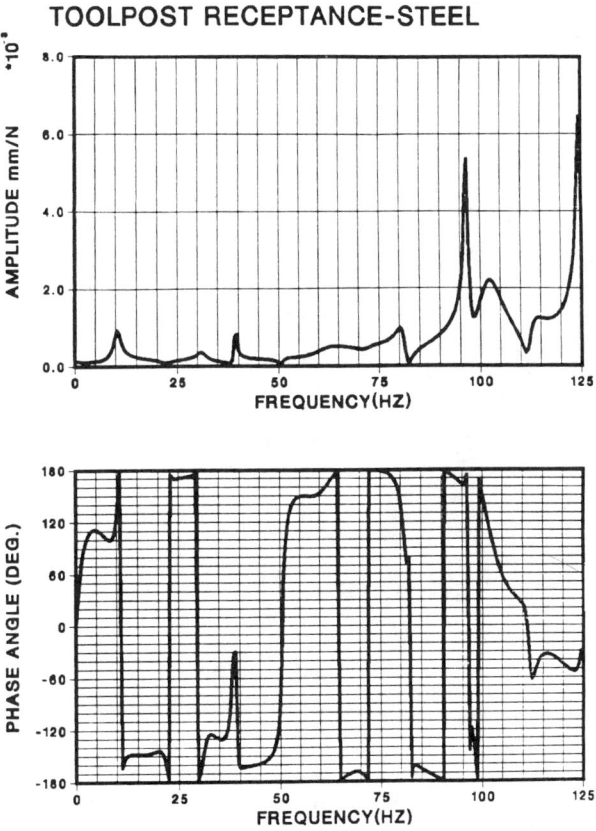

Figure 6.10 Toolpost receptance—lathe cutting. (Reproduced from Okafor and Pandit (1987), Courtesy of the Society of Manufacturing Engineers, Copyright 1987.)

TABLE 6.2 Characteristics of Major Modes of Tool Post Receptance–Steel Workpiece: Natural Frequency (Hz)|Damping Ratio

Condition	Rotation Freq.	Identified Mode Number								Max. Receptance (mm/N)
		1	2	3	4	5	6	7	8	
Idle	10.51	3.0121			39.99 −0.0031	57.07 0.2118		99.33 0.0073		4.6529
Cutting	10.51	9.972 0.0565	21.98 0.1424	30.84 0.0421	39.26 −0.0034	62.72 0.0848	77.45 0.0002	103.00 0.0137	112.00 0.0154	0.0065

rotation appears as the first mode in the tool post when cutting is in progress but does not appear when the machine tool is idling and hence the tool is not contacting the workpiece. This has been confirmed at other cutting conditions as well. A more detailed study of more complete tables similar to Table 6.2 has revealed the relative stability of the three machine-tool units and the optimum cutting conditions over the experimental range.

6.5 MULTICHANNEL MODELING EXAMPLE—FORGE HAMMER NOISE

The simple extension of the univariate single series ARMA($n,n-1$) modeling strategy to multiple series outlined in Section 6.4.1 suffices when the number of inputs and outputs is small. It uses the same autoregressive order $n = n_i$ for all series; this order is successively increased until the decrease in the residual sum of squares of each series is statistically insignificant. For high-order models and/or large number of series, such a "uniform" autoregressive order increase could be quite time consuming. Hence a modeling strategy that takes advantage of the knowledge of the eigenvalues λ_i of the single-series model of each series to rapidly converge on the adequate multiple-series model orders, was outlined in Pandit and Ma (1989) and illustrated by application to forge hammer noise analysis. We will follow this reference for the example in this section.

6.5.1 Modeling Strategy

The strategy rests on the simple concept that the total number of eigenvalues of the state matrices from all the single-series models cannot be less than the number of system eigenvalues of the state matrix of the multivariate (transformed) model and hence provides an upper bound. This upper bound can be further reduced by omitting from the total count the number of eigenvalues repeating between the series, but not within the same series, because the system itself may have repeated eigenvalues. The ratio of this total count of eigenvalues with the number of series, rounded to the next smallest integer, provides an upper bound for the autoregressive order of all series in the multiple series model.

The three-step modeling procedure based on this concept may be outlined as follows:

Step 1: Fit ARMA($n,n-1$) model to each series by increasing n until the residuals are uncorrelated and their sum of squares does not decrease significantly, as judged by the F test.

Step 2: Find the total count of eigenvalues of all series, counting repeated eigenvalues (if any) between series only once. Divide the total count by the number of series and find the next smallest integer n_u.

336 SPECTRUM ANALYSIS

Step 3: Fit the multiple series models (6.4.1) with $n_i = n_u$ for all i. If the residuals are correlated, increase n_i by the usual $(n, n-1)$ strategy; if they are uncorrelated, reduce n_i by 1 successively until residual sum of squares increases significantly for every series at least once. Select the lowest order model with residual sum of squares justified by the F test and uncorrelated residuals for each i.

The extended ARMA models (6.4.1) so obtained are then transformed to the ARMAV models by using (6.4.2). The multivariate spectrum analysis can then be conducted by using the desired expressions from Section 6.3.

6.5.2 Illustration by Forge Hammer Noise Data

The strategy was applied to the forge hammer noise data reported in Tretheway and Evensen (1981), which may be referred to for more details of the experimental procedure and the instrumentation configuration reproduced in Figure 6.11. The inputs are from the ram (X_{1t}), column (X_{2t}), and anvil (X_{3t})

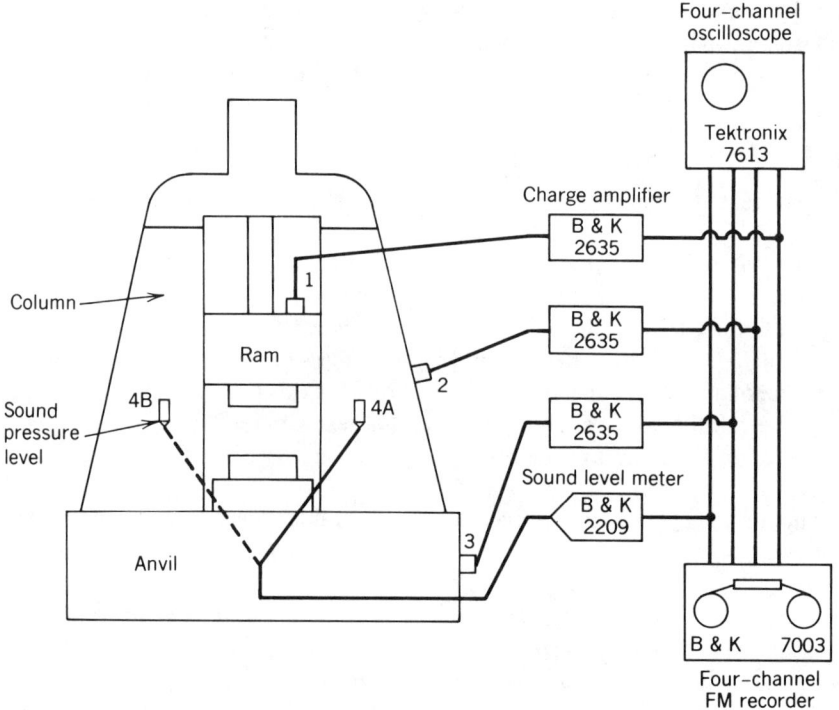

Figure 6.11 Instrumentation configuration for the forge hammer noise data collection. (From Trethewey and Evensen (1981), reproduced with permission.)

acceleration and the output is from sound pressure level—SPL (X_{4t}) measurements. Each series consists of 512 observations sampled at 6.4 kHz sampling frequency. The data are listed in Appendix B.2.

Modeling Results

The results of applying the modeling strategy step by step are as follows:

Step 1: ARMA$(n, n-1)$ model n values

	Ram X_{1t}	Column X_{2t}	Anvil X_{3t}	SPL X_{4t}
Set A	29	38	49	25
Set B	43	44	47	25

Step 2:

Set A: $n_u = 36 \simeq \dfrac{29 + 38 + 49 + 25}{4}$

Set B: $n_u = 40 \simeq \dfrac{43 + 44 + 47 + 25}{4}$

Step 3: Starting from $n_i = 36$ for set A and $n_i = 40$ for set B and reducing by 1 until the residual sum of squares of every series increases significantly at least once, gives the following selected orders.

	n_1	n_2	n_3	n_4
Set A	34	31	31	34
Set B	36	34	34	30

Note that although in theory the number of eigenvalues repeated between the series are supposed to be subtracted from the total count in step 2, this was not done in the present case to illustrate that the procedure is robust enough to correct for it in step 3 with minimal additional computation. In practice this is desirable unless the additional computation is critical, because deciding on an eigenvalue repeated between series is difficult. How small should the difference between real and imaginary parts of two eigenvalues from two different series be for the eigenvalues to be considered equal depends upon the statistical nature of data, computational errors, and the correct (unknown) model. Therefore, it is much more expedient to let the selected model orders in step 3 drop these eigenvalues, as seen in the illustration used above. This is in fact one of the strengths of the strategy as it eliminates one more subjective trial-and-error-based decision.

338 SPECTRUM ANALYSIS

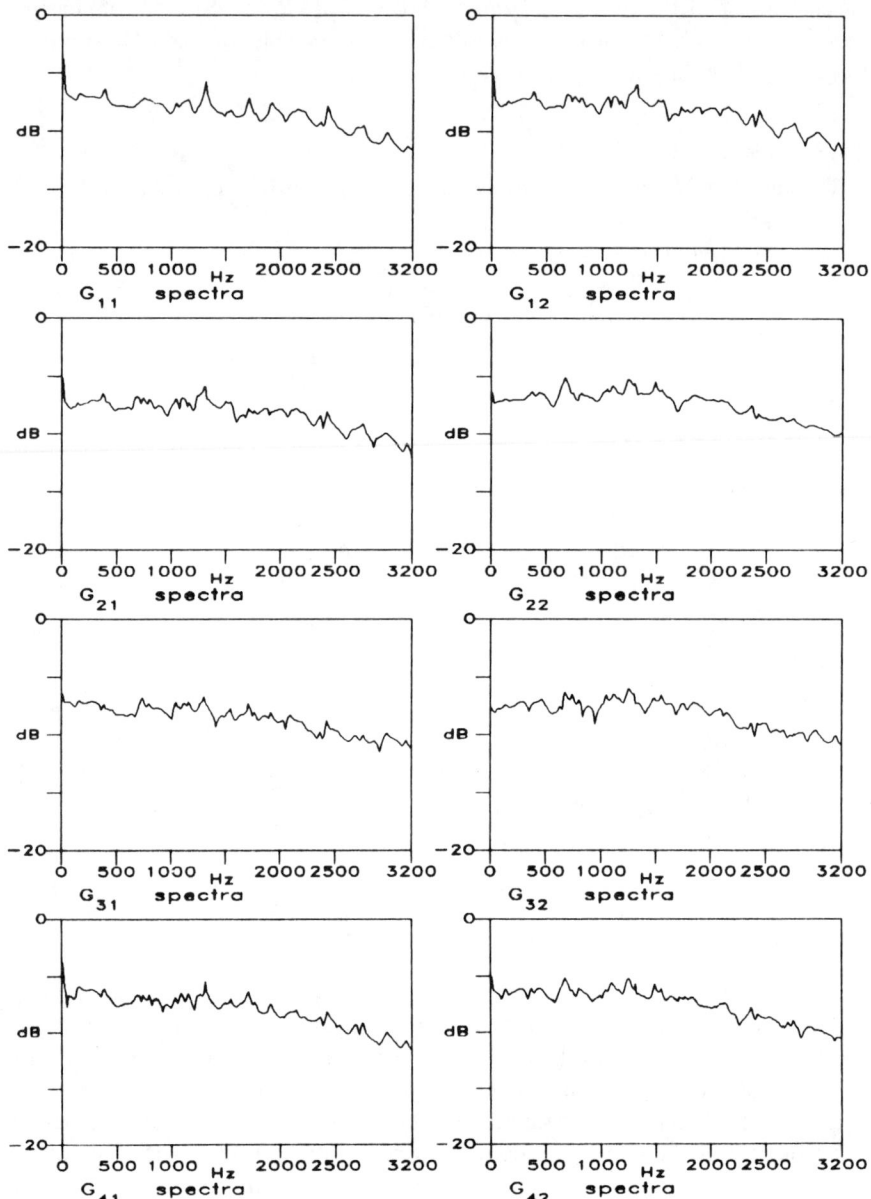

These models were transformed using (6.4.2) and the spectral expressions in Section 6.3 were evaluated. Some of these for set A are plotted in Figures 6.12–6.15 for illustrative purposes.

Auto- and Cross-Spectra
A comparison of the DDS and FFT autospectra (not reproduced here) has shown that the DDS autospectra obtained from a single sample are smoother

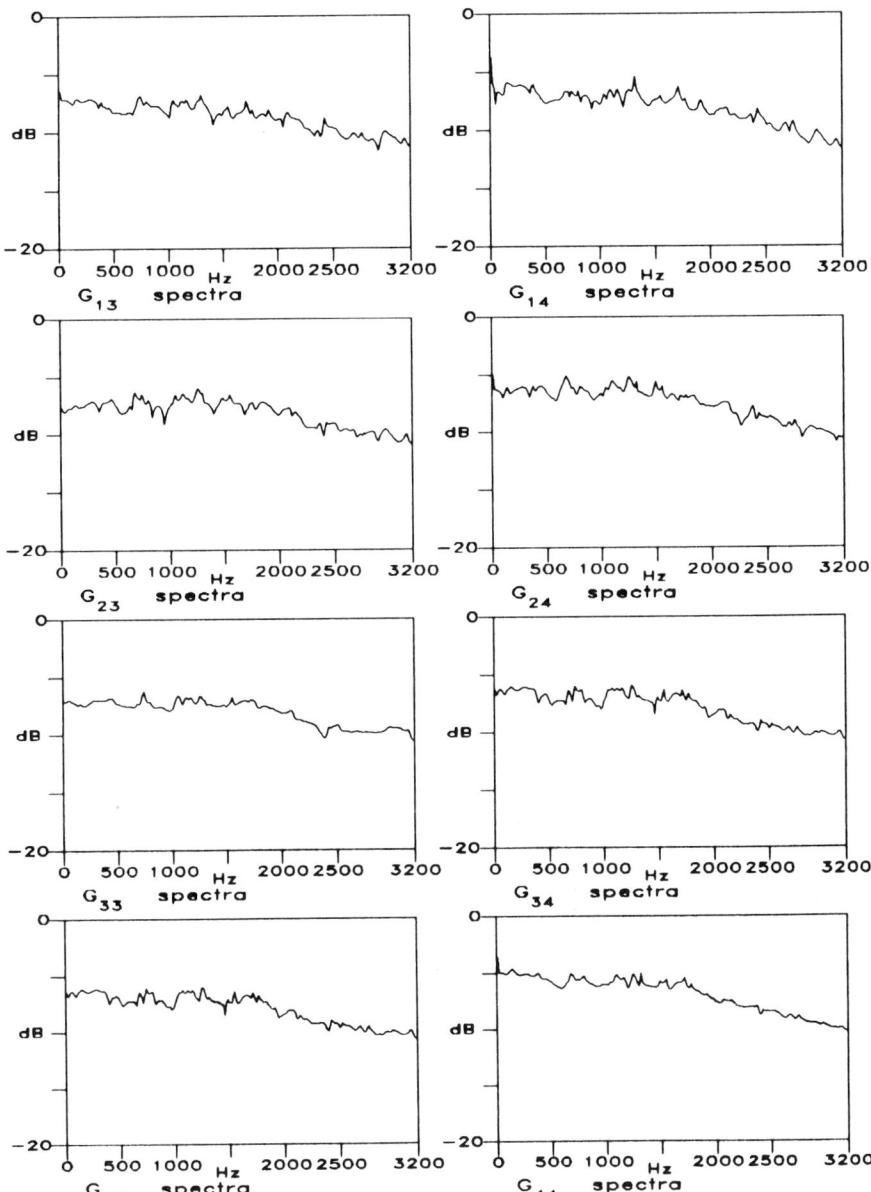

Figure 6.12 Auto- and cross-spectra from forging hammer noise models.

than a 50-sample-averaged FFT. Since the DDS model has 34×4 eigenvalues there are at most $34 \times 2 = 68$ complex conjugate pairs, as many of them can be real. Thus the number of peaks/frequencies is less than 68 as, unlike FFT, only statistically significant peaks are retained.

SPECTRUM ANALYSIS

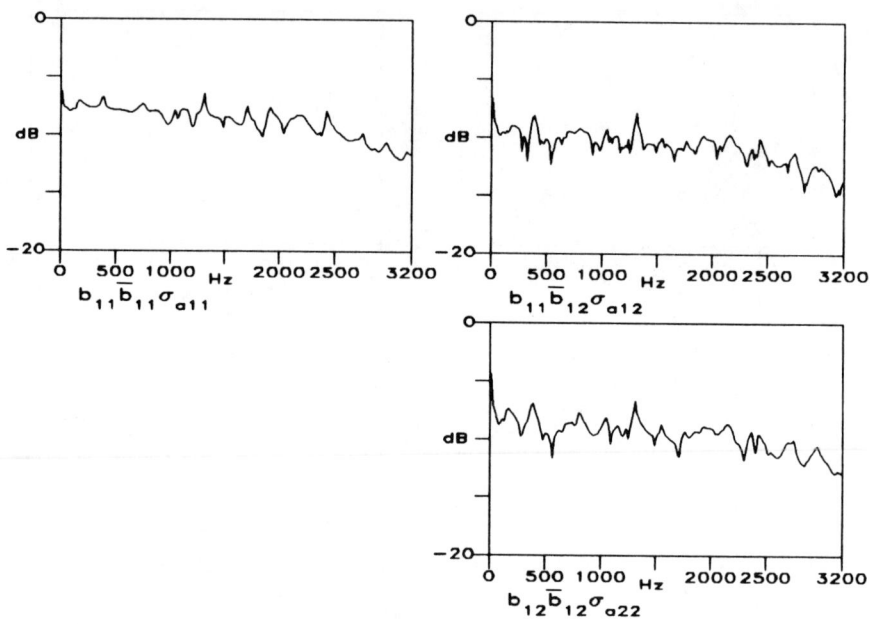

Full Bandwidth Power 0.147166949E-03   %

0.380E-05	0.303E-06	0.953E-07	0.173E-05	2.58	0.21	0.06	1.17
	0.357E-04	0.459E-05	0.392E-05		24.29	3.12	2.66
		0.258E-04	0.571E-05			17.53	3.88
			0.655E-04				44.49

Full Bandwidth Positive Power

0.380E-05	0.314E-06	0.109E-06	0.180E-05	2.58	0.21	0.07	1.23
	0.357E-04	0.465E-05	0.393E-05		24.29	3.16	2.67
		0.258E-04	0.581E-05			17.53	3.95
			0.655E-04				44.49

Full Bandwidth Negative Power

0.000E+00	-0.112E-07	-0.140E-07	-0.762E-07	0.00	-0.01	-0.01	-0.05
	0.000E+00	-0.553E-07	-0.136E-07		0.00	-0.04	-0.01
		0.000E+00	-0.925E-07			0.00	-0.06
			0.000E+00				0.00

The sampling frequency of the data is 6400 Hz, so the Nyquist frequency is 3200 Hz. However, the FFT requires that the entire spectrum must decay before the Nyquist frequency to avoid aliasing. Hence commercial Fourier analyzers, to allow for antialiasing filters, provide frequencies usually up to half the Nyquist frequency that is 1600 Hz in the present case. The DDS spectra do not

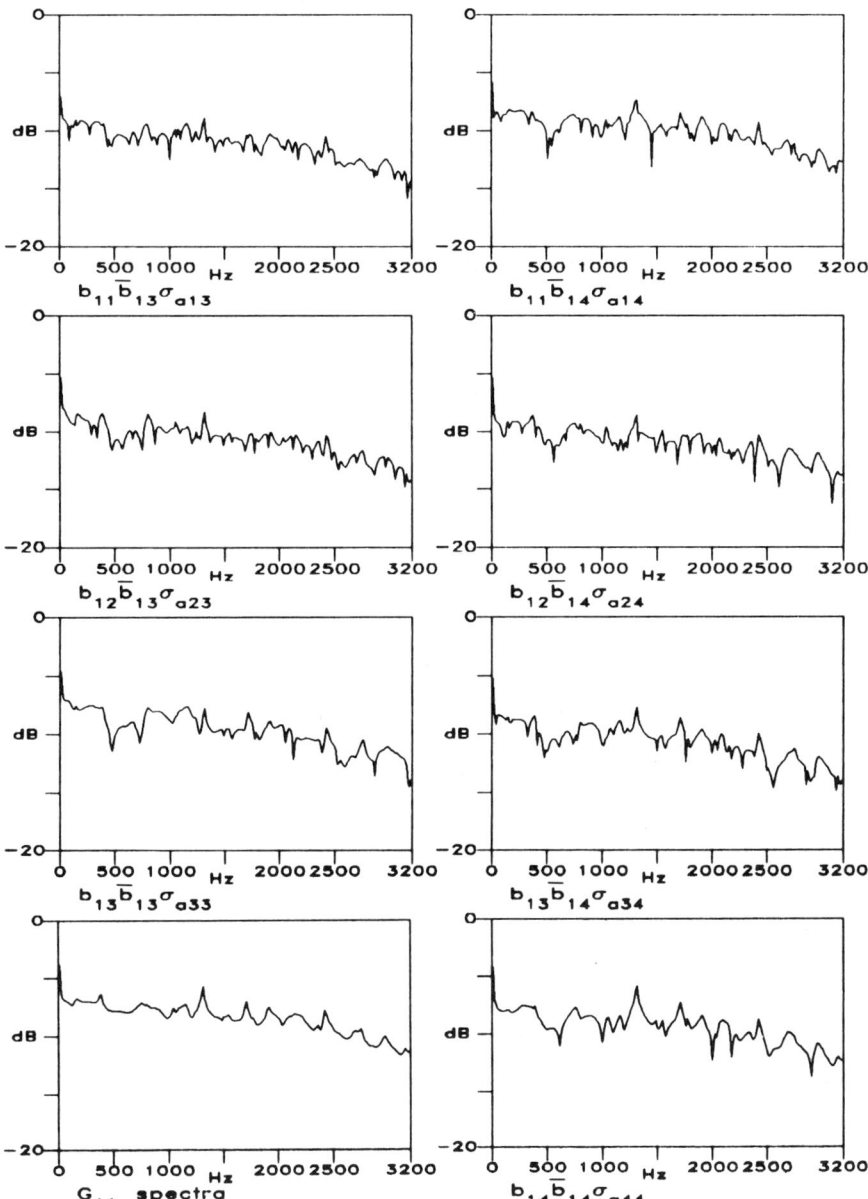

Figure 6.13 White noise variance–covariance decomposition of series 1 (ram).

have this limitation as they require only the peak frequencies to be less than the Nyquist frequency to avoid aliasing. Therefore, as shown in Figure 6.12, the DDS auto- and cross-spectra provide a full range of frequencies up to the Nyquist frequency of 3200 Hz. Moreover, the frequency and damping ratio of

342 SPECTRUM ANALYSIS

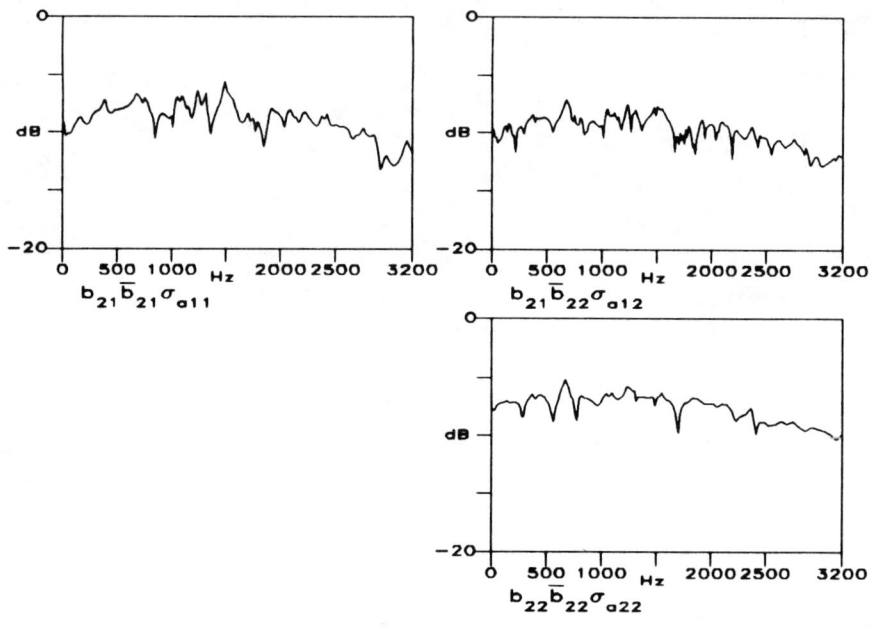

```
              Full Bandwidth Power  0.861655844E-04                         %

0.876E-05   0.292E-06  -0.501E-07  -0.121E-07        10.16      0.34      -0.06     -0.01
            0.510E-04   0.437E-06   0.443E-06                  59.17       0.51      0.51
                        0.756E-05  -0.171E-06                               8.78     -0.20
                                    0.179E-04                                        20.81

              Full Bandwidth Positive Power

0.876E-05   0.451E-06   0.218E-07   0.690E-06        10.16      0.52       0.03      0.80
            0.510E-04   0.884E-06   0.715E-06                  59.17       1.03      0.83
                        0.756E-05   0.213E-06                               8.78     0.25
                                    0.179E-04                                        20.81

              Full Bandwidth Negative Power

0.000E+00  -0.159E-06  -0.719E-07  -0.703E-06         0.00     -0.18      -0.08     -0.82
           0.000E+00   -0.447E-06  -0.271E-06                   0.00      -0.52     -0.31
                       0.000E+00   -0.384E-06                              0.00     -0.45
                                   0.000E+00                                          0.00
```

each peak is directly available from the eigenvalues; there is no need for further processing such as curve fitting. Since the multivariate spectrum matrix is Hermitian, the magnitude spectrum matrix plotted in Figure 6.12 is symmetric.

Decomposition by White Noise Variance–Covariance

The decomposition of Section 6.3.2 by white-noise variance–covariance, given by (6.3.23) but computed by integrating the b_{ij} operators illustrated in (6.4.7a

MULTICHANNEL MODELING EXAMPLE—FORGE HAMMER NOISE   343

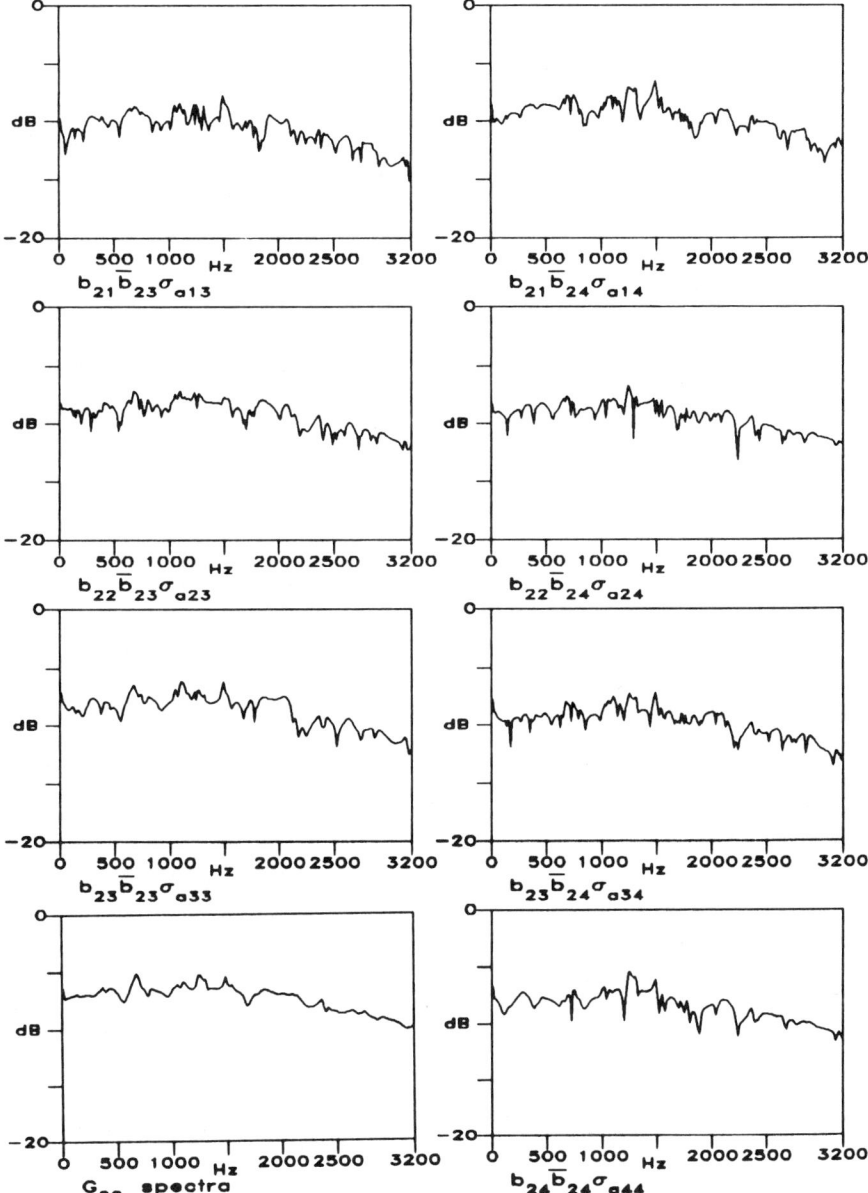

Figure 6.14 White noise variance–covariance decomposition of series 2 (column).

and b), was computed for the set A models. It is illustrated graphically in Figures 6.13–6.16 for the four series, each together with a table showing numerical values after integration on the left and percent values on the right; the layout of the elements matches with the format of the plots. Note that for the off-diagonal

SPECTRUM ANALYSIS

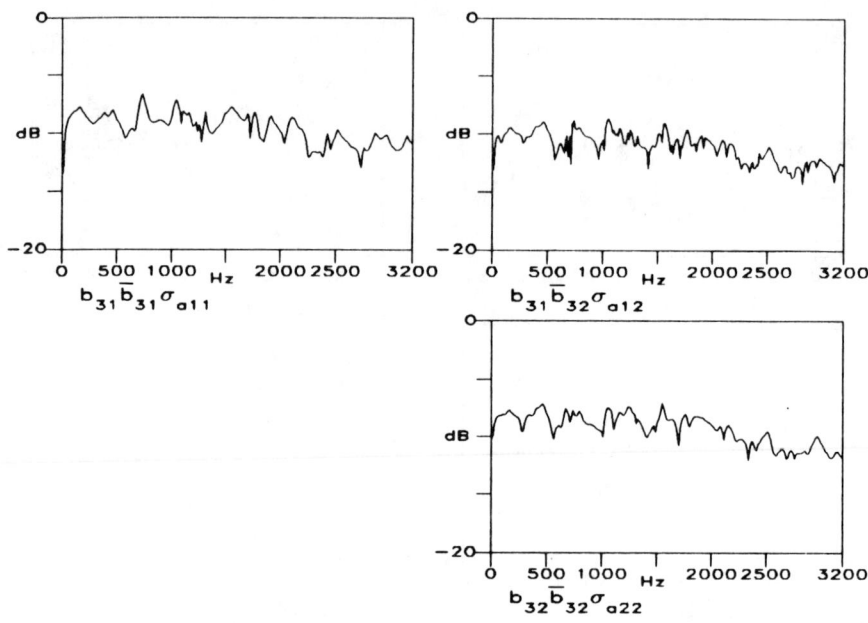

```
            Full Bandwidth Power  0.111254007E-04                    %

0.114E-05  -0.286E-08   0.296E-08   0.520E-07    10.25    -0.03     0.03      0.47
            0.127E-05   0.466E-07  -0.320E-07             11.39     0.42     -0.29
                        0.661E-05  -0.811E-07                      59.45     -0.73
                                    0.212E-05                                19.04

            Full Bandwidth Positive Power

0.114E-05   0.125E-07   0.169E-07   0.119E-06    10.25     0.11     0.15      1.07
            0.127E-05   0.130E-06   0.164E-07             11.39     1.17      0.15
                        0.661E-05   0.900E-07                      59.45      0.81
                                    0.212E-05                                19.04

            Full Bandwidth Negative Power

0.000E+00  -0.153E-07  -0.139E-07  -0.667E-07     0.00    -0.14    -0.13     -0.60
            0.000E+00  -0.835E-07  -0.484E-07              0.00    -0.75     -0.44
                        0.000E+00  -0.171E-06                       0.00     -1.54
                                    0.000E+00                                 0.00
```

complex expressions, we always use twice the real part as illustrated in Figure 6.4 as well as (6.4.7a and b).

As emphasized earlier, since the decomposition can provide positive as well as negative contributions, which enhance as well as diminish the variance or power, the two more tables splitting each contribution into positive and

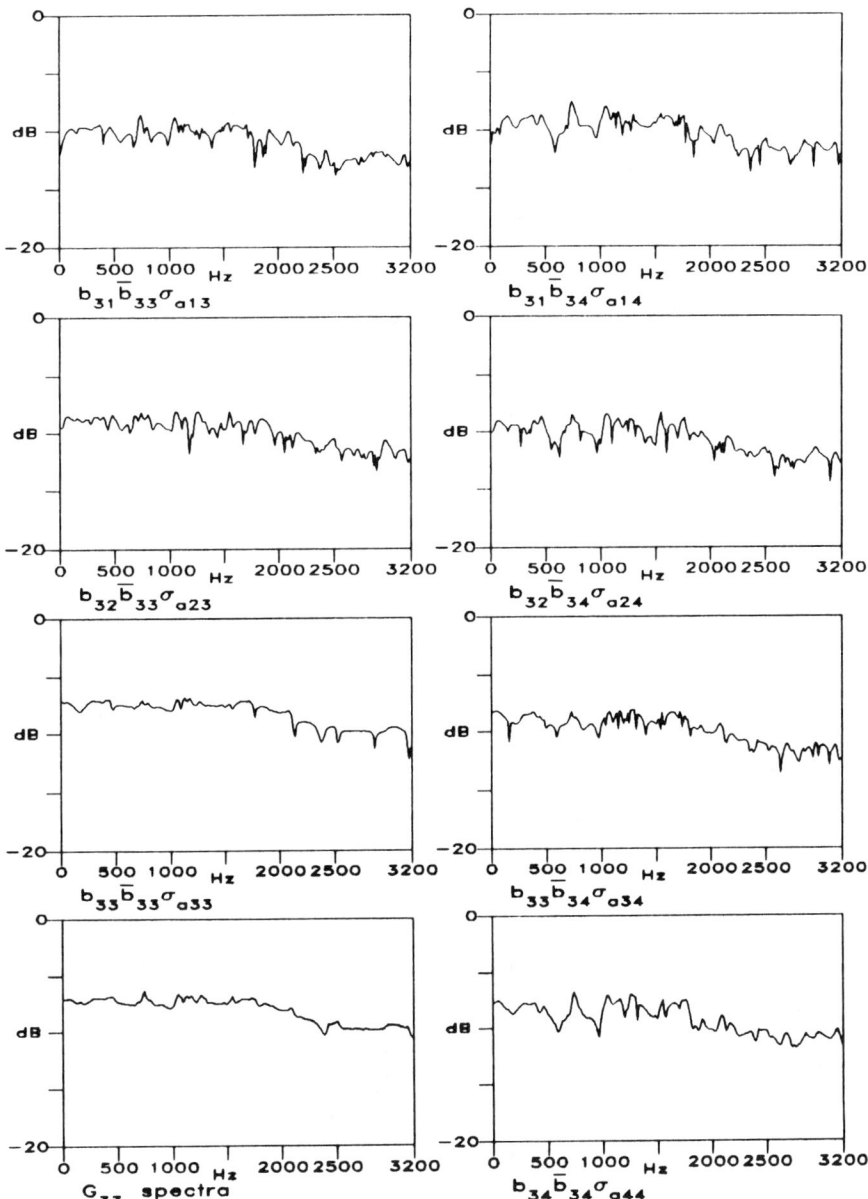

Figure 6.15 White noise variance–covariance decomposition of series 3 (anvil).

negative components are provided. Negligible contribution from the negative components shows that there are no variance-diminishing components in the system that can help reduce the noise.

It is also seen that in most cases the white noise variance from each series is the largest contributor to the variance of that series: 59, 59, and 56% for the

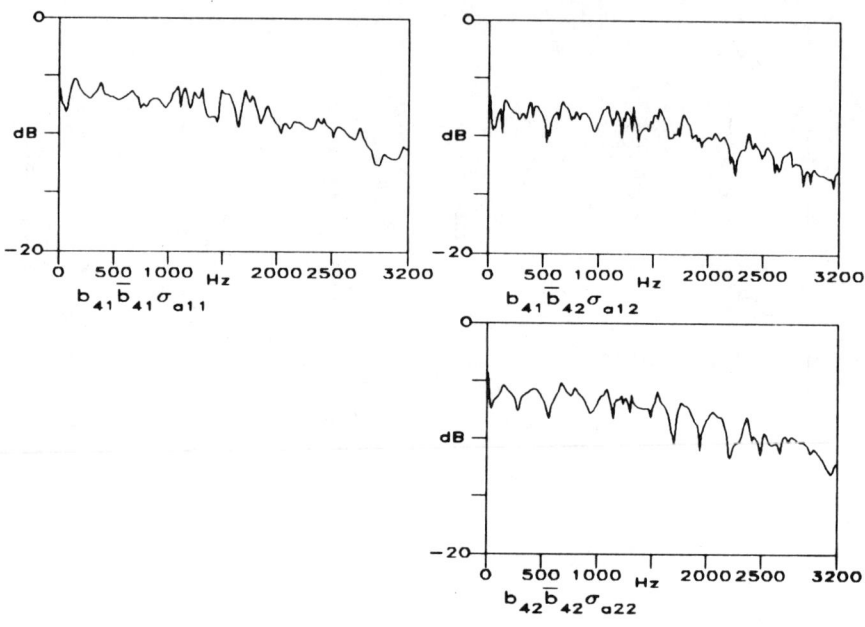

Full Bandwidth Power 0.820963685E-03   %

0.493E-04	0.119E-05	0.369E-06	-0.341E-06	6.01	0.14	0.04	-0.04
	0.127E-03	0.958E-05	0.754E-05		15.50	1.17	0.92
		0.157E-03	0.127E-04			19.06	1.55
			0.457E-03				55.65

Full Bandwidth Positive Power

0.493E-04	0.161E-05	0.713E-06	0.831E-05	6.01	0.20	0.09	1.01
	0.127E-03	0.124E-04	0.104E-04		15.50	1.51	1.26
		0.157E-03	0.196E-04			19.06	2.39
			0.457E-03				55.65

Full Bandwidth Negative Power

0.000E+00	-0.423E-06	-0.343E-06	-0.865E-05	0.00	-0.05	-0.04	-1.05
	0.000E+00	-0.283E-05	-0.283E-05		0.00	-0.34	-0.34
		0.000E+00	-0.695E-05			0.00	-0.85
			0.000E+00				0.00

series 2 (Column), 3 (Anvil), and 4 (SPL), respectively. The only exception is the series 1 (Ram) in which the major contribution 44.5% comes from σ_{a44} of SPL. The table for series 4 (SPL) shows that apart from the 56% contribution from its own σ_{a44}, the largest contribution of 19% comes from σ_{a33} of the anvil, and the next largest of 15.5% from the column. Redesign of these two elements would

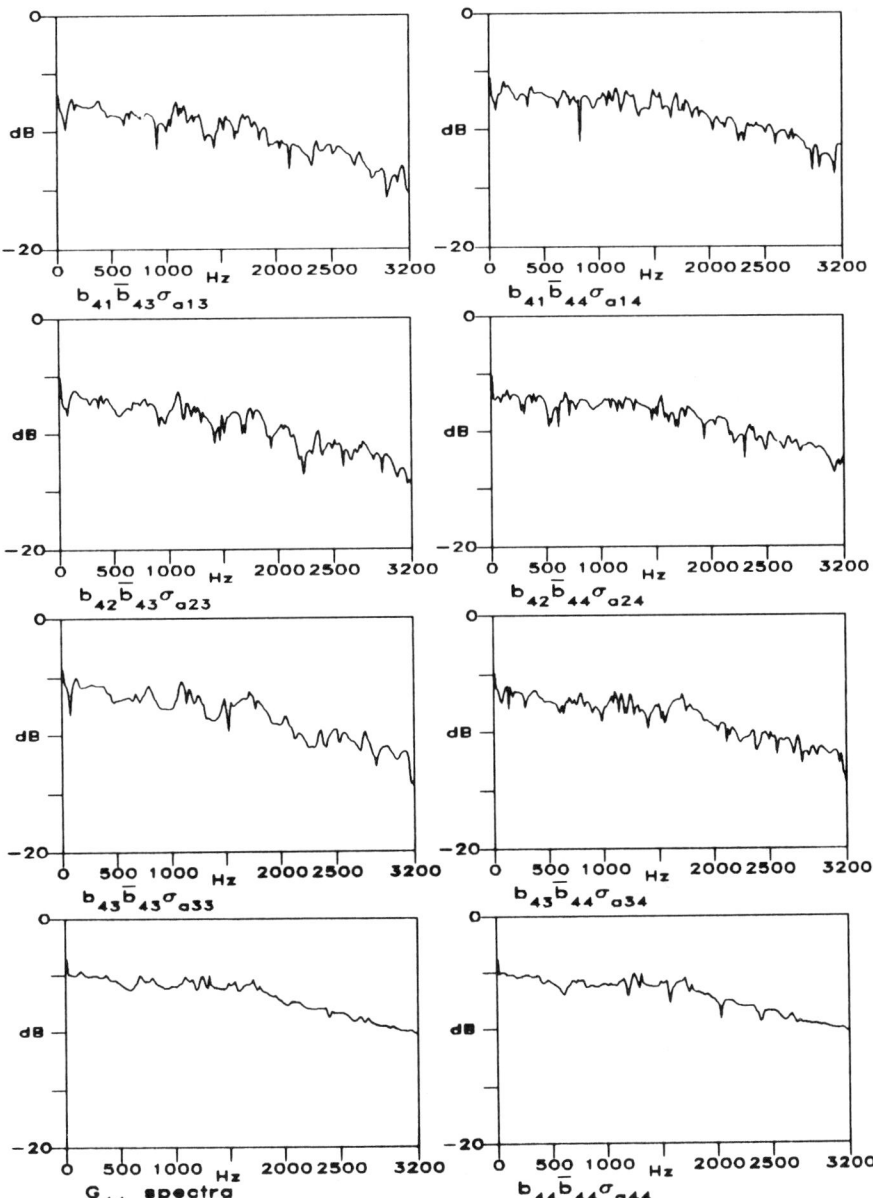

Figure 6.16 White noise variance–covariance decomposition of series 4 (SPL).

therefore help most in reducing the noise. The plots in Figure 6.16 essentially confirm this.

Conditioned Transfer Functions and Spectra
The conditioned spectra and the transfer functions from the fitted models of set

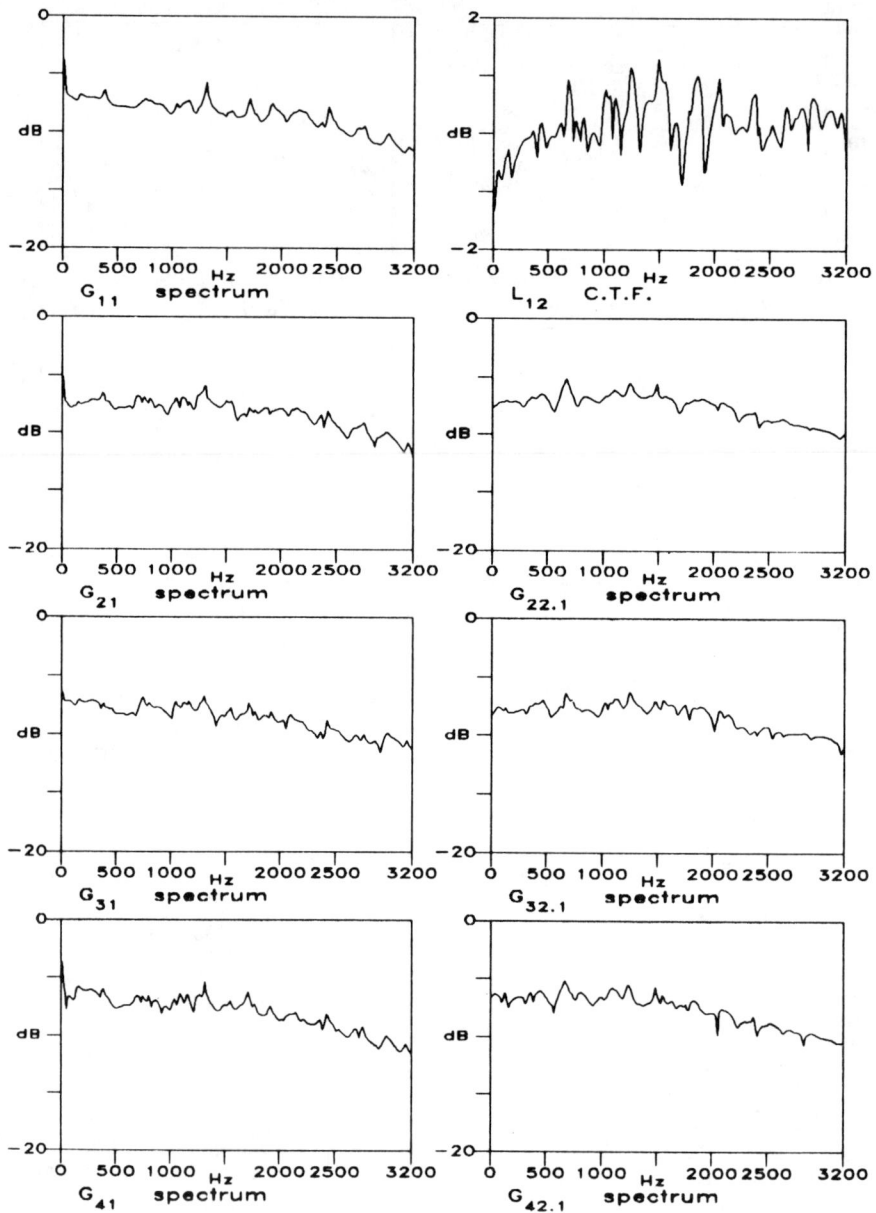

A are shown in Figure 6.17. The conditioned spectra are laid out in the lower triangular matrix, whereas the conditioned transfer functions occupy the remaining upper triangular matrix; the autospectra are on the diagonal and the cross-spectra are on off-diagonal elements as usual. This plotting scheme is also

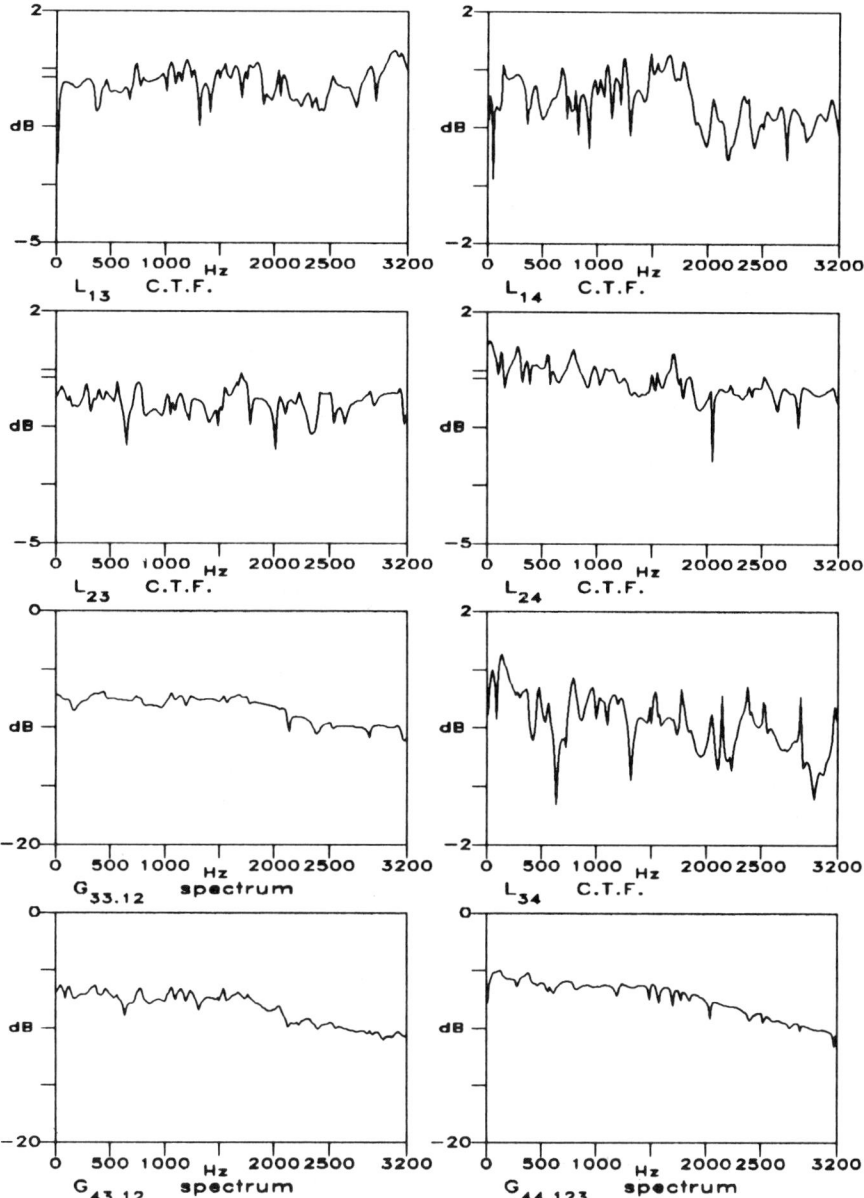

Figure 6.17 Conditioned transfer functions and auto- and cross-spectra.

the storage scheme adopted in the computation, and illustrates the saving in memory achieved by overwriting S_{ik} with L_{ik} for $i > k$ in the computation algorithm following (6.3.29) and (6.3.30).

The S_{ik} used in the computation were obtained by the DDS approach and

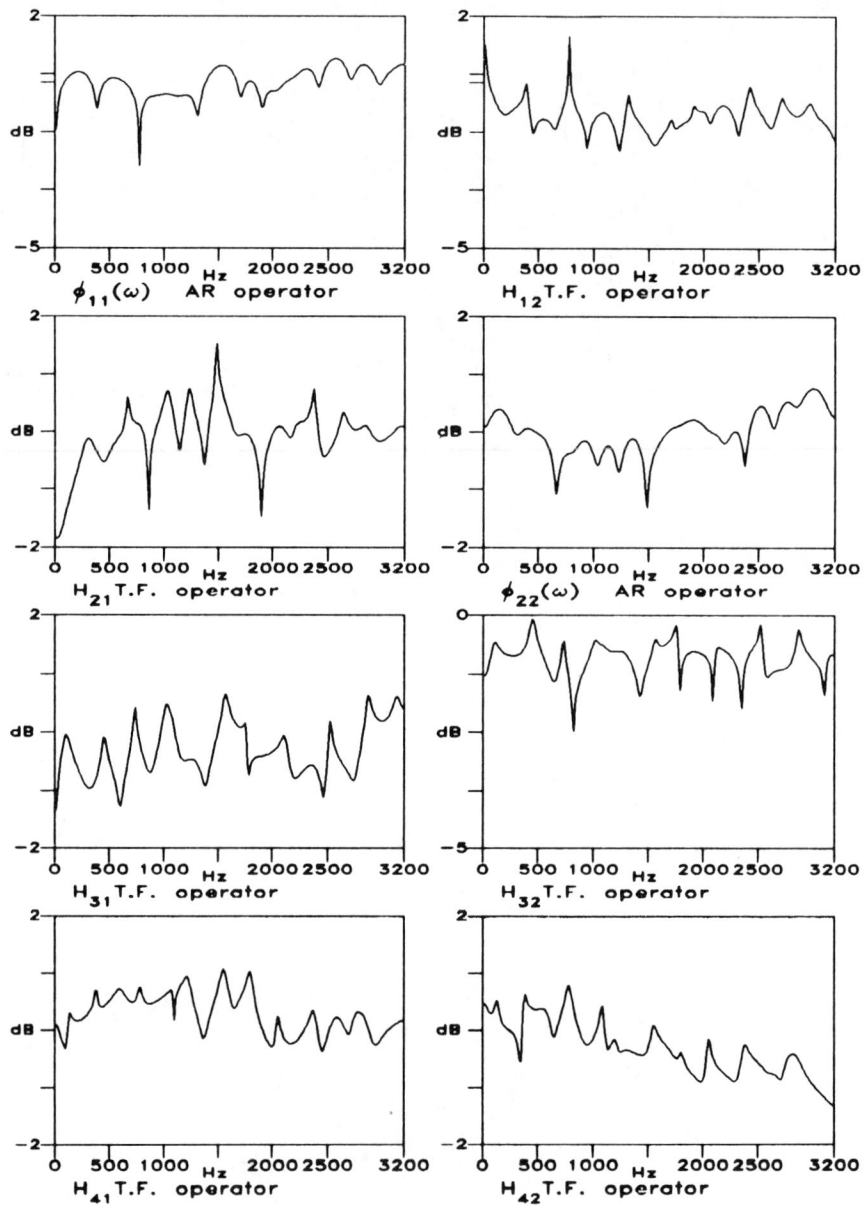

were already smoother than those obtained using FFT, illustrated in Tretheway and Evensen (1981). In spite of this, the conditioned spectra computed from these relatively smooth spectra are still rough, as seen from Figure 6.17. This explains the practical difficulty of applying the residual spectrum technique

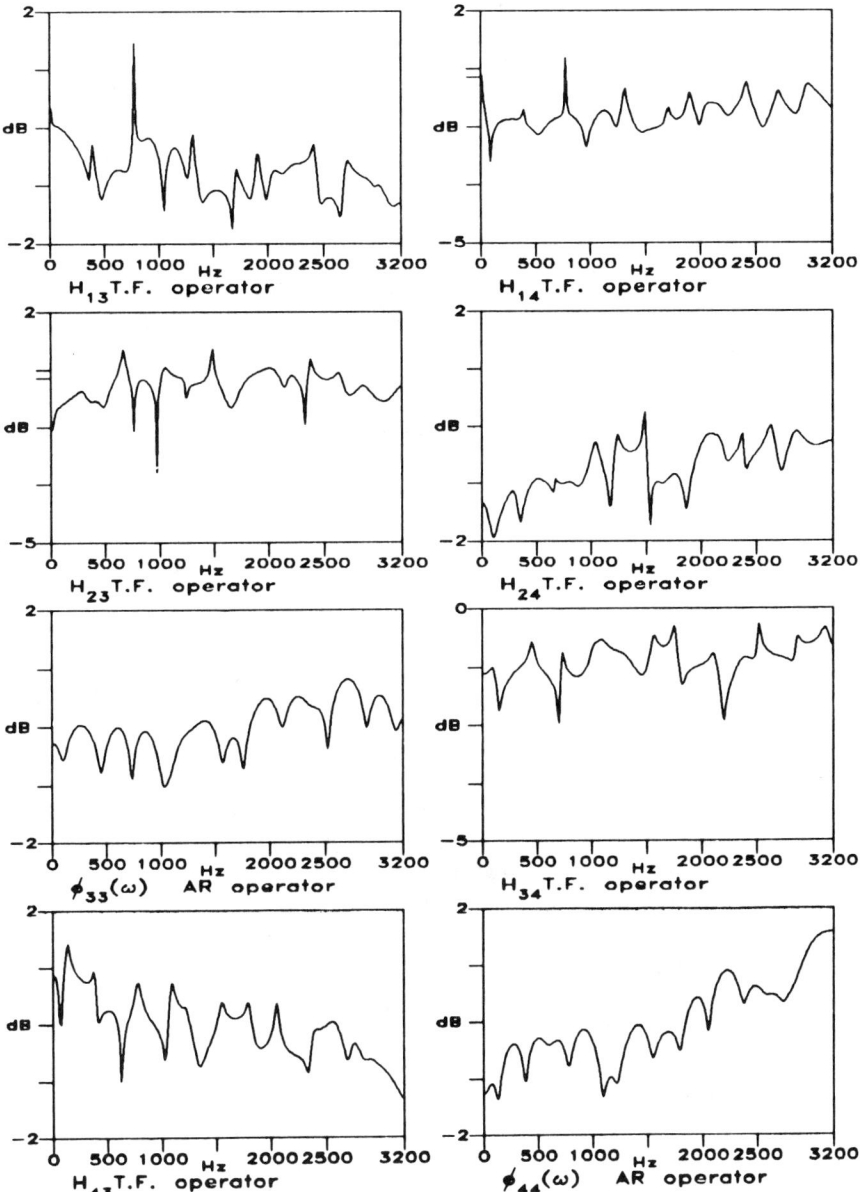

Figure 6.18 Autoregressive operators (diagonal) and ordinary transfer functions (off-diagonal).

using commercial FFT-based Fourier analyzers. The resemblance of the residual noise term in Figure 6.17 with the σ_{a44} term in Figure 6.16 (both at the right bottom) is not surprising. Both these plots represent the residual noise in SPL after the effect of ram, column, and anvil has been removed.

352 SPECTRUM ANALYSIS

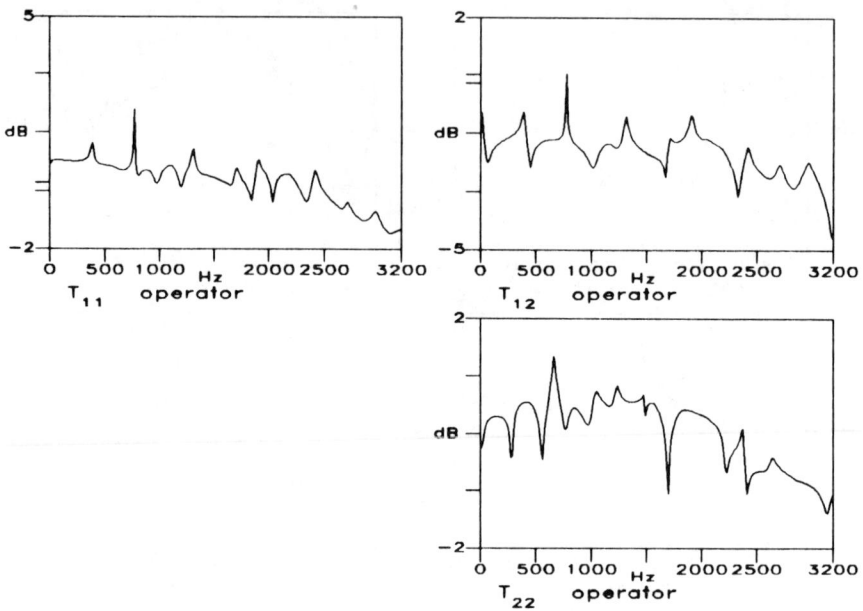

Autoregressive Moving Average Operators

Since all the DDS spectra have been obtained from the ARMAV models, it is helpful to see the plots of the ARMA operators in the frequency domain; these are given in Figures 6.18–6.20. Figure 6.18 gives the autoregressive operator $\phi_{ii}(f)$ on the diagonal and the transfer functions $H_{ij}(f) = \phi_{ij}(f)/\phi_{ii}(f)$, $i \neq j$, on the off-diagonal elements. Figure 6.19 gives the transfer functions relating the series i with residual a_{jt}, $T_{ij}(f) = \theta_{ij}(f)/\phi_{ii}(f)$; this figure is upper triangular because the θ matrices of the ARMAV model transformed from the extended ARMA models are upper triangular. Figure 6.20 gives the elements of the $\phi^{-1}(f)\theta(f)$ matrix, whereas Figure 6.21 gives the elements of the $\phi^{-1}(f)\theta(f)\sigma_a$ matrix which finally provides the $S(f)$ matrix by (6.3.20a).

Figures 6.18 and 6.19 show that the transfer function of the hammer constituents with each other and with random noise sources have only a few well-defined poles (peaks) and zeros (troughs). Their interaction in Figure 6.20 mixes up all these peaks and troughs and this mixing is continued in Figure 6.21 when the σ_a matrix is multiplied from the right. Therefore peaks and troughs in Figures 6.18 and 6.19 would be much easier to deal with in finding their sources rather than their jumbled-up picture in the auto- and cross-spectra. Moreover, the clear-cut peaks and troughs in Figures 6.18 and 6.19 are merely a visual guide; their exact frequencies and damping ratios are known from the eigenvalues directly and can be used to identify and eliminate them individually in order to reduce the overall noise levels in any of the existing frequency bands.

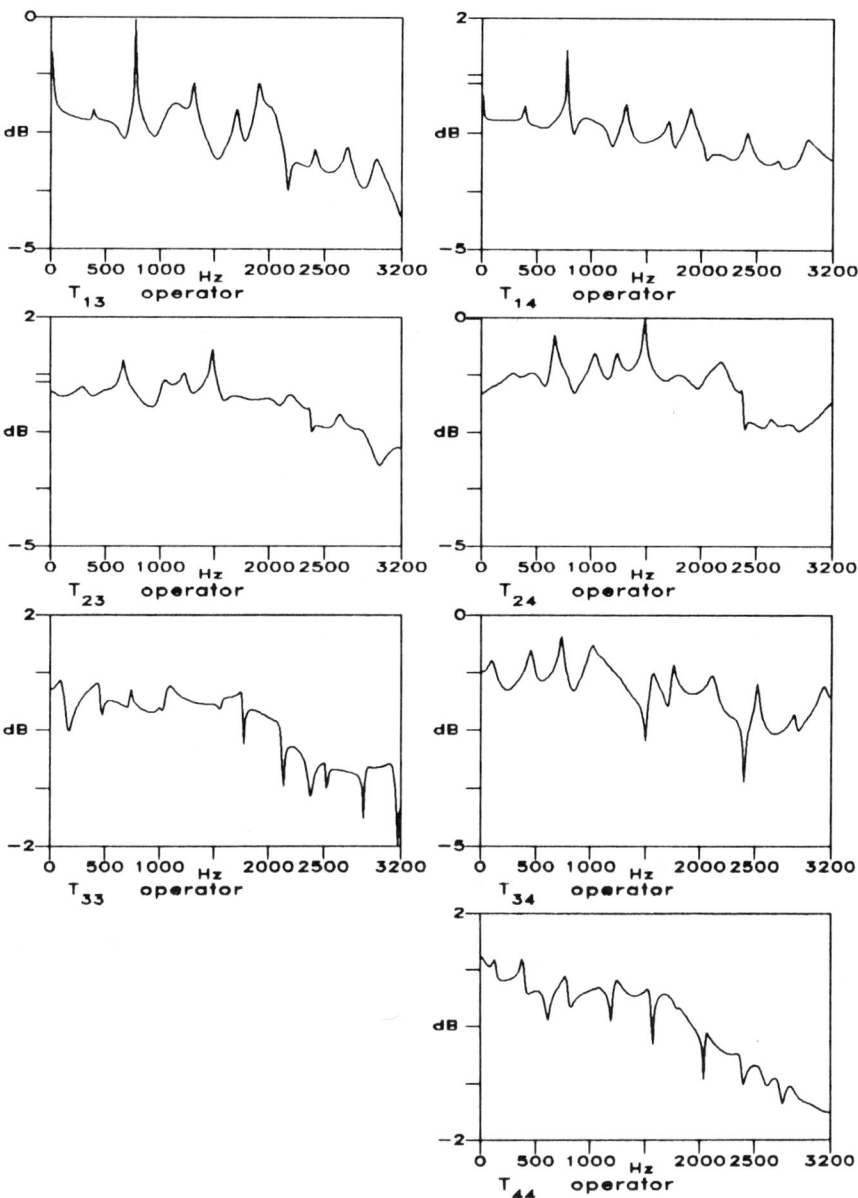

Figure 6.19 Transfer functions for white noise inputs.

This gets us back to the modal analysis of the Green's function (matrix) given by (6.3.14) and (6.3.18), which provides contribution $\mathbf{d}_i$ for a real λ_i and $\mathbf{d}_i + \mathbf{d}_{i+1}$ for complex pair λ_i, λ_{i+1} to the covariance matrix γ_0 by (6.3.19a). Such analysis can be further refined by separating the deterministic and stochastic parts and analyzing them separately, as discussed in Sections 5.3.1 and 5.3.2.

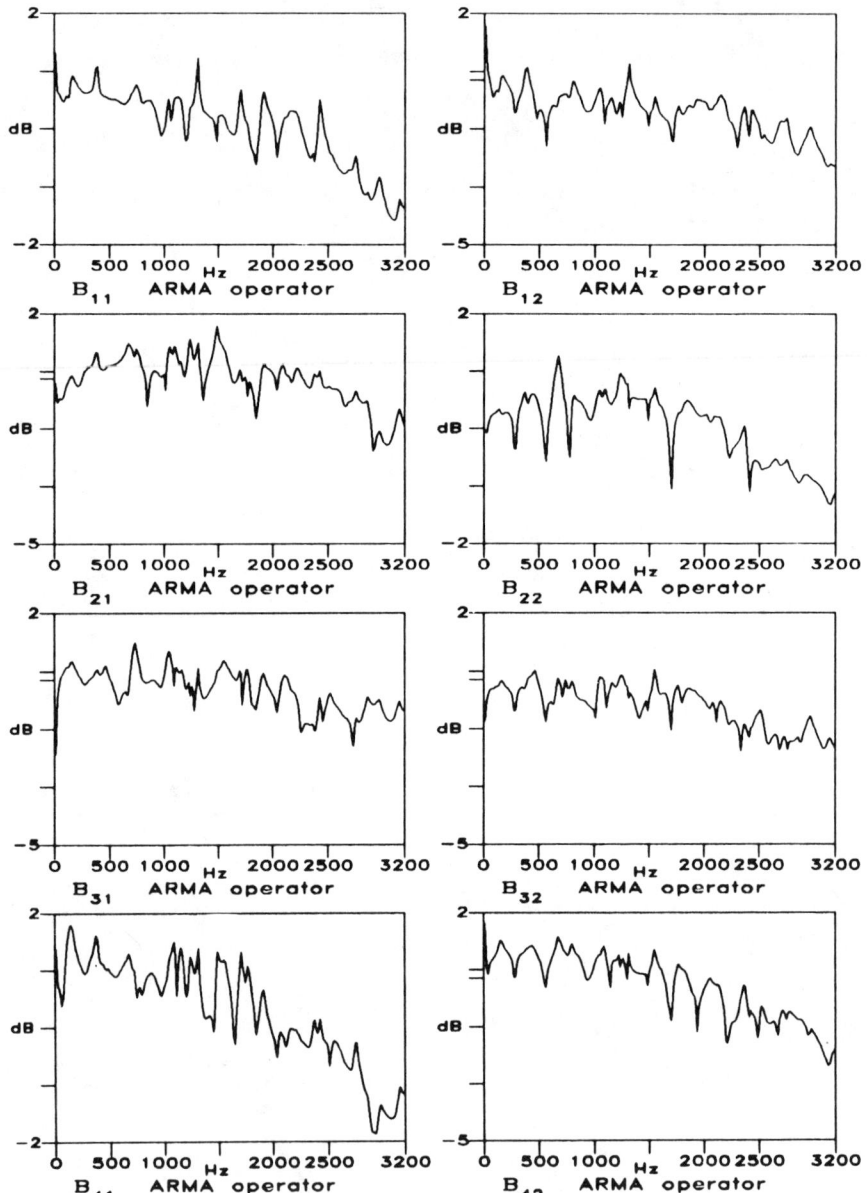

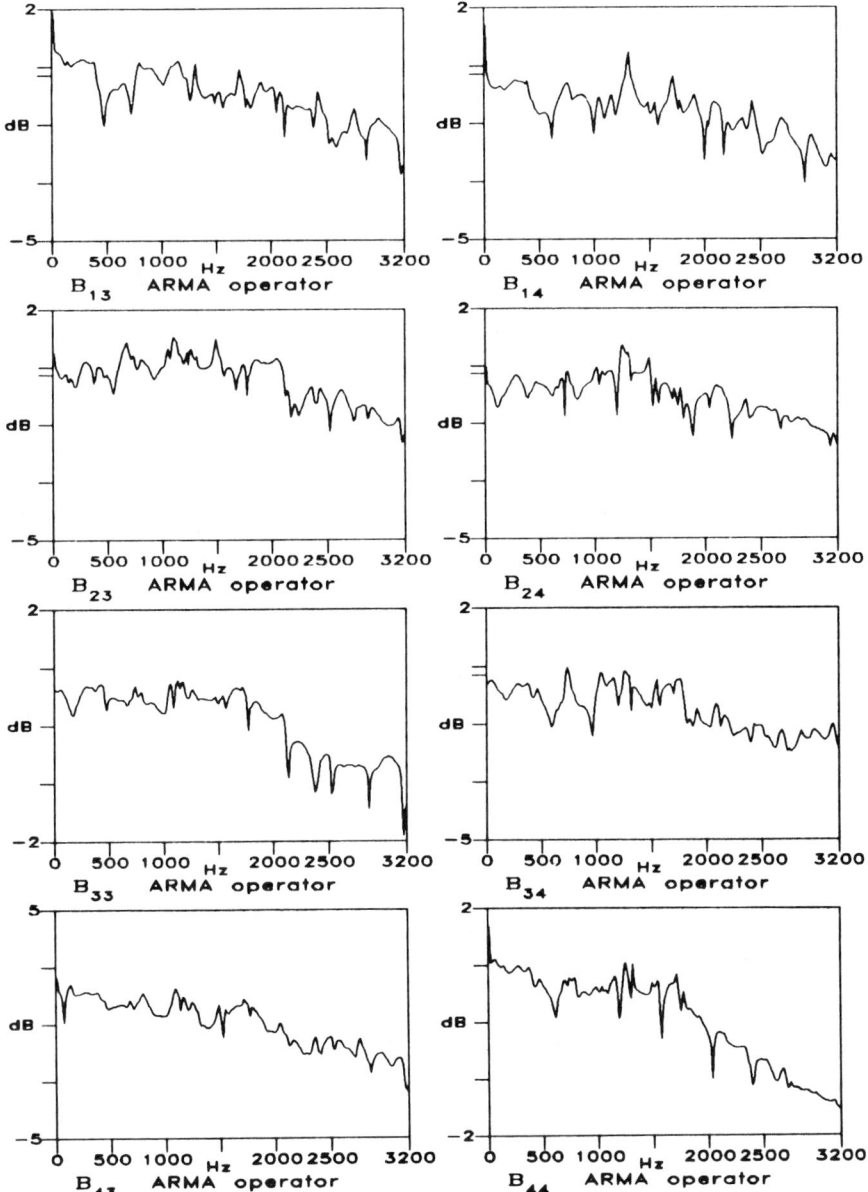

Figure 6.20 Elements of the ARMA operator matrix $\phi^{-1}(f)\,\theta(f)$.

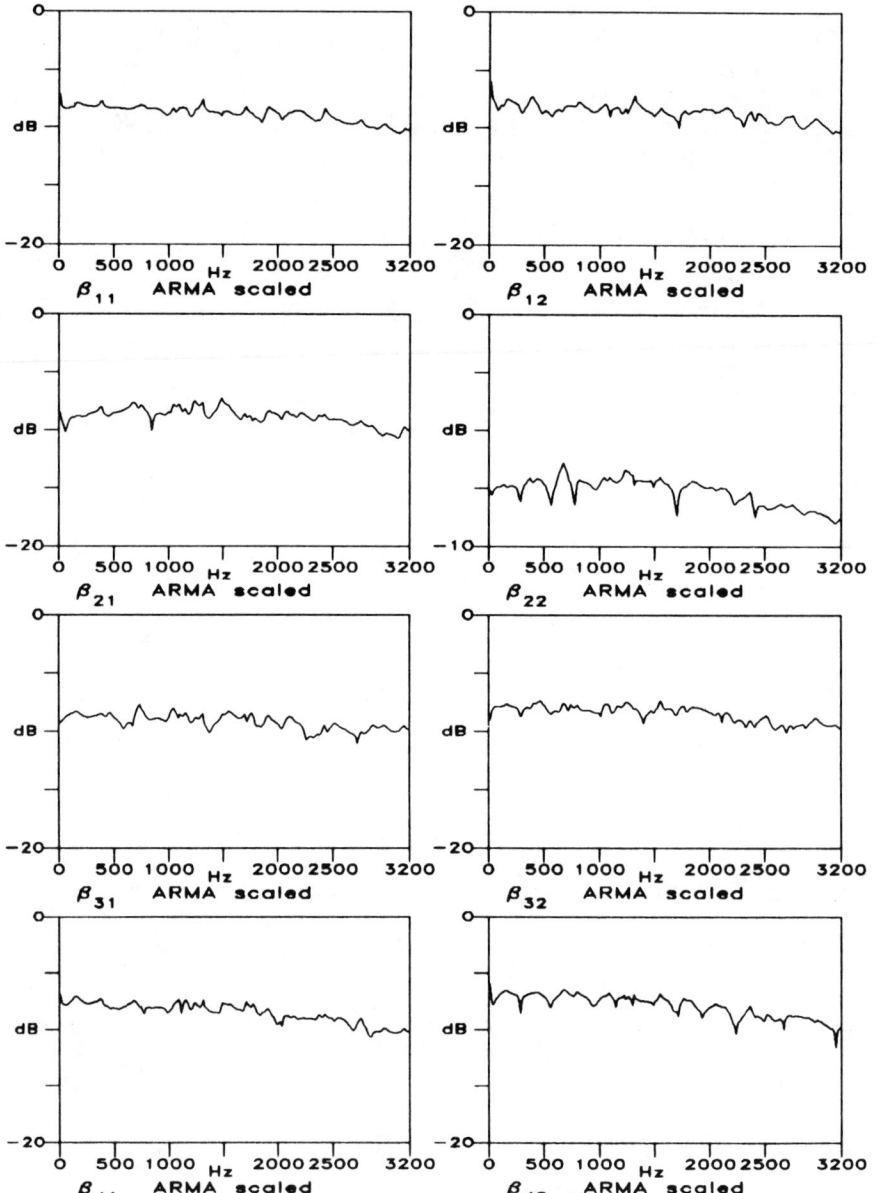

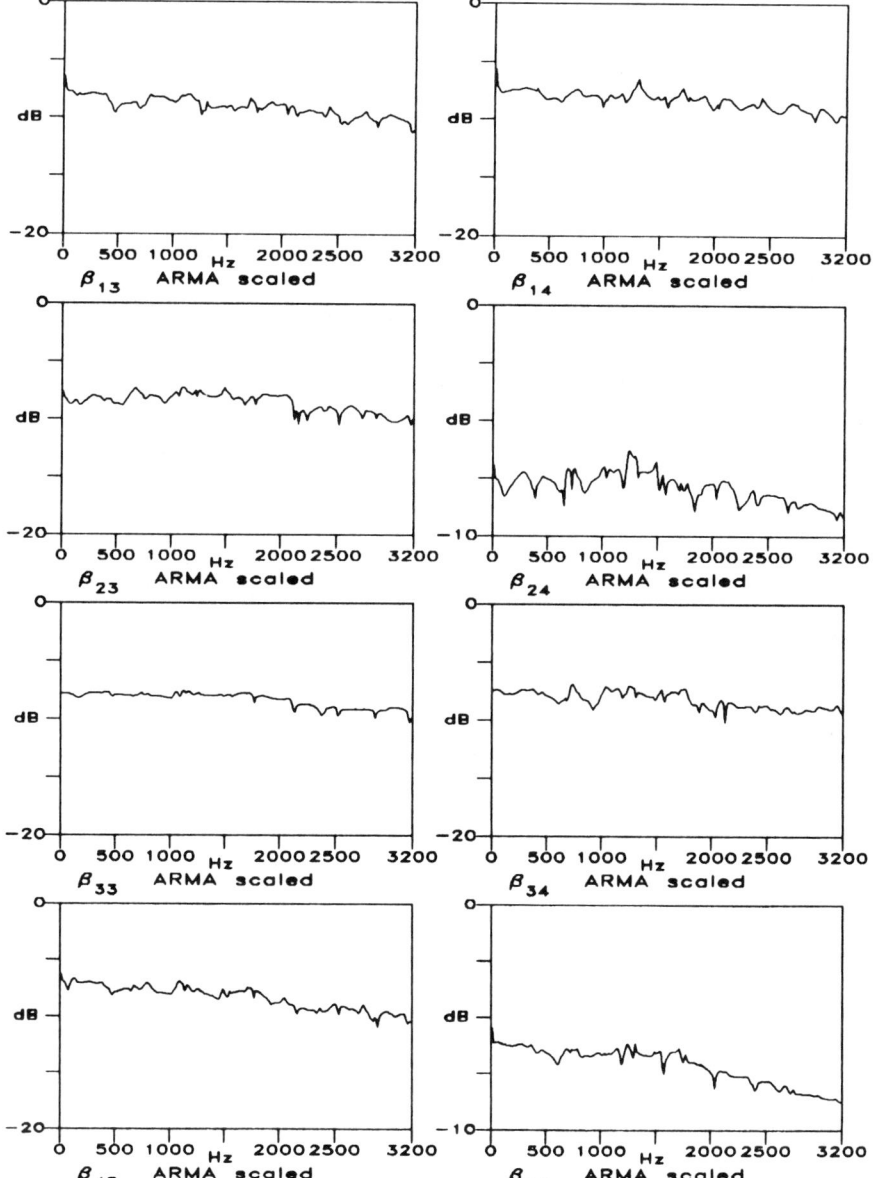

Figure 6.21 Elements of the scaled ARMA operator matrix $\phi^{-1}(f)\,\theta(f)\sigma_a$.

Remarks

This case study has shown that the DDS approach to noise analysis and source identification has the following advantages:

1. A single record suffices, no averaging is needed.
2. Smooth and easily understandable plots are available from analytical expressions.
3. Full range of frequencies up to the Nyquist frequency are available.
4. Only statistically significant peaks are retained and their frequencies and damping ratios can be directly obtained from the eigenvalues of the model without any further processing of spectral plots.
5. Most of the subjective decisions and trial-and-error methods are eliminated, thus minimizing the expertise needed in applications.
6. Both global and local indicators of contributions from physical sources and individual frequencies are available. These are helpful in identifying and eliminating the sources of undesirable noise.

APPENDIX A

STATISTICAL PRELIMINARIES

Observed data is usually random or stochastic in nature. Its quantitative analysis by the commonly used statistical techniques is summarized in this appendix, just enough to provide a background for the material in the main text. This appendix concentrates on the basic concepts, primarily dealing with the statistics of independent data. Statistical texts, such as Rao (1965) should be referred for a more advanced exposition. The statistical theory of time series, dealing with dependent data by means of sample correlations and sample spectra, gets unnecessarily complicated. The DDS approach provides a simpler and practically more useful alternative; a review of statistics from this angle and its contrast with the DDS approach may be found in Pandit and Wu (1983).

A.1 Mean, Variance, Covariance, and Correlation of Random Variables and Their Estimators

If an observed quantity can be considered as the outcome of a controlled experiment and can give different values under the repetition of the same experiment, then the quantity may be called a random variable. The individual outcomes are called the values of the random variable or sampled realizations of the random variable. Random variables are generally denoted by capital letters, say X, and its value by small letters, say x. If we make N observations on the same random variable, say $x_1, x_2, \ldots, x_N$, the simplest 'statistic" of this observed data is its average,

$$\bar{x} = \frac{1}{N}(x_1 + x_2 + \cdots + x_N) \qquad (A.1.1)$$

which characterizes the location around which the data values are spread. If this spread does not increase as the number of observations N is increased, the average will tend to a limiting value called the mean or the expected value of the random variable X, denoted by μ_x:

$$\text{Mean of } X: \mu_x = E(X) = \lim_{N \to \infty} \frac{1}{N}(x_1 + x_2 + \cdots + x_N) \quad \text{(A.1.2)}$$

Note that the expectation is a linear operator, so that if X and Y are two random variables with means μ_X and μ_Y, and a and b are constants, then

$$\mu_{aX+bY} = E(aX + bY) = aE(X) + bE(Y) = a\mu_X + b\mu_Y \quad \text{(A.1.3)}$$

The finite spread of the random variable is characterized by the expected value of the square of its deviation from the mean, called the variance, denoted by σ_x^2

$$\text{Variance of } X: \text{Var}(X) = \sigma_x^2 = E(X - \mu_x)^2 \quad \text{(A.1.4)}$$

σ_x, the positive square root of variance, is called the standard deviation and provides a measure of spread of the random variable from its mean without regard to its sign.

When the variance, which is always nonnegative by definition, is zero, that is, $\sigma_x^2 = 0$, X is a degenerate random variable or the usual deterministic variable taking on a fixed value such as 10.1. Then $E(X) = 10.1$, $\sigma_x^2 = 0$, and we simply write $X = 10.1$. Thus a deterministic variable is a (somewhat artificial) limiting case of a stochastic or random variable. Note that for a "genuine" random variable X with $\sigma_x^2 > 0$, the statement "$X = 10.1$" is meaningless, although a particular realization or sample may give the value $x = 10.1$. When the standard deviation is negligibly small, $\sigma_x \approx 0$, then the statement "$X = 10.1$" is acceptable as a deterministic approximation.

The concept of mean readily extends to a vector random variable, say a p-variate column vector $\mathbf{X} = [X_1, X_2, \ldots, X_p]^T$:

$$\text{Mean of } \mathbf{X}: E(\mathbf{X}) = \boldsymbol{\mu}_x = [E(X_1), E(X_p), \ldots, E(X_p)]^T$$
$$= [\mu_{x1}, \mu_{x2}, \ldots, \mu_{xp}]^T \quad \text{(A.1.5)}$$

The deviation of the random variable $\mathbf{X}$ from its mean $\boldsymbol{\mu}_x$ is then also a p vector $(\mathbf{X} - \boldsymbol{\mu}_x)$. Since the expression $(\mathbf{X} - \boldsymbol{\mu}_x)^2$ is meaningless, and the expression

$$E(\mathbf{X} - \boldsymbol{\mu}_x)^T(\mathbf{X} - \boldsymbol{\mu}_x) = \sigma_{x1}^2 + \sigma_{x2}^2 + \cdots + \sigma_{xp}^2 \quad \text{(A.1.6)}$$

is a scalar containing the sum of the variance of the elements of $\mathbf{X}$, they are not useful in extending the concept of variance to a vector random variable. A proper extension of the concept of variance to a vector random variable turns out to be

$$V(\mathbf{X}) = \sigma_x^2 = E(\mathbf{X} - \mu_x)(\mathbf{X} - \mu_x)^T \qquad (A.1.7)$$

This is called the dispersion or variance–covariance matrix of the random variable $\mathbf{X}$, because, considering the bivariate case with $p = 2$ for illustration,

$$V(\mathbf{X}) = \begin{bmatrix} E(X_1 - \mu_{x1})^2 & E(X_1 - \mu_{x1})(X_2 - \mu_{x2}) \\ E(X_2 - \mu_{x2})(X_1 - \mu_{x1}) & E(X_2 - \mu_{x2})^2 \end{bmatrix}$$

$$= \begin{bmatrix} \text{Var}(X_1) & \text{Cov}(X_1, X_2) \\ \text{Cov}(X_1, X_2) & \text{Var}(X_2) \end{bmatrix}$$

and the covariance of two scalar random variables say X and Y, is defined as

$$\text{Covariance of } X \text{ and } Y: \text{Cov}(X, Y) = \gamma_{xy} = E(X - \mu_x)(Y - \mu_Y) \qquad (A.1.8)$$

Since the square matrix $\mathbf{A}\bar{\mathbf{A}}^T$ is nonnegative-definite for an arbitrary matrix $\mathbf{A}$ (see property iv of Section 3.5.1),

$$\pm \text{Cov}(X, Y) \le \sqrt{\text{Var}(X)\text{Var}(Y)} \qquad (A.1.9)$$

The covariance expresses the extent of relation or dependence between a pair of random variables and is zero when they are statistically independent. Unlike the variance, the covariance can be positive or negative. It can also take arbitrarily large values because it is affected by scale

$$\text{Cov}(cX, cY) = c^2 \text{Cov}(X, Y)$$

for an arbitrary constant c. This is also true for variance, which is the "covariance of the variable with itself." Hence it is sometimes convenient to indicate the extent of relation by defining the "scaleless" covariance called correlation:

$$\text{Correlation of } X \text{ and } Y: \text{Cor}(X, Y) = \frac{\text{Cov}(X, Y)}{\sqrt{\text{Var}(X)\text{Var}(Y)}} \qquad (A.1.10)$$

$$\rho_{XY} = \frac{\gamma_{XY}}{\sigma_X \sigma_Y}$$

It follows from (A.1.9) that

$$|\rho_{XY}| \le 1 \qquad (A.1.11)$$

which is particularly convenient for plotting purposes. For a vector of statistically independent random variables, their dispersion or variance–covariance matrix is diagonal, $V(\mathbf{X}) = \text{diag}\{\sigma_{xi}^2\}$.

Therefore, in general, for an arbitrary column vector of constants $\ell = [\ell_1, \ell_2, \ldots, \ell_p]^T$, the mean and variance of a linear combination $\ell^T \mathbf{X}$ are given by

$$E(\ell^T \mathbf{X}) = \ell^T E(\mathbf{X}) = \ell^T \mu_x \qquad (A.1.12)$$

$$\mathrm{Var}(\ell^T \mathbf{X}) = \ell^T E(\mathbf{X} - \mu_x)(\mathbf{X} - \mu_x)^T \ell = \ell^T V(\mathbf{X}) \ell \qquad (A.1.13)$$

and only for the special case of statistically independent X_i:

$$\mathrm{Var}(\ell^T \mathbf{X}) = \ell^T \mathrm{diag}\{\sigma_{xi}^2\} \ell = \ell_1^2 \sigma_{x1}^2 + \ell_2^2 \sigma_{x2}^2 + \cdots + \ell_p^2 \sigma_{xp}^2 \qquad (A.1.13a)$$

Hence the importance of statistical independence in elementary statistics.

Covariance matrix of two vector random variables is readily defined as

Covariance of $\mathbf{X}$ and $\mathbf{Y}$: $\mathrm{Cov}(\mathbf{X}, \mathbf{Y}) = \gamma_{XY} = E(\mathbf{X} - \mu_X)(\mathbf{Y} - \mu_Y)^T \qquad (A.1.14)$

which need not be symmetric or nonnegative-definite. These definitions readily extend to complex random variables when ordinary transpose is replaced by complex conjugate transpose; the property of being symmetric is then replaced by the Hermitian $\mathbf{A} = \bar{\mathbf{A}}^T$. Thus, in particular, for zero mean complex, vector random variables $\mathbf{X}$ and $\mathbf{Y}$, and complex ℓ,

$$V(\mathbf{X}) = E(\mathbf{X}\bar{\mathbf{X}}^T), \quad V(\mathbf{Y}) = E(\mathbf{Y}\bar{\mathbf{Y}}^T), \quad \mathrm{Cov}(\mathbf{X}, \mathbf{Y}) = E(\mathbf{X}\bar{\mathbf{Y}}^T) \qquad (A.1.15)$$

and even with nonzero means

$$\mathrm{Var}(\ell^T \mathbf{X}) = \ell^T V(\mathbf{X}) \bar{\ell} \qquad (A.1.16)$$
$$= |\ell_1|^2 \sigma_{x1}^2 + |\ell_2|^2 \sigma_{x2}^2 + \cdots + |\ell_p|^2 \sigma_{xp}^2, \quad \text{for independent } X_i \qquad (A.1.16a)$$

We will continue to use only real variables and constants in the rest of the appendix for simplicity.

As seen from (A.1.2), the expectation operation is a weighted summation over a possibly infinite "population." In practice one needs to estimate the characteristics such as mean and variance from the available finite data. The easiest way to accomplish this is by replacing the expectation operation by the usual summation, as in (A.1.1). Thus $\bar{x}$ is an estimate of the mean μ_x. However, if a sample of N observations is repeated, we would possibly get different values $\bar{x}$. In this sense, the average itself is a random variable and should be denoted by the capital letter $\bar{X}$; it is then called an estimator. The estimators of mean, variance, covariance, and correlation based on N statistically independent observations can thus be written by replacing E with the finite average $1/N \Sigma$, where Σ denotes summation over $i = 1, 2, \ldots, N$, as

$$\bar{X} = \frac{1}{N}\Sigma X_i \tag{A.1.17}$$

$$\hat{\sigma}_x^2 = \frac{1}{N}\Sigma(X_i - \bar{X})^2 \tag{A.1.18}$$

$$\hat{\gamma}_{XY} = \frac{1}{N}\Sigma(X_i - \bar{X})(Y_i - \bar{Y}) \tag{A.1.19}$$

$$\hat{\rho}_{XY} = \frac{\hat{\gamma}_{XY}}{\hat{\sigma}_X \hat{\sigma}_Y} \tag{A.1.20}$$

Since an estimator is a random variable, it has its own mean and variance indicating how far its values in a particular sample may spread from the mean. For a good estimator, it is desirable that its mean or the expected value is the characteristic being estimated; then it is said to be unbiased. Another desirable property is that the variance of the estimator tends to zero as N is increased; then it is said to be consistent. Thus, for large enough N, an unbiased and consistent estimator practically coincides with the deterministic characteristic it is estimating.

Since each of the independent observations X_i has mean μ_x and variance σ_x^2 and $\bar{X} = \ell^T \mathbf{X}$ with $\ell = [1/N, 1/N, \ldots, 1/N]^T$ it is left as an exercise to show that

$$E(\bar{X}) = \mu_x, \quad \text{Var}(\bar{X}) = \frac{\sigma_x^2}{N} \tag{A.1.21}$$

Thus $\bar{X}$ is an unbiased and consistent estimator of the mean μ_x.

Similar manipulation shows that

$$E(\hat{\sigma}_x^2) = \left(\frac{N-1}{N}\right)\sigma_x^2 \tag{A.1.22}$$

Thus $\hat{\sigma}_x^2$ is *not* an unbiased estimator, it underestimates σ_x^2; the bias, however, reduces with N and is negligible for large N. However, since

$$E\left[\frac{N}{(N-1)}\hat{\sigma}_x^2\right] = \sigma_x^2$$

the estimator

$$s^2 = \frac{N}{(N-1)}\hat{\sigma}_x^2 = \frac{1}{(N-1)}\Sigma(X_i - \bar{X})^2 \tag{A.1.23}$$

is an unbiased estimator and hence is used, particularly for small N.

A.2 Mean and Variance–Covariance of Estimators in the Linear Least Squares Model and Sample Auto Correlations

The commonly used model in linear least squares theory can be succinctly written as

$$\mathbf{Y} = \mathbf{X}\boldsymbol{\beta} + \boldsymbol{\varepsilon}, \qquad E(\boldsymbol{\varepsilon}) = \mathbf{0}, \qquad V(\boldsymbol{\varepsilon}) = \sigma_\varepsilon^2 \mathbf{I} \qquad (A.2.1)$$

where $\mathbf{Y}$ is an $N \times 1$ column vector of statistically independent observations, $\mathbf{X}$ is an $N \times n$ design matrix of (functionally) independent deterministic variables, $\boldsymbol{\beta}$ is an $n \times 1$ column vector of parameters, and $\boldsymbol{\varepsilon}$ is an $N \times 1$ column vector of uncorrelated errors, each with zero mean and variance σ_ε^2. For $n = 1$ it represents simple regression of a line through the origin.

Then

$$E(\mathbf{Y}) = \mathbf{X}\boldsymbol{\beta} \quad \text{and} \quad V(\mathbf{Y}) = \sigma_\varepsilon^2 \mathbf{I} \qquad (A.2.2)$$

It is shown in Chapter 3 that the least squares estimates are given by the normal equations

$$\mathbf{X}^T \mathbf{X} \boldsymbol{\beta} = \mathbf{X}^T \mathbf{Y} \qquad (A.2.3)$$

as

$$\hat{\boldsymbol{\beta}} = (\mathbf{X}^T \mathbf{X})^{-1} \mathbf{X}^T \mathbf{Y} \qquad (A.2.4)$$

assuming $\mathbf{X}$ is full rank. [For rank-deficient $\mathbf{X}$ one can use generalized inverse, see Rao (1965)]. Since

$$E(\hat{\boldsymbol{\beta}}) = (\mathbf{X}^T \mathbf{X})^{-1} \mathbf{X}^T E(\mathbf{Y}) = (\mathbf{X}^T \mathbf{X})^{-1} (\mathbf{X}^T \mathbf{X}) \boldsymbol{\beta} = \boldsymbol{\beta} \qquad (A.2.5)$$

the least squares estimates are unbiased. Their variance–covariance matrix follows from (A.1.13) and (A.2.2) as

$$V(\boldsymbol{\beta}) = (\mathbf{X}^T \mathbf{X})^{-1} \mathbf{X}^T V(\mathbf{Y}) \mathbf{X} (\mathbf{X}^T \mathbf{X})^{-1} = (\mathbf{X}^T \mathbf{X})^{-1} \sigma_\varepsilon^2 \qquad (A.2.6)$$

These results are not exactly applicable to the stationary zero mean AR(1) model

$$X_t = \phi_1 X_{t-1} + a_t \qquad (A.2.7)$$

even though a_t's are uncorrelated because X_{t-1} is a random variable. However, at time t, the observation X_{t-1} is known or fixed and deterministic, and then the conditional least squares estimator of ϕ_1 is obtained as

$$\hat{\phi} = \frac{\sum X_t X_{t-1}}{\sum X_{t-1}^2} \qquad (A.2.8)$$

and its approximate variance is given by

$$\text{Var}(\hat{\phi}_1) = \frac{\sigma_a^2}{\Sigma X_{t-1}^2} \simeq \frac{\sigma_a^2}{N \, \text{Var}(X_t)} = \frac{(1-\phi_1^2)}{N} \quad \text{(A.2.9)}$$

showing that it is unbiased and consistent. Under the actual unconditional least squares set-up these results are only approximately true for large N.

Since a stationary X_t has the same mean μ_x, variance γ_0, and covariance γ_k with X_{t-k} for any t, its variance covariance expressions simplify to

Autocovariance of X_t and X_{t-k}: $\gamma_k = E(X_t - \mu_x)(X_{t-k} - \mu_x)$ \quad (A.2.10)

Autocorrelation of X_t and X_{t-k}: $\rho_k = \dfrac{\gamma_k}{\gamma_0}$ \quad (A.2.11)

with the corresponding sample estimators

$$\hat{\gamma}_k = \frac{1}{N} \Sigma (X_t - \bar{X})(X_{t-k} - \bar{X}) \quad \text{(A.2.12)}$$

$$\hat{\rho}_k = \frac{\hat{\gamma}_k}{\hat{\gamma}_0} \quad \text{(A.2.13)}$$

It is a common practice to treat the average subtracted stationary data as X_t with zero mean; then it is clear that $\hat{\rho}_1 = \hat{\phi}_1$ and $\rho_1 = \phi_1$ for an AR(1) model.

For statistically independent data, X_t has an AR(1) model with $\phi_1 = 0$ and (1.2.9) gives

$$\text{Var}(\hat{\rho}_1) = \text{Var}(\hat{\phi}_1) \simeq \frac{1}{N} \quad \text{(A.2.14a)}$$

and in general

$$\text{Var}(\hat{\rho}_k) \simeq \frac{1}{N} \quad \text{(A.2.14)}$$

For statistically dependent data, if X_t is stationary and its $\rho_k \to 0$ as $|k| \to \infty$, it can be shown that

$$\text{Cov}(\hat{\rho}_r, \hat{\rho}_{r+s}) \simeq \frac{1}{N} \sum_{m=-\infty}^{\infty} \rho_m \rho_{m+s} \quad \text{(A.2.15)}$$

and in particular,

$$\text{Var}(\hat{\rho}_r) = \frac{1}{N} \sum_{m=-\infty}^{\infty} \rho_m^2 \quad \text{(A.2.15a)}$$

if $\rho_s \approx 0$ for $|s| \geq r$, see, for example, Priestley (1981, p. 333). Thus the autocorrelation estimator has large variance and is more highly autocorrelated than the original X_t. This is also true of cross-correlation estimators. Hence we have avoided modeling procedures based on estimated autocorrelations. For the autocorrelations of the residual a_t's, however, their whiteness or uncorrelatedness can be checked by the unified autocorrelation defined by

$$\text{Unified autocorrelation} = \frac{\hat{\rho}_k}{\sqrt{\text{Var}(\hat{\rho}_k)}} \simeq \sqrt{N} \hat{\rho}_k^a \qquad (A.2.16)$$

Using the normal distribution discussed in the next section, we can conclude, with only 5% chance of error, that the autocorrelations are zero and residuals are white if all the unified autocorrelations are less than 1.96 in absolute value.

A.3 Probability Distributions—Normal, Chi-Square, and F

The definition of expectation by (A.1.2) as a limit with $N \to \infty$ is not rigorous because the limit is not well defined. A rigorous definition can be given using a probability distribution associated with the random variable. If a value x_i of a discrete random variable repeats n_i times in a sample of N observations, its frequency is (n_i/N), with $\Sigma(n_i/N) = 1$. For a population of large N this frequency is defined as the probability $p_i = \text{Prob}(X = x_i)$ and then a proper definition of expectation for a discrete random variable is

$$E(X) = \sum_{i=-\infty}^{\infty} x_i p_i, \qquad \sum_{i=-\infty}^{\infty} p_i = 1$$

For a continuous random variable X that can take continuous values x, the probability that it has values in an infinitesimal interval dx is defined $f(x)\,dx$ and the expectation is given by

$$E(X) = \mu_x = \int_{-\infty}^{\infty} x f(x)\,dx, \quad \text{or} \quad \int_{-\infty}^{\infty} (x - \mu_x) f(x)\,dx = 0 \qquad (A.3.1)$$

where

$$\int_{-\infty}^{\infty} f(x)\,dx = 1 \qquad (A.3.2)$$

Hence the nonnegative valued function $f(x)$ is called the probability density function and the nondecreasing function

$$F(x) = \int_{-\infty}^{x} f(y)\,dy = \text{Prob}[X \le x] \qquad (A.3.3)$$

is called the cumulative probability function. The expected value of any function of a random variable, say $g(X)$, is then given by

$$E[g(X)] = \int_{-\infty}^{\infty} g(x)f(x)\,dx \qquad (A.3.4)$$

and in particular

$$\sigma_x^2 = \int_{-\infty}^{\infty} (x-\mu)^2 f(x)\,dx \qquad (A.3.5)$$

For two random variables, say X and Y, their joint probability is the product of the conditional probability of X given $Y(X/Y)$ and the probability of Y, so that

$$f(x, y) = f(x|y)f(y) \qquad (A.3.6)$$

Two random variables are statistically independent when their conditional probability is the same as unconditional probability, that is, if X and Y are independent

$$f(x|y) = f(x) \qquad (A.3.7a)$$

and

$$f(x, y) = f(x)f(y) \qquad (A.3.7)$$

Therefore, for independent random variables

$$\begin{aligned}\operatorname{Cov}(X, Y) &= E(X - \mu_x)(Y - \mu_Y) \\ &= \int_{-\infty}^{\infty} \int_{-\infty}^{\infty} (x - \mu_x)(y - \mu_Y) f(x, y)\,dx\,dy \\ &= \int_{-\infty}^{\infty} (x - \mu_x)f(x)\,dx \int_{-\infty}^{\infty} (y - \mu_Y)f(y)\,dy \\ &= 0 \end{aligned} \qquad (A.3.8)$$

and hence they are uncorrelated. The converse is, however, not always true.

Normal Distribution and Confidence Intervals

The most commonly used distribution is the Gaussian or normal distribution, in which the probability mass is distributed around the mean by a bell-shaped curve with its spread determined by the variance. If a scalar or univariate random variable X with mean μ_x and variance σ_x^2 has normal distribution, written in short as $X \sim N(\mu_x, \sigma_x^2)$, then the probability density of the values x of X is given by

$$f(x) = \frac{1}{\sqrt{2\pi}\,\sigma_x} e^{-[(x-\mu_x)^2/2\sigma_x^2]}, \qquad -\infty < x < \infty \qquad (A.3.9)$$

One of the most useful properties of the normal distribution is that a linear combination of normally distributed random variables is again normal. If

$$X \sim N(\mu_x, \sigma_x^2), \qquad Y \sim N(\mu_Y, \sigma_Y^2)$$

then

$$aX + bY \sim N[(a\mu_x + b\mu_Y), (a^2\sigma_x^2 + 2ab\gamma_{xy} + b^2\sigma_Y^2)] \qquad (A.3.10)$$

In particular, if $X_1, X_2, \ldots, X_N$ are independent observations from $X \sim N(\mu_x, \sigma_x^2)$, that is, $X_i \sim \text{NID}(\mu_x, \sigma_x^2)$, then, using (A.1.21),

$$\bar{X} \sim N\left(\mu_x, \frac{\sigma_x^2}{N}\right) \qquad (A.3.11)$$

and

$$Z \sim \frac{(\bar{X} - \mu_x)}{\sigma_x/\sqrt{N}} \sim \frac{(X - \mu_x)}{\sigma_x} \sim N(0, 1) \qquad (A.3.12)$$

where Z is called a standard normal random variable.

Since the normal distribution is completely determined by the mean and variance of a random variable, for a p-variate vector random variable $\mathbf{X}$ its multivariate normal distribution is completely determined by its mean $\boldsymbol{\mu}_x$ and variance–covariance matrix σ_x^2. A p-variate random variable $\mathbf{X}$ has multivariate normal distribution if and only if ($\leftrightarrow$) every linear combination of $\mathbf{X}$ has a univariate normal distribution

$$\mathbf{X} \sim \text{MVN}(\boldsymbol{\mu}_x, \sigma_x^2) \leftrightarrow \boldsymbol{\ell}^T \mathbf{X} \sim N(\boldsymbol{\ell}^T \boldsymbol{\mu}_x, \boldsymbol{\ell}^T \sigma_x^2 \boldsymbol{\ell}) \qquad (A.3.13)$$

Using this definition, many of the multivariate generalizations of the univariate results can be easily found. Thus it is easy to show that if $\mathbf{X}_1, \mathbf{X}_2, \ldots, \mathbf{X}_n$ are independent observations from $\mathbf{X} \sim \text{MVN}(\boldsymbol{\mu}_x, \sigma_x^2)$ then

$$\bar{\mathbf{X}} \sim \text{MVN}\left(\boldsymbol{\mu}_x, \frac{1}{N}\sigma_x^2\right) \qquad (A.3.14)$$

The normal density function (A.3.9), which is thus applicable for both scalar and vector random variables, is not useful in practice because it does not possess a closed form integral. The probability computation

$$\text{Prob}[a \leq X \leq b] = F(b) - F(a) \qquad (A.3.15)$$

must be done with the help of tables giving the cumulative probability computed by numerical integration. The tables are generally available for the standard normal variable Z. The values most used in practice are

$$F(1.65) = 0.95053, \qquad F(1.96) = 0.975, \qquad F(2.58) = 0.99506 \qquad (A.3.16)$$

Due to the symmetry of the normal distribution

$$\text{Prob}(-1.65 \leq Z \leq 1.65) = F(1.65) - F(-1.65)$$
$$= F(1.65) - [1 - F(1.65)]$$
$$\simeq 0.90 = 90\%$$

Thus 90, 95, and 99% probability limits on Z are ± 1.65, ± 1.96, and ± 2.58. Therefore, using the definition of Z by (A.3.12), 90, 95, and 99% probability limits on X are respectively

$$\mu_x \pm 1.65\sigma_x, \qquad \mu_x \pm 1.96\sigma_x, \qquad \text{and} \qquad \mu_x \pm 2.58\sigma_x \qquad (A.3.17)$$

Another important property of the normal distribution is that a sum of a large number of random variables with most nonnormal distributions also tends to be normal. Therefore, knowing the mean and variance of such a sum, (A.3.17) can be used to determine probability limits. These limits provide an interval, known as a *confidence interval*, in which the true parameter is likely to be with given probability. For example, suppose we know the variance σ_x from past experience and want to estimate a possibly changed mean. Then using (A.3.11) and (A.3.17)

$$\text{Prob}\left(\mu_x - 1.96\frac{\sigma_x}{\sqrt{N}} \leq \bar{X} \leq \mu_x + 1.96\frac{\sigma_x}{\sqrt{N}}\right) = 0.95$$

so that 95% confidence interval on the mean μ_x is $(\bar{x} \pm 1.96\,\sigma_x/\sqrt{N})$. Similarly using (A.2.8) and (A.2.9), an approximate 95% confidence interval on the parameter $\hat{\phi}_1$ of an AR(1) model for large N is

$$\hat{\phi}_1 \pm 1.96\sqrt{(1 - \hat{\phi}_1^2)/N} \qquad (A.3.18)$$

For vector estimates, such as the linear least squares estimates discussed in Section A.2, one needs multidimensional confidence *regions*, which are difficult to interpret. An easier shortcut is to only consider an individual element in the vector or a linear combination of elements and use the variance to specify the confidence interval. This unfortunately ignores the correlation with other linear combinations. To illustrate, the 95% confidence interval on $\ell^T\beta$ is obtained from (A.2.4)–(A.2.6) and (A.1.17) as

$$\ell^T\hat{\beta} \pm 1.96[\ell^T(\mathbf{X}^T\mathbf{X})^{-1}\ell]\hat{\sigma}_\varepsilon \qquad (A.3.19)$$

where, similar to (A.1.23),

$$\hat{\sigma}_\varepsilon^2 = (\mathbf{Y} - \mathbf{X}\hat{\beta})^T(\mathbf{Y} - \mathbf{X}\hat{\beta})/(N - r) \qquad (A.3.20)$$

Note that in (A.3.19) and (A.3.20) **X** is an $N \times n$ *matrix* and not a p vector, as mostly used in this section. For an individual parameter, we take ℓ with 1 in its place and zero everywhere else, for example, the confidence interval on β_1 is obtained by taking $\ell = [1, 0, \ldots, 0]^T$.

Although this method can be used to find confidence intervals on the individual parameters ϕ_{ijk} of the ARV models estimated by least squares as discussed in Chapter 5, these are of limited practical use. First, there are too many ϕ_{ijk}'s. Second, the covariances of their estimators may be even more important, and finally the parameters of real interest such as natural frequencies, damping ratios, and mode shapes are complicated nonlinear functions of the ϕ_{ijk}'s. Hence the condition number of the matrix $\mathbf{X}^T\mathbf{X}$ seems to be a good overall indicator of the estimation situation. When the condition number is low, the estimation situation is good and the confidence interval would be mostly narrow.

Chi-Square and F Distribution

If $Z_1, Z_2, \ldots, Z_v$ are *independent* standard normal variables, then their sum of squares has a distribution called Chi-square with v degrees of freedom, denoted as χ_v^2

$$\chi_v^2 \sim Z_1^2 + Z_2^2 + \cdots + Z_v^2, \qquad Z_i \sim \text{NID}(0,1) \qquad (A.3.21)$$

Since Z_i has mean zero and variance 1

$$E(\chi_v^2) = v$$

that is, the mean of a Chi-square variable is its degrees of freedom. It can be shown that

$$\frac{(N-1)s^2}{\sigma_x^2} = \frac{\Sigma(X_i - X)^2}{\sigma^2} \sim \chi_{(N-1)}^2$$

so that

$$E\left[\frac{(N-1)s^2}{\sigma_x^2}\right] = (N-1) \qquad \text{or} \qquad E(s^2) = \sigma_x^2$$

used in Section A.1.

A ratio of two independent Chi-square variables divided by their degrees of freedom has F distribution:

$$F_{v1,v2} \sim \frac{\chi_{v1}^2/v_1}{\chi_{v2}^2/v_2}, \qquad \chi_{v1}^2 \text{ independent of } \chi_{v2}^2 \qquad (A.3.22)$$

The values $F_{v1, v2, p}$ of the random variables $F_{v1, v2}$, for which the cumulative Prob $(0 < F_{v1, v2} \leq F_{v1, v2, p}) = P = 0.95, 0.99$, and 0.999 are given in Table A following this section. The distribution is used to check that one estimate of variance is significantly larger than another. If the computed $F_{v1,v2} > F_{v1,v2,p}$ from the table, then the hypothesis that the variances are equal is rejected, with a $(1 - P) \times 100\%$ risk of the decision being wrong. An illustrative example of its application in the F test for DDS modeling is illustrated in Section 2.5.4 and its multivariate approximation is given in Section 5.2.1.

Table A Percentage Points of the F-Distribution[a]

$$P = \frac{1}{B(v_1/2, v_2/2)} \int_0^{v_1 F_{v_1,v_2}.P/v_2} g^{v_1/2-1}(1+g)^{-(v_1+v_2)/2} \, dg$$

$P = 0.95$

v_2 \ v_1	1	2	3	4	5	6	7	8	9
1	161.4	199.5	215.7	224.6	230.2	234.0	236.8	238.9	240.5
2	18.51	19.00	19.16	19.25	19.30	19.33	19.35	19.37	19.38
3	10.13	9.55	9.28	9.12	9.01	8.94	8.89	8.85	8.81
4	7.71	6.94	6.59	6.39	6.26	6.16	6.09	6.04	6.00
5	6.61	5.79	5.41	5.19	5.05	4.95	4.88	4.82	4.77
6	5.99	5.14	4.76	4.53	4.39	4.28	4.21	4.15	4.10
7	5.59	4.74	4.35	4.12	3.97	3.87	3.79	3.73	3.68
8	5.32	4.46	4.07	3.84	3.69	3.58	3.50	3.44	3.39
9	5.12	4.26	3.86	3.63	3.48	3.37	3.29	3.23	3.18
10	4.96	4.10	3.71	3.48	3.33	3.22	3.14	3.07	3.02
11	4.84	3.98	3.59	3.36	3.20	3.09	3.01	2.95	2.90
12	4.75	3.89	3.49	3.26	3.11	3.00	2.91	2.85	2.80
13	4.67	3.81	3.41	3.18	3.03	2.92	2.83	2.77	2.71
14	4.60	3.74	3.34	3.11	2.96	2.85	2.76	2.70	2.65
15	4.54	3.68	3.29	3.06	2.90	2.79	2.71	2.64	2.59
16	4.49	3.63	3.24	3.01	2.85	2.74	2.66	2.59	2.54
17	4.45	3.59	3.20	2.96	2.81	2.70	2.61	2.55	2.49
18	4.41	3.55	3.16	2.93	2.77	2.66	2.58	2.51	2.46
19	4.38	3.52	3.13	2.90	2.74	2.63	2.54	2.48	2.42
20	4.35	3.49	3.10	2.87	2.71	2.60	2.51	2.45	2.39
21	4.32	3.47	3.07	2.84	2.68	2.57	2.49	2.42	2.37
22	4.30	3.44	3.05	2.82	2.66	2.55	2.46	2.40	2.34
23	4.28	3.42	3.03	2.80	2.64	2.53	2.44	2.37	2.32
24	4.26	3.40	3.01	2.78	2.62	2.51	2.42	2.36	2.30
25	4.24	3.39	2.99	2.76	2.60	2.49	2.40	2.34	2.28
26	4.23	3.37	2.98	2.74	2.59	2.47	2.39	2.32	2.27
27	4.21	3.35	2.96	2.73	2.57	2.46	2.37	2.31	2.25
28	4.20	3.34	2.95	2.71	2.56	2.45	2.36	2.29	2.24
29	4.18	3.33	2.93	2.70	2.55	2.43	2.35	2.28	2.22
30	4.17	3.32	2.92	2.69	2.53	2.42	2.33	2.27	2.21
40	4.08	3.23	2.84	2.61	2.45	2.34	2.25	2.18	2.12
60	4.00	3.15	2.76	2.53	2.37	2.25	2.17	2.10	2.04
120	3.92	3.07	2.68	2.45	2.29	2.17	2.09	2.02	1.96
∞	3.84	3.00	2.60	2.37	2.21	2.10	2.01	1.94	1.88

[a] Reproduced with permission from Pearson and Hartley, *Biometrika Tables for Statisticians*, Vol. 1 (1958), pp. 159–163.

10	12	15	20	24	30	40	60	120	∞
241.9	243.9	245.9	248.0	249.1	250.1	251.1	252.2	253.3	254.3
19.40	19.41	19.43	19.45	19.45	19.46	19.47	19.48	19.49	19.50
8.79	8.74	8.70	8.66	8.64	8.62	8.59	8.57	8.55	8.53
5.96	5.91	5.86	5.80	5.77	5.75	5.72	5.69	5.66	5.63
4.74	4.68	4.62	4.56	4.53	4.50	4.46	4.43	4.40	4.36
4.06	4.00	3.94	3.87	3.84	3.81	3.77	3.74	3.70	3.67
3.64	3.57	3.51	3.44	3.41	3.38	3.34	3.30	3.27	3.23
3.35	3.28	3.22	3.15	3.12	3.08	3.04	3.01	2.97	2.93
3.14	3.07	3.01	2.94	2.90	2.86	2.83	2.79	2.75	2.71
2.98	2.91	2.85	2.77	2.74	2.70	2.66	2.62	2.58	2.54
2.85	2.79	2.72	2.65	2.61	2.57	2.53	2.49	2.45	2.40
2.75	2.69	2.62	2.54	2.51	2.47	2.43	2.38	2.34	2.30
2.67	2.60	2.53	2.46	2.42	2.38	2.34	2.30	2.25	2.21
2.60	2.53	2.46	2.39	2.35	2.31	2.27	2.22	2.18	2.13
2.54	2.48	2.40	2.33	2.29	2.25	2.20	2.16	2.11	2.07
2.49	2.42	2.35	2.28	2.24	2.19	2.15	2.11	2.06	2.01
2.45	2.38	2.31	2.23	2.19	2.15	2.10	2.06	2.01	1.96
2.41	2.34	2.27	2.19	2.15	2.11	2.06	2.02	1.97	1.92
2.38	2.31	2.23	2.16	2.11	2.07	2.03	1.98	1.93	1.88
2.35	2.28	2.20	2.12	2.08	2.04	1.99	1.95	1.90	1.84
2.32	2.25	2.18	2.10	2.05	2.01	1.96	1.92	1.87	1.81
2.30	2.23	2.15	2.07	2.03	1.98	1.94	1.89	1.84	1.78
2.27	2.20	2.13	2.05	2.01	1.96	1.91	1.86	1.81	1.76
2.25	2.18	2.11	2.03	1.98	1.94	1.89	1.84	1.79	1.73
2.24	2.16	2.09	2.01	1.96	1.92	1.87	1.82	1.77	1.71
2.22	2.15	2.07	1.99	1.95	1.90	1.85	1.80	1.75	1.69
2.20	2.13	2.06	1.97	1.93	1.88	1.84	1.79	1.73	1.67
2.19	2.12	2.04	1.96	1.91	1.87	1.82	1.77	1.71	1.65
2.18	2.10	2.03	1.94	1.90	1.85	1.81	1.75	1.70	1.64
2.16	2.09	2.01	1.93	1.89	1.84	1.79	1.74	1.68	1.62
2.08	2.00	1.92	1.84	1.79	1.74	1.69	1.64	1.58	1.51
1.99	1.92	1.84	1.75	1.70	1.65	1.59	1.53	1.47	1.39
1.91	1.83	1.75	1.66	1.61	1.55	1.50	1.43	1.35	1.25
1.83	1.75	1.67	1.57	1.52	1.46	1.39	1.32	1.22	1.00

TABLE A(Continued)

$$P = \frac{1}{B(v_1/2, v_2/2)} \int_0^{v_1 F_{v_1,v_2}\cdot P/v_2} g^{v_1/2-1}(1+g)^{-(v_1+v_2)/2} dg$$

$P = 0.99$

v_2 \ v_1	1	2	3	4	5	6	7	8	9
1	4052	4999.5	5403	5625	5764	5859	5928	5982	6022
2	98.50	99.00	99.17	99.25	99.30	99.33	99.36	99.37	99.39
3	34.12	30.82	29.46	28.71	28.24	27.91	27.67	27.49	27.35
4	21.20	18.00	16.69	15.98	15.52	15.21	14.98	14.80	14.66
5	16.26	13.27	12.06	11.39	10.97	10.67	10.46	10.29	10.16
6	13.75	10.92	9.78	9.15	8.75	8.47	8.26	8.10	7.98
7	12.25	9.55	8.45	7.85	7.46	7.19	6.99	6.84	6.72
8	11.26	8.65	7.59	7.01	6.63	6.37	6.18	6.03	5.91
9	10.56	8.02	6.99	6.42	6.06	5.80	5.61	5.47	5.35
10	10.04	7.56	6.55	5.99	5.64	5.39	5.20	5.06	4.94
11	9.65	7.21	6.22	5.67	5.32	5.07	4.89	4.74	4.63
12	9.33	6.93	5.95	5.41	5.06	4.82	4.64	4.50	4.39
13	9.07	6.70	5.74	5.21	4.86	4.62	4.44	4.30	4.19
14	8.86	6.51	5.56	5.04	4.69	4.46	4.28	4.14	4.03
15	8.68	6.36	5.42	4.89	4.56	4.32	4.14	4.00	3.89
16	8.53	6.23	5.29	4.78	4.44	4.20	4.03	3.89	3.78
17	8.40	6.11	5.18	4.67	4.34	4.10	3.93	3.79	3.68
18	8.29	6.01	5.09	4.58	4.25	4.01	3.84	3.71	3.60
19	8.18	5.93	5.01	4.50	4.17	3.94	3.77	3.63	3.52
20	8.10	5.85	4.94	4.43	4.10	3.87	3.70	3.56	3.46
21	8.02	5.78	4.87	4.37	4.04	3.81	3.64	3.51	3.40
22	7.95	5.72	4.82	4.31	3.99	3.76	3.59	3.45	3.35
23	7.88	5.66	4.76	4.26	3.94	3.71	3.54	3.41	3.30
24	7.82	5.61	4.72	4.22	3.90	3.67	3.50	3.36	3.26
25	7.77	5.57	4.68	4.18	3.85	3.63	3.46	3.32	3.22
26	7.72	5.53	4.64	4.14	3.82	3.59	3.42	3.29	3.18
27	7.68	5.49	4.60	4.11	3.78	3.56	3.39	3.26	3.15
28	7.64	5.45	4.57	4.07	3.75	3.53	3.36	3.23	3.12
29	7.60	5.42	4.54	4.04	3.73	3.50	3.33	3.20	3.09
30	7.56	5.39	4.51	4.02	3.70	3.47	3.30	3.17	3.07
40	7.31	5.18	4.31	3.83	3.51	3.29	3.12	2.99	2.89
60	7.08	4.98	4.13	3.65	3.34	3.12	2.95	2.82	2.72
120	6.85	4.79	3.95	3.48	3.17	2.96	2.79	2.66	2.56
∞	6.63	4.61	3.78	3.32	3.02	2.80	2.64	2.51	2.41

10	12	15	20	24	30	40	60	120	∞
6056	6106	6157	6209	6235	6261	6287	6313	6339	6366
99.40	99.42	99.43	99.45	99.46	99.47	99.47	99.48	99.49	99.50
27.23	27.05	26.87	26.69	26.60	26.50	26.41	26.32	26.22	26.13
14.55	14.37	14.20	14.02	13.93	13.84	13.75	13.65	13.56	13.46
10.05	9.89	9.72	9.55	9.47	9.38	9.29	9.20	9.11	9.02
7.87	7.72	7.56	7.40	7.31	7.23	7.14	7.06	6.97	6.88
6.62	6.47	6.31	6.16	6.07	5.99	5.91	5.82	5.74	5.65
5.81	5.67	5.52	5.36	5.28	5.20	5.12	5.03	4.95	4.86
5.26	5.11	4.96	4.81	4.73	4.65	4.57	4.48	4.40	4.31
4.85	4.71	4.56	4.41	4.33	4.25	4.17	4.08	4.00	3.91
4.54	4.40	4.25	4.10	4.02	3.94	3.86	3.78	3.69	3.60
4.30	4.16	4.01	3.86	3.78	3.70	3.62	3.54	3.45	3.36
4.10	3.96	3.82	3.66	3.59	3.51	3.43	3.34	3.25	3.17
3.94	3.80	3.66	3.51	3.43	3.35	3.27	3.18	3.09	3.00
3.80	3.67	3.52	3.37	3.29	3.21	3.13	3.05	2.96	2.87
3.69	3.55	3.41	3.26	3.18	3.10	3.02	2.93	2.84	2.75
3.59	3.46	3.31	3.16	3.08	3.00	2.92	2.83	2.75	2.65
3.51	3.37	3.23	3.08	3.00	2.92	2.84	2.75	2.66	2.57
3.43	3.30	3.15	3.00	2.92	2.84	2.76	2.67	2.58	2.49
3.37	3.23	3.09	2.94	2.86	2.78	2.69	2.61	2.52	2.42
3.31	3.17	3.03	2.88	2.80	2.72	2.64	2.55	2.46	2.36
3.26	3.12	2.98	2.83	2.75	2.67	2.58	2.50	2.40	2.31
3.21	3.07	2.93	2.78	2.70	2.62	2.54	2.45	2.35	2.26
3.17	3.03	2.89	2.74	2.66	2.58	2.49	2.40	2.31	2.21
3.13	2.99	2.85	2.70	2.62	2.54	2.45	2.36	2.27	2.17
3.09	2.96	2.81	2.66	2.58	2.50	2.42	2.33	2.23	2.13
3.06	2.93	2.78	2.63	2.55	2.47	2.38	2.29	2.20	2.10
3.03	2.90	2.75	2.60	2.52	2.44	2.35	2.26	2.17	2.06
3.00	2.87	2.73	2.57	2.49	2.41	2.33	2.23	2.14	2.03
2.98	2.84	2.70	2.55	2.47	2.39	2.30	2.21	2.11	2.01
2.80	2.66	2.52	2.37	2.29	2.20	2.11	2.02	1.92	1.80
2.63	2.50	2.35	2.20	2.12	2.03	1.94	1.84	1.73	1.60
2.47	2.34	2.19	2.03	1.95	1.86	1.76	1.66	1.53	1.38
2.32	2.18	2.04	1.88	1.79	1.70	1.59	1.47	1.32	1.00

TABLE A(Continued)

$$P = \frac{1}{B(v_1/2, v_2/2)} \int_0^{v_1 F_{v_1,v_2,P/v_2}} g^{v_1/2-1}(1+g)^{-(v_1+v_2)/2} dg$$

$P = 0.999$

$F_{v_1,v_2,0.999}$

v_1 \ v_2	1	2	3	4	5	6	7	8	9
1	4051*	5000*	5404*	5625*	5764*	5859*	5929*	5981*	6023*
2	998.5	999.0	999.2	999.2	999.3	999.3	999.4	999.4	999.4
3	167.0	148.5	141.1	137.1	134.6	132.8	131.6	130.6	129.9
4	74.14	61.25	56.18	53.44	51.71	50.53	49.66	49.00	48.47
5	47.18	37.12	33.20	31.09	29.75	28.84	28.16	27.64	27.24
6	35.51	27.00	23.70	21.92	20.81	20.03	19.46	19.03	18.69
7	29.25	21.69	18.77	17.19	16.21	15.52	15.02	14.63	14.33
8	25.42	18.49	15.83	14.39	13.49	12.86	12.40	12.04	11.77
9	22.86	16.39	13.90	12.56	11.71	11.13	10.70	10.37	10.11
10	21.04	14.91	12.55	11.28	10.48	9.92	9.52	9.20	8.96
11	19.69	13.81	11.56	10.35	9.58	9.05	8.66	8.35	8.12
12	18.64	12.97	10.80	9.63	8.89	8.38	8.00	7.71	7.48
13	17.81	12.31	10.21	9.07	8.35	7.86	7.49	7.21	6.98
14	17.14	11.78	9.73	8.62	7.92	7.43	7.08	6.80	6.58
15	16.59	11.34	9.34	8.25	7.57	7.09	6.74	6.47	6.26
16	16.12	10.97	9.00	7.94	7.27	6.81	6.46	6.19	5.98
17	15.72	10.66	8.73	7.68	7.02	6.56	6.22	5.96	5.75
18	15.38	10.39	8.49	7.46	6.81	6.35	6.02	5.76	5.56
19	15.08	10.16	8.28	7.26	6.62	6.18	5.85	5.59	5.39
20	14.82	9.95	8.10	7.10	6.46	6.02	5.69	5.44	5.24
21	14.59	9.77	7.94	6.95	6.32	5.88	5.56	5.31	5.11
22	14.38	9.61	7.80	6.81	6.19	5.76	5.44	5.19	4.99
23	14.19	9.47	7.67	6.69	6.08	5.65	5.33	5.09	4.89
24	14.03	9.34	7.55	6.59	5.98	5.55	5.23	4.99	4.80
25	13.88	9.22	7.45	6.49	5.88	5.46	5.15	4.91	4.71
26	13.74	9.12	7.36	6.41	5.80	5.38	5.07	4.83	4.64
27	13.61	9.02	7.27	6.33	5.73	5.31	5.00	4.76	4.57
28	13.50	8.93	7.19	6.25	5.66	5.25	4.93	4.69	4.50
29	13.39	8.85	7.12	6.19	5.59	5.18	4.87	4.64	4.45
30	13.29	8.77	7.05	6.12	5.53	5.12	4.82	4.58	4.39
40	12.61	8.25	6.60	5.70	5.13	4.73	4.44	4.21	4.02
60	11.97	7.76	6.17	5.31	4.76	4.37	4.09	3.87	3.69
120	11.38	7.32	5.79	4.95	4.42	4.04	3.77	3.55	3.38
∞	10.83	6.91	5.42	4.62	4.10	3.74	3.47	3.27	3.10

* Multiply these entries by 100.

10	12	15	20	24	30	40	60	120	∞
6056*	6107*	6158*	6209*	6235*	6261*	6287*	6313*	6340*	6366*
999.4	999.4	999.4	999.4	999.5	999.5	999.5	999.5	999.5	999.5
129.2	128.3	127.4	126.4	125.9	125.4	125.0	124.5	124.0	123.5
48.05	47.41	46.76	46.10	45.77	45.43	45.09	44.75	44.40	44.05
26.92	26.42	25.91	25.39	25.14	24.87	24.60	24.33	24.06	23.79
18.41	17.99	17.56	17.12	16.89	16.67	16.44	16.21	15.99	15.75
14.08	13.71	13.32	12.93	12.73	12.53	12.33	12.12	11.91	11.70
11.54	11.19	10.84	10.48	10.30	10.11	9.92	9.73	9.53	9.33
9.89	9.57	9.24	8.90	8.72	8.55	8.37	8.19	8.00	7.81
8.75	8.45	8.13	7.80	7.64	7.47	7.30	7.12	6.94	6.76
7.92	7.63	7.32	7.01	6.85	6.68	6.52	6.35	6.17	6.00
7.29	7.00	6.71	6.40	6.25	6.09	5.93	5.76	5.59	5.42
6.80	6.52	6.23	5.93	5.78	5.63	5.47	5.30	5.14	4.97
6.40	6.13	5.85	5.56	5.41	5.25	5.10	4.94	4.77	4.60
6.08	5.81	5.54	5.25	5.10	4.95	4.80	4.64	4.47	4.31
5.81	5.55	5.27	4.99	4.85	4.70	4.54	4.39	4.23	4.06
5.58	5.32	5.05	4.78	4.63	4.48	4.33	4.18	4.02	3.85
5.39	5.13	4.87	4.59	4.45	4.30	4.15	4.00	3.84	3.67
5.22	4.97	4.70	4.43	4.29	4.14	3.99	3.84	3.68	3.51
5.08	4.82	4.56	4.29	4.15	4.00	3.86	3.70	3.54	3.38
4.95	4.70	4.44	4.17	4.03	3.88	3.74	3.58	3.42	3.26
4.83	4.58	4.33	4.06	3.92	3.78	3.63	3.48	3.32	3.15
4.73	4.48	4.23	3.96	3.82	3.68	3.53	3.38	3.22	3.05
4.64	4.39	4.14	3.87	3.74	3.59	3.45	3.29	3.14	2.97
4.56	4.31	4.06	3.79	3.66	3.52	3.37	3.22	3.06	2.89
4.48	4.24	3.99	3.72	3.59	3.44	3.30	3.15	2.99	2.82
4.41	4.17	3.92	3.66	3.52	3.38	3.23	3.08	2.92	2.75
4.35	4.11	3.86	3.60	3.46	3.32	3.18	3.02	2.86	2.69
4.29	4.05	3.80	3.54	3.41	3.27	3.12	2.97	2.81	2.64
4.24	4.00	3.75	3.49	3.36	3.22	3.07	2.92	2.76	2.59
3.87	3.64	3.40	3.15	3.01	2.87	2.73	2.57	2.41	2.23
3.54	3.31	3.08	2.83	2.69	2.55	2.41	2.25	2.08	1.89
3.24	3.02	2.78	2.53	2.40	2.26	2.11	1.95	1.76	1.54
2.96	2.74	2.51	2.27	2.13	1.99	1.84	1.66	1.45	1.00

APPENDIX B.1
% Silicon Content of the Blast Furnace Output
(200 Observations, read rowwise)

.53	1.00	.78	.80	1.13	.99	.79	1.01	.94	.79
.91	1.01	.90	.86	.84	.61	.83	.87	.74	.74
1.05	.70	.95	1.44	1.39	1.46	1.07	1.24	1.18	1.12
1.40	.86	1.02	.91	1.26	1.35	1.31	.66	1.33	.82
.71	1.36	1.29	1.03	1.21	1.25	1.37	.97	.67	.78
.95	.83	.78	1.21	.95	.95	1.00	1.17	1.38	1.61
.95	.94	.98	.89	1.00	1.07	1.49	1.00	1.12	.90
1.00	1.21	.80	1.03	1.30	1.05	1.14	1.18	1.11	1.20
1.12	1.10	.99	.89	1.17	1.12	1.41	.93	1.00	.82
.68	.87	.97	1.04	1.12	1.03	1.00	1.05	.94	1.12
1.31	1.23	.99	.95	1.08	1.12	1.21	1.17	1.38	1.09
.86	1.02	1.33	1.07	1.27	1.33	1.15	1.11	1.30	.96
1.00	1.11	1.09	1.02	.90	.88	.96	1.03	.94	1.07
1.15	.96	1.20	1.14	1.09	.87	.84	.75	1.13	1.15
1.08	.89	.66	.67	.93	1.05	1.46	1.29	1.16	1.22
.74	.77	1.03	1.34	1.54	1.34	.80	.81	1.13	.75
.90	1.51	1.26	1.22	1.21	1.02	1.08	.96	1.24	1.17
.86	1.05	1.03	1.27	1.35	1.26	.90	1.11	1.35	1.00
1.24	1.50	1.02	1.12	1.10	1.01	1.01	.85	.91	.76
.58	.54	.92	1.36	1.55	1.26	1.22	1.33	1.13	1.14

Source: Data from Ali A. N. Hashemi, "Dynamic Behavior and Control of Blast Furnace via Time Series Approach," MS Thesis, University of Wisconsin, Madison, 1976.

APPENDIX B.2
Forge Hammer Data Set A: Ram Acceleration
Sampling Interval: $\Delta = 156.25\,\mu$ Second
(512 observations, read rowwise)

.0482940700	.0145111100	−.0288076400	−.0442028000
−.0733757000	−.1685715000	−.2696686000	−.3243561000
−.3922577000	−.4485931000	−.3804474000	−.2598724000
−.1528702000	−.0112299900	.0965881300	.0811615000
.0214233400	−.0018920300	.0069427490	.0039367670
−.0052030090	−.0149378800	−.0188741700	−.0096888540
.0087585450	.0272522000	.0209960900	.0157775900
.0455322300	.0735015900	.0867767300	.0941467300
.0748748800	.0403595000	.0048675540	−.0243063000
−.0369243600	−.0263509700	−.0145869300	−.0064237120
−.0166616400	−.0179281200	−.0069119930	−.0014647840
.0404510500	.0540161100	.0325927700	.0523529100
.0690765400	.0354766800	.0088653560	.0186309800
.0101928700	−.0029448270	−.0160665500	−.0471935300
−.0546550700	−.0132594100	.0271759000	.0299987800
.0191345200	.0033874510	.0151367200	.0721740700
.0710296600	.0236816400	.0213012700	.0365753200
.0458679200	.0395050000	−.0010680560	−.0346965800
−.0204000500	.0004272460	.0159149200	.0124816900
.0086669920	.0443115200	.0674133300	.0663757300
.0680084200	.0838165300	.0985717800	.0934753400
.0785522500	.0504913300	.0319519000	.0385589600

APPENDIX B.2
Forge Hammer Data Set A: Ram Acceleration
(*Continued*)

.0123291000	−.0220174800	−.0295705800	−.0446453100
−.0345287300	−.0004882365	−.0047147270	−.0099940300
.0137634300	.0385589600	.0516510000	.0449829100
.0421752900	.0369415300	.0507354700	.0700225800
.0371246300	.0050964350	.0183563200	.0300750700
.0256958000	.0199585000	.0361328100	.0732574500
.0857086200	.0792083700	.0664367700	.0662536600
.0940399200	.0843505900	.0625000000	.0557403600
.0358581500	.0418396000	.0454101600	.0072479250
−.0217428200	−.0064084530	.0260772700	.0330810500
.0190582300	.0180206300	.0219116200	.0351104700
.0570983900	.0555572500	.0439605700	.0333557100
.0366668700	.0578765900	.0292053200	−.0120234500
−.0043179990	−.0013427140	.0005035400	−.0115809400
−.0293874700	.0074005130	.0378570600	.0376739500
.0399169900	.0293273900	.0252685500	.0449218700
.0491638200	.0267333900	−.0060880180	−.0045926570
.0155181900	−.0084376330	−.0390605900	−.0244588900
.0068206790	.0238800000	.0032958980	−.0157308600
−.0038603540	.0134582500	.0378875700	.0477294900
.0339355400	.0171356200	.0302887000	.0570068400
.0380401600	−.0123591400	−.0129084600	.0134124800

APPENDIX B.2
Forge Hammer Data Set A: Ram Acceleration
(Continued)

.0126190200	.0197601300	.0276336700	.0221405000
.0167541500	.0201721200	.0424957300	.0616607700
.0505523700	.0489807100	.0622558500	.0495452900
.0385284400	.0523681600	.0549621600	.0333557100
.0133514400	-.0105128300	-.0163564700	.0173950200
.0374145500	.0423584000	.0578765900	.0408477800
.0127258300	.0167846700	.0040893550	-.0050504210
.0381012000	.0769195600	.0658569300	.0482788100
.0338592500	.0359344500	.0390930200	.0339813200
.0598907500	.0698852500	.0503082300	.0505065900
.0386962900	.0046234130	.0003814697	.0228576700
.0363922100	.0489807100	.0529480000	.0368957500
.0396881100	.0426483200	.0036773680	-.0047452450
.0278778100	.0189208900	.0112915000	.0336761500
.0264129600	.0234832800	.0378723100	.0538787800
.0635986300	.0359802200	.0179443400	.0370178200
.0417327900	.0187225300	-.0052640440	-.0106959300
-.0104517900	-.0158529300	-.0189046900	-.0146021800
.0063781730	.0186004600	.0017242430	-.0064237120
.0056152340	.0163269000	.0337524400	.0303344700
-.0000915490	-.0127863900	.0021209720	.0103149400
-.0149989100	-.0397014600	-.0313549000	-.0136408800

APPENDIX B.2
Forge Hammer Data Set A: Ram Acceleration
(Continued)

-.0060422420	.0030364990	.0077514640	.0048980710
.0144958500	.0453643800	.0509643600	.0160980200
-.0028380160	.0089263920	.0195770300	.0186767600
-.0041959290	-.0339489000	-.0268392600	-.0124049200
-.0189046900	-.0184164000	-.0142359700	-.0148768400
.0038757320	.0176239000	.0009307861	-.0100550700
.0098419190	.0366058300	.0246429400	.0025024410
-.0015563370	.0000915527	-.0021666290	-.0114588700
-.0203847900	-.0238790500	-.0063931940	.0190277100
.0129241900	-.0050199030	.0057373040	.0292205800
.0445098900	.0429534900	.0161590600	-.0072629450
.0087585450	.0247802700	.0034790030	-.0251607900
-.0220327400	-.0104823100	-.0057065490	.0117187500
.0025024410	-.0093836780	.0135192900	.0236206100
.0116424600	.0038604740	.0118713400	.0169525100
.0027465820	.0066070560	.0223846400	.0092315670
-.0016326310	-.0045316220	-.0293569600	-.0370311700
-.0099635120	.0081787100	.0111694300	.0077514640
.0055847160	.0089874270	.0278015100	.0356750500
.0068664550	-.0083308220	.0087890620	.0237274200
.0164489700	-.0023955110	-.0144038200	-.0192556400
-.0105738600	.0163726800	.0191650300	.0026550290

APPENDIX B.2
Forge Hammer Data Set A: Ram Acceleration
(*Continued*)

.0151519800	.0311431900	.0240631100	.0170898400
.0188903800	.0180816700	.0151214600	.0203094500
.0082244870	−.0166463900	−.0084528920	.0030975340
−.0096583370	−.0113062900	−.0032500030	.0087738040
.0307159400	.0308075000	.0039672850	−.0088343620
.0105438200	.0326232900	.0232849100	.0048980710
.0057373040	.0137634300	.0203552200	.0154266400
−.0125575100	−.0356731400	−.0109858500	.0141296400
−.0073239800	−.0271902100	−.0088038440	.0197296100
.0289917000	.0252533000	.0159454300	.0028076170
.0068206790	.0271759000	.0194702100	−.0170583700
−.0184469200	.0047149660	−.0019835230	−.0101466200
−.0065305230	−.0064542290	−.0018004780	−.0007629096
−.0113673200	−.0170736300	−.0033720730	.0015563960
.0004882812	.0009460449	−.0126490600	−.0146937400
−.0048825740	−.0102381700	−.0301809300	−.0431652100
−.0281972900	.0011749270	.0046081540	−.0103297200
−.0149378800	−.0149378800	−.0111384400	.0060729980
.0138244600	−.0037230250	−.0132746700	.0028991690
.0003814697	−.0250539800	−.0231618900	−.0101313600
−.0131831200	−.0190115000	−.0254812200	−.0278616000
−.0156698200	−.0015715960	.0011749270	−.0038450960

APPENDIX B.2
Forge Hammer Data Set A: Ram Acceleration
(*Continued*)

-.0024412870	.0040283200	.0059509270	.0032806400
-.0148921000	-.0267477000	-.0122523300	-.0032805200
-.0179128600	-.0356426200	-.0281972900	-.0102686900
-.0111231800	-.0119776700	-.0101923900	-.0229177500
-.0183553700	.0041198730	.0051574700	-.0119776700
-.0216360100	-.0145869300	-.0072171690	-.0203390100
-.0318889600	-.0341320000	-.0406932800	-.0385723100
-.0338420900	-.0351085700	-.0304861100	-.0215902300
-.0121607800	-.0092768660	-.0214223900	-.0305013700
-.0176992400	-.0075528620	-.0206747100	-.0272054700
-.0156850800	-.0160055200	-.0293264400	-.0377330800
-.0431499500	-.0380382500	-.0203695300	-.0060117240
-.0053403380	-.0212698000	-.0270528800	-.0060422420
.0108947800	-.0044400690	-.0235128400	-.0175008800
-.0113825800	-.0210256600	-.0349559800	-.0384349800
-.0321483600	-.0249776800	-.0199117700	-.0204916000
-.0259847600	-.0271902100	-.0201101300	-.0153651200
-.0196065900	-.0169515600	-.0045926570	-.0083765980

APPENDIX B.2
Forge Hammer Data Set A: Column Acceleration
Sampling Interval: $\Delta = 156.25\,\mu$ Second
(512 observations, read rowwise)

.0021209720	.0514221200	.0305786100	.0492248500
.0601654100	.0362396200	.0100250200	-.0499401100
-.2242203000	-.3369904000	-.1980972000	-.0765647900
-.1189690000	-.0972404500	-.0328502700	-.0696525600
-.0957756000	.0804595900	.2374115000	.1565399000
.1116333000	.2240295000	.2492523000	.1459656000
.0375213600	-.0422496800	-.0613231700	-.0505199400
-.1067772000	-.2001266000	-.2005081000	-.0329265600
.1020660000	.0206298800	-.0111842200	.0068206790
-.0361766800	-.1004906000	-.1268692000	-.0451641100
.0650024400	.0283813500	-.0207662600	.1736908000
.3474731000	.2037354000	-.1099968000	-.2841034000
-.1858902000	.0932769800	.1234131000	-.0143427800
.0142517100	.1225739000	.0768280000	-.0109095600
-.0398082700	-.1153221000	-.1294632000	-.0583782200
-.0065000060	.1313477000	.2691345000	.1176300000
-.0488719900	.0188751200	.0029296870	-.1338272000
-.1776505000	-.0444774600	.1565399000	.2107849000
.0585479700	-.0463390300	.0151825000	.0547943100
-.0702629100	-.2436447000	-.1951981000	.0306854200
.1740570000	.1384430000	-.0044400690	.0160827600
.0811615000	-.0243520700	-.1427383000	-.1122398000

APPENDIX B.2
Forge Hammer Data Set A: Column Acceleration
(Continued)

.0024108890	.0418090800	.0354309100	.0831909200
.1000366000	-.0873680100	-.2399979000	-.0771904000
.1026764000	.1106720000	.0162048300	-.0803947400
.0344390900	.2388458000	.1611023000	-.1216545000
-.1860580000	-.0264730500	.0354309100	.0381622300
.0785827600	.0032043460	-.0748405500	.0300750700
.1559906000	.1296539000	.0292205800	-.1196556000
-.1466751000	.0294952400	.0647430400	-.0766563400
-.0902061500	.0340576200	.1187897000	.0796814000
-.0502605400	-.1132469000	-.0801658600	.0438537600
.0843963600	-.0109858500	-.0237569800	.0552215600
.1244202000	.0867157000	-.0600414300	-.1257706000
-.0813407900	-.0125727700	.1119995000	.1569214000
.0358581500	-.0813865700	-.0028990510	.1039734000
.0303039600	-.1296005000	-.1756973000	-.0823020900
.0887146000	.1826935000	.0628051800	.0042419430
.1498413000	.1799011000	.0292053200	-.0576000200
-.0701866100	-.1247978000	-.0781974800	.0621337900
.1044006000	-.0050351620	-.1321487000	-.1157646000
-.0686607400	-.0660057100	-.0575542400	.0341491700
.1287384000	.1601257000	.0949401900	-.0274801300
-.0476970700	-.0538158400	-.1112785000	-.1270828000

APPENDIX B.2
Forge Hammer Data Set A: Column Acceleration
(Continued)

-.1032372000	-.0112605100	.1345520000	.1335907000
-.0206441900	-.1048698000	-.1223106000	-.0701561000
.0977020300	.1412811000	.0183868400	-.0164327600
.0714263900	.0668335000	.0367279100	.0196991000
-.1441269000	-.1787491000	.0052948000	.0543212900
.0008087158	-.0061948300	-.0037230250	.0007934570
-.0014800430	-.0709800700	-.1394730000	-.0162496600
.1431580000	.1376648000	.0484314000	.0391998300
.0937652600	.0453033400	-.0758018500	-.0903892500
-.0334911300	-.0232229200	.0148010300	.0797882100
.0826263400	.0234527600	.0099487300	.0386810300
.0051727290	-.0406322500	-.0164938000	.0196685800
.0632476800	.0739898700	.0044555660	-.0917320200
-.1370316000	-.0810508700	.0322876000	.0415344200
-.0508403800	-.0275106400	.0611419700	.0768585200
.0437011700	-.0105128300	-.0546703300	-.0258779500
.0468597400	.0409698500	-.0282125500	-.0285787600
.0135803200	.0099029540	.0118713400	.0054931640
-.0259237300	-.0012359020	.0429687500	.0313110400
.0178527800	-.0148615800	.0202331500	.1020203000
.0087127690	-.1193047000	-.0952262900	-.0090785030
.0342559800	-.0065305230	-.0107722300	.0159912100

APPENDIX B.2
Forge Hammer Data Set A: Column Acceleration
(Continued)

.0005187988	.0421752900	.1127625000	.0634918200
-.0815849300	-.1453629000	-.0745658900	.0657959000
.1427765000	.0707092300	-.0083918570	.0155181900
.0480041500	-.0192098600	-.1004906000	-.0489635500
.0503082300	.0718383800	.0426940900	.0197448700
.0611419700	.0764923100	-.0324535400	-.0923728800
-.0289449700	.0340728800	.0321655300	.0241851800
.0438537600	.0512542700	.0240783700	-.0230398200
-.0358409900	-.0007476509	.0351257300	-.0179433800
-.0808067300	.0181579600	.1095581000	.0337829600
-.0774650600	-.0598735800	.0167541500	.0065460200
-.0815849300	-.0997581500	.0228729200	.1186829000
.0530395500	-.0557537100	-.0871696500	-.0099482540
.0492401100	-.0041654110	-.0371685000	.0239715600
.0515899700	-.0083613400	-.0484447500	-.0221090300
.0036315910	-.0173482900	-.0380229900	-.0018767710
.0717010500	.0862121600	.0242157000	-.0234365500
-.0241842300	-.0272207300	-.0635795600	-.0539531700
.0239563000	.0732421800	.0424499500	-.0180501900
-.0285024600	.0139007600	.0390930200	-.0218191100
-.0717277500	-.0267477000	.0370483400	.0451355000
.0359191900	.0391082800	.0178833000	-.0177145000

APPENDIX B.2
Forge Hammer Data Set A: Column Acceleration
(Continued)

-.0419597600	-.0347881300	-.0054776670	.0161743200
.0173797600	.0303497300	.0420837400	.0205078100
.0156402600	.0083312990	-.0189046900	-.0212087600
-.0325145700	-.0300130800	.0296020500	.0688934300
.0529937700	.0096282960	-.0213613500	-.0199880600
-.0110316300	-.0045773980	.0077056880	-.0023192170
-.0011290910	.0348815900	.0502166700	.0100097700
-.0317363700	-.0274038300	.0187225300	.0354614300
-.0083613400	-.0442485800	-.0217733400	.0535430900
.0888061500	.0396575900	-.0205831500	-.0385265300
-.0175924300	.0172271700	.0431823700	.0385742100
.0220031700	.0015411380	-.0041196350	.0243530300
.0257720900	-.0062711240	-.0380382500	-.0280447000
.0268554600	.0491180400	.0297546400	-.0048062800
-.0157003400	.0101623500	.0069580070	-.0261373500
-.0176382100	.0137023900	.0311889600	.0257873500
-.0034483670	-.0123896600	.0075073240	.0077819820
-.0134272600	-.0104060200	.0032501220	.0256958000
.0493011500	.0331420900	.0086669920	-.0112757700
-.0336284600	-.0291891100	-.0038298370	.0181274400
.0275116000	.0229034400	.0137939500	.0098114010
.0005493164	-.0131068200	-.0131220800	-.0065915580

APPENDIX B.2
Forge Hammer Data Set A: Column Acceleration
(Continued)

-.0008849800	.0092926030	.0226898200	.0163574200
.0005035400	.0013732910	-.0027464630	-.0161733600
-.0229635200	-.0133051900	.0191497800	.0362854000
.0284271200	.0257720900	.0228118900	-.0050199030
-.0406322500	-.0369548800	-.0036467310	.0180053700
.0150146500	.0071411130	.0212402300	.0415191600
.0306701700	-.0181264900	-.0456523900	-.0130305300
.0131530800	-.0021361110	-.0016021130	.0374298100
.0474700900	.0128631600	-.0166158700	-.0170278500
.0025482170	.0048522940	-.0106043800	-.0024718050
.0303955100	.0456695600	.0317077600	.0052337650
-.0037077670	.0060577390	-.0067288880	-.0198812500
.0146942100	.0484314000	.0379791300	-.0001525804
-.0151209800	.0211486800	.0456237800	.0120697000
-.0322856900	-.0256643300	.0184936500	.0380554200
.0200195300	.0026092520	.0084838860	.0216980000
.0146179200	.0002746582	.0042724600	.0114746000
.0036010740	-.0029600860	.0205078100	.0463867100

APPENDIX B.2
Forge Hammer Data Set A: Anvil Acceleration
Sampling Interval: $\Delta = 156.25\,\mu$ Second
(512 observations, read rowwise)

.0210418700	.0324249300	.0668335000	.1179047000
.0904083200	.0375366200	.0982360800	.1952972000
.1524048000	-.0635337800	-.1953659000	-.1655807000
-.2044144000	-.2530670000	-.0849418600	.1654053000
.2442017000	.2601624000	.2370148000	.1254120000
.0801391600	.0682830800	-.1342697000	-.3847504000
-.3976593000	-.2245865000	-.0419139900	-.0185689900
-.0618419600	.0469512900	.0967102000	.0022277830
.0387725800	.1736908000	.1723633000	.0695800800
.0005187988	-.0187673600	-.1227226000	-.2910919000
-.3576508000	-.2326889000	-.1054802000	-.0938224800
.0621337900	.2814636000	.1905365000	.0314636200
.0634918200	.0908966100	-.0158834500	-.0850944500
-.0775260900	-.0615825700	-.0753135700	-.1234398000
-.1629105000	-.1624069000	-.1380386000	-.0566082000
.0867767300	.1624146000	.1710205000	.1955719000
.1519928000	.0236053500	-.1031303000	-.1445084000
-.1402664000	-.0897483800	.0051422120	.0710754400
.0709075900	.0882568400	.0822143600	.0199890100
-.0167226800	-.0583019300	-.0163717300	.0333709700
.0367889400	-.0190572700	-.0487957000	-.0780143700
-.0970573400	-.0794181800	-.0253133800	.0527496300

APPENDIX B.2
Forge Hammer Data Set A: Anvil Acceleration
(*Continued*)

.0972290000	.0624084500	.0615997300	.1263428000
.0764770500	-.0942649800	-.1494980000	-.0532360100
.0139617900	.0097808840	.0135955800	.0280914300
.0401763900	.0451965300	.0182800300	-.0108180000
.0175781200	.0562133800	.0456085200	.0141143800
-.0048062800	-.0155329700	-.0532360100	-.0829429600
-.0642814600	-.0078883170	.0162963900	.0426178000
.1162720000	.1378174000	.0530700700	-.0220022200
-.0332927700	-.0487346600	-.0703086900	-.0321025800
.0127563500	.0054321280	-.0250692400	-.0409679400
-.0286703100	-.0221853300	-.0280904800	.0185546900
.0804901100	.0776519800	.0568237300	.0407714800
-.0007934271	-.0560588800	-.1050835000	-.1175346000
-.0880546600	-.0093531610	.0627746600	.0585632300
.0043945310	-.0302877400	-.0057065490	.0218353300
-.0032957790	-.0348186500	-.0029143090	.0173492400
.0029449460	-.0314617200	-.0628471400	-.0846977200
-.0779991100	-.0564098400	-.0146632200	.0528259300
.0798797600	.0427093500	.0049591060	-.0059659480
-.0319500000	-.0420360600	-.0237112000	.0056304930
.0187377900	.0011444090	-.0165243100	-.0143427800
-.0016021130	-.0088343620	-.0135645900	.0003051757

APPENDIX B.2
Forge Hammer Data Set A: Anvil Acceleration
(Continued)

.0150909400	.0383148200	.0675659200	.0610504100
.0064392080	−.0493602800	−.0677452100	−.0513134000
−.0234823200	−.0136866600	−.0173330300	−.0014037490
.0197143600	.0209808300	.0095672610	−.0017089250
−.0070950980	−.0009917617	.0167694100	.0476837200
.0498046800	.0023803710	−.0445385000	−.0456676500
−.0182333000	−.0106654200	.0011444090	.0337371800
.0716247600	.0814819300	.0456085200	−.0179128600
−.0463085200	−.0217733400	.0111541700	.0072326660
−.0156850800	−.0366344500	−.0314006800	.0069427490
.0185241700	−.0110926600	−.0292959200	−.0083765980
.0321502700	.0511016800	.0234832800	−.0151362400
−.0284566900	−.0288991900	−.0214376400	−.0100550700
−.0063779350	−.0023039580	.0092620850	.0133361800
.0061187740	.0022583000	.0041656490	−.0024870630
−.0225668000	−.0195760700	.0069580070	.0024414060
−.0361614200	−.0598888400	−.0413799300	−.0224294700
−.0397777600	−.0465374000	−.0132441500	.0267791700
.0218505900	−.0117335300	−.0193471900	−.0044248100
−.0028532740	−.0223989500	−.0247793200	−.0035551790
−.0011596080	−.0218038600	−.0305318800	−.0172109600
.0036773680	.0024719230	−.0188436500	−.0220937700

APPENDIX B.2
Forge Hammer Data Set A: Anvil Acceleration
(Continued)

.0032653800	.0277557400	.0222320600	.0061492920
-.0089564320	-.0438823700	-.0612316100	-.0220480000
.0093078610	.0023345950	-.0148310700	-.0161428500
-.0016021130	.0090484610	-.0046384330	-.0326366400
-.0382518800	-.0165548300	.0145568800	.0257568400
.0112304700	-.0109553300	-.0159139600	-.0070340630
.0047454830	.0104827900	.0038757320	-.0071256160
-.0022429230	.0108184800	.0063171380	-.0122065500
-.0287618600	-.0242605200	-.0040738580	.0000610352
-.0016936660	.0148010300	.0247345000	.0087890620
-.0127253500	-.0207815200	-.0203695300	-.0149378800
-.0032194850	.0066986080	.0082855220	.0048065190
.0002441406	-.0025328400	-.0040433410	-.0091853140
-.0085749630	.0005493164	.0074157710	.0074920650
.0015411380	-.0023039580	-.0067899230	-.0213766100
-.0261678700	-.0054318900	.0129241900	.0098266600
-.0055539610	-.0166158700	-.0126490600	-.0040128230
-.0019529460	-.0040128230	-.0032194850	.0015106200
.0032501220	.0006103515	-.0028990510	-.0086207390
-.0126795800	-.0063931940	.0075988760	.0150909400
.0062561040	-.0034636260	-.0045773980	-.0005340278
.0048522940	.0071258540	.0054168700	.0017089840

APPENDIX B.2
Forge Hammer Data Set A: Anvil Acceleration
(Continued)

-.0000305148	.0009002685	.0011138920	-.0056760310
-.0071713920	.0028686520	.0062866210	-.0038298370
-.0073087220	-.0033110380	-.0019225480	-.0038450960
-.0080409050	-.0114131000	-.0069425110	-.0055539610
-.0109095600	-.0100245500	-.0055081840	-.0006713569
.0029602050	.0000305176	-.0102381700	-.0198354700
-.0171804400	-.0084223750	-.0032957790	-.0022581820
-.0032500030	-.0076444150	-.0118556000	-.0137476900
-.0114741300	-.0038450960	.0006103515	.0025329580
.0016937260	-.0078425410	-.0174703600	-.0138087300
-.0073544980	-.0084834100	-.0109400700	-.0091700550
-.0083155630	-.0129237200	-.0157003400	-.0096278190
-.0041654110	-.0039517880	-.0068662170	-.0096430780
-.0088496210	-.0110163700	-.0136866600	-.0066831110
.0021514890	.0016937260	-.0078883170	-.0153956400
-.0080866810	.0059814450	.0087280270	.0011596670
-.0025328400	-.0012511610	-.0086665150	-.0189962400
-.0168447500	-.0013579730	.0125885000	.0097198490
-.0005645454	-.0052182670	-.0036772490	-.0030211210
-.0071866510	-.0055081840	.0028991690	.0078125000
.0032196040	-.0072934630	-.0095667840	-.0017394420
.0056152340	.0077819820	.0022430420	-.0051419730

APPENDIX B.2
Forge Hammer Data Set A: Anvil Acceleration
(Continued)

-.0091700550	-.0051114560	.0045318600	.0068817140
-.0006866157	-.0047910210	-.0015563370	.0034027090
.0037078860	.0014495840	-.0005187690	.0000152588
.0032653800	.0052490230	.0050354000	.0065307610
.0087585450	.0055847160	-.0022886990	-.0062253480
-.0012969370	.0054626460	.0081024170	.0054779050
.0011291500	-.0025786160	-.0050351620	-.0077970030
-.0076291560	-.0016936660	.0015716550	-.0034178500
-.0076138970	-.0045316220	-.0021055940	-.0029143090
-.0025938750	.0030670170	.0027313230	-.0031126740
-.0045163630	-.0035399200	-.0055387020	-.0050199030
.0010223390	.0027771000	-.0001220591	-.0033720730
-.0038145780	-.0017547010	-.0010375380	-.0037382840
-.0069730280	-.0101466200	-.0095362660	-.0058133600
-.0043332580	-.0049283500	-.0065000060	-.0066831110
-.0044095520	-.0052182670	-.0080561640	-.0095210080
-.0120997400	-.0125575100	-.0079035760	-.0015258190
-.0019072890	-.0101923900	-.0137019200	-.0096278190

APPENDIX B.2
Forge Hammer Data Set A: Sound Pressure Level
Sampling Interval: $\Delta = 156.25\ \mu$ Second
(512 observations, read rowwise)

-.0060727600	.0896301300	.1809082000	.2495422000
.3028564000	.3382568000	.4371033000	.6207275000
.7094727000	.6428833000	.6750488000	.8222656000
.8107910000	.6341553000	.4673157000	.3939819000
.2893982000	.1011047000	-.0966453600	-.1719894000
-.1600571000	-.2066269000	-.2126694000	-.1460800000
-.1179771000	-.1604233000	-.2069321000	-.2620087000
-.3497162000	-.2626495000	-.0571270000	-.0499858900
-.1690292000	-.1632614000	-.0764732400	-.1270065000
-.2960358000	-.3973236000	-.3976288000	-.3934174000
-.3199310000	-.0621318800	.1674805000	.1929016000
.2343750000	.3822021000	.3750000000	.2125244000
.1366272000	.0759582500	-.1161461000	-.3779755000
-.4303131000	-.2299423000	-.1963120000	-.3415680000
-.2372360000	.1065674000	.3092041000	.3211975000
.3551941000	.3314209000	.1196289000	-.0203847900
.1200562000	.2969666000	.3090515000	.2250977000
.0621032700	-.2214584000	-.5440369000	-.6841125000
-.4578094000	-.0371074700	.1928101000	.1956482000
.1606445000	.2273254000	.2976074000	.0949707000
-.2245407000	-.3315582000	-.3055267000	-.1906357000
-.0508709000	-.1087303000	-.2555695000	-.2038498000

APPENDIX B.2
Forge Hammer Data Set A: Sound Pressure Level
(Continued)

-.0299673100	.1090088000	.2315979000	.3085937000
.3637695000	.5254517000	.6655273000	.5272827000
.3077698000	.2685852000	.3065186000	.3289185000
.2803345000	.1673889000	.0272827100	-.0552349100
.0252380400	.0456237800	-.1671982000	-.2830963000
-.1621323000	-.1320724000	-.3076935000	-.4676666000
-.4653778000	-.3674164000	-.2487717000	-.0647239700
-.0340862300	-.2854767000	-.4773712000	-.4173431000
-.3056183000	-.4177399000	-.4838409000	-.3674469000
-.4020538000	-.6319275000	-.6702576000	-.4301910000
-.2585297000	-.2479782000	-.1928329000	-.0707054100
-.0246877700	-.0268240000	.0559387200	.1621094000
.0608520500	-.0951499900	-.0946311900	-.0737266500
-.2590485000	-.5356140000	-.5552063000	-.3238678000
-.1718063000	-.0019225480	.4045715000	.6515503000
.3091125000	-.1267624000	-.0526409100	.1965637000
.0841064500	-.1226463000	-.0504131300	.0327453600
-.0857200600	-.1827011000	-.2059250000	-.3285980000
-.4216156000	-.2219162000	.1299744000	.2838745000
.2526245000	.2250977000	.1263733000	-.1118126000
-.2425461000	-.1531296000	-.1187706000	-.2317734000
-.1531601000	.1327820000	.1008606000	-.1682968000

APPENDIX B.2
Forge Hammer Data Set A: Sound Pressure Level
(*Continued*)

-.1996078000	-.1761093000	-.3247833000	-.3360138000
.0131225600	.3086548000	.1177368000	-.2024460000
-.1344833000	.0587463400	.0136413600	-.1709824000
-.2854767000	-.2588959000	-.0867271400	.1440125000
.1822815000	.0513610800	.0050659170	.0131225600
-.0203542700	.0616760300	.2687378000	.2929993000
.1097107000	.0973205600	.3330078000	.4199829000
.2199707000	.0799560500	.1981812000	.2067566000
-.0194082300	-.1097374000	.0290222200	.1680908000
.1005249000	-.0830650300	-.1058617000	.0465393100
.1382751000	.0644531200	.0413818400	.1203003000
.1125183000	.1300659000	.2351990000	.1448059000
-.0761070300	-.0926170300	.0572509800	.1718140000
.1904602000	.1956177000	.2449951000	.2641907000
.2470398000	.2420349000	.1611633000	-.0384197200
-.0885582000	.1275024000	.2592468000	.1126099000
-.1070518000	-.1389389000	-.0453167000	-.0795249900
-.1787338000	-.1797104000	-.1220970000	-.1625290000
-.1972885000	-.0867576600	-.0005798042	-.0221853300
-.0777244600	-.1594467000	-.2288437000	-.2230148000
-.1313705000	.0448303200	.1152954000	-.0411968200
-.2021103000	-.2185287000	-.1112022000	-.0054013730

APPENDIX B.2
Forge Hammer Data Set A: Sound Pressure Level
(Continued)

-.0031737090	-.0976829500	-.1731186000	-.0503521000
.1553040000	.1371460000	-.0659141500	-.1659775000
-.0839500400	.0259704600	.1064453000	.1230164000
-.0182790800	-.1936264000	-.1963730000	-.0538005800
.0307922400	-.0340862300	-.1411362000	-.1997604000
-.1304245000	-.0143122700	-.0358257300	-.0468730900
.0895690900	.1891785000	.1850281000	.1105652000
-.0036314730	-.0618267100	-.0347881300	.0262756300
.0352478000	.0005798339	-.0204153100	-.0289907500
-.0198049500	.0029907230	.0299682600	.0882568400
.1523132000	.2067566000	.2250671000	.1780090000
.0799560500	-.0252981200	-.0905418400	-.1017723000
-.0438823700	.0027465820	.0166626000	.0670471200
.1103210000	.0725708000	.0285034200	.0292053200
.0138549800	-.0445537600	-.0646934500	.0044555660
.0329284700	-.0401287100	-.1517258000	-.1503525000
-.0369854000	.0514831500	.0234375000	-.0997276300
-.1496811000	-.1002159000	-.0543499000	-.0306692100
-.0494670900	-.0965843200	-.0604839300	.0493774400
.0619506800	-.0022276640	-.0515422800	-.0642662000
-.0186147700	.0576782200	.1116638000	.0457153300
-.0637474100	-.0710716200	-.0752830500	-.1393967000

APPENDIX B.2
Forge Hammer Data Set A: Sound Pressure Level
(Continued)

-.2089157000	-.1736984000	-.0740013100	-.0304250700
-.0291433300	-.0723228500	-.0693016100	.0393981900
.0656433100	.0133361800	.0233764600	.0488281200
.0221557600	-.0148921000	-.0493450200	-.0927391100
-.1138573000	-.1202965000	-.1431808000	-.1285019000
.0051269530	.1332397000	.0808715800	-.0739707900
-.1560287000	-.1165428000	-.0488567400	.0345459000
.1347046000	.1511841000	.0806274400	.0159301800
.0189514200	.0518493700	.0164184600	-.0951805100
-.1919785000	-.1706772000	.0047302240	.2241821000
.2836914000	.1498718000	.0666503900	.1009827000
.0899658200	.0180053700	-.0031431910	.0622558500
.1020813000	.0552673300	-.0036619900	-.0541362800
-.2005539000	-.3915253000	-.3941193000	-.2401047000
-.1514511000	-.0745506300	.0172424300	.0271301300
.0125732400	.0587768600	.1134338000	.0796203600
-.0175466500	-.0500469200	-.0082392690	-.0162649200
-.0800437900	-.0779075600	-.0571270000	-.0846519500
-.0939903300	-.0482769000	.0102233900	.0524292000
.1430054000	.2872620000	.3301086000	.2068481000
.1431885000	.2385864000	.2574463000	.1041870000
-.0500164000	-.0481853500	.0710449200	.0928649900

APPENDIX B.2
Forge Hammer Data Set A: Sound Pressure Level
(Continued)

.0400390600	.0791320800	.1510315000	.1749878000
.1963501000	.2309265000	.1773987000	.1192932000
.1671448000	.2224426000	.1761475000	.0607910200
.0170288100	.0680236800	.0782470700	.0287780800
.0195617700	-.0132441500	-.0968284600	-.0762596100
.0422973600	.0802612300	.0430908200	.0914611800
.1598206000	.1125793000	.0461731000	.0713501000
.0662841800	-.0605449700	-.1665878000	-.1152306000
-.0109858500	-.0301809300	-.1271591000	-.1457138000
-.0672569300	.0097656250	.0240783700	.0060729980
-.0128169100	-.0111384400	.0199585000	.0610961900
.0887146000	.0778808600	.0233764600	-.0487041500
-.0874595600	-.0795249900	-.0746421800	-.1072655000
-.1423874000	-.0938377400	-.0297231700	-.0373210900
-.0645103500	-.0483989700	.0051269530	.0231018100
-.0176687200	-.0754051200	-.0627098100	.0672302200
.1343994000	.0344848600	-.0698509200	-.0837669400
-.0747337300	-.1315536000	-.2181625000	-.2351608000

REFERENCES

Adock, J. and Potter, R. (1985). "A Frequency Domain Curve Fitting Algorithm with Improved Accuracy," *Proc. 3rd International Modal Analysis Conference*, pp. 541–547.

Akaike, H. (1976). "Canonical Correlation Analysis of Time Series and the Use of an Information Criterion," in R. Mehra and D. Lainiotis (Eds.), *System Identification: Advances and Case Studies*, Academic Press, New York.

Aoki, M. (1987). *State Space Modeling of Time Series*, Springer-Verlag, New York.

Bendat, J. S. and Piersol, A. G. (1980). *Engineering Applications of Correlation and Spectral Analysis*, Wiley Interscience, New York.

Brogan, W. L. (1985). *Modern Control Theory*, 2nd ed., Prentice Hall, Englewood Cliffs, NJ.

Brown, D. L., Allemang, R. J., Zimmerman, R., and Mergay, M. (1979). "Parameter Estimation Technique for Modal Analysis," S.A.E. Paper #790221.

d'Abro, A. (1939). *The Rise of the New Physics*, D. Van Nostrand, revised Dover edition, 1952.

Dabrowski, S. (1985). "Multivariate System Analysis Using Data Dependent Systems and Fourier Transform Methods," Ph.D. Dissertation, Michigan Technological University, Houghton, MI.

Dodds, C. J. and Robson, J. D. (1975). "Partial Coherence in Multivariate Random Processes," *J. Sound and Vibration*, Vol. 42, No. 2, pp. 243–249.

Ewins, D. J. (1985). *Modal Testing: Theory and Practice*, John Wiley, New York.

Ewins, D. J. and Gleeson, P. T. (1982). "A Method For Modal Identification of Lightly Damped Structures," *J. Sound and Vibration*, Vol. 84, No. 1, pp. 57–79.

Gersch, W., Brotherton, T., and Braun S. (1980). "Parametric Time Domain Analysis of the Multiple Input/Scalar Output Problem: The Source Identification Problem," *J. Sound and Vibration*, Vol. 69, No. 3, pp. 441–460.

Golub, G. H. and Reinsch, C. (1970). "Singular Value Decomposition and Least Squares Solutions," *Numerical Math.*, **14**, 403–420.

Golub, G. H. and VanLoan, C. F. (1987). *Matrix Computations*, The Johns Hopkins University Press, Baltimore, MD.

Grenander, U. (1951). "On Empirical Spectral Analysis of Stochastic Processes," *Ark. Mat.*, Vol. 1, pp. 503–531.

Helsel, R. J., Evensen, H. A., and Pandit, S. M. (1988). "Experimental Confirmation of Time-Domain Analysis via Autoregressive Models in State Space," *Proc. 6th International Modal Analysis Conference*, pp. 243–249.

Hewlett Packard (1989). *The Fundamentals of Signal Analysis*, Application Note 243.

Ho. B. L. and Kalman, R. E. (1965). "Effective Construction of Linear State-Variable Models From Input/Output Data," *Proc. 3rd Annual Allerton Conf. Circuit and System Theory*, pp. 449–459.

Horn, R. A. and Johnson, C. R. (1985). *Matrix Analysis*, Cambridge University Press, New York.

Ibrahim, S. R. (1978). "Modal Confidence Factor in Vibration Testing," *J. Spacecraft and Rockets*, Vol. 15, pp. 313–316.

Ibrahim, S. R. and Mikulcik, E. C. (1977). "A Method for the Direct Identification of Vibration Parameters from Free Response," *Shock and Vibration Bulletin*, No. 47, pp. 183–198.

Jenkins, G. M. and Watts, D. G. (1968). *Spectral Analysis and Its Applications*, Holden-Day, San Francisco, CA.

Jones, R. and Kobayashi, Y. (1986). "Global Parameter Estimation Using Rational Fractional Polynomials," *Proc. 4th International Modal Analysis Conference*, pp. 864–869.

Juang. J. N. and Pappa, R. S. (1984). "An Eigensystem Realization Algorithm (ERA) for Modal Parameter Identification and Model Reduction," *Proc. 4th International Conference on Applied Numerical Modeling*, Dec. 27–29, pp. 491–501.

Juang, J. N. (1986). "Identification of Modal Characteristics for Flexible Structures," Section 5 in E. Denman et al., *Identification of Large Space Structures on Orbit*, Report No. AFRPLTR-86-054, American Society of Civil Engineers, New York.

Kalman, R. E., Falb, P. L., and Arbib, M. A. (1969). *Topics in Mathematical System Theory*, McGraw Hill, New York.

Kay, S. M. (1988). *Modern Spectral Estimation*, Prentice-Hall, Englewood Cliffs, NJ.

Kendall, M. G. (1945). "On the Analysis of Oscillatory Time Series," *J. Royal Statistical Soc.*, Vol. 108, p. 93.

Lang, G. F. (1983). "Modal Density—A Limiting Factor in Analysis." *Sound and Vibration*, March 1983, pp. 20–22.

Lee, R. C. K. (1964). *Optimal Estimation, Identification and Control*, Research Monograph #28, MIT Press.

Marple. L. M. (1987). *Digital Spectral Analysis With Applications*, Prentice-Hall, Englewood Cliffs, NJ.

Mehta, N. P. and Pandit, S. M. (1989). "Modal Analysis of Multiple Eigenvalue Systems by Data Dependent Systems," *Vibration Analysis—Techniques and Applications*, T. S. Sankar et al. (Eds.), DE-Vol. 18-4, The American Society of Mechanical Engineers, Bk. #H0508D, pp. 311–320, revised and abridged version in *ASME Journal of Vibration and Accoustics*, 1991, to appear.

Mitchell, L. D. (1986). "Signal Processing and the Fast-Fourier-Transform (FFT) Analyzer—A Survey", *The International Journal of Analytical and Experimental Modal Analysis*, Vol. 1, No. 1, pp. 24–36.

Okafor, A. C. and Pandit, S. M. (1987). "A Comparative Study of Identification of Dynamic Characteristics of the Headstock, Toolpost and Tailstock of a Lathe," Proc. 15th North American Manufacturing Research Conference, pp. 565–572.

Ogata, K. (1967). *State Space Analysis of Control Systems*, Prentice-Hall, Englewood Cliffs, NJ.

Pandit, S. M. (1970). "A Bayesian Approach to Time Series Analysis in Adaptive Control," M.S. Thesis, The Pennsylvania State University, State College, PA.

Pandit, S. M. (1973). "Data Dependent Systems: Modeling, Analysis and Optimal Control Via Time Series". Ph.D. Dissertation, University of Wisconsin, Madison, WI.

Pandit, S. M. (1977a). "Analysis of Vibration Records by Data Dependent Systems." *The Shock & Vibration Bulletin*, No. 47, pp. 161–174.

Pandit, S. M. (1977b). "Stochastic Linearization by Data Dependent Systems." *ASME Journal of Dynamic Systems Measurement and Control*, 99G, pp. 221–226.

Pandit, S. M., Helsel, R. J., and Evensen, H. A. (1986). "Modal Estimation of Lumped Parameter Systems Using Vector Data Dependent System Models," *Proc. 4th International Modal Analysis Conference*, pp. 414–421.

Pandit, S. M., Hu, Z. and Schilke, P. A. (1990). "Modal Analysis of V-6 Engine Using In-Situ Impulse Test Data." *Proc. 8th International Modal Analysis Conference*, pp. 871–877.

Pandit, S. M. and Jacobson, E. N. (1988). "Data Dependent Systems Approach to Modal Analysis, Part II: Application to Structural Modification of a Disc-Brake Rotor," *J. Sound and Vibration*, **122**(3), 423–432.

Pandit, S. M., Jacobson, E. N. and Shapton, W. R. (1985). "Modal Parameters of Disc Brake Rotor by Data Dependent Systems," *Proc. 3rd International Modal Analysis Conference*, pp. 850–856.

Pandit, S. M. and Lin, G. (1991). "Data Dependent Systems Methodology for Validation of Simulation Models Illustrated by End-Milling," *International Journal of Production Research*, to appear.

Pandit, S. M. and Ma, A. (1989). "A Time Domain Modeling Strategy Applied to Forge Hammer Noise," *Proc. 7th International Modal Analysis Conference*, pp. 260–267.

Pandit, S. M. and Mehta, N. P. (1984). "Data Dependent Systems Approach to Modal Parameter Identification," *Proc. 2nd International Modal Analysis Conference*, pp. 35–43.

Pandit, S. M. and Mehta, N. P. (1985). "Data Dependent Systems Approach to Modal Analysis Via State Space," *Transactions ASME, Journal of Dynamic Systems, Measurement and Control*, Vol. 107, pp. 132–137.

Pandit, S. M. and Mehta, N. P. (1988). "Data Dependent Systems Approach to Modal Analysis, Part I: Theory," *J. Sound and Vibration*, Vol. 122, No. 3. pp. 413–422.

Pandit, S. M. and Weber, C. R. (1990). "Image Decomposition by Data Dependent Systems," *Proc. USA–Japan Symposium on Flexible Automation*, Vol. II, 1988, pp. 414–419, revised version, *ASME Transactions, Journal of Engineering for Industry*, Vol. 112, 1990, pp. 286–292.

Pandit, S. M. and Wu, S. M. (1977). "Modeling and Analysis of Closed Loop Systems from Operating Data," *Technometrics*, Vol. 19. pp. 485–487.

Pandit, S. M. and Wu, S. M. (1983). *Time Series and System Analysis with Applications*, John Wiley, New York. Reprinted by Krieger Publishing Co., Melbourne, FL, 1990.

Pandit, S. M. and Yao, Y. (1991). "Improved Frequency Response Function Estimation by Data Dependent Systems, "*Proc. 9th International Modal Analysis Conference*, pp. 651–656.

Papoulis, A. (1977). *Signal Analysis*, McGraw Hill, New York.

Pernebo, L. and Silverman, L. M. (1982). "Model Reduction via Balanced State Representations", *IEEE Transactions on Automatic Control*, Vol. 27, pp. 382–387.

Priestley, M. B. (1981). *Spectral Analysis and Time Series*, Academic Press, New York.

Rao, C. R. (1965). *Linear Statistical Inference and Its Applications*, 2nd ed., John Wiley, 1973.

Trethewey, M. W. and Evensen, H. A. (1981). "Identification of Noise Sources of Forge Hammers During Production: An Application of Residual Spectrum Techniques to Transients," *J. Sound and Vibration*, Vol. 77, No. 3, pp. 357–374.

Tse, F. S., Morse, I. E., and Hinkle, R. T. (1978). *Mechanical Vibrations Theory and Applications*, Allyn and Bacon, Boston, MA.

Vold, H., Kundrat, J., Rocklin, G. T. and Russel, R. (1982). "A Multi-Input Modal Estimation Algorithm for Mini-Computers," S.A.E. Paper #820194.

Zeiger, H. P. and McEwen, A. J. (1974). "Approximate Linear Realizations of Given Dimension Via Ho's Algorithm," *IEEE Transactions in Automatic Control*, Vol. 19, p. 153.

INDEX

{ } Notation for diagonal matrix, 136, 140
{ } Notation for matrix, 34, 304, 305

Accelerance, 175
Acceleration, 246, 277, 337
Accelerometer, 246, 269
Admittance, 175
Algebra, 65, 204
Algebraic multiplicity, 112
Aliasing, 71, 211–214
   controlled, 207
   effect on DDS, 212
Amplitude, 151
Analytic function, 10, 292, 295, 308, 312
Angle between vectors, 83
Antialiasing filter, 212, 269, 317
Antiresonance, 212–213.
Apparent mass, 175
"AR(0)", 35
AR(1), 14–24
   application of Gram–Schmidt/QR, 93
   application of Householder, 108–109
   nonstationary, 23–24
   in one dimensional state space, 36–37
   stationary, 22–23, 288–292
ARMA(2,1), 29–31
   autocovariance and spectrum, 293
   Green's function, 138, 292
   in two-dimensional state space, 37–39
ARMA($n, n - 1$), 32–33, 135–141, 294–296
   extended, 316
ARMAV, 13, 164
   transformed from extended ARMA, 316
ARMAV($n, n - 1$), 33–35, 161–165, 306–308
ARV(1), 29, 153–161
ARV(2), 162, 250, 267
ARV(n) estimation, 225–229
Autocorrelation, 287, 365
   unified, 40, 366
Autocovariance, 74, 122, 286, 304, 365
   decomposition, 294–296
Autocovariance matrix, *see* Covariance matrix, lagged
Autoregressive matrix polynomial, *see* Matrix polynomial
Autoregressive moving average, *see* ARMA
Autoregressive moving average operators, 352–357
Autospectrum, 10, 286
Average modal amplitude, AM, CM, 237, 279
Averaging noise, 73

Backshift operator B, 58–59, 135, 155
Backward identity matrix, 111
Backward substitution, 91
   for AR estimation, 91
   for ARV estimation, 228
   for least squares by Cholesky, 110
Bandwidth, 181
   on signal analyzer, 317

Bartlett window, 74
Basis, 80–82
  change of, 102–103, 144
    application to AR, 103
Bias:
  of estimators, 363
  of spectral windows, 74–75
  vs. variance (random error), 74–75
Bivariate models:
  ARMAV(2,1), 33
  ARV(1), 29, 153
Bode plot, 176
Boundary conditions, 64
Boundary values, 8
Brake squeal, 260

Calculus, 65, 204
Cauchy–Schwartz inequality, 83
Chapter outline, 11
Characteristic polynomial (equation), 1
  of a matrix, 112
  for natural frequencies, 188
  of nth order difference equation, 136
  of nth order differential equation, 126
Characteristic roots, see Eigenvalues
Chi-square, 76, 370
Cholesky factorization, 110, 225–226
Circle-fit method, see Modal circle
Circulant matrix, 70
Classically damped system, 249, 255
Classical mechanics:
  origin of state space, 2, 44–51
  vibration theory of, 187–197
Classical system modeling, 2
Cofactor, 132, 136
Coherence (squared), 10, 297, 305
Commuting matrices, 185, 190, 192
Companion matrix, 52
  for nth order differential equation, 125
Complex exponential method, 199–200
Compliance, 175
Computer, 4, 43, 65, 260, 317
Conditional sum of squares and product matrix, see CSP matrix
Conditioned characteristics, see Partial characteristics
Condition of a matrix, 111
  in ARV estimation, 226
  and determinant, 121
  in lieu of confidence interval, 370
  number, 120
Confidence interval, 41, 43, 369
Confidence region, 369
Conservative system, 46

Consistency:
  of estimators, 363
  of linear equations, 81
  of sample spectrum, 75
Constrained system, 45
Controllability, 167
Control system design, 303
Control theory:
  and ERA, 203
  modern, 1, 168–169
Coordinates:
  Cartesian, 45
  generalized, 45, 144
  normal, 188
  principal, 189
  and residual a_t's, 35
Correlation, 361
Covariance, 361
  of vector random variables, 362
Covariance matrix, 121
  decomposition, 307–308
  lagged, 304
Cross-correlation, 305
Cross-covariance, 304
Cross-spectrum, 10, 305
CSP matrix, 217
Curve fitting, 197–203
Cutting stiffness, 332

Damped forced system, 49–51
Damping coefficient (rate), 182
Damping (damper) force, 49
Damping matrix, see Mass, stiffness and damping matrices
Damping ratio (factor), 9
  from data, 230, 233
  definition, 176
  overestimation by FFT, 182, 208, 266
Dashpot, see Mass-spring-dashpot system
Data dependent systems:
  approximated by deterministic, 4
  definition, 4
  estimated by SVD, 229
  methodology, 4, 6–7
  modeling:
    approach illustrated, 35–44
    as exponential expansion, 59
    strategy 26–27, 78, 167–172
  philosophy, 2, 3–6
  point of view (approach), 1
  related to Fourier methods, 66
Data vector, 6, 8, 79, 90, 92
DDS, see Data dependent systems
Decoupling, 135, 144

Degeneracy of a matrix, 82
  full, 115
  simple, 126
Degrees of freedom (DOF):
  of Chi-square, 370
  system, definition, 45
    $n(2)$, 50–52
    one, 46–50
Δ, see Sampling interval
Delta function, see Impulse function
Design matrix, 98
Determinant:
  of a Householder matrix, 108
  partitioning, 146
  related to eigenvalues, 112
  of a unitary (orthogonal) matrix, 105
Deterministic:
  component (part), 4, 8, 19
    and ARMA model, 63
    as a limit of stochastic, 5, 18, 21, 360
  mode manifestation, 2
  part and average $\overline{X}$, 55–59
  system, 20
    criterion for determining, 44
DFS (DFT), 69
Diagonalizable, 114
  simultaneously, 185, 190
Diagonal matrix, 113
Difference equation:
  first order, see AR(1)
  homogenous linear, 16
  for mode separation, 1
  scalar, see AR(1); ARMA (n,n − 1)
  and sum of exponentials, 60–63
  vector, see ARMAV (n,n − 1); ARV(1)
Differential equation:
  first order, 24–26
  for mode separation, 1
  nth order, 125–135
  scalar, 47, 49, 52
  and sum of exponentials, 60–62
  vector, 51
Dimension:
  of state space, see State space
  of vector space, see Vector space
Dirichlet conditions, 64
Disc-brake rotor, 260–266
Discrete Fourier series, see DFS (DFT)
Discrete Fourier Transform, see DFS (DFT)
Dispersion matrix, see Variance–covariance matrix
Displacement (position), 45–46, 246
Dissipation energy, 46, 49, 50

Distance between vectors, 83
Dynamic flexibility, 175
Dynamics, 44
Dynamic stiffness, 175

Economics, 1
Eigenfunction expansion, 64
Eigenspace, 116
Eigensystem realization algorithm, see ERA
Eigenvalues:
  complex, 147–151, 159–160
  definition, 112
  of differential/difference equations, 7
  discrete to continuous, 233
  multiple, 143, 156–159
Eigenvalue–eigenvector decomposition, see Spectral decomposition
Eigenvectors:
  of continuous time state matrix $\mathbf{A}$, 126, 147, 233
  definition, 112
  of discrete time state matrix $\mathbf{\Phi}$, 9, 136, 154, 232
  generalized, 118, 143, 232
Einstein, 3
Electronic revolution, 3
Energy method, see Lagrange, equations of motion
Equations of motion, 45–52
ERA, 121, 201–203, 223
Ergodic(ity), 73, 75, 288
Estimators, 362–366
Excitation, 173, 230
  burst-random, 216
  chirp, 216
  general, 241–243
  impulse, 243–245
  random, 243–245
  sinusoidal, 245–246
  step, 247
Expected value, see Mean
Experimental modal analysis, 173
Exponential(s):
  discrete, 16
  matrix, 28, 127, 129, 148, 155
  sum of, 60–63
Exponential expansion:
  DDS as, see Data dependent systems
  difference/differential equations, 61–62
  and variety of data, 61
Exponentially weighted moving average, 54
Exponential window, 209
Extended ARMA models (for multiple series), 316

Fast Fourier Transform, see FFT
F distribution, 370–371
   tables, 372–377
Feedback:
   of a machine-tool, 330–335
   present or absent, 154–157
   transfer function, 299, 302
Feedforward transfer function, 299–302
FFT, 9
   computational efficiency of, 67–71
   limitations, 204–216
Finite element, 235
   validation, 237–238
Flop, 71
   count for ARV, 226
Folding, 211
Forced response (solution), 19, 240–247
Forced system (vibration), 49
Forcing function, 19
Forecasting, 98–100
Forge hammer noise, 336–358
Fourier analyzer, 260, 277
Fourier methods:
   in complex plane, 65–67
      related to DDS, 66
   computational efficiency, see FFT
Fourier series, 64–65
   complex (exponential), 64
   discrete, 68–71
Fourier transform, 65
   limitations in modal analysis, 178
   and random function, 72–75
Free conservative system (vibration), 45
Frequency resolution, 205–207
Frequency response function, 10. See also FRF
FRF:
   estimation, 216, 230–231, 296–304
   MDOF system, matrix:
      from modal model, 234
      spectral representation, 186–187
      undamped, 185
   SDOF, 174
F-test:
   scalar, 39–43, 371
   vector, 218
Fundamental matrix, 129

Gain, 130–131, 140, 300
Galileo, 2
   spacecraft, 121
Gaussian distribution, see Normal distribution
Geometric multiplicity, 115
Gram–Schmidt procedure, 84–85
   for AR models, 86–94

   for conditioned characteristics, 314
   for least squares, 110–111
   modified, 90
Gravitation, theory of, 3
Green's function, 129
   matrix, 155
      explicit form, 162, 306–307
      implicit method, 163
      without matrix inversion, 157–159
      physical interpretation, 161

Half power frequencies, 180
Hamilton, equations of motion, 45–49
Hamiltonian, 46
Hankel matrix, 110–111
   and ARV estimation, 220
   covariance matrix, 224
Hermitian matrix:
   definition and properties, 112
   eigenvalues same as singular values, 120
Heterodyning, see Aliasing, controlled
Ho–Kalman algorithm, 121, 223
Homogenous equations, 80
Homogenous responses (solution), 19
Householder transformation (reflection, matrix), 106–111
   determinant, 108
   for least squares, 110–111
Hysteretic damping, see Structural damping

Idempotent matrix, 107
Identifiability, 165–166
Ill-conditioning of a matrix, see Condition of a matrix
Image processing, 1. See also Machine vision
Impedance, mechanical, 175
Impulse function, 128
Impulse response:
   continuous time system, scalar, 128–129
   continuous time system, vector, 144–145, 148–149
   discrete time system, scalar, 139
   discrete time system, vector, 152–153
   modal decomposition, 299–303
Independent variable matrix, see Design matrix
Inertance, 175
Inertia matrix, 272
Influence coefficient, see Compliance
Initial value(s)/conditions, 8, 19
   and ARMA model, 63
Inner product, $<\mathbf{x}, \mathbf{y}>$, 83
In situ measurements, 277
Integrating factor, 25
Inverse Fourier transform, 199

Inverse function, 94
Inverse of a matrix:
  eigenvalues, 112
  generalized, 123
Inverse method, 184
ITD method, 200–201

Jordan form, 116, 119

Kinetic energy, 46–50
  MDOF system, 187

Lagrange, equations of motion, 45–51
Lagrangian, 45
λ, autoregressive roots:
  from difference operator, 59, 62
  as eigenvalues of Φ, 136, 154, 162, 232
  equal to moving average roots, 63
Laplace transform:
  DDS as generalized, 65–67
  Fourier transform as a cut of, 179, 180, 181
Leakage, 207–210
Least squares estimates, 17, 97–98
  computation, 110–111
  geometry of, 124
  nonlinear, 95, 232
  relative error, 228–229
  by SVD, 123–124
  unbiased, 364
  weighted, 98
Least squares estimators, 364
Levinson–Derbin algorithms, 225
Linear algebra, 78
Linear averaging, 214
Linear damping, see Viscous damping
Linear equations, 80–82
Linearization, 9, 303
Linearly dependent, 80
Linearly independent, 80
Linear transformation, 102, 125
  simple, 116
L_1 norm, 6
Loss factor, 184
LU decomposition, 310, 314

MA, see Moving average
Machined surfaces, 1
Machine-tool dynamics identification, 330
Machine vision, 296
Magnification factor, 177
Mass normalized, 188
Mass-spring-dashpot system, 49–50, 127–132
Mass, stiffness and damping matrices, 50, 184, 190. See also Spatial (physical) model

Matrix decomposition, 111–124
Matrix exponential, see Exponential(s), matrix
Matrix methods:
  of spectral decomposition, 79
  for state space, 1
Matrix polynomial:
  nth order, 164, 232, 302
  second order, 162
    complex, 190
Maxwell's reciprocity theorem, 187
MDOF system:
  damped, 190–197
  undamped, 184–190
Mean, 16, 21–23
  definition, 360, 366
  effect of estimation, 55–59
Mechanical impedance, 175
Minimal realization, see Realization
Minimum mean squared error, 99
Minimum variance:
  control, 100
  forecast, 99
Mixing, see Aliasing, controlled
Mobility, 175
Modal analysis:
  broadly interpreted, 1
  continuous time scalar systems, 134
  from data, 173–283
Modal assurance criterion, MAC, 239
Modal circle, 182–184
Modal confidence factor, MCF, 239, 279–281
Modal constant, see Residue
Modal decomposition:
  concept, 5
  of FRF, 299–303
  of multivariate spectrum, 307
  scalar continuous-time, 125–135
  scalar discrete-time, 135–137
  of spectrum, 293, 295
  vector continuous-time, 142–152
  vector discrete-time, 152–165
Modal density, 255
Modal mass, 189
  complex, 193
Modal matrix of classical vibration theory, 151, 258
Modal matrix **M**, 113
  from discrete Φ, 232–233
  of two-mass undamped system, 147
Modal model, 10, 230–234
  from scalar models, 249, 255, 262
  from vector models, 267–277
Modal participation factor, see Residue
Modal power, 237, 263

Modal signal-to-noise ratio, MSN, 238
Modal and spectrum analysis, 2, 9–11, 118
Modal stiffness, 189
  complex, 193
Modal testing, experimental, 10, 173
Modal variance, MV, 238
Modal vector, 151, 186, 232
Modes:
  bending, 274, 276–277
  closely-spaced, 249–255
  computational, 235
  normal, 188, 192
  patterns, 1
  principal, 189, 192
  pseudo, 280
  selection, 234–240
  strength of, *see* Average modal amplitude, AM, CM
Mode shape, 151
  animated, 230
  complex, 193, 262
  mass-normalized, 188
  scaled, 10, 232, 261
Model adequacy, 6, 8–9, 26–27
Model decomposition, 5. *See also* Modal decomposition
Moving average, 30
  misnomer, 53–54
  need for, 53
Multiple-degree-of-freedom system, *see* MDOF system
Multiple-input-multiple-output (MIMO) system, 7, 301–304
Multiple-input-single-output (MISO) system, 300–301, 322–330
Multiple (squared) coherence, 315
Multiple series, *see* Extended ARMA models (for multiple series)
Multivariate frequency response, *see* FRF, MDOF system, matrix
Multivariate normal distribution, 368
Multivariate spectrum, 304–315

NASA, 121
Natural frequency, 9, 47, 175–176
  squared, as eigenvalue, 187
  complex, 193
Near-rank deficiency, 120
Newton, 2, 3
NID, 16
Nodal line, 261
Nonclassically damped system, 247–249
Nonconservative system, 49

Noise floor, 58, 219
Nonhomogenous equation, 81
  consistent, 82
Nonlinear systems, 9, 303
Nonnegative definite matrix, 110
  definition and eigenvalues, 112
  variance–covariance and spectrum, 305
Nonstationarity, 23
  in the mean and variance, 56
Nonstationary, 7, 8, 9
  stochastic process, 21, 145–146, 153
Norm:
  of a matrix, 120
  of a vector, 83
Normal distribution, 16, 367–368
Normal equations, 97–98, 364
  computation and accuracy, 110–111
Normal matrix, 115
Null spaces:
  of a matrix, 82, 118, 124
  v, moving average roots:
    definition, 94
    equal to autoregressive, 63
  of a transformation, 116
Nyquist:
  frequency, 211, 340
  plot, 176

Observability, 167
O.n.b., *see* Orthonormal, basis
One sided spectra, 319
One-to-one transformation, 102
Operation count:
  in ARV estimation, 226–229
  in FFT, 70–71
  in least squares, 110–111
Optimal control, 7, 100–101
  strategy, 101
Orthogonal:
  decomposition, 85, 94
    applications of, 95–101
    of stationary models, 95
  projection, 96–97
  transformation (matrix), 105
  vectors, 83
Orthogonality:
  of Fourier methods, 64–65
  and parsimony, 67
Orthonormal:
  basis by Gram–Schmidt, 84–85
  basis by Householder, 110–111
  transformation, 105
  vectors, 83
Outer product, $> <$, 106

Parseval relation, 295
Parsimony, 67
Partial characteristics, 309–315, 323–330, 347–351
   recursive expressions for, 312–313, 323–325
Partial fractions, 125, 190
Particular response (solution), 19
Peak-amplitude method, *see* Peak-picking method
Peak-picking method, 180
Periodogram, *see* Sample spectrum
Phase angle:
   FRF, 177
   and MAC, 239
   mode shapes, 248, 250, 252–253, 256–259, 271, 282
   plots, 318, 326–329, 333
Poles, 125
   as squared frequencies, 190
Polyreference method, 200
Positive definite matrix, 50
   definition and eigenvalues, 112
   variance–covariance and spectrum, 310, 313
   for vibration energies, 187
Positive semidefinite matrix, *see* Nonnegative definite matrix
Potential energy, 46–50
   MDOF system, 187
Power averaging, *see* Spectral averaging
Power spectral density (PSD), 285, 286
Prediction, *see* Forecasting; Response prediction
Probability distributions, 366–371
Projection, 106–107
Prony method, 200
Proportional damping, 191–192, 249, 255
Ptolemaic epicycle theory, 3

QR decomposition (factorization), 85
   for ARV estimation, 226–229
   by Householder, 108
Quadratic form, 50
Quantum mechanics, 2
   and DDS, 4

Ramp, 130, 140
Random, *see also* Stochastic
   component (part), 4, 8
   forcing function, 10
Random error, *see* Variance of spectral estimators
Random function, 16, 72–75
Random variable, 21, 359
   complex, 362
   degenerate, 21, 360
Random vibrations, 10, 244, 303
Random walk, 23
Range space of a matrix, 82
Rank:
   deficiency and ERA, 203
   for real data, 224
   dimension of vector space, 80
   of a matrix, 82
   and SVD, 119–121
Rank-deficient matrix, 87
Rational fraction polynomial method, 200
Realization:
   canonical, 203
   least squares, 203
   minimal, 121, 203
Receptance:
   of machine-tool, 332–335
   MDOF system, 184–197
   SDOF system, 174–184
Reciprocal basis, 86, 117
Reciprocity theorem, Maxwell's 187
Reference point method, 234
   illustrated, 278
Reflection, *see* Householder transformation
Relative amplitude, *see* Mode shape
Relativity, theory of, 3
Relativistic mechanics and DDS, 4
Repeated roots, 260. *See also* Eigenvalues, multiple
Residual:
   concept, 5
   as pseudo-input, 54
   source of random, 4, 7
   sum of squares, 17
   white, 4, 8
Residual characteristics, *see* Partial characteristics
Residual modes, 198
Residue, 125, 189, 263
Response model:
   MDOF system, 192, 230
   SDOF system, 174
Response prediction, 232
   error, 277
Right-hand side dynamics, 53
Rigid body, 235
RMS averaging, *see* Spectral averaging
Rotational inertia, 272
RSS, *see* Residual, sum of squares

Sample autocorrelation, 364–365
Sample autocovariance, 74–75, 288, 365
   as a poor estimator, 75, 122

Sample covariance matrix, 121
Sample spectrum, 74–75
  advantage of, 75–76
  discouraging properties, 75
  for transfer function, 297
  vs. spectrum, 287–288
Sampling frequency, 211, 337, 340
Sampling interval, 8
Scalar systems, 125–141
Scaling factor, 234–240
Scientific method, 2
SDOF system:
  damped, 176–182
  undamped, 174–176
Signal-to-noise:
  indicators, 11
  ratio, 238, 249, 274, 276–277, 327, 330
Similarity transformation, 104
Simulation:
  ACSL, 247
  3-DOF system, 267
  validation, 237
Single-degree-of-freedom system, see SDOF system
Single-input–single-output (SISU) system, 298–300, 317–322
Singular value decomposition, see SVD
Singular values, see SVD
Singular vectors, see SVD
Sound pressure level, 336–337
Spatial (physical) model, 10, 176, 230, 240–247
  from scalar models, 255–260
  from vector models, 272–277
Spectral averaging, 214
Spectral decomposition, 9, 111–119
Spectral density function, 287
Spectral representation:
  as crux of system analysis, 117–118
  of nth order difference equation, 136–137
  of nth order differential equation, 125–135
Spectrum analysis, 284–358
  for modal information, 2, 230
Spring:
  constant k, 46
  force, 46
  matrix, see Mass, stiffness and damping matrices
Stable, 7, 8, 9
  asymptotically, 130, 140, 145–146, 153
Standard deviation, 360
Standard normal variable, 368
State, 45
State matrix, continuous $\mathbf{A}$:
  definition, 10
  for nth order scalar system, 125
  in response (solution), 28
State matrix, discrete $\boldsymbol{\phi}$:
  for ARMA($n, n - 1$), 135
  definition, 9
  in response (solution), 28
  singularity, 164
State model, continuous:
  n-degrees-of-freedom, 50–52
  one-degree-of-freedom, 46–50
  with solution, 28
  standard form, 28
State model, discrete:
  for ARMA($n, n - 1$), 135–141
  with solution, 28–35
  standard form, 30
State space:
  definition, 45
  dimension as rank, 88–89
  dimension and SVD, 121–123
  formalism, 1, 7–9
  one dimensional, 36–37
  origin in mechanics, 44–55
  two dimensional, 37–39
Stationarity condition, 8, 9, 145–146
Stationary, 7, 8, 9, 73
  wide-sense, 21, 73
Statistic, 359
Statistically dependent data, 14
  reduction to independent, 76–77
Statistically independent, 367
Steady-state response, 245–246
Step response:
  continuous time system, scalar, 130–132
  continuous time system, vector, 145, 149–150
  discrete time system, scalar, 139–141
  discrete time system, vector, 153
Stiffness matrix, see Mass, stiffness and damping matrices
Stochastic, see also Random
  component (part), 4, 8, 19, 238
  and ARMA model, 63
  mode manifestation, 1
  system (process), 20–21
Stopping criterion:
  deterministic systems, 43–44
  stochastic systems, 39–43
Structural damping:
  MDOF system, 192–194
  SDOF system, 182–184
Structural modification, 260–266

Subspace, 80
SVD, 98, 119–124
　for ARV estimation, 229

Time constant, 276
Time series, 121, 359
Toeplitz matrix, 70, 110
　and ARV estimation, 220
Toolpost receptance, 332–333
Torsional damping, 272
Torsional stiffness, 272
Torsional system, 268–277
Total least squares, 63, 98
Trace of a matrix, 112
Transfer function:
　for Green's function, 155
　MDOF system, matrix, undamped, 185
　for modal information, 2, 230
　SDOF system, 175
Transient response (solution), 19, 240
Transformation, *see* Linear transformation
Triangular reduction:
　by Gram–Schmidt, 84–94
　by Householder, 108–109
Trigonometric polynomial, 67, 207

Unbiased, 17, 363, 364
Uncertainty, 4
　and determinism, 5
　and relativity, 5
Uncoupled coordinates, *see* Decoupling
Undamped system, 46–48
　two-degree-of-freedom, 146–152
Unitarily diagonalizable, 114, 115
Unitary transformation (matrix), 105
Univariate systems, *see* Scalar systems
Unstable, 7, 8, 9, 130, 140, 153

Vandermonde matrix, 70, 127
　determinant and inverse, 132–133, 158
Variance, 16, 21–23, 286
　decomposition, *see* Autocovariance, decomposition
　definition, 360
　of a linear combination, 362
　rapid computation by integration, 296
　of spectral estimators, 74–75
Variance–covariance matrix, 304
　definition, 361
　modal decomposition, 307
Vector autoregressive moving average, *see* ARMAV
Vector differential equation, 51
Vector space, 80–86
Velocity, 45–48
Viscous damping, 182
　MDOF system, 194–197
　SDOF system, 182
V-6 engine, 277

White noise:
　continuous time, 26
　discrete time, 8
　multivariate, 305
　properties, 287
　sampled, 289–292
　variance–covariance decomposition, 308–309
　illustrated, 318–322, 342–347
Window carpentry, 10, 210
Windows, 74–75, 208–210

Yule–Walker equations, 122, 307

Zoom 207
z transform, 297